Lecture Notes in Computer Science   8878

Commenced Publication in 1973
Founding and Former Series Editors:
Gerhard Goos, Juris Hartmanis, and Jan van Leeuwen

Marcos K. Aguilera   Leonardo Querzoni
Marc Shapiro (Eds.)

Principles of Distributed Systems

18th International Conference, OPODIS 2014
Cortina d'Ampezzo, Italy, December 16-19, 2014
Proceedings

 Springer

Volume Editors

Marcos K. Aguilera
Mountain View, CA, USA
E-mail: mkaguilera@gmail.com

Leonardo Querzoni
Sapienza University of Rome, Italy
E-mail: querzoni@dis.uniroma1.it

Marc Shapiro
Inria Paris-Rocquencourt
and Sorbonne Universités
UPMC Univ Paris 06, LIP6
Paris, France
E-mail: marc.shapiro@acm.org

ISSN 0302-9743   e-ISSN 1611-3349
ISBN 978-3-319-14471-9   e-ISBN 978-3-319-14472-6
DOI 10.1007/978-3-319-14472-6
Springer Cham Heidelberg New York Dordrecht London

Library of Congress Control Number: 2014958596

LNCS Sublibrary: SL 1 – Theoretical Computer Science and General Issues

Typesetting: Camera-ready by author, data conversion by Scientific Publishing Services, Chennai, India

Printed on acid-free paper

Springer is part of Springer Science+Business Media (www.springer.com)

Preface

On behalf of the Technical Committee of the International Conference on Principles of Distributed Systems, we are very pleased to present in this volume the proceedings of the 18th edition of the conference, which was held during 16 to 19 December 2014, in Cortina d'Ampezzo, Italy.

OPODIS, the International Conference on Principles of Distributed Systems, is an international forum for the exchange of state-of-the-art knowledge on distributed computing and systems among researchers from around the world. All aspects of distributed systems are within the scope of OPODIS, including theory, specification, design, performance, and system building. OPODIS has traditionally been strong in the theoretical aspects of distributed systems. Since the 2013 edition, OPODIS is seeking to expand its coverage to include the overlap between theoretical solutions and practical implementations, as well as experimentation and quantitative assessments.

Papers were sought soliciting original research contributions to the theory, specification, design, and implementation of distributed systems. In response to the call for papers, 98 submissions were received. In a first round, each paper was reviewed by at least three members of the Program Committee (PC), sometimes with the help of external reviewers. The 69 top papers went through a second round of reviewing and received more PC reviews. After some online exchanges, the PC held a physical meeting, to select the 32 papers of this edition. Rigorous reviewing and vigorous discussion ensured a high-quality selection process. We would like to thank the Programme Committee members, as well as the external reviewers, for their fundamental contribution in selecting the best papers.

In addition to the technical papers, the program included a keynote presentation from Lorenzo Alvisi (University of Texas at Austin, USA) and two tutorial presentations from Yann Busnel (Crest - Ensai, Rennes and LINA - University of Nantes, France) and Christian Cachin (IBM Research Zürich, Switzerland). This edition also had a co-located workshop, the Second Workshop on Distributed Computing: Computability and Complexity. We would like to thank the tutorial and workshop chair Étienne Rivière (University of Neuchâtel, Switzerland) for his help in attracting tutorial proposals and for helping with the workshop co-location.

This event would not have been possible without the technical and administrative support of Gabriella Caramagno and Giuliana Bottaro. We also wish to express our gratitude to our supporting institutions: la Sapienza University of Rome, the Research Center for Cyber Intelligence and Information Security (CIS), and CINI.

November 2014

Marcos K. Aguilera
Leonardo Querzoni
Marc Shapiro

Organization

General Chair

Leonardo Querzoni   Sapienza University of Rome, Italy

Program Co-chairs

Marcos K. Aguilera   Unaffiliated
Marc Shapiro   Inria and UPMC-LIP6, France

Steering Committee

Roberto Baldoni   Sapienza University of Rome, Italy
Antonio Fernandez Anta   Institute IMDEA Networks, Spain
Paola Flocchini   University of Ottawa, Canada
Giuseppe Prencipe   University of Pisa, Italy
Binoy Ravindran   Virginia Tech, USA
Nicola Santoro   Carleton University, Ottawa, Canada
Maarten van Steen   VU University Amsterdam, The Netherlands

Strategic Advisory Committee

Alain Bui   University of Versailles S.Q., France
Marc Bui   LAISC, EPHE, Paris, France
Nicola Santoro   Carleton University, Canada
Philippas Tsigas   Chalmers University of Technology, Sweden

Program Committee

Marco Aiello   University of Groningen, The Netherlands
François Bonnet   JAIST, Japan
Allen Clement   Google and Max Planck Institute for Software
             Systems, Switzerland and Germany
Paolo Costa   Microsoft Research Cambridge, UK
Carole Delporte   Université Paris Denis Diderot, France

Murat Demirbas   University at Buffalo, USA
Sameh Elnikety   Microsoft Redmond, USA
Hugues Fauconnier   Université Denis Diderot LIAFA, France
Pascal Felber   Université de Neuchâtel, Switzerland
Pierre Fraigniaud   CNRS and University Paris Diderot, France
Alexey Gotsman   IMDEA Software Institute, Spain
Tim Harris   Oracle Labs, UK
Maurice Herlihy   Brown University, USA
Konrad Iwanicki   University of Warsaw, Poland
Márk Jelasity   University of Szeged, Hungary
Alex Kogan   Oracle Labs, USA
Tamer Öszu   University of Waterloo, Canada
Boaz Patt-Shamir   Tel Aviv University, Israel
Fernando Pedone   University of Lugano, Switzerland
Maria Potop-Butucaru   Université Pierre et Marie Curie-LIP6, France
Nuno Preguiça   Universidade Nova de Lisboa, Portugal
Sergio Rajsbaum   UNAM, Mexico
Luís Rodrigues   INESC-ID, Universidade de Lisboa, Portugal
Paolo Romano   INESC-ID and IST, Universidade de Lisboa,
                               Portugal
Marco Serafini   Qatar Computing Research Institute, Qatar
Liuba Shrira   Brandeis, USA
Peter Van Roy   Université Catholique de Louvain, Belgium
Jennifer Welch   Texas A&M University, USA
Yukiko Yamauchi   Kyushu University, Japan

Additional Reviewers

Ailidani Ailijiang   Ilche Georgievski   Iulian Moraru
Hagit Attiya   Emmanuel Godard   Calvin Newport
Mor Baruch   Lukasz Golab   Fukuhito Ooshita
Samuel Benz   Heerko Groefsema   Leandro Pacheco
Giovanni Bernardi   Urs Hengartner   Ricardo Padilha
Eduardo Bezerra   Taisuke Izumi   Giuliano Andrea Pagani
Frank Blaauw   Tomoko Izumi   Roberto Palmieri
Manuel Bravo   Vasiliki Kalavri   Marcelo Pasin
Doina Bucur   Sayaka Kamei   Ruma Paul
Zuhal Can   Sven Kohler   Erez Petrank
Hyun Chul Chung   Amos Korman   Seth Pettie
Christian Colombo   Saptaparni Kumar   Étienne Rivière
Stefano Coniglio   Shay Kutten   Stéphane Rovedakis
Shantanu Das   Anissa Lamani   Daniele Sciascia
Diego Didona   Alexander Lazovik   Pierre Sutra
Nuno Diegues   Zhongmiao Li   Edward Talmage
Sérgio Duarte   Ying Liu   Serafettin Tasci

Ando Emerencia   José Legatheaux Martins   Jeremy Topolski
Guy Even   Hugues Mercier   Corentin Travers
Leszek A. Gasieniec   Sayan Mitra   Jons-Tobias Wamhoff

Organizing Committee

Silvia Bonomi
  (*Publicity Chair*)   Sapienza University of Rome, Italy
Gabriella Caramagno   Sapienza University of Rome, Italy
Leonardo Querzoni
  (*General Chair*)   Sapienza University of Rome, Italy
Étienne Rivière
  (*Workshop and*   Université de Neuchâtel, Switzerland
Tutorials Chair)

Salt: Combining ACID and BASE in a Distributed Database (Invited Talk)

Lorenzo Alvisi

The University of Texas at Austin, USA

Abstract. What is the right abstraction to support scalable and available storage and retrieval of data in a distributed database? Today's options—ACID transactions and BASE implementations—force developers to compromise either ease of programming or performance. This talk will discuss Salt, a new database that allows the ACID and BASE paradigms to *coexist* in order to combine the desirable qualities of both. Salt is based on the observation, rooted in Pareto' s principle, that, when an application outgrows the performance and availability offered by an ACID implementation, it is often because of the requirements of only a few transactions: most transactions never test the limits of what ACID can offer. Through the new abstraction of BASE transactions, Salt allows to safely "BASE-ify" only those few performance-critical ACID transactions, without compromising the ACID guarantees enjoyed by the remaining transactions: in so doing, Salt can reap most of the performance benefits of the BASE paradigm, without unleashing the cost and complexity that traditionally come with it.

Distributed Large-Scale
Data Stream Analysis
(Tutorial)

Yann Busnel

Crest - Ensai, Rennes, France
LINA - University of Nantes, France

Abstract. This tutorial aims to survey some existing algorithms that process huge amount of data inline, efficiently in term of space and time complexity. The interest of estimating metrics or identify specific patterns between several (a.k.a. distributed) data streams is important in data intensive applications. Many different domains are concerned by such analyses including machine learning, data mining, databases, information retrieval, and network monitoring. In all these applications, it is necessary to quickly and precisely process a huge amount of data. For instance, in IP network management, the analysis of input streams allows to rapidly detect the presence of anomalies or intrusions when changes in the communication patterns occur. The problem of extracting pertinent information in a data stream is similar to the problem of identifying patterns that do not conform to the expected behaviour, which has been an active area of research for many decades. For instance, depending on the specificities of the domain considered and the type of outliers considered, different methods have been designed, namely classification-based, clustering-based, nearest neighbour based, statistical, spectral, and information theory. We aim to propose a comprehensive survey of these techniques, their advantages and their drawbacks in this tutorial. A common feature of these techniques is their space complexity and their computational cost, as they rely on small space approximation algorithms for analysing their data.

Integrity, Consistency, and Verification of Remote Computation
(Tutorial)

Christian Cachin

IBM Research Zurich, Switzerland

Abstract. With the advent of cloud computing, many clients have outsourced computation and data storage to remote servers. This has led to prominent concerns about the privacy of the data and computation placed outside the control of the clients. On the other hand, the integrity of the responses from the remote servers has been addressed in depth only recently. Violations of correctness are potentially more dangerous, however, in the sense that the safety of a service is in danger and that the clients rely on the responses. Incidental computation errors as well as deliberate and sophisticated manipulations on the server side are nearly impossible to discover with today's technology. Over the last few years, there has been rising interest in technology to verify the results of a remote computation and to check the consistency of responses from a cloud service. These advances rely on recently introduced cryptographic techniques, including authenticated data types (ADT), probabilistically checkable proofs (PCPs), fully-homomorphic encryption (FHE), quadratic programs (QP), and more. With multiple clients accessing the remote service, a further dimension is added to the problem in the sense that clients isolated from each other need to guarantee that their verification operations relate to the same "version" of the server's computation state. This tutorial will survey the recent work in this area and provide a broad introduction to some of the key concepts underlying verifiable computation, towards single and multiple verifiers. The aim is to give a systematic survey of techniques in the realm of verifiable computation, remote data integrity, authenticated queries, and consistency verification.

The approaches rely on methods from cryptography and from distributed computing. The presentation will introduce the necessary background techniques from these fields, describe key results, and illustrate how they ensure integrity in selected cases.

The tutorial consists of three parts: i) verifiable computation, ii) authenticated data types and iii) distributed consistency enforcement.

Table of Contents

Models

Radio Networks

Robots

Self-Stabilization

Shared Data Structures

Shared Memory

Synchronization and Universal Construction

Verifying the Consistency of Remote Untrusted Services with Commutative Operations

Christian Cachin[1] and Olga Ohrimenko[2],*

[1] IBM Research - Zurich, Switzerland
cca@zurich.ibm.com
[2] Microsoft Research, Cambridge, United Kingdom
oohrim@microsoft.com

Abstract. A group of mutually trusting clients outsources a computation service to a remote server, which they do not fully trust and that may be subject to attacks. The clients do not communicate with each other and would like to verify the correctness of the remote computation and the consistency of the server's responses. This paper first presents the *Commutative-Operation verification Protocol (COP)* that ensures linearizability when the server is correct and preserves fork-linearizability in any other case. All clients that observe each other's operations are consistent, in the sense that their own operations and those operations of other clients that they see are linearizable. Second, this work extends COP through authenticated data structures to *Authenticated COP* , which allows consistency verification of outsourced services whose state is kept only remotely, by the server. This yields the first fork-linearizable consistency verification protocol for generic outsourced services that (1) *relieves* clients from *storing the state*, (2) supports *wait-free* client operations, and (3) handles *sequences* of arbitrary *commutative operations*.

Keywords: cloud computing, fork-linearizability, data integrity, verifiable computation, commutative operations, Byzantine emulation.

1 Introduction

With the advent of *cloud computing*, most computations run in remote data centers and no longer on local devices. As a result, users are bound to trust the service provider for the confidentiality and the correctness of their computations. This work addresses the *integrity* of outsourced data and computations and the *consistency* of the provider's responses. Consider a group of mutually trusting clients who want to collaborate on a resource that is provided by a remote partially trusted server. This could be a wiki containing data of a common project, an archival document repository, or a groupware tool running in the cloud. A subtle change in the remote computation, whether caused inadvertently by a bug or deliberately by a malicious adversary, may result in wrong responses to the clients. The clients trust the provider only partially, hence, they would like to assess the integrity of the computation, to verify that responses are correct, and to check that they all get consistent responses.

---

* Work done at IBM Research - Zurich and at Brown University.

M.K. Aguilera et al. (Eds.): OPODIS 2014, LNCS 8878, pp. 1–16, 2014.
© Springer International Publishing Switzerland 2014

In an asynchronous network model without communication among clients such as considered here, the server may perform a *forking attack* and omit the effects of operations by some clients in her responses to other clients. Not knowing which operations other clients execute, the forked clients cannot detect such violations. The best achievable consistency guarantee in this setting is captured by *fork-linearizability*, introduced by Mazières and Shasha [23] for storage systems. Fork-linearizability ensures that whenever the server in her responses to a client C_1 has ignored an operation executed by a client C_2, then C_1 can never again observe an operation by C_2 afterwards and vice versa. This property ensures clearly defined service semantics in the face of an attack and allows clients to detect server misbehavior easily.

Several conceptual [8, 21, 5, 6] and practical advances [29, 13, 20, 27] have recently been made that improve consistency checking and verification with fork-linearizability and related notions. The resulting protocols ensure that when the server is correct, the service is linearizable and (ideally) the algorithm is *wait-free*, that is, every client's operations complete independently of other clients. It has been recognized, however, that read/write conflicts cause such protocols to block; this applies to fork-linearizable semantics [23, 8] and to other forking consistency notions [5, 6].

In this paper, we go beyond storage services and verify the consistency of remote *computation* on a Byzantine server. The *Commutative-Operation verification Protocol* or *COP* imposes fork-linearizable semantics for arbitrary functionalities, exploits commuting operations, and allows clients to operate concurrently without blocking unless operations conflict. Furthermore, the extension to *Authenticated COP* also relieves clients from storing the computation state and from executing all operations. Fork-linearizability makes it easy to expose Byzantine behavior of the server. For instance, the clients may exchange a message outside the model over a low-bandwidth channel and thereby verify the correctness of a service in an end-to-end way.

Efficient handling of wait-free operations is a key feature for collaboration with remote coordination, as geographically separated clients may operate at different speed. Consequently, previous work has devoted a lot of attention to identifying and avoiding blocking [23, 8, 19]. For example, read operations in a storage service commute and do not lead to a conflict. On the other hand, when a client writes a data item concurrently with another client who reads it, the reader has to wait until the write operation completes; otherwise, fork-linearizability is not guaranteed [8]. If all operations are to proceed without blocking, though, it is necessary to weaken the consistency guarantees to weak fork-linearizability [6], for instance. COP is wait-free and never blocks because it aborts non-commuting operations that cannot proceed.

The *Blind Stone Tablet (BST)* protocol [29], the closest predecessor of this work, supports an encrypted remote database hosted by an untrusted server that is accessed by multiple clients. Its consistency checking algorithm allows some commuting client operations to proceed concurrently, but only to a limited extent, as we explain below. Every client has to maintain the complete service state and to execute all operations, in contrast to this work. Furthermore, the BST protocol guarantees fork-linearizability only for database state updates, but does not ensure it for all responses output by a client.

SPORC [13] considers a groupware collaboration service whose operations may not commute, but can be made to commute by applying operational transformations. Through this mechanism, different execution orders still converge to the same state. All SPORC operations are wait-free but respect only fork-* linearizability, which is weaker than fork-linearizability.

Contributions. This paper considers a generic service executed by an untrusted server and provides new protocols for consistency verification through fork-linearizable semantics. More concretely, it introduces the Commutative-Operation verification Protocol (COP) and its extension to Authenticated COP (called ACOP) with the following properties:

1. COP is the first wait-free, abortable consistency verification protocol that emulates an arbitrary functionality on a Byzantine server with fork-linearizability and exploits commuting operation sequences. (See Sect. 3.)
2. ACOP is the first wait-free fork-linearizable consistency verification protocol for services, where the state is maintained by the server and the clients do not execute every operation. (See Sect. 4.)
3. COP comes with a formal analysis that proves fork-linearizable semantics for generic service execution; previous work did not establish this notion.

COP and ACOP follow the general pattern of most previous fork-linearizable emulation protocols. For determining when to proceed with concurrent operations, we consider *sequences* of operations that jointly commute and the state of the service, in contrast to earlier protocols, which considered only isolated operations. For lack of space, some definitions and the formal analysis are contained in the full version [7].

For computations supported by suitable authenticated data structures, ACOP enables *authenticated remote computation*, where operations are executed by the server and the clients no longer need to maintain the state of the computation. In contrast to previous work, this enables ACOP to handle services with large state.

1.1 Related Work

Storage protocols. Fork-linearizability has been introduced (under the name of *fork consistency*) together with the SUNDR storage system [23, 18]. Conceptually SUNDR operates on storage objects with simple read/write semantics. Subsequent work of Cachin et al. [8] improves the efficiency of untrusted storage protocols. A lock-free storage protocol with abortable operations, which lets all operations complete in the absence of step contention, has been proposed by Majuntke et al. [21].

FAUST [6] and Venus [27] go beyond the fork-linearizable consistency guarantee and model occasional message exchanges among the clients. This allows FAUST and Venus to obtain stronger semantics, in the sense that they eventually reach consistency (i.e., linearizability) or detect server misbehavior. In the model considered here, fork-linearizability is the best possible guarantee [23]. The relation of these protocols and others to COP is summarized in Tab. 1.

Table 1. Summary of related protocols. In this table under *function*, the BST protocol supports only a *single* commuting operation and does not achieve wait-freedom (as indicated by the parentheses in the first column); SPORC is wait-free for generic functions that have *operational transforms*; COP and ACOP are wait-free for generic *commuting* operation sequences. *Weak fork-linearizability* (or *fork-* consistency*) allows the last operation of a client to be inconsistent compared to *fork-linearizability*; however, BST and SPORC do not guarantee their consistency notion for client responses, only for state changes that may occur much later (as indicated by the parentheses). The *execution* column indicates whether the *clients* compute operations and maintain state or whether this is done by the *server*.

Protocol	Wait-free	Function	Consistency	Execution
SUNDR [23, 18]	—	storage	fork-lin.	server
FAUST & Venus [6, 27]	✓	storage	weak fork-lin.	server
BST [29]	(✓)	single comm. op.	(fork-lin.)	clients
SPORC [13]	✓	generic o.-t. op.	(weak fork-lin.)	clients
COP (Sec. 3)	✓	generic comm. op.	fork-lin.	clients
ACOP (Sec. 4)	✓	generic comm. op.	fork-lin.	server

Blind Stone Tablet (BST). The BST protocol [29] considers transactions on a database, coordinated by the remote server. A client first *simulates* a transaction on its own copy, potentially generating *local output*, then coordinates with the server for ordering the transaction. From the server's response it determines if a transaction commutes with other, pending transactions invoked by different clients that were reported by the server. If they conflict, the client undoes the transaction and basically aborts; otherwise, he commits the transaction and relays it via the server to other clients. When a client receives such a relayed transaction, the client *applies* the transaction to its database copy.

BST has several limitations: First, because a client applies his own transactions only when all pending transactions by other clients have been applied to his own state, updates induced by his transactions are delayed in dependence on other clients. Thus, he cannot always execute his next transaction from the modified state and produce the correct output. This implies the client is blocked and the protocol is not "wait-free" as claimed [29]. Second, the notion of "trace consistency" in the analysis of BST considers only transactions that have been applied to the local state, not the responses as required to satisfy fork-linearizability. However, a transaction may be applied long after its response was output, hence, client operations might not be fork-linearizable. In contrast, the analysis of COP shows it is fork-linearizable for all *responses* output by clients. Finally, every client in BST maintains a copy of the database and replays all operations locally, which is not necessary in ACOP.

COP extends BST and allows one client to execute multiple operations independently of the other clients, as long as his *sequence* of operations jointly commutes with the *sequence* of pending operations by other clients, considering the current service *state*. BST considers only the commutativity of individual operations. Note that two operations o_1 and o_2 may independently commute with an operation o_3 from a particular starting state, but their concatenation, $o_1 \circ o_2$, may not commute with o_3. Operation sequences and state-based commutativity have recently been exploited for building scalable services on multicore systems [10].

Non-blocking protocols. SPORC [13] is a group collaboration system where operations do not need to be executed in the same order at every client by virtue of employing *operational transforms*. The latter concept allows for shifting operations to a different position in an execution by transforming them according to properties of the skipped operations. Differently ordered and transformed variants of a common sequence converge to the same end state. SPORC is claimed to provide fork-* linearizability [19], which is almost the same as weak fork-linearizability [6]; both notions are strict relaxations of fork-linearizability that permit concurrent operations to proceed without blocking, such that protocols become wait-free. The increased concurrency is traded for weaker consistency, as up to one diverging operation may exist between two clients. Moreover, there is no formal analysis for SPORC. As in BST, SPORC addresses only the updates of client states and does *not* consider *local outputs*; however, for showing linearizability, one has to consider the respones of operations.

FAUST [6], mentioned before, never blocks clients and enjoys eventual consistency, but guarantees only weak fork-linearizability. Abortable operations have been introduced in this context by Majuntke et al. [21] for data storage.

In contrast to SPORC and FAUST, COP ensures the stronger fork-linearizability condition, where every operation is consistent as soon as it completes. In terms of expressiveness, SPORC is neither weaker nor stronger than COP: On one hand, SPORC seems more general as it never blocks clients even for operations that do not appear to commute; on the other hand, SPORC is limited to functions with transformable operations and does not address conflicting operations (which exist in some functions [8]); COP, however, works for arbitrary functions.

In BST and SPORC, all clients execute all operations. ACOP eliminates this drawback and shifts the state and the computation to the server by exploiting the notion of authenticated data structures, as suggested by Cachin [3] in a more restricted setting. In storage protocols (SUNDR and FAUST), clients do not "execute" each other's operations due to the limited functionality.

Last but not least, the protocol of Cachin [3] provides also fork-linearizable execution for generic services like COP. However, the protocol is inherently blocking.

2   Definitions

System model. We consider an asynchronous distributed system with n clients, $C_1, \ldots, C_n$ and a server S, modeled as processes. Each client is connected to the server through an asynchronous, reliable communication channel that respects FIFO order. A protocol specifies the operations of the processes. All clients are *correct* and follow the protocol, whereas S operates in one of two modes: either she is *correct* and follows the protocol or she is *Byzantine* and may deviate arbitrarily from the specification.

Functionality. We consider a deterministic *functionality* F (also called a type) defined over a set of *states* $\mathcal{S}$ and a set of *operations* $\mathcal{O}$. F takes as arguments a state $s \in \mathcal{S}$ and an operation $o \in \mathcal{O}$ and returns a tuple (s', r), where $s' \in \mathcal{S}$ is a state that reflects any changes that o caused to s and $r \in \mathcal{R}$ is a response to o i.e., $(s', r) = F(s, o)$. This is also called the *sequential specification* of F.

We extend this notation for executing a sequence of operations $\langle o_1, \ldots, o_k \rangle$, starting from an initial state s_0, and write $(s', r) = F(s_0, \langle o_1, \ldots, o_k \rangle)$ for $(s_i, r_i) = F(s_{i-1}, o_i)$ with $i = 1, \ldots, k$ and $(s', r) = (s_k, r_k)$. Note that an operation in $\mathcal{O}$ may represent a batch of multiple application-level operations.

Commutative Operations. Commutative operations of F play a role in protocols that may execute multiple operations concurrently. Two operations $o_1, o_2 \in \mathcal{O}$ are said to *commute in a state* s if and only if these operations, when applied in different orders starting from s, yield the same respective states and responses. Formally, if $(s', r_1) \leftarrow F(s, o_1)$, $(s'', r_2) \leftarrow F(s', o_2)$; and $(t', q_2) \leftarrow F(s, o_2)$, $(t'', q_1) \leftarrow F(t', o_1)$, then $r_1 = q_1$, $r_2 = q_2$, and $s'' = t''$. Furthermore, we say two operations $o_1, o_2 \in \mathcal{O}$ *commute* when they commute in any state of $\mathcal{S}$.

Also sequences of operations can commute. Suppose two sequences ρ_1 and ρ_2 consisting of operations in $\mathcal{O}$ are mixed together into one sequence π such that the partial order among the operations from ρ_1 and from ρ_2 is retained in π, respectively. If executing π starting from a state s gives the same respective responses and the same final state as for every other such mixed sequence, in particular for $\rho_1 \circ \rho_2$ and for $\rho_2 \circ \rho_1$, where $\circ$ denotes concatenation, we say that ρ_1 and ρ_2 *commute in state* s. Analogously, we say that ρ_1 and ρ_2 *commute* if they commute in any state.

Operations that do not commute are said to *conflict*. We define a Boolean predicate $commute_F(s, \rho_1, \rho_2)$ that is true if and only if ρ_1 and ρ_2 commute in s according to F. W.l.o.g. we assume all operations of F and $commute_F$ are efficiently computable.

Abortable services. When operations of F conflict, a protocol may either decide to block or to abort. Aborting and giving the client a chance to retry the operation at his own rate often has advantages compared to blocking, which might delay an application in unexpected ways.

As in previous work that permitted aborts [1, 21], we allow operations to abort and augment F to an *abortable* functionality F' accordingly. F' is defined over the same set of states $\mathcal{S}$ and operations $\mathcal{O}$ as F, but returns a tuple defined over $\mathcal{S}$ and $\mathcal{R} \cup \{\bot\}$. F' may return the same output as F, but F' may also return $\bot$ and leave the state unchanged, denoting that a client is not able to execute F. Hence, F' is a non-deterministic relation and satisfies $F'(s, o) = \{(s, \bot), F(s, o)\}$. Since F' is not deterministic, a sequence of operations no longer uniquely determines the resulting state and response value.

Abortable functionalities may be seen as obstruction-free objects [1, 15] and vice versa; such objects guarantee that every client operation completes assuming the client eventually runs in isolation.

Operations, histories, and consistency properties. Clients interact with F via operations. Every operation at a client C_i is associated with an *invocation* and a *response* event that occurs at C_i. We say that C_i *executes* an operation between the corresponding invocation and response events. We use the standard notions of events, precedence, and histories.

The condition of *linearizability* [16] requires that the operations of all clients appear to execute atomically in one sequence, and its extension to *fork-linearizability* [23, 8], which relaxes the condition of one sequence to permit multiple "forks" of an execution. Under fork-linearizability, every client observes a linearizable history and when some operation is observed by multiple clients, the history of events up to this operation is the same.

Our protocol provides a *fork-linearizable Byzantine emulation* [8] of the service on an untrusted server. This notion ensures two dual properties: first, when the server is correct, then the service should guarantee the standard notion of linearizability; otherwise, the protocol should ensure fork-linearizability to the clients. Formal definitions appear in the full version [7].

Cryptography. We make use of two cryptographic primitives, namely a collision-free hash function *hash* and a digital signature scheme, with operations denoted by $sign_i$ and $verify_i$ for signatures computed by C_i. As our focus lies on concurrency and correctness and not on cryptography, we model both as ideal, deterministic functionalities implemented by a trusted entity (see [4]).

3 The Commutative-Operation Verification Protocol

Notation. The function *length*(a) for a list a denotes the number of elements in a and $\|$ denotes concatenation of strings. Several variables are *dynamic arrays* or *maps*, which associate keys to values. A value is stored in a map H by assigning it to a key, denoted $H[k] \leftarrow v$; if no value has been assigned to a key, the map returns $\bot$. Recall that F' is the abortable extension of functionality F.

Overview. COP, presented in Algorithms 1–3, adopts the structure of previous protocols that guarantee fork-linearizable semantics [23, 29, 3]. It aims at obtaining a globally consistent order for the operations of all clients, as determined by the server.

When a client C_i invokes an operation o, he sends an INVOKE message to the server S. He expects to receive a REPLY message from S telling him about the position of o in the global sequence of operations. The message contains the operations that are *pending* for o, that is, operations that C_i may not yet know and that are ordered before o by a correct S. (A Byzantine S may introduce consistency violations here.) We distinguish between *pending-other* operations invoked by other clients and *pending-self* operations, which are operations executed by C_i up to o.

Client C_i then verifies that the data from the server is consistent. If this or any other verification step fails, the formal protocol simply halts; in practice, the clients would then recover the service state, abandon the faulty S, and switch to another provider. In order to ensure fork-linearizability for the response values, the client first simulates the pending-self operations and tests if o *commutes* with the pending-other operations. If the test succeeds, he declares o to be *successful*, executes o, and computes the response r according to F; otherwise, O is *aborted* and the response is $r = \bot$. According to this, the *status* of o is a value in $\mathcal{Z} = \{\text{SUCCESS}, \text{ABORT}\}$. Through these steps the client *commits* o. Then he sends a corresponding COMMIT message to S and outputs r.

The (correct) server records the committed operation and relays it to all clients via a BROADCAST message. When the client receives such a broadcast operation, he verifies that it is consistent with everything the server told him so far. If this verification succeeds, we say that the client *confirms* the operation. If the operation's status was SUCCESS, then the client executes it and *applies* it to his local state.

Data structures. Every client locally maintains a set of variables during the protocol. The state $s \in S$ is the result of applying all successful operations, received in BROADCAST messages, to the initial state s_0. Variable c stores the sequence number of the last operation that the client has confirmed. H is a map containing a *hash chain* computed over the global operation sequence as announced by S. The contents of H are indexed by the sequence number of the operations. Entry $H[l]$ is computed as $hash(H[l-1]\|o\|l\|i)$, with $H[0] = $ NULL, and represents an operation o with sequence number l executed by C_i. (The notation $\|$ stands for concatenating values as bit strings.) A variable u is set to o whenever the client has invoked an operation o but not yet completed it; otherwise u is $\bot$. Variable Z maps the sequence number of every operation that the client has executed himself to the status of the operation. The client only needs the entries in Z with index greater than c.

The (correct) server also keeps several variables locally. She stores the invoked operations in a map I and the completed operations in a map O, both indexed by sequence number. Variable t determines the global sequence number for the invoked operations. Finally, variable b is the sequence number of the last broadcast operation and ensures that S disseminates operations to clients in the global order.

Protocol. When client C_i invokes an operation o, he stores it in u and sends an INVOKE message to S containing o, c, and τ, a digital signature computed over o and i. In turn, a correct S sends a REPLY message with the list ω of pending operations; they have a sequence number greater than c. Upon receiving a REPLY message, the client checks that ω is consistent with any previously sent operations and uses ω to assemble the successful pending-self operations μ and the pending-other operations γ. He then determines whether o can be executed or has to be aborted.

In particular, during the loop in Algorithm 1, for every operation o in ω, C_i determines its sequence number l and verifies from the digital signature that o was indeed invoked by C_j. He computes the entry of o in the hash chain from o, l, j, and $H[l-1]$. If $H[l] = \bot$, then C_i stores the hash value there. Otherwise, $H[l]$ has already been set and C_i verifies that the hash values are equal; this means that o is consistent with the pending operation(s) that S has sent previously with indices up to l.

If operation o is his own and its saved status in $Z[l]$ was SUCCESS, then he appends it to μ. The client remembers the status of his own operations in Z, since *commute$_F$* depends on the state and that could have changed if he applied operations after committing o.

Finally, when C_i reaches the end of ω (i.e., when C_i considers $o = u$), he checks that ω is not empty and that it contains u at the last position. He then creates a temporary state a by applying μ to the current state s, and tests whether u commutes with the pending-other operations γ in a. If they do, he records the status of u as SUCCESS in $Z[l]$ and computes the response r by executing u on state a. If u does not commute with γ, he sets status of u to ABORT and $r \leftarrow \bot$. Then C_i signs u together with its sequence number, status, and hash chain entry $H[l]$ and includes all values in the COMMIT message sent to S.

Algorithm 1. Commutative-operation verification protocol (client C_i)

State

    $u \in \mathcal{O} \cup \{\bot\}$: the operation being executed currently or $\bot$ if no operation runs, initially $\bot$

    $c \in \mathbb{N}_0$: sequence number of the last operation that has been confirmed, initially 0

    $H : \mathbb{N}_0 \to \{0,1\}^*$: hash chain (see text), initially containing only $H[0] = \text{NULL}$

    $Z : \mathbb{N}_0 \to \mathcal{Z}$: status map (see text), initially empty

    $s \in \mathcal{S}$: current state, after applying operations, initially s_0

upon invocation o do

    $u \leftarrow o$

    $\tau \leftarrow sign_i(\text{INVOKE}\|o\|i)$

    send message $[\text{INVOKE}, o, c, \tau]$ to S

upon receiving message $[\text{REPLY}, \omega]$ from S **do**

    $\gamma \leftarrow \langle \rangle$ // list of pending-other operations

    $\mu \leftarrow \langle \rangle$ // list of successful pending-self operations

    $k \leftarrow 1$

    while $k \leq length(\omega)$ **do**

        $(o, j, \tau) \leftarrow \omega[k]$

        $l \leftarrow c + k$ // promised sequence number of o

        if not $verify_j(\tau, \text{INVOKE}\|o\|j)$ **then**

            halt

        if $H[l] = \bot$ **then**

            $H[l] \leftarrow hash(H[l-1]\|o\|l\|j)$ // extend hash chain

        else if $H[l] \neq hash(H[l-1]\|o\|l\|j)$ **then**   // server replies are inconsistent

            halt

        if $j = i \wedge Z[l] = \text{SUCCESS} \wedge k < length(\omega)$ **then**

            $\mu \leftarrow \mu \circ \langle o \rangle$

        else if $j \neq i$ **then**

            $\gamma \leftarrow \gamma \circ \langle o \rangle$

        $k \leftarrow k + 1$

    if $k = 1 \vee o \neq u \vee j \neq i$ **then**   // variables o, j, and $l = c + length(\omega)$ keep their values

        halt   // last pending operation must equal the current operation

    $(a, r) \leftarrow F(s, \mu)$ // compute temporary state with successful pending-self operations

    if $commute_F(a, \langle u \rangle, \gamma)$ **then**   // $u = o$ is the current operation

        $(a, r) \leftarrow F(a, u)$ // compute response to u

        $Z[l] \leftarrow \text{SUCCESS}$

    else

        $r \leftarrow \bot$

        $Z[l] \leftarrow \text{ABORT}$

    $\phi \leftarrow sign_i(\text{COMMIT}\|u\|l\|H[l]\|Z[l])$

    send message $[\text{COMMIT}, u, l, H[l], Z[l], \phi]$ to S

    $u \leftarrow \bot$

    return r

Algorithm 2. Commutative-operation verification protocol (client C_i, continued)

---

upon receiving message [BROADCAST, o, q, h, z, ϕ, j] from S **do**

    if not $\big(q = c + 1$ **and** $verify_j(\phi, \text{COMMIT}\|o\|q\|h\|z)\big)$ **then**   // server replies are not consistent

        halt

    if $H[q] = \bot$ **then**   // operation has not been pending at client

        $H[q] \leftarrow hash(H[q-1]\|o\|q\|j)$

    if $h \neq H[q]$ **then**

        halt   // server replies are not consistent

    if $z = \text{SUCCESS}$ **then**   // at this point, the operation is confirmed

        $(s, r) \leftarrow F(s, o)$ // apply the operation and ignore response

    $c \leftarrow c + 1$

Algorithm 3. Commutative-operation verification protocol (server S)

State

    $t \in \mathbb{N}_0$: sequence number of the last invoked operation, initially 0

    $b \in \mathbb{N}_0$: sequence number of the last broadcast operation, initially 0

    $I : \mathbb{N} \to \mathcal{O} \times \mathbb{N}_0 \times \{0,1\}^*$: invoked operations (see text), initially empty

    $O : \mathbb{N} \to \mathcal{O} \times \{0,1\}^* \times \mathcal{Z} \times \{0,1\}^* \times \mathbb{N}$: committed operations (see text), initially empty

upon receiving message [INVOKE, o, c, τ] from C_i **do**

    $t \leftarrow t + 1$

    $I[t] \leftarrow (o, i, \tau)$

    $\omega \leftarrow \langle I[b+1], \ldots, I[t] \rangle$ // include non-committed operations and o

    send message [REPLY, ω] to C_i

upon receiving message [COMMIT, o, q, h, z, ϕ] from C_i **do**

    $O[q] \leftarrow (o, h, z, \phi, i)$

    while $O[b+1] \neq \bot$ **do**   // broadcast operations ordered by their sequence number

        $b \leftarrow b + 1$

        $(o', h', z', \phi', j) \leftarrow O[b]$

        send message [BROADCAST, o', b, h', z', ϕ', j] to all clients

---

Upon receiving a COMMIT message for an operation o with sequence number q, the (correct) server records its content as $O[q]$ in the map of committed operations. Then she is supposed to send a BROADCAST message containing $O[q]$ to the clients. She waits with this until she has received COMMIT messages for all operations with sequence number less than q and broadcast them. This ensures that completed operations are disseminated in the global order to all clients. Waiting here leads to blocking in BST, as mentioned in the Introduction. In COP, this does not forbid clients from progressing with their own operations as we explain below.

In a BROADCAST message received by client C_i, the committed operation is represented by a tuple (o, q, h, z, ϕ, j). The client conducts several verification steps; if successful, we say o is *confirmed*. Subsequently he *applies* o to his state s. In more detail, the client first verifies that the sequence number q is the next operation according to c; hence, o follows the global order and the server did not omit any operations. Second, he uses the digital signature ϕ on the message to verify that C_j indeed committed o. Lastly, C_i computes his own hash-chain entry $H[q]$ for o and confirms that it is equal to the hash-chain value h from the message. This ensures that C_i and C_j have received consistent operations from S up to o. Once the verification succeeds, the client applies o to his state s only if its status z was SUCCESS, that is, when C_j has not aborted o.

Commuting operation sequences. Consider the following example F of a counter restricted to non-negative values: Its state consists of an integer s; an $add(x)$ operation adds x to s and returns TRUE; a $dec(x)$ operation subtracts x from s and returns TRUE if $x \leq s$, but does nothing and returns FALSE if $x > s$. Suppose the current state s at C_i is 7 and C_i executes $dec(4)$ and subsequently $dec(6)$. During both operations of C_i, the server announces that $add(2)$ by another client is pending. Note that C_i executes $dec(4)$ successfully but aborts $dec(6)$ because $dec(6)$ does not commute with $add(2)$ from 3, the temporary state (a in Algorithm 1) computed by C_i after the pending-self operation. However, the latter two operations, $add(2)$ and $dec(6)$, do commute in the current state 7. This shows why the client executes the pending-self operations before testing the current operation for a conflict.

Suppose now the current state s is again 7 and C_i executes $dec(4)$. The server reports the pending sequence $\langle dec(2), dec(3) \rangle$. Thus, C_i aborts $dec(4)$. Even though $dec(4)$ commutes with $dec(2)$ and with $dec(3)$ individually in state 7, it does not commute with their sequence. This illustrates why COP checks for a conflict with the sequence of pending operations.

Memory requirements. For saving storage space, the client may garbage-collect entries of H and Z with sequence numbers smaller than c. The server can also save space by removing the entries in I and O for the operations that she has broadcast. However, if new clients are allowed to enter the protocol, the server should keep all operations in O and broadcast them to new clients upon their arrival.

With the above optimizations the client has to keep only pending operations in H and pending-self operations in Z. The same holds for the server: the maximum number of entries stored in I and O is proportional to the number of pending operations at any client.

Communication. Every operation executed by a client requires him to perform one roundtrip to the server: send an INVOKE message and receive a REPLY. For every executed operation the server simply sends a BROADCAST message. Clients do not communicate with each other in the protocol. However, as soon as they do, they benefit from fork-linearizability and can easily discover a forking attack by comparing their hash chains.

Messages INVOKE, COMMIT, and BROADCAST are independent of the number of clients and contain only a description of one operation, while the REPLY message contains the list of pending operations ω. If even one client is slow, then the length of ω for all other clients grows proportionally to the number of further operations they are executing. To reduce the size of REPLY messages, the client can remember all pending operations received from S, and S can send every pending operation only once.

Aborts and wait-freedom. Every client executing COP can proceed with an operation o for F as long as it does not conflict with pending operations of other clients. Observe that the state used by the client for executing o reflects all of his own operations executed so far, even if he has not yet confirmed or applied them to his state because operations of other clients have not yet completed. After successfully executing o, the client outputs the response immediately after receiving the REPLY message from S. A conflict arises when o does not commute with the pending operations of other clients. In this case, the client aborts o and outputs $\perp$, according to F'.

Hence, for F where all operations and operation sequences commute, COP is wait-free. For arbitrary F, however, no fork-linearizable Byzantine emulation can be wait-free [8]. COP avoids blocking via the augmented functionality F'. Clients complete every operation in the sense of F', which includes aborts; therefore, COP is wait-free for F'. In other words, regardless of whether an operation aborts or not, the client may proceed executing further operations.

To mitigate the risk of conflicts, the clients may employ a synchronization mechanism such as a contention manager, scheduler, or a simple random waiting strategy. Such synchronization is common for services with strong consistency demands. If one considers also clients that may crash (outside our formal model), then the client group has to be adjusted dynamically or a single crashed client might hold up progress of other clients forever. Previous work on the topic has explored how a group manager or a peer-to-peer protocol may control a group membership protocol [18, 27]; these methods apply also to COP.

Analysis. COP emulates the abortable functionality F' on a Byzantine server with fork-linearizability. Furthermore, all histories of COP where the clients execute operations sequentially are fork-linearizable w.r.t. F (no operations abort), and if, additionally, the server is correct, then all such histories are also linearizable w.r.t. F. Here we give only a brief summary of this result; the details appear in the full version [7].

There are two points to consider. First, with a correct S, we show that the output of every client satisfies F' also in the presence of many pending-self operations. The check for commutativity, applied after simulating the client's pending-self operations, ensures that the client's response is the same as if the pending-other operations would have been executed before the operation itself.

The second main innovation lies in the construction of a view for every client that includes all operations that he has executed or applied, together with those of his operations that some other clients have confirmed. Since these operations may have changed the state at other clients, they must be considered. More precisely, some C_k may have confirmed an operation o executed by C_i that C_i has not yet confirmed or applied. In order to be fork-linearizable, the view of C_i must include o as well, including all operations that were "promised" to C_i by S in the sense that they were announced by S as pending for o. It follows from the properties of the hash chain that the view of C_k up to o is the same as C_i's view including the promised operations. The view of C_i further includes all operations that C_i has executed after o. Taken together this demonstrates that every execution of COP is fork-linearizable w.r.t. F'.

4 Authenticated Computation

In this section, we introduce *Authenticated COP* or *ACOP* , which shifts state maintenance and service execution to the server and lets clients only perform verification. ACOP extends COP with an authenticated data structure [24] for the service functionality. It enables *authenticated remote computation* for many realistic services with complex interfaces[12, 25, 9, 17], such as indexed databases, search trees, document processing services, and generic storage schemes; typically their operations permit queries and updates. Recent advances in cryptographic tools for verifying remote computation suggest that it may even become feasible to construct authenticators for generic computations while preserving the privacy of the inputs [14, 2].

4.1 Authenticated COP

We consider a server that stores shared state and executes operations of the functionality F invoked by clients. When F supports an *authenticated data structure* [24], the clients may verify the integrity of a response to an operation from a cryptographic proof in the form of an authenticator for the response. ACOP results from integrating the authenticated data structure into COP and ensures the fork-linearizability of the service, retaining all other benefits of COP.

More formally, suppose S maintains the state of F in variable x, called the *server's state*; when S receives an operation o from a client, she should update the state by executing $(x', r) \leftarrow F(x, o)$ and send the response r to the client. For adding authentication, the server's state is extended to include authentication data, and an authenticator α is computed with the response as $(x', \alpha, r) \leftarrow$ authexec$_F(x, o)$. The server sends r together with α to the client. The client maintains a *digest* d between operations, which authenticates the (potentially large) state of F maintained by S. For checking the correctness of the response, the client computes $(d', r') \leftarrow$ verify$_F(d, \alpha, o, r)$, whereby $r' = \bot$ indicates that the verification failed, and otherwise, $r' = r$ is the correct response. The authexec$_F$ and verify$_F$ operations encapsulate the authenticated data structure; more information can be found in the rich literature on the subject [28, 22]. For practical authentication techniques such as hash trees and authenticated dictionaries, α is usually much smaller than the full state.

We now describe how to extend Algorithms 1–3 for ACOP.

4.2 Server

We start with the changes for S. As part of her state, S additionally maintains a state map $X : \mathbb{N}_0 \to \{0, 1\}^*$ indexed by operations, where $X[0] = s_0$ is the initial state. Entry $X[b]$ is assigned when the server broadcasts an operation with sequence number b such that $X[b]$ contains the result of executing the operations with sequence numbers from $1, \ldots, b$.

When the server receives the INVOKE message from C_i with an operation o, she increments the index t and considers the pending operations ω with index between b and t. Then S executes the pending-self operations ν of C_i, which include o, to obtain the response and authenticator for o as $(x', \alpha, r) \leftarrow$ authexec$_F(X[b], \nu)$; she sends ω

and r to C_i together with α. Note that x' is discarded and that S uses $X[b]$ to compute the result using the operation sequence ν, which includes o, as C_i has only applied the operations with sequence numbers $1, \ldots, b$ at the time when he invokes o.

In COP the client checks for commutativity between an invoked operation and the pending operations by himself. With the above modification, S also needs to abort operations as the client would determine from $commute_F$ when computing r and α, and S must include additional information that allows the client to execute $commute_F$. In practice, the server may store only the latest state $X[b]$ and the changes induced by the operations with lower sequence numbers. Moreover, once S learns from INVOKE messages that all clients have received and applied all operations with sequence number q, then she may discard the state changes for q as well.

4.3 Client

The clients no longer maintain state s and instead store a digest map $G : \mathbb{N}_0 \to \{0, 1\}^*$ indexed by operations, where $G[q]$ authenticates the state resulting from executing the operations with index up to q, starting from s_0. The client uses G to verify the server's responses to his operations in a REPLY message. In particular, for operation o, client C_i runs Algorithm 1, executes its pending-self operations (μ) upon input $G[c]$ to obtain a temporary state a and a corresponding digest g, performs the commutativity check, and, if successful, computes $(d', r') \leftarrow verify_F(g, \alpha, o, r)$. The client halts if the original algorithm halts or if $r' = \bot$; otherwise, the response is $r \leftarrow r'$. The client augments the COMMIT message with α and r' and signs the entire message. Note that d' is again used only temporarily for verifying the pending-self operations and is discarded when the method returns.

Upon receiving a BROADCAST message when the last confirmed operation has index c, the client verifies the signature from client C_j that invoked the operation and the hash value as before. Then C_i intends to verify that the response and digest are consistent (between him and C_j) and to compute the next digest $G[c + 1]$. Note that C_i cannot use α, however, to update the digest, as α authenticates o in the state where C_j committed it, but this state may differ from the state at index c, which is current for C_i. We therefore require that S sends an additional authenticator α' for o in state $X[c]$. The client verifies that α' and r correspond to o by executing $(G[c + 1], r') \leftarrow verify_F(G[c], \alpha', o, r)$, and verifying that $r' \neq \bot$. The client may garbage-collect entries in G in a similar way as for the hash chain in COP.

5   Conclusion

This paper has introduced COP and ACOP, two variants of the Commutative-Operation verification Protocol, which allow a group of clients to execute a generic service coordinated by a remote untrusted server. COP ensures fork-linearizability and allows clients to easily verify the consistency and integrity of the service responses. In contrast to previous work, COP is wait-free and supports commuting operation sequences (but may sometimes abort conflicting operations); ACOP extends COP by shifting state and operation execution from the clients to the server.

Given the popularity of outsourced computation and cloud computing, the problem of checking the results of remote computations cryptographically has received a lot of attention recently [11, 26, 14, 2]. However, these protocols typically address only a two-party model and, with some exceptions [2], do not support state changes. An important direction for future work lies in integrating these verifiable computation protocols into COP and related protocols for guaranteeing cryptographic integrity in the sense of fork-linearizability for multiple clients.

Acknowledgments. We thank Marcus Brandenburger for interesting discussions and valuable comments.

This work has been supported in part by the European Union's Seventh Framework Programme (FP7/2007–2013) under grant agreement number ICT-257243 TCLOUDS.

References

[1] Aguilera, M.K., Frølund, S., Hadzilacos, V., Horn, S.L., Toueg, S.: Abortable and query-abortable objects and their efficient implementation. In: Proc. 26th ACM Symposium on Principles of Distributed Computing (PODC) (2007)

[2] Braun, B., Feldman, A.J., Ren, Z., Setty, S.T.V., Blumberg, A.J., Walfish, M.: Verifying computations with state. In: Proc. 24th ACM Symposium on Operating Systems Principles (SOSP), pp. 341–357 (2013)

[3] Cachin, C.: Integrity and consistency for untrusted services. In: Černá, I., Gyimóthy, T., Hromkovič, J., Jefferey, K., Královič, R., Vukolić, M., Wolf, S. (eds.) SOFSEM 2011. LNCS, vol. 6543, pp. 1–14. Springer, Heidelberg (2011)

[4] Cachin, C., Guerraoui, R., Rodrigues, L.: Introduction to Reliable and Secure Distributed Programming, 2nd edn. Springer (2011)

[5] Cachin, C., Keidar, I., Shraer, A.: Fork sequential consistency is blocking. Information Processing Letters 109(7), 360–364 (2009)

[6] Cachin, C., Keidar, I., Shraer, A.: Fail-aware untrusted storage. SIAM Journal on Computing 40(2), 493–533 (2009), preliminary version appears In: Proc. DSN 2009

[7] Cachin, C., Ohrimenko, O.: Verifying the consistency of remote untrusted services with commutative operations. Report arXiv:1302.4808v2, CoRR (December 2013),
http://arxiv.org/abs/1302.4808v2

[8] Cachin, C., Shelat, A., Shraer, A.: Efficient fork-linearizable access to untrusted shared memory. In: Proc. 26th ACM Symposium on Principles of Distributed Computing (PODC), pp. 129–138 (2007)

[9] Canetti, R., Paneth, O., Papadopoulos, D., Triandopoulos, N.: Verifiable set operations over outsourced databases. In: Krawczyk, H. (ed.) PKC 2014. LNCS, vol. 8383, pp. 113–130. Springer, Heidelberg (2014)

[10] Clements, A.T., Kaashoek, M.F., Zeldovich, N., Morris, R.T., Kohler, E.: The scalable commutativity rule: Designing scalable software for multicore processors. In: Proc. 24th ACM Symposium on Operating Systems Principles (SOSP), pp. 1–17 (2013)

[11] Cormode, G., Mitzenmacher, M., Thaler, J.: Practical verified computation with streaming interactive proofs. In: Proc. 3rd Conference on Innovations in Theoretical Computer Science (ITCS), pp. 90–112 (2012)

[12] Crosby, S.A., Wallach, D.S.: Authenticated dictionaries: Real-world costs and trade-offs. ACM Transactions on Information and System Security 14(2) (2011)

[13] Feldman, A.J., Zeller, W.P., Freedman, M.J., Felten, E.W.: SPORC: Group collaboration using untrusted cloud resources. In: Proc. 9th Symp. Operating Systems Design and Implementation (OSDI) (2010)

[14] Gennaro, R., Gentry, C., Parno, B., Raykova, M.: Quadratic span programs and succinct NIZKs without PCPs. In: Johansson, T., Nguyen, P.Q. (eds.) EUROCRYPT 2013. LNCS, vol. 7881, pp. 626–645. Springer, Heidelberg (2013)

[15] Herlihy, M., Luchangco, V., Moir, M.: Obstruction-free synchronization: Double-ended queues as an example. In: Proc. 23rd Intl. Conference on Distributed Computing Systems, (ICDCS) (2003)

[16] Herlihy, M.P., Wing, J.M.: Linearizability: A correctness condition for concurrent objects. ACM Transactions on Programming Languages and Systems 12(3), 463–492 (1990)

[17] Kosba, A.E., Papadopoulos, D., Papamanthou, C., Sayed, M.F., Shi, E., Triandopoulos, N.: TRUESET: Nearly practical verifiable set computations. In: Proc. 23rd USENIX Security Symposium (2014)

[18] Li, J., Krohn, M., Mazières, D., Shasha, D.: Secure untrusted data repository (SUNDR). In: Proc. 6th Symp. Operating Systems Design and Implementation (OSDI), pp. 121–136 (2004)

[19] Li, J., Mazières, D.: Beyond one-third faulty replicas in Byzantine fault-tolerant systems. In: Proc. 4th Symp. Networked Systems Design and Implementation (NSDI) (2007)

[20] Mahajan, P., Setty, S., Lee, S., Clement, A., Alvisi, L., Dahlin, M., Walfish, M.: Depot: Cloud storage with minimal trust. In: Proc. 9th Symp. Operating Systems Design and Implementation (OSDI) (2010)

[21] Majuntke, M., Dobre, D., Serafini, M., Suri, N.: Abortable fork-linearizable storage. In: Abdelzaher, T., Raynal, M., Santoro, N. (eds.) OPODIS 2009. LNCS, vol. 5923, pp. 255–269. Springer, Heidelberg (2009)

[22] Martel, C., Nuckolls, G., Devanbu, P., Gertz, M., Kwong, A., Stubblebine, S.G.: A general model for authenticated data structures. Algorithmica 39, 21–41 (2004)

[23] Mazières, D., Shasha, D.: Building secure file systems out of Byzantine storage. In: Proc. 21st ACM Symposium on Principles of Distributed Computing (PODC) (2002)

[24] Naor, M., Nissim, K.: Certificate revocation and certificate update. IEEE Journal on Selected Areas in Communications 18(4), 561–570 (2000)

[25] Papamanthou, C., Tamassia, R., Triandopoulos, N.: Optimal verification of operations on dynamic sets. In: Rogaway, P. (ed.) CRYPTO 2011. LNCS, vol. 6841, pp. 91–110. Springer, Heidelberg (2011)

[26] Setty, S., Vu, V., Panpalia, N., Braun, B., Blumberg, A.J., Walfish, M.: Taking proof-based verified computation a few steps closer to practicality. In: Proc. 21st USENIX Security Symposium (2012)

[27] Shraer, A., Cachin, C., Cidon, A., Keidar, I., Michalevsky, Y., Shaket, D.: Venus: Verification for untrusted cloud storage. In: Proc. Cloud Computing Security Workshop (CCSW). ACM (2010)

[28] Tamassia, R.: Authenticated data structures. In: Di Battista, G., Zwick, U. (eds.) ESA 2003. LNCS, vol. 2832, pp. 2–5. Springer, Heidelberg (2003)

[29] Williams, P., Sion, R., Shasha, D.: The blind stone tablet: Outsourcing durability to untrusted parties. In: Proc. Network and Distributed Systems Security Symposium (NDSS) (2009)

Logical Physical Clocks

Sandeep S. Kulkarni[1], Murat Demirbas[2],
Deepak Madappa[2], Bharadwaj Avva[2], and Marcelo Leone[1]

[1] Computer Science & Engineering, Michigan State University
[2] Computer Science & Engineering, University at Buffalo, SUNY

Abstract. There is a gap between the theory and practice of distributed systems in terms of the use of time. The theory of distributed systems shunned the notion of time, and introduced "causality tracking" as a clean abstraction to reason about concurrency. The practical systems employed physical time (NTP) information but in a best effort manner due to the difficulty of achieving tight clock synchronization. In an effort to bridge this gap and reconcile the theory and practice of distributed systems on the topic of time, we propose a hybrid logical clock, HLC, that combines the best of logical clocks and physical clocks. HLC captures the causality relationship like logical clocks, and enables easy identification of consistent snapshots in distributed systems. Dually, HLC can be used in lieu of physical/NTP clocks since it maintains its logical clock to be always close to the NTP clock. Moreover HLC fits in to 64 bits NTP timestamp format, and is masking tolerant to NTP kinks and uncertainties. We show that HLC has many benefits for *wait-free* transaction ordering and performing snapshot reads in multiversion globally distributed databases.

1 Introduction

1.1 Brief History of Time

Logical Clock (LC). LC [15] was proposed in 1978 by Lamport as a way of timestamping and ordering events in a distributed system. LC is divorced from physical time (e.g., NTP clocks): the nodes do not have access to clocks, there is no bound on message delay and on the speed/rate of processing of nodes. The causality relationship captured, called happened-before (**hb**), is defined based on passing of information, rather than passing of time [15]. While being beneficial for the theory of distributed systems, LC is impractical for today's distributed systems: 1) Using LC, it is not possible to query events in relation to physical time. 2) For capturing **hb**, LC assumes that all communication occurs in the present system and there are no backchannels. This is obsolete for today's integrated, loosely-coupled system of systems.

In 1988, the vector clock (VC) [9,22] was proposed to maintain a vectorized version of LC. VC maintains a vector at each node which tracks the knowledge this node has about the logical clocks of other nodes. While LC finds one consistent snapshot (that with same LC values at all nodes involved), VC finds all possible consistent snapshots, which is useful for debugging applications. In Figure 1, while LC would find (a,w) as a consistent cut, VC would also identify (b,w), (c,w) as consistent cuts. Unfortunately,

M.K. Aguilera et al. (Eds.): OPODIS 2014, LNCS 8878, pp. 17–32, 2014.

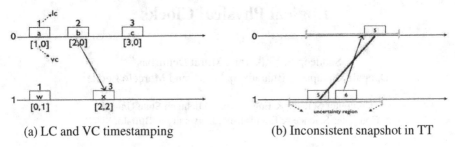

(a) LC and VC timestamping   (b) Inconsistent snapshot in TT

Fig. 1. LC/VC timestamping and TT timestamping

the space requirement of VC is on the order of nodes in the system, and is prohibitive, and it stays prohibitive with optimizations (e.g., [25]) that reduce the size of VC.

Physical Time (PT). PT leverages on physical clocks at nodes that are synchronized using the Network Time Protocol (NTP) [23]. Since perfect clock synchronization is infeasible for a distributed system [24], there are uncertainty intervals associated with PT. While PT avoids the disadvantages of LC by using physical time for timestamping, it introduces new disadvantages: 1) When the uncertainty intervals are overlapping, PT cannot order events. NTP can usually maintain time to within tens of milliseconds over the public Internet, and can achieve one millisecond accuracy in local area networks under ideal conditions, however, asymmetric routes and network congestion can occasionally cause errors of 100 ms or more. 2) PT has several kinks such as leap seconds [16, 17] and non-monotonic updates to POSIX time [10] which may cause the timestamps to go backwards.

TrueTime (TT). TrueTime is proposed recently by Google for developing Spanner [2], a multiversion distributed database. TT relies on a well engineered tight clock synchronization available at all nodes thanks to GPS clocks and atomic clocks made available at each cluster. However, TT introduces new disadvantages: 1) TT requires special hardware and a custom-build tight clock synchronization protocol, which is infeasible for many systems (e.g., using leased nodes from public cloud providers). 2) If TT is used for ordering events that respect causality then it is essential that if e hb f then $tt.e < tt.f$. Since TT is purely based on clock synchronization of physical clocks, to satisfy this constraint, Spanner delays event f when necessary. Such delays and reduced concurrency are prohibitive especially under looser clock synchronization.

HybridTime (HT). HT, which combines VC and PT clocks, was proposed for solving the stabilizing causal deterministic merge problem [13]. HT maintains a VC at each node which includes knowledge this node has about the PT clocks of other nodes. HT exploits the clock synchronization assumption of PT clocks to trim entries from VC and reduces the overhead of causality tracking. In practice the size of HT at a node would only depend on the number of nodes that communicated with that node within the last ϵ time, where ϵ denotes the clock synchronization uncertainty. Recently, Demirbas and Kulkarni [3] explored how HT can be adopted to solve the consistent snapshot problem in Spanner [2].

1.2   Contributions of This Work

In this paper we aim to bridge the gap between the theory (LC) and practice (PT) of timekeeping and timestamping in distributed systems and to provide guarantees that generalize and improve that of TT.

- We present a logical clock version of HT, which we name as Hybrid Logical Clocks (HLC). HLC refines both the physical clock (similar to PT and TT) and the logical clock (similar to LC). HLC maintains its logical clock to be always close to the NTP clock, and hence, HLC can be used in lieu of physical/NTP clock in several applications such as snapshot reads in distributed key value stores and databases. Most importantly, HLC preserves the property of logical clocks (e hb $f \Rightarrow hlc.e < hlc.f$) and as such HLC can identify and return consistent global snapshots without needing to wait out clock synchronization uncertainties and without needing prior coordination, in a posteriori fashion.
- HLC is backwards compatible with NTP, and fits in the 64 bits NTP timestamp. Moreover, HLC works as a superposition on the NTP protocol (i.e., HLC only reads the physical clocks and does not update them) so HLC can run alongside applications using NTP without any interference. Furthermore HLC is general and does not require a server-client architecture. HLC works for a peer-to-peer node setup across WAN deployment, and allows nodes to use different NTP servers.[1] In Section 3, we present the HLC algorithm and prove a tight bound on the space requirements of HLC and show that the bound suffices for HLC to capture the LC property for causal reasoning.
- HLC provides masking tolerance to common NTP problems (including nonmonotonous time updates) and can make progress and capture causality information even when time synchronization has degraded. HLC is also self-stabilizing fault-tolerant [4] and is resilient to arbitrary corruptions of the clock variables, as we discuss in Section 4.
- We implement HLC and provide experiment results of HLC deployments under various deployment scenarios. In Section 5, we show that even under stress-testing, HLC is bounded and the size of the clocks remain small. These practical bounds are much smaller than the theoretical bounds proved in our analysis. Our HLC implementation is made available in an anonymized manner at `https://github.com/AugmentedTimeProject`
- HLC has direct applications in identifying consistent snapshots in distributed databases [2, 14, 18, 19, 27, 29]. It is also useful in many distributed systems protocols including causal message logging in distributed systems [1], Byzantine fault-tolerance protocols [11], distributed debugging [26], distributed filesystems [21], and distributed transactions [30]. In Section 6, we showcase the benefits of HLC for snapshot reads in distributed databases. An open source implementation of Spanner [2] that uses HLC is available at `https://github.com/cockroachdb/cockroach`.

---

[1] HLC can also work with ad hoc clock synchronization protocols [20] and is not bound to NTP.

2 Preliminaries

A distributed system consists of a set of nodes whose number may change over time. Each node can perform three types of actions, a send action, a receive action, and a local action. The goal of a timestamping algorithm is to assign a timestamp to each event. We denote a timestamping algorithm with an all capital letters name, and the timestamp assigned by this algorithm by the corresponding lower case name. E.g., we use LC to denote the logical clock algorithm by Lamport [15], and use $lc.e$ to denote the timestamp assigned to event e by this algorithm.

The notion of happened before hb captures the causal relation between events in the system. As defined in [15], event e happened before event f (denoted by e hb f) is a transitive relation that respects the following: e and f are events on the same node and e occurred before f, or e is a send event and f is the corresponding receive event. We say that e and f are concurrent, denoted by $e\|f$, iff $\neg(e$ hb $f) \wedge \neg(f$ hb $e)$. Based on the existing results in the literature, the following are true:

$$e \text{ hb } f \Rightarrow lc.e < lc.f$$
$$lc.e = lc.f \Rightarrow e\|f$$
$$e \text{ hb } f \Leftrightarrow vc.e < vc.f$$

3 HLC: Hybrid Logical Clocks

In this section, we introduce our HLC algorithm starting with a naive solution first. We then prove correctness and tight bounds on HLC. We also elaborate on the useful features of the HLC for distributed systems.

3.1 Problem Statement

The goal of HLC is to provide one-way causality detection similar to that provided by LC, while maintaining the clock value to be always close to the physical/NTP clock. The formal problem statement for HLC is as follows.

Given a distributed system, assign each event e a timestamp, $l.e$, such that

1. e hb $f \Rightarrow l.e < l.f$,
2. Space requirement for $l.e$ is $O(1)$ integers,
3. $l.e$ is represented with bounded space,
4. $l.e$ is *close* to $pt.e$, i.e., $|l.e - pt.e|$ is bounded.

The first requirement captures one-way causality information provided by HLC. The second requirement captures that the space required for $l.e$ is $O(1)$ integers. To prevent encoding of several integers into one large integer, we require that any update of $l.e$ is achieved by $O(1)$ operations. The third requirement captures that the space required to represent $l.e$ is bounded, i.e., it does not grow in an unbounded fashion. In practice, we like $l.e$ to be the size of $pt.e$, which is 64 bits in the NTP protocol.

Finally, the last requirement states that $l.e$ should be close to $pt.e$. This enables us to utilize HLC in place of PT. To illustrate this consider the case where the designer wants to take a snapshot at (physical) time t. Given that physical clocks are not perfectly synchronized, it is not possible to get a consistent snapshot by just reading state at different nodes at time t as shown in Figure 1. On the other hand, using HLC we can obtain such a snapshot by taking the snapshot of every node at *logical time t*. Such a snapshot is guaranteed to be consistent, because from the HLC requirement 1 we have $l.e = l.f \Rightarrow e \| f$. In Section 6, we discuss in more detail how HLC enables users to take uncoordinated a-posteriori consistent snapshots of the distributed system state.

3.2 Description of the Naive Algorithm

Given the goal that $l.e$ should be close to $pt.e$, in the naive algorithm we begin with the rule: *for any event e, $l.e \geq pt.e$.* We design our algorithm as shown in Figure 2. This algorithm works similar to LC. Initially all l values are set to 0. When a send event, say f, is created on node j, we set $l.f$ to be $max(l.e+1, pt.j)$, where e is the previous event on node j. This ensures $l.e < l.f$. It also ensures that $l.f \geq pt.f$. Likewise, when a receive event f is created on node j, $l.f$ is set to $max(l.e + 1, l.m + 1, pt.j)$, where $l.e$ is the timestamp of the previous event on j, and $l.m$ is the timestamp of the message (and, hence, the send event). This ensures that $l.e < l.f$ and $l.m < l.f$.

Initially $lc.j := 0$

Send or local event
$l.j := max(l.j + 1, pt.j)$
Timestamp with $l.j$

Receive event of message m
$l.j := max(l.j + 1, l.m + 1, pt.j)$
Timestamp with $l.j$

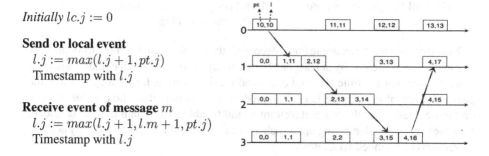

Fig. 2. Naive HLC algorithm for node j **Fig. 3.** Counterexample for Naive Algorithm

It is easy to see that the algorithm in Figure 2 satisfies the first two requirements in the problem statement. However, this naive algorithm violates the fourth requirement, which also leads to a violation of the third requirement for bounded space representation. To show the violation of the fourth requirement, we point to the counterexample in Figure 3 which shows how $|l.e - pt.e|$ grows in an unbounded fashion. The messaging loop among nodes 1, 2, and 3 can be repeated forever, and at each turn of the loop the drift between logical clock and physical clock (the $l - pt$ difference) will keep growing.

The root of the unbounded drift problem is due to the naive algorithm using l to maintain both the maximum of pt values seen so far and the logical clock increments from new events (local, send, receive). This makes the clocks lose information: it becomes unclear if the new l value came from pt (as in the message from node 0 to node 1) or from causality (as is the case for the rest of messages). As such, there is no suitable place to reset l value to bound the $l - pt$ difference, because resetting l may lead to losing the hb relation, and, hence, a violation of requirement 1.

Initially $l.j := 0; c.j := 0$

Send or local event
$l'.j := l.j;$
$l.j := max(l'.j, pt.j);$
If $(l.j = l'.j)$ then $c.j := c.j + 1$
    Else $c.j := 0;$
Timestamp with $l.j, c.j$

Receive event of message m
$l'.j := l.j;$
$l.j := max(l'.j, l.m, pt.j);$
If $(l.j = l'j = l.m)$
        then $c.j := max(c.j, c.m) + 1$
Elseif $(l.j = l'.j)$ then $c.j := c.j + 1$
Elseif $(l.j = l.m)$ then $c.j := c.m + 1$
Else $c.j := 0$
Timestamp with $l.j, c.j$

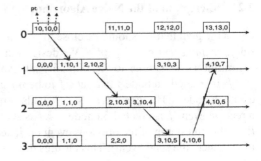

Fig. 4. HLC algorithm for node j **Fig. 5.** Fixing the Counterexample in Figure 3 with Algorithm in Figure 4

Note that the counterexample holds even with the requirement that the physical clock of a node is incremented by at least one between any two events on that node. However, if we assume that the time for send event and receive event is long enough so that the physical clock of *every* node is incremented by at least one, then the counterexample on Figure 3 fails, and the naive algorithm would be able to maintain $|l - pt|$ bounded. However, instead of depending upon such assumption, we show how to properly achieve correctness of bounded HLC, next.

3.3 HLC Algorithm

We use our observations from the counterexample to develop the correct HLC algorithm. In this algorithm, the $l.j$ in the naive algorithm is expanded to two parts: $l.j$ and $c.j$. The first part $l.j$ is introduced as a level of indirection to maintain the maximum of pt information learned so far, and c is used for capturing causality updates only when l values are equal.

In contrast to the naive algorithm where there was no suitable place to reset l without violating hb , in the HLC algorithm, we can reset c when the information heard about maximum pt catches up or goes ahead of l. Since l denotes the maximum pt heard among nodes and is not continually incremented with each event, within a bounded time, either one of the following is guaranteed to occur: 1) a node receives a message with a larger l, and its l is updated and c is reset to reflect this, or 2) if the node does not hear from other nodes, then its l stays the same, and its pt will catch up and update its l, and reset the c.

The HLC algorithm is as shown in Figure 4. Initially, l and c values are set to 0. When a new send event f is created, $l.j$ is set to $max(l.e, pt.j)$, where e is the previous event on j. Similar to the naive algorithm, this ensures that $l.j \geq pt.j$. However, because we have removed the "+1", it is possible that $l.e$ equals $l.f$. To deal with this, we utilize the value of $c.j$. By incrementing $c.j$, we ensure that $\langle l.e, c.e \rangle < \langle l.f, c.f \rangle$ is true with lexicographic comparison.[2] If $l.e$ differs from $l.f$ then $c.j$ is reset, and this allows us to guarantee that c values remain bounded. When a new receive event is created, $l.j$ is set to $max(l.e, l.m, pt.j)$. Now, depending on whether $l.j$ equals $l.e$, $l.m$, both or neither, $c.j$ is set.

Let's reconsider the counterexample to the naive algorithm. This example replayed with the HLC algorithm is shown in Figure 5. When we continue the loop among nodes 1, 2, 3, we see that pt at nodes 1, 2 and 3 catches up and exceeds $l = 10$ and resets c to 0. This keeps the c variable bounded at each node.

To prove the correctness of the HLC algorithm as well as to prove that it satisfies requirement 4 (closeness between HLC value and PT), we present the following two theorems, whose proofs follow trivially from HLC implementation. (Proofs of other theorems is presented in [12].)

Theorem 1. *For any two events e and f, e hb $f \Rightarrow (l.e, c.e) < (l.f, c.f)$*   □

Theorem 2. *For any event f, $l.f \geq pt.f$*   □

Theorem 3. *$l.f$ denotes the maximum clock value that f is aware of. In other words,*
$l.f > pt.f \Rightarrow (\exists g : g$ hb $f \wedge pt.g = l.f)$

Physical clocks are synchronized to be within ϵ. Hence, we cannot have two events e and f such that e hb f and $pt.e > pt.f + \epsilon$. Hence, combining this with Theorem 3, we have

Corollary 1. *For any event f, $|l.f - pt.f| \leq \epsilon$*

Finally, we prove requirement 3, by showing that c value of HLC is bounded as well. To this end, we extend Theorem 3 to identify the relation of c and events created at a particular time. As we show in Theorem 4, $c.f$ captures information regarding events created at time $l.f$.

Theorem 4. *For any event f,*
$c.f = k \wedge k > 0$
$\Rightarrow \quad (\exists g_1, g_2, \cdots, g_k :$
$\quad\quad (\forall j : 1 \leq j < k : g_i$ hb $g_{i+1}) \wedge (\forall j : 1 \leq j \leq k : l.(g_i) = l.f) \wedge g_k$ hb $f)$

From Theorem 4, the following corollary follows.

Corollary 2. *For any event f $c.f \leq |\{g : g$ hb $f \wedge l.g = l.f)\}|$.*

Theorem 5. *For any event f, $c.f \leq N * (\epsilon + 1)$*

We note that the above bound is almost tight and can be shown to be so with an example similar to that in Figure 3. However, if we assume that message transmission delay is large enough so that the physical clock of every process is increased by at least d, where d is a given parameter, we can reduce the bound on c further. For reasons of space, the proof of this claim is relegated to [12].

---

[2] $(a, b) < (c, d)$ iff $((a < c) \quad \vee \quad ((a = c) \wedge (b < d)))$.

3.4 Properties of HLC

HLC algorithm is designed for arbitrary distributed architecture and is also readily applicable to other environments such as the client-server model.

We intentionally chose to implement HLC as a superposition on NTP. In other words, HLC only reads the physical clock but does not update it. Hence, if a node receives a message whose timestamp is higher, we maintain this information via l and c instead of changing the physical clock. This is crucial in ensuring that other programs that use NTP alone are not affected. This also avoids the potential problem where clocks of nodes are synchronized with each other even though they drift substantially from *real wall-clock*. Furthermore, there are impossibility results showing that accepting even tiny unsynchronization to adjust the clocks can lead to diverging clocks [8]. Finally, while HLC utilizes NTP for synchronization, it does not depend on it. In particular, even when physical clocks utilize any ad hoc clock synchronization algorithm [20], HLC can be superposed on top of such a service, so can also be used in ad hoc networks.

4 Resilience of HLC

4.1 Self-stabilization

Here we discuss how we design self-stabilizing [4] fault-tolerance to HLC, which enables HLC to be eventually restored to a legitimate state, even when HLC is perturbed/corrupted to an arbitrary state.

Stabilization of HLC rests on the superposition property of HLC on NTP clocks. Since HLC does not modify the NTP clock, it does not interfere with the NTP correcting/synchronizing the physical clock of the node. Once the physical/NTP clock stabilizes, HLC can be corrected based on observations in Theorem 2 and Corollaries 5 and 2. These results identify the maximum permitted value of $l - pt$ and the maximum value of c. In the event of extreme clock errors by NTP or transient memory corruption, the application may reach a state where these bounds are violated. In that case, we take the physical clock as the authority, and reset l and c values to pt and 0 respectively. In other words the stabilization of HLC follows that of stabilization of pt via NTP clock.

In order to contain the spread of corruptions due to bad HLC values, we have a rule to ignore out of bounds messages. We simply ignore reception of messages that cause l value to diverge too much from pt. This prevention action fires if the sender of the message is providing a clock value that is significantly higher suggesting the possibility of corrupted clock. In order to contain corruptions to c, we make its space bounded, so that even when it is corrupted, its corruption space is limited. This way c would in the worst case roll over, or more likely, c would be reset to an appropriate value as a result of l being assigned a new value from pt or from another l received in a message.

Note that both the reset correction action and the ignore out-of-bounds message action are local correction actions at a node. If HLC fires either of these actions, it also logs the offending entries for inspection and raises an exception to notify the administrator.

4.2 Masking of Synchronization Errors

In order to make HLC resilient to common NTP synchronization errors, we assign sufficiently large space to $l - pt$ drift so that most (99.9%) NTP kinks can be masked smoothly. While Theorem 2 and Corollaries 5 and 2 state that $l - pt$ stay within ϵ the clock synchronization uncertainty (crudely two times the NTP offset value), we set a very conservative value, Δ, on the $l - pt$ bound. The bound Δ can be set to a constant factor of ϵ, and even on the order of seconds depending on the application semantics. This way we tolerate and mask common NTP clock synchronization errors within normal operation of HLC. And when Δ bound is violated, the local reset correction action and the ignore message prevention action fire as discussed in the previous subsection.

Using this approach, HLC is robust to stragglers, nodes with pt stuck slightly in the past. Consider a node that lost connection to its NTP server and its clock started drifting behind the NTP time. Such a straggler can still keep up with the system for some time and maintain up-to-date and bounded HLC time: As long as it receives messages from other nodes, it will learn new/higher l values and adopt them. This node will increment its c by 1 when it does not adopt a new l value, but this does not cause the c rise excessively for the other nodes in the system. Even if this node sends a message with high c number, the other nodes will have up-to-date time and ignore that c and will use $c = 0$. Similarly, HLC is also robust to the rushers, nodes with pt slightly ahead of others. The masking tolerance of HLC makes it especially useful for last write wins (LWW) database systems like Cassandra [10, 17]. We investigate this tolerance empirically in the next section.

5 Experiments

5.1 AWS Deployment Results

The experiments used Amazon AWS xlarge instances running Ubuntu 14.04. The machines were synchronized to a stratum 2 NTP server, 0.ubuntu.pool.ntp.org. In our basic setup, we programmed all the instances to send messages to each other continuously using TCP sockets, and in a separate thread receive messages addressed to them. The total messages sent range from 75,000 to 425,000.

Using the basic setup (all nodes are senders and sending to each other) within the same AWS region, we get the following results. The value "c" indicates that the value of the c component of the HLC at the nodes. The remaining columns show the frequency: the percentage of times the HLC at the nodes had the corresponding c values out of the total number of events. For each setup, we collected data with two different NTP synchronization levels, indicated by the average offset of nodes' clocks from NTP. When we allow the NTP daemons at the nodes more time (a couple hours) to synchronize, we get lower NTP offset values. We used "ntpdc -c loopinfo" and "ntpdc -c kerninfo" calls to obtain the NTP offset information at the nodes.

The experiments with 4 nodes show that the value of c remains very low, less than 4. This is a much lower bound than the worst case possible theoretical bound we proved in Section 3. We also see that the improved NTP synchronization helps move the c distribution toward lower values, but this effect becomes more visible in the 8 and 16 node experiments. With the looser NTP synchronization, with average offset 5 ms, the maximum $l - pt$ difference was observed to be 21.7 ms. The 90th percentile of $l - pt$ values correspond to 7.8 ms, with their average value computed to be 0.2 ms. With the tighter NTP synchronization, with average offset 1.5 ms, the maximum $l - pt$ difference was observed to be 20.3 ms. The 90th percentile of $l - pt$ values correspond to 8.1 ms, with their average value computed to be 0.2 ms.

The experiments with 8 nodes highlights the lowered c values due to improved NTP synchronization. For the experiments with average NTP offset 9ms, the maximum $l - pt$ difference was observed to be 107.9 ms. The 90th percentile of $l - pt$ values correspond to 41.4 ms, with their average value computed to be 4.2 ms. For the experiments with average NTP offset 3ms, the maximum $l - pt$ difference was observed to be 7.4 ms. The 90th percentile of $l - pt$ values correspond to 0.1 ms, with their average value computed to be 0 ms.

Using 8 m1.xlarge nodes

c	offset=9ms	offset=3ms
0	65.56 %	91.18 %
1	15.39 %	8.82 %
2	8.14 %	0 %
3	5.90 %	
4	2.74 %	
5	1.39 %	
6	0.56 %	
7	0.20 %	
8	0.08 %	
9	0.03 %	

Using 4 m1.xlarge nodes

c	offset=5ms	offset=1.5ms
0	83.90 %	83.66 %
1	12.12 %	12.03 %
2	3.37 %	4.09 %
3	0.24 %	0.21 %

Using 16 m1.xlarge nodes

c	offset=16ms	offset=6ms
0	66.96 %	75.43 %
1	19.40 %	18.51 %
2	7.50 %	3.83 %
3	4.59 %	1.84 %
4	1.76 %	0.32 %
5	0.61 %	0.06 %
6	0.14 %	0.01 %
7	0.02 %	

The 16 node experiments also showed very low c values despite all nodes sending to each other at practically at the wire speed. For the experiments with average NTP offset 16ms, the maximum $l - pt$ difference was observed to be 90.5 ms. The 90th percentile of $l - pt$ values correspond to 25.2 ms, with their average value computed to be 2.3 ms. For the experiments with average NTP offset 6ms, the maximum $l - pt$ difference was observed to be 46.8 ms. The 90th percentile of $l - pt$ values correspond to 8.4 ms, with their average value computed to be 0.3 ms.

WAN Deployment Results. We deployed our HLC testing experiments on a WAN environment as well. Specifically, we used 4 m1.xlarge instances each one located at a different AWS region: Ireland, US East, US West and Tokyo. Our results show that with 3ms NTP offset, the $c = 0$ values constitute about 95% of the cases and $c = 1$ constitute the remaining 5%. These values are much lower than the corresponding values for the single datacenter deployment. The maximum $l - pt$ difference remained extremely low, about 0.02 ms, and the 90th percentile of $l - pt$ values corresponded to 0. These values are again much lower than the corresponding values for the single datacenter deployment.

The reason for seeing very low $l - pt$ and c values in the WAN deployment is because the message communication delays across WAN are much larger than the ϵ, the clock synchronization uncertainty. As a result, when a message is received, its l timestamp is already in the past and is smaller than the l value at the receiver which is updated by its pt. Since the single cluster deployment with short message delays is the most demanding scenario in terms of HLC testing we focused on those results in our presentation.

5.2   Stress Testing and Resilience Evaluation in Simulation

To further analyze the resiliency of HLC, we evaluated it in scenarios where it will be stressed, e.g., where the event rate is too high and where the clock synchronization is significantly degraded. In our simulations, we considered the case where the event creation rate was 1 event per millisecond and clock drift varies from $10ms$ to $100ms$. Given the relation between l and pt from Theorem 2, the drift between l and pt is limited to the clock drift. Hence, we focus on values of c for different events.

In these simulations, a node is allowed to advance its physical clock by 1ms as long as its clock drift does not exceed beyond ϵ. If a node is allowed to advance its physical clock then it increases it with a 50% probability. When it advances its clock, it can send a message with certain probability (All simulations in this section correspond to the case where this probability is 100%). We deliver this message at the earliest possible feasible time, essentially making delivery time to be 0. The results are as shown in Figure 6. As shown in these figures, the distribution of c values was fairly independent of the value of ϵ. Moreover, for more than 99% of events, the c value was 4 or less. Less than 1% of events had c values of 5-8.

To evaluate HLC in the presence of degraded clock synchronization, we added a straggler node to the system. This node was permitted to violate clock drift constraints by always staying behind. We consider the case where the straggler just resides at the end of permissible boundary, i.e., its clock drift from the highest clock is ϵ. We also consider the case where straggler violates the clock drift constraints entirely and it is upto 5ϵ behind the maximum clock. The results are as shown in Figures 7 and 8. Even with the straggler, the c value for 99% events was 4 or less. However, in these simulations, significantly higher c values were observed for some events. In particular, for the case where the straggler remained just at the end of permissible boundary, events with c value of upto 97 were observed at the straggler node. For the case where the straggler was permitted to drift by 5ϵ, c value of upto 514 was observed again only at the straggler node. The straggler node did not raise the c values of other nodes in the system.

We also conducted the experiments where we had a rusher, a node that was excessively ahead. Figures 9 and 10 demonstrate the results. The maximum c value observed in these experiments was 8. And, the number of events with c value greater than 3 is less than 1%.

As a result of these experiments we conclude that the straggler node affects the c value more than the rusher node, but only for itself. In our experiments, each node selects the sender randomly with uniform distribution. Hence, messages sent by the rusher node do not have a significant cumulative effect. However, messages sent by all nodes to the straggler node causes its c value to grow.

6   Discussion

In this section, we discuss application of HLC for finding consistent snapshots in distributed databases, compact representations of l and c, and other related work.

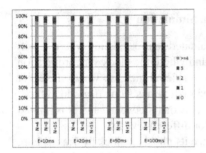

Fig. 6. c value distribution for varying ϵ

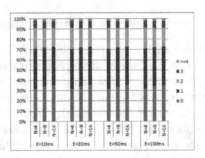

Fig. 7. c value distribution with ϵ straggler

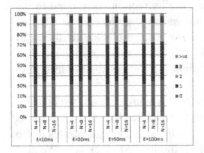

Fig. 8. c value distribution with a 5ϵ straggler

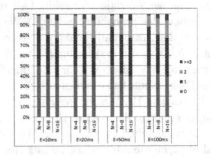

Fig. 9. c value distribution with a ϵ rusher

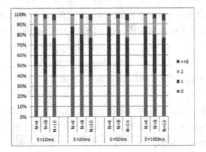

Fig. 10. c value distribution with a 5ϵ rusher

6.1 Snapshots

In snapshot read, the client is interested in obtaining a snapshot of the data at a given time. HLC can be used to perform snapshot read similar to that performed by TrueTime. In other words, with HLC, each process simply needs to choose the values with a given timestamp (as described below) to obtain a consistent snapshot. Unlike approaches with VC where checking concurrency of the chosen events is necessary, the events chosen by our approach are guaranteed to be concurrent with each other. Moreover, unlike TT, there is no need to delay any transaction due to uncertainty in the clock values.

To describe our approach more simply, we introduce the concept of virtual dummy events. Let e and f be two events on the same node such that $l.e < l.f$. In this case, we introduce dummy (internal) events whose l value is in the range $[l.e + 1, l.f]$ and $c.f = 0$. (If $c.f = 0$ then the last event in the sequence is not necessary.) Observe that introducing such dummy events does not change timestamps of any other events in the system. However, this change ensures that for any time t, there exists an event on every node where l value equals t and c value equals 0. With the virtual dummy events adjustment, given a request for snapshot read at time t, we can obtain the values at timestamp $\langle l = t, c = 0 \rangle$.[3] Our adjustment ensures that such events are guaranteed to exist. And, by the logical clock hb relationship mentioned in requirement 2, we have $hlc.e = hlc.f \Rightarrow e \| f$ and so we can conclude that the snapshots taken at this time are consistent with each other and form a consistent global snapshot. Moreover, based on Theorem 3 and Corollary 2, this snapshot corresponds to the case where the global time is in the window $[t - \epsilon, t]$. We refer the reader to Figure 11 for an example of finding consistent snapshot read at time $t = 10$.

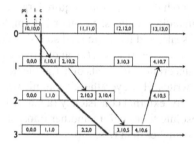

Fig. 11. Consistent snapshot for $t = 10$ in HLC trace

6.2 Compact Timestamping Using l and c

NTP uses 64-bit timestamps which consist of a 32-bit part for seconds and a 32-bit part for fractional second. (This gives a time scale that rolls over every 2^{32} seconds—136 years— and a theoretical resolution of 2^{-32} seconds—233 picoseconds.) Using a single 64-bit timestamp to represent HLC is also very desirable for backwards compatibility with NTP. This is important because many distributed database systems and distributed key-value stores use NTP clocks to timestamp and compare records.

There are, however, several challenges for representing HLC as a single 64-bit timestamp. Firstly, the HLC algorithm maintains l and c separately, to differentiate between increases due to the physical clock versus send/receive/local events. Secondly, by tracking the pt, the size of l is by default 64-bits as the NTP timestamps.

We propose the following scheme for combining l and c and storing it in single 64 bit timestamp. This scheme involves restricting l to track only the most significant 48 bits of pt in the HLC algorithm presented in Figure 4. Rounding up pt values to 48 bits l values still gives us microsecond granularity tracking of pt. Given NTP synchronization

---

[3] Actually we can obtain snapshot reads for any $\langle l = t, c = K \rangle$ and not just at $\langle l = t, c = 0 \rangle$.

levels, this is sufficient granularity to represent NTP time. The way we round up pt is to always take the ceiling to the 48th bit. In the HLC algorithm in Figure 4, l is updated similarly but is done for 48 bits. When the l values remain unchanged in an event, we capture that by incrementing c following the HLC algorithm in Figure 4. 16 bits remain for c and allows it room to grow up to 65536, which is more than enough as we show in our experiments in Section 5.

Using this compact representation, if we need to timestamp (message or data item for database storage), we will concatenate c to l to create the HLC timestamp. The distributed consistent snapshot finding algorithm described above is unaffected by this change to the compact representation. The only adjustment to be made is to round up the query time t to 48 bits as well.

6.3 Other Related Work

Dynamo [28] adopts VC as version vectors for causality tracking of updates to the replicas. Orbe [5] uses dependency matrix along with physical clocks to obtain causal consistency. In the worst case, both these solutions require large timestamps. Cassandra uses PT and LWW-rule for updating replicas. Spanner [2] employs TT to order distributed transactions at global scale, and facilitate read snapshots across the distributed database. In order to ensure e hb $f \Rightarrow tt.e < tt.f$ and provide consistent snapshots, Spanner requires waiting-out uncertainty intervals of TT at the transaction commit time which restricts throughput on writes. However, these "commit-waits" also enable Spanner to provide a stronger property, external consistency (a.k.a, strict serializability): if a transaction t1 commits (in absolute time) before another transaction t2 starts, then t1's assigned commit timestamp is smaller than t2's.

HLC does not require waiting out the clock uncertainty, since it is able to record causality relations within this uncertainty interval using the HLC update rules. HLC can also be adopted for providing external consistency and still keeping the throughput on writes unrestricted by introducing client-notification-wait after a transaction ends.

An alternate approach for ordering events is to establish explicit relation between events. This approach is exemplified in the Kronos system [7], where each event of interest is registered with the Kronos service, and the application explicitly identifies events that are of interest from causality perspective. This allows one to capture causality that is application-dependent at the increased cost of searching the event dependency relation graph. By contrast, LC/VC/PT/HLC assume that if a node performs two consecutive events then the second event causally depends upon the first one. Thus, the ordering is based solely on the timestamps assigned to the events.

Clock-SI [6] work considers the snapshot isolation problem for distributed databases/data stores. In contrast to the conventional snapshot isolation implementations that use a centralized timestamp authority for consistent versioning, Clock-SI proposes a way to use NTP-synchronized clocks to assign snapshot and commit timestamps to transactions. HLC improves the Clock-SI solution if it is used instead of NTP-clocks in Clock-SI. HLC avoids incurring the clock-uncertainty wait-out delay in Figure 1 of Clock-SI work [6], because HLC also uses hb information as encoded in HLC clocks.

7 Conclusion

In this paper, we introduced the hybrid logical clocks (HLC) that combines the benefits of logical clocks (LC) and physical time (PT) while overcoming their shortcomings. HLC guarantees that (one way) causal information is captured, and hence, it can be used in place of LC. Since HLC provides nodes a logical time that is within possible clock drift of PT, HLC is substitutable for PT in any application that requires it. HLC is strictly monotonic and, hence, can be used in place of applications in order to tolerate NTP kinks such as non-monotonic updates. HLC can be implemented using 64 bits space, and is backwards compatible with NTP clocks. Moreover, HLC only reads NTP clock values but does not change it. Hence, applications using HLC do not affect other applications that only rely on NTP. HLC is highly resilient. Since its space requirement is bounded by theoretical analysis and is shown to be even more tightly bounded by our experiments, we use this as a foundation to design stabilizing fault tolerance to HLC. snapshot read. Moreover, since the drift between HLC and physical clock is less than the clock drift, a snapshot taken with HLC is an acceptable choice for a snapshot at a given physical time. Thus, HLC is especially useful as a timestamping mechanism in multiversion distributed databases. For example in Spanner, HLC can be used in place of TrueTime (TT) to overcome one of the drawbacks of TT that requires events to be delayed/blocked in the *clock synchronization uncertainty window*. An open source implementation of Spanner that uses HLC is available at https://github.com/cockroachdb/cockroach.

Acknowledgment. The material is based upon work supported by National Science Foundation awards NS-1329807 and CNS-1318678.

References

1. Bhatia, K., Marzullo, K., Alvisi, L.: Scalable causal message logging for wide-area environments. Concurrency and Computation: Practice and Experience 15(10), 873–889 (2003)
2. Corbett, J., Dean, J., et al.: Spanner: Google's globally-distributed database. In: Proceedings of OSDI (2012)
3. Demirbas, M., Kulkarni, S.: Beyond truetime: Using augmentedtime for improving google spanner. In: Workshop on Large-Scale Distributed Systems and Middleware (LADIS) (2013)
4. Dijkstra, E.W.: Self-stabilizing systems in spite of distributed control. Communications of the ACM 17(11) (1974)
5. Du, J., Elnikety, S., Roy, A., Zwaenepoel, W.: Orbe: Scalable causal consistency using dependency matrices and physical clocks. In: Proceedings of the 4th Annual Symposium on Cloud Computing, SOCC 2013, pp. 11:1–11:14. ACM, New York (2013), http://doi.acm.org/10.1145/2523616.2523628
6. Du, J., Elnikety, S., Zwaenepoel, W.: Clock-SI: Snapshot isolation for partitioned data stores using loosely synchronized clocks. In: IEEE Symposium on Reliable Distributed Systems (SRDS), pp. 173–184 (2013)
7. Escriva, R., Dubey, A., Wong, B., Sirer, E.: Kronos: The design and implementation of an event ordering service. In: EuroSys (2014)
8. Fan, R., Lynch, N.: Gradient clock synchronization. In: PODC, pp. 320–327 (2004)
9. Fidge, J.: Timestamps in message-passing systems that preserve the partial ordering. In: Proceedings of the 11th Australian Computer Science Conference, vol. 10(1), pp. 56–66 (1988)

10. Kingsbury, K.: The trouble with timestamps,
    `http://aphyr.com/posts/299-the-trouble-with-timestamps`
11. Kotla, R., Alvisi, L., Dahlin, M., Clement, A., Wong, E.: Zyzzyva: Speculative byzantine fault tolerance. SIGOPS Oper. Syst. Rev. 41(6), 45–58 (2007)
12. Kulkarni, S., Demirbas, M., Madeppa, D., Avva, B., Leone, M.: Logical physical clocks and consistent snapshots in globally distributed databases. Tech. Rep. 2014-04, SUNY Buffalo (May 2014)
13. Kulkarni, S., Ravikant: Stabilizing causal deterministic merge. J. High Speed Networks 14(2), 155–183 (2005)
14. Lakshman, A., Malik, P.: Cassandra: Structured storage system on a p2p network. In: Proceedings of the 28th ACM Symposium on Principles of Distributed Computing, PODC 2009, p. 5 (2009)
15. Lamport, L.: Time, clocks, and the ordering of events in a distributed system. Communications of the ACM 21(7), 558–565 (1978)
16. The future of leap seconds,
    `http://www.ucolick.org/~sla/leapsecs/onlinebib.html`
17. Another round of leapocalypse, `http://www.itworld.com/security/288302/another-round-leapocalypse`
18. Li, C., Porto, D., Clement, A., Gehrke, J., Preguiça, N., Rodrigues, R.: Making geo-replicated systems fast as possible, consistent when necessary. In: Symposium on Operating Systems Design and Implementation (OSDI 2012), pp. 265–278 (2012)
19. Lloyd, W., Freedman, M., Kaminsky, M., Andersen, D.: Don't settle for eventual: Scalable causal consistency for wide-area storage with cops. In: SOSP, pp. 401–416 (2011)
20. Maroti, M., Kusy, B., Simon, G., Ledeczi, A.: The flooding time synchronization protocol. In: SenSys (2004)
21. Mashtizadeh, A., Bittau, A., Huang, Y., Mazières, D.: Replication, history, and grafting in the ori file system. In: SOSP, pp. 151–166 (2013)
22. Mattern, F.: Virtual time and global states of distributed systems. In: Parallel and Distributed Algorithms, pp. 215–226 (1989)
23. Mills, D.: A brief history of ntp time: Memoirs of an internet timekeeper. ACM SIGCOMM Computer Communication Review 33(2), 9–21 (2003)
24. Patt-Shamir, B., Rajsbaum, S.: A theory of clock synchronization (extended abstract). In: ACM Symposium on Theory of Computing (STOC), pp. 810–819 (1994)
25. Saito, Y.: Unilateral version vector pruning using loosely synchronized clocks. Tech. rep., HP Labs (2002)
26. Sigelman, B., Barroso, L., Burrows, M., Stephenson, P., Plakal, M., Beaver, D., Jaspan, S., Shanbhag, C.: Dapper, a large-scale distributed systems tracing infrastructure. Tech. rep., Google, Inc. (2010),
    `http://research.google.com/archive/papers/dapper-2010-1.pdf`
27. Sovran, Y., Power, R., Aguilera, M., Li, J.: Transactional storage for geo-replicated systems. In: SOSP, pp. 385–400 (2011)
28. Vogels, W.: Eventually consistent. Communications of the ACM 52(1), 40–44 (2009)
29. Wu, Z., Butkiewicz, M., Perkins, D., Katz-Bassett, E., Madhyastha, H.: Spanstore: Cost-effective geo-replicated storage spanning multiple cloud services. In: SOSP, pp. 292–308 (2013)
30. Zhang, Y., Power, R., Zhou, S., Sovran, Y., Aguilera, M., Li, J.: Transaction chains: Achieving serializability with low latency in geo-distributed storage systems. In: SOSP, pp. 276–291 (2013)

Be General and Don't Give Up Consistency in Geo-Replicated Transactional Systems

Alexandru Turcu, Sebastiano Peluso, Roberto Palmieri, and Binoy Ravindran

Virginia Tech, ECE Department, Blacksburg VA 24061, USA
{talex,peluso,robertop,binoy}@vt.edu

Abstract. We present ALVIN, a system for managing concurrent transactions running on a set of geographically distributed sites. ALVIN supports general-purpose transactions, and guarantees strong consistency criteria. Through a novel partial order broadcast protocol, ALVIN maximizes the parallelism of ordering and local transaction processing. ALVIN processes read-only transactions either locally or globally, according to the selected consistency criterion, and orders only conflicting transactions across all sites. We built ALVIN in the Go language and conducted an evaluation study relying on the Amazon EC2 infrastructure and Paxos- and EPaxos-based state machine replication protocols as competitors. Our experimental results reveal that ALVIN provides significant speed up for read-dominated TPC-C workloads and on 7 datacenters by as much as 4.8x when compared to EPaxos, and up to 26% in write-intensive workloads.

Keywords: Geo-Replication, Transaction, Distributed System.

1 Introduction

In the recent years, transaction processing on geographically distributed computer systems (or "GDS") received significant research interest [22,12,23,5,17]. Geo-replicated concurrency control protocols can be classified in two approaches. The first approach ensures high consistency, but restricts the type of transactions that are allowed [23,17]. This enables exploiting specific protocol optimizations to achieve high performance. The second approach allows general-purpose transactions, but weakens the consistency criterion for better performance [2,22]. This has the negative effect of reduced programmability, as programmers must cope with potential inconsistent states in application behaviors.

Motivated by this gap between strong consistency/poor performance and weak consistency/good performance, we propose a geo-replicated transactional system called ALVIN, which finds an effective tradeoff between performance and strong consistency. At the core of ALVIN is a novel Partial Order Broadcast protocol (*POB*) that globally orders only conflicting transactions and minimizes the number of communication steps for non-conflicting transactions. While the idea of defining the agreement of consensus on the basis of message semantics is not new and has been previously introduced in Generalized Consensus [13] or Generic

M.K. Aguilera et al. (Eds.): OPODIS 2014, LNCS 8878, pp. 33–48, 2014.
© Springer International Publishing Switzerland 2014

Broadcast [19], POB encompasses a novel approach for ordering transactions' commits that overcomes the limitations of existing single leader-based solutions (i.e., Generalized Paxos [13]) when deployed in GDS. POB does not rely on a designated leader to either order transactions or support conflict resolution in case of conflicting concurrent transactions.

POB has been designed to inherit the benefits of state-of-the-art, multi-leader, state machine replication protocols specifically proposed for GDS such as *Mencius* [16] and *EPaxos* [17], and, at the same time, to overcome their drawbacks. In particular, POB, like Mencius [16], has the advantages of defining the final order of messages on the sender nodes. Typically, this technique avoids expensive distributed decisions by determining an a priori assignment of delivered positions to messages. This approach suffers from potentially expensive waiting conditions that are needed to ensure that the delivery of a message in position p does not precede the delivery of a message in position $p' < p$. However, POB, unlike Mencius, relies on a quorum of replies, instead of waiting for the information about delivered positions from all nodes. This makes POB's performance robust even in scenarios where nodes are far apart (as is often the case in GDS), or when the message sending rate is unbalanced among nodes.

On the other hand, POB, like EPaxos [17], may adjust the order of a message that has been already proposed, according to its dependencies, to reduce communication steps in scenarios of no conflicting proposals of dependent messages. However, unlike EPaxos, POB does not need to build a dependency graph of received messages and avoids the execution of complex tasks on that graph. Such housekeeping operations can be significantly expensive in transaction processing: the number of dependencies in the dependency graph can rapidly grow when a transaction's size and data contention increases.

Roughly, in POB, each node is the leader of transactions originating on it and is responsible for assigning a final position to those transactions. A node has a predefined and exclusive subset of positions that can be used for the assignment. As in Mencius, transactions can be delivered in the order defined by their position numbers. However, unlike Mencius, the delivery of a transaction at a certain position does not need to wait for the notification of all previous positions. This is because, besides a position, a transaction T is associated with a set of dependencies, namely, the set of transactions conflicting with T that must precede T in the order defined by POB. T's leader computes the position and the dependencies of T on the basis of a partial view of the system built by means of quorums. POB ensures that for any pair of transactions T_1 and T_2, if T_1 is in T_2's dependencies, then the position of T_1 is less than the position of T_2. Therefore, a transaction T is delivered on a node after all transactions in T's dependencies have been delivered on that node.

POB's advantages are fully exploited by P-CC, a local parallel concurrency control layer that we propose. P-CC commits non-conflicting transactions in parallel with conflicting transactions, thereby increasing the parallelism.

ALVIN's processing model allows clients to execute transactions locally on the spawning site, whose execution is globally certified against concurrent transactions at other sites. To this goal, POB disseminates transactions and P-CC locally validates and commits them according to the delivery order provided by POB using a timestamp-based multi-versioning scheme. This combination allows all transactions, including those aborted, to always observe a consistent state. This property is mandatory for in-memory deployment in order to avoid unexpected failures due to inconsistent memory accesses [8].

In addition to these features, ALVIN exports design choices to programmers to customize the POB and P-CC according to the needs of the application and system at hand. As an example, ALVIN offers two strong consistency criteria that programmers can select, namely, Serializability (SR) [3] and Extended Update Serializability (EUS) [1,20] (i.e., PL-3U [1]). With the former, transactions that never write (i.e., read-only) must be broadcast through POB. In contrast, with the latter, such transactions execute locally at the cost of generating some non-serializable schedules, which, however, are usually silent to the application. Another example is the potential for computing a fast decision on the transaction delivery order, at the cost of quorum bigger than that for a classic decision.

We built ALVIN in the *Go* programming language and evaluated on the Amazon EC2 infrastructure using up to 7 sites, and benchmarks including Bank [11] and TPC-C [6]. As competitors, we implemented two certification-based transactional systems [18] that rely on MultiPaxos [14] and EPaxos [17] for their ordering layer. Our experiments reveal that ALVIN provides significant speed up for TPC-C workloads and 7 datacenters by as much as 4.8× when compared to EPaxos and configured for exploiting EUS. This significant gain is due to a more efficient execution of read-only workload, which is enabled by EUS's semantics. Rather, if ALVIN runs under SR, it gains up to 26% over EPaxos because it does not pay the cost of graph analysis needed by EPaxos for delivering transactions. On Bank, due to its small transactions and trivial dependency graphs, that cost is not significant, thus EPaxos behaves similarly to ALVIN. MultiPaxos highlights the drawbacks of having a single leader in GDS, thus its performance is lower than other (multi-leader) competitors.

The paper makes the following contributions: *(1)* ALVIN, the first geo-replicated transactional system that guarantees a strong consistency level and supports the execution of general-purpose transactions in classic asynchronous environments; *(2)* a novel multi-leader protocol for partially ordering transactions, enabling high scalability in geo-replicated environments. In addition, the protocol does not need complex local processing for determining the final delivery order, yielding reduced client-perceived latency; *(3)* a publicly available prototype[1], which can be customized for coping with different execution environments.

2   Related Work

Many modern transactional systems employ geo-replication as a means to reduce data access latency and to provide fault-tolerance and disaster recovery.

---

[1] http://www.hyflow.org/software.html

Spanner [5] is Google's globally-replicated database. It provides externally-consistent transactions, but its architecture is complex: it relies on the TrueTime API, which exposes the absolute time and the uncertainty of the time measurement. ALVIN's architecture is more general and suited for easier deployment.

Walter [22] and MDCC [12] are two solutions designed for geo-replicated transactional systems. Walter ensures Parallel Snapshot Isolation, which allows non-conflicting write transactions that span multiple sites to commit even if they observed incompatible histories. ALVIN ensures that all update transactions are serializable. On the other hand, MDCC commits transactions by using one instance of Multi-Paxos [14] (or Generalized Paxos [13] to exploit commutative operations) per replication group containing the accessed data items and, if a transaction touches multiple replication groups, an additional phase is required to reach a consensus among the leaders of the various groups.

Lynx [23] is a geo-distributed transactional storage that works by chopping transactions into sequences of pieces. Each piece executes at a different datacenter, and the system usually replies to clients after the first hop. Lynx's drawback is that it does not tolerate aborts after a chain's first segment.

Finally, we consider EPaxos [17] and Mencius [16] as the closest approaches to ALVIN. EPaxos [17] proposes a partial order protocol for ordering conflicting commands and it uses a per-command leader to avoid the designated leader of (Generalized) Paxos. It considers two types of quorums for executing a command: one is used for implementing a fast-path of one round-trip of communication in case the command does not conflict with other concurrent commands; the other is used in case two phases of communication are required to agree on the order.

EPaxos yields high performance but it has several drawbacks when plugged in transactional processing or in the presence of read operations. In fact, after having agreed on the dependency set for a command, each node adds that command to a dependency graph and its execution is in accordance with an order computed over the strongly connected components of that graph. In case a command represents a transaction or even a read operation, the client has to wait until the command's outcome is available, thus putting the graph analysis into the execution's critical path. ALVIN is not based on graph analysis because dependencies are already available when the transaction attempts to commit, thus resulting in better performance.

At the core of Mencius' [16] ordering protocol there is the fixed assignment of sending slots to nodes. A sender can decide the order of a message only after hearing from all nodes. This approach results in poor performance in case there is a slow or faraway node, as in geo-replication.

3 Assumptions and System Model

We assume a set of geographically distributed sites $\Pi = \{P_1, P_2, \ldots, P_N\}$ that cooperate to synchronize their activities on common shared data. They rely on a wide area network as the communication infrastructure, therefore we assume an asynchronous distributed system. We do not assume any specific distribution of network delays and we do not upper-bound them either. Every message may experience an arbitrarily large, although finite, delay.

Each site (or node) can be seen as a logical representation of a datacenter. Managing the synchronization within each datacenter is an orthogonal problem which we scope out in this paper. Each site is equipped with the entire shared data set, thus transactions running on that site can access data locally.

We assume that the total number of sites is equal to N, where at most $f < \lceil \frac{N}{2} \rceil$ of them can be faulty at any time, thus at least a majority of nodes is always correct. In this paper we assume sites fail according to the crash-stop failure model [3] and we scope out any malicious behavior. In any ordering communication step, a node contacts all the sites and waits for a *quorum Q* of replies. We define two types of quorum size: a *classic quorum* (CQ) size and a *fast quorum* (FQ) size. We assume that both CQ and FQ are at least equal to $\lfloor \frac{N}{2} \rfloor + 1$. This way any two quorums always intersect, thus ensuring that, even though f failures happen, there is always at least one site with the last updated information that we can use for recovering the system. The values assumed by CQ and FQ are configuration-dependent, and they will be specified throughout the presentation of the communication layer.

In order to eventually reach an agreement on the order of transactions when sites are faulty (e.g., a datacenter is unreachable), we assume that the system can be enhanced with the weakest type of unreliable failure detector [10] that is necessary to implement a leader election service [9].

4 Alvin: Geo-Replicated Transactional System

We propose simple object-oriented interfaces, where all accesses (READ, WRITE) to shared objects are enclosed between BEGIN and COMMIT operations.

ALVIN bases its benefits on the exploitation of a partial order of transactions rather than a total order. In fact, ordering all the transactions' commits on all nodes is sufficient to guarantee that all nodes execute the same state transitions, but it is too strong as a condition, especially in GDS, because it enforces that the finalization of a transaction is delayed by the completion of even non-conflicting transactions, thus hampering the system's scalability. On the contrary, enforcing that only conflicting transactions are ordered on all nodes (as in ALVIN) has a twofold benefit: it still guarantees that all nodes eventually converge on a common state, and it allows a degree of parallelism needed for scaling in low inter-datacenter conflict scenarios (which are the expected workloads in GDS).

The software architecture of ALVIN includes two fundamental layers: the Partial Order Broadcast layer (POB) and the Parallel Concurrency Control layer (P-CC). POB is in charge of broadcasting transactions to certify and commit them according to the certification-based approach [18] and in a way such that conflicting transactions are always delivered in the same order on all nodes. P-CC is responsible for optimistically executing transactions by always providing a consistent view of the transactional state, and applying the updates of write transactions that can commit. This makes ALVIN a geo-replication solution also suitable for in-memory transactional systems, which require that all transactions (even those aborted) do not observe incorrect states. This requirement has been defined to be desirable for non-sandboxed environments [8] because reading from an inconsistent snapshot could generate an application's unrecoverable failure.

The transactional application executing on top of the platform is composed of multiple threads balanced on all nodes. According to the certification-based replication scheme [18], each thread activates and executes a transaction T at the same node where it is running, recording objects read from and written to in private spaces called the read-set ($T.RS$) and the write-set ($T.WS$) respectively.

T is optimistically executed under the control of P-CC and, when it reaches the stage where all of its operations have been executed, the executing thread broadcasts T via the POB layer and waits until T is globally validated and either aborted or committed. In the former case the application thread has to re-issue T from its very beginning; in the latter case T's updates are applied to the transactional shared state after the commit of any other transaction preceding T in the order defined by POB. During the optimistic execution of a transaction, in fact, the updates of write operations are only buffered in the transaction's write-set and they cannot be directly applied to the shared state because the transaction could abort later on.

The POB layer provides two interfaces to send and receive a transaction T: POBROADCAST(T), used for broadcasting a transaction T along with its read-set and write-set; PODELIVER($T, \{T_1, \cdots, T_m\}$), used for delivering a transaction T to nodes, along with the set of transactions $\{T_1, \cdots, T_m\}$, defined as $deps_T$, which conflict with T and must be processed (i.e., certified and possibly committed) before T. Formally, two transactions T and T' are conflicting if at least one of the following three conditions are verified: (i) $T.WS \cap T'.WS \neq \emptyset$, (ii) $T.WS \cap T'.RS \neq \emptyset$, (iii) $T.RS \cap T'.WS \neq \emptyset$.

4.1 Partial Order Broadcast Layer

The core idea behind the design of POB is guaranteeing that all nodes agree on the same delivery order for conflicting transactions. This is because, if two transactions do not conflict, then they can be validated and committed (or aborted) in any order (i.e., all the orders are equivalent due to the absence of conflicts). Formally, POB guarantees that any pair of conflicting transactions – i.e., two transactions that access at least one common object, where at least one of the accesses is a write operation – are not delivered in different orders on two nodes.

Therefore POB guarantees the following properties:
- P1: *Strong Uniform Conflicting Order.* If some node delivers message $m = [T, deps_T]$ before message $m' = [T', deps_{T'}]$ and transactions T and T' conflict, then every node delivers m' only after m.
- P2: *Local Dependency.* For any node that delivers message $m = [T, deps_T]$ before message $m' = [T', deps_{T'}]$ and T and T' conflict, then $T \in deps_{T'}$ and $T' \notin deps_T$ (i.e., no circular dependency between conflicting transactions).

Property *P1* is defined as strong because it does not allow omission of messages. It is in contrast with the weak order property that, instead, allows the omission of messages despite the fact that the order of delivery on all nodes is still preserved. In particular, POB does not allow a scenario in which a node P_i delivers m before m' while a node P_j delivers m' without delivering m, where m and m' contain two conflicting transactions. We need the strong version of this property because in transaction processing, even if the partial order is not violated, the aforementioned scenario can generate two different outcomes for the same transaction T', enclosed in m', on the nodes P_i and P_j. As an example, the P-CC on P_i could abort T' because its execution has been invalidated by transaction T contained in m, while P_j commits T'.

The property *P2* regards the semantics of the interfaces exposed to P-CC. In particular, when POB delivers a message $m' = [T', deps_{T'}]$ to P-CC, transaction T' has to wait for the completion of all the transactions in $deps_{T'}$ before determining its outcome. This condition is sufficient for ensuring that all transactions are processed in accordance with the partial order defined by POB. In addition, POB also guarantees the typical properties of a reliable broadcast service (Validity, Integrity, Uniform Agreement) [7].

Due to space constraints we report the detailed correctness proofs of POB in the technical report.

Overview. The idea of enforcing an order only among conflicting commands has already been specified by the Generalized Consensus [13] and Generic Broadcast [19] problems and followed by a set of implementations, e.g., Generalized Paxos [13], EPaxos [17]. POB improves the above proposals by relying on a fully decentralized design without leveraging on a stable leader to establish the order of transactions and without expensive housekeeping computations before issuing the delivery of a transaction.

The main idea behind POB is to define a deterministic scheme for the assignment of *delivery slots* (i.e., positions in the final order that are associated with positive integers) to submitted transactions, by following the general design of *communication history*-based total order broadcast protocols [7,16] in which the delivery order of messages is determined by the senders. In POB, for each transaction T we define a unique transaction leader tl_T that establishes the final delivery position of T by applying the following rules:

- *Rule 1.* If a node P_i is T's leader (i.e., tl_T), then T can only be delivered in unused positions numbered with pos_T, such that $pos_T \bmod N = i$.
- *Rule 2.* Transaction T' is delivered in position $pos_{T'}$ if and only if, for each conflicting transaction T delivered in position $pos_T > pos_{T'}$, $T' \in deps_T$ and $T \notin deps_{T'}$, where $deps_T$ (respectively $deps_{T'}$) is the set of transactions which T (respectively T') depends on.

Rule 1 guarantees that two transactions from different leaders cannot occupy the same position. However, ALVIN is also able to concurrently broadcast multiple requests from the same node and, as it will be clear later, this could cause two transactions from the same leader to be assigned the same position number. Such transactions are deterministically ordered using the transaction identifier. On the other hand, *Rule 2* is specifically defined for satisfying property *P2*.

The transaction leader tl_T for a transaction T is either the sender of T, or any other elected node if T's sender is suspected as crashed by the failure detector.

Protocol. A transaction T, that is submitted to the POB service via the PO-BROADCAST(T) interface, goes through four phases: *Proposal phase, Decision phase, Accept phase* and *Delivery phase.*

Proposal phase. The node P_i, acting as the leader of T (i.e., tl_T), selects the next available position number for T to be proposed to all the other nodes. This position, named pos_T, is the smallest number among the ones allowed by *Rule 1* and greater than any other position that P_i has observed as already used. P_i also selects the set $deps_T$ of dependencies, namely all transactions T' conflicting with T and having a (even temporary) position less than pos_T.

Subsequently, P_i broadcasts a PROPOSE message with the tuple $\langle T, pos_T, deps_T, e \rangle$ to all nodes. By broadcasting a transaction T we mean broadcasting T's identifier ($T.tid$), read-set ($T.RS$) and write-set ($T.WS$).

The e value is an epoch number associated with transaction T and the messages containing T. It identifies the epoch in which messages for T can be exchanged. A transition to a new epoch is forced by T's new elected leader when T's old leader is suspected as crashed. Messages associated with an epoch e_1 cannot be processed by nodes that have already executed a transition to an epoch e_2, with $e_2 > e_1$. In the PROPOSE message, the epoch number is 0 since it identifies the initial epoch of T in which T's sender is recognized by default as the initial leader tl_T of T.

A node P_j receiving a PROPOSE message for T, replies with an ACKPROPOSE message in order to update P_i with the set of transactions conflicting with T and observed by P_j so far, i.e., $newDeps_T^j$, and a possibly new position to be chosen for T, i.e., $newPos_T^j$. In particular, let us define $temp_T^j$ as the smallest number among the ones allowed by *Rule 1* for P_i and greater than any other position used by transactions conflicting with T and already received by P_j. Then $newPos_T^j$ is equal to $temp_T^j$ in case $temp_T^j$ is greater than pos_T (the position proposed by P_i); otherwise it is equal to pos_T. On the other hand, $newDeps_T^j$ is the set of all transactions T' conflicting with T and having a (even temporary) position less than $newPos_T^j$.

A transaction T received during this phase is marked as PENDING and it is inserted in a data structure named *delivery queue* (DQueue). On each node P_j, DQueue is a queue storing the transactions received by P_j as tuples $\langle T, pos_T, deps_T, status \rangle$, where *status* has values in {PENDING, ACCEPTED, STABLE}. The tuples in the DQueue are totally ordered according to their pos_T's values.

Decision phase. Transaction T's leader P_i waits for a quorum of FQ replies from the previous phase. It then computes the final position pos_T and final dependencies $deps_T$ that are used for the delivery of T in the next phases as follows: pos_T is the maximum position among the proposals ($newPos_T^j$) in the quorum, while $deps_T$ is the union among the dependency sets ($newDeps_T^j$) proposed in the quorum. Afterwards, P_i broadcasts an ACCEPT message for T with the final position and dependencies in order to request to other nodes to accept the delivery of T. The value of FQ in the base configuration of POB is equal to $f + 1$. Section 4.1 shows how to enable a so called *fast transaction decision* by changing the value of FQ.

Accept phase. A node P_j receiving T updates its DQueue accordingly. This means changing the status of T to ACCEPTED and replacing the old values of pos_T and $deps_T$ with the ones received in this phase. Then P_j replies with an ACKACCEPT message by including pos_T and a possibly new set of dependencies $newDeps_T^j$. In fact, in this phase P_j can also attach an additional set $deltadeps_T$ to $deps_T$, if it detects that it received transactions T^δ conflicting with T and having a position in between the old and the new values of pos_T in DQueue. This is because, P_j could have been received T^δ after that $deps_T$ was computed in the *Proposal phase*. More formally, $newDeps_T^j$ is equal to $deps_T \cup deltadeps_T$, where $deltadeps_T$ is the set of all transactions $T' \notin deps_T$ conflicting with T and having a (even temporary) position less than pos_T.

Delivery phase. T's leader P_i waits for a quorum of CQ replies from the previous phase, where CQ is equal to $f + 1$, to be sure that its decision will be stable even if f failures (including itself) occur. After that, it broadcasts its decision via a STABLE message including pos_T, which was already decided in the *Decision phase*, and $deps_T$, which is computed as the union of the $newDeps_T^j$ collected during the previous phase.

A node receiving the STABLE message for T marks T as STABLE in its DQueue by also replacing the old values of pos_T and $deps_T$ with the ones received in this phase. Then, the node can deliver the message $[T, deps_T]$ to the concurrency control when all transactions in $deps_T$ have been already delivered by triggering PODELIVER$(T, deps_T)$.

Since the position of a transaction T' can change throughout the execution of the POB protocol, there might be scenarios in which a transaction $T'' \in deps_{T'}$ becomes STABLE with a position $pos_{T''}$ greater than the final position of T', which would lead T' to wait infinitely for a conflicting transaction that is actually ordered after it. To address this problem, in such a case T'' is removed from the $deps_{T'}$ set. Note that, when this condition is true, T' is guaranteed to be already present in $deps_{T''}$.

Failure Recovery. When a node P_k detects that T's current leader P_i crashed, and P_k has not yet marked T as STABLE, it attempts to become T's new leader by executing a classic Paxos *Prepare phase* [14]. Therefore, P_k broadcasts an epoch number e for T greater than the last one observed for T. Then it waits for a PROMISE from a quorum Q of $f+1$ nodes, meaning that they will not participate in any new *Prepare phase* or *Proposal/Accept phases* for T associated with an epoch number less than e. The nodes in Q also send back the latest status known for T and identified by the most recent tuple $\langle T, pos_T, deps_T, status \rangle$ they have in their DQueue. This allows P_k to take a final decision that cannot differ from the one P_i took (if any).

Therefore, we distinguish three cases depending on the value of *status*:

- At least one $\langle T, pos_T, deps_T, \text{STABLE} \rangle$ is received from Q. In this case, P_k starts a *Delivery phase* by broadcasting a STABLE message for T with pos_T and $deps_T$.
- At least one $\langle T, pos_T, deps_T, \text{ACCEPTED} \rangle$ is received from Q and no STABLE status is present. In this case, P_k starts an *Accept phase* by broadcasting an ACCEPT message for T with pos_T and $deps_T$.
- Neither ACCEPTED nor STABLE value is received from Q. In this case, P_k selects a new position available for T by restarting a new instance of the protocol starting from the *Proposal phase* for T.

Fast Transaction Decision. POB can be configured to allow a so called *fast transaction decision* about the order of a transaction if there are no concurrent conflicting transactions. The idea is the same as adopted in [15,17] and entails that a transaction leader can determine the final position of a transaction early, i.e., after only two communication steps, because it has received all equal ACK-PROPOSE messages from a quorum of nodes in the *Decision phase*. Enabling the fast decision introduces a trade-off. On the one hand, the leader can define the order of a transaction in fewer communication delays, but on the other hand, quorum sizes become bigger and the recovery phase more complex.

When fast decisions are enabled, POB must use a size FQ greater than CQ, i.e., fast quorums bigger than classic quorums, otherwise a fast ordering decision by a transaction leader P_i could be irrecoverable after the fault of P_i. Specifically, the new leader of a transaction T, e.g., P_k, has to decide in the same way the old leader of T, e.g., P_i, decided.

First of all, we have to notice that in case P_i had a fast decision for T by including (respectively not including) a concurrent and conflicting transaction T' in its dependencies, it would be impossible that the leader of T' also had a fast decision by including (respectively not including) T in its dependencies, due to the definition of quorums. Therefore, a trivial recovery of P_k would be contacting the leader of T' to know the final decision for T' with respect to T. On the contrary, in case the new leader of T, i.e., P_k, is not able to contact the current leaders of the transactions conflicting with and concurrent to T, it must take a decision by analyzing collected replies.

If P_i had a fast decision for T, i.e., it collected all equal proposals for T hence it decided in two communication steps, we have to enforce a deterministic behavior on the quorum of replies collected by P_k during recovery. Specifically, in that case, we want P_k to have a majority (i.e., $\left\lfloor \frac{CQ}{2} \right\rfloor + 1$) of values equal to the fast decision in the quorum of replies collected during recovery. In other words, when the new leader of T collects a classic quorum in the recovery phase, then the number of replies different from a possible fast decision of the old leader (and that do not include the reply from the leader of a generic conflicting transaction), i.e., $N - FQ - 1$, has to be less than the majority in the quorum, i.e., $\left\lfloor \frac{CQ}{2} \right\rfloor + 1$. Equation 1 follows.

In addition to the above, another constraint is needed to avoid two new leaders of two conflicting and concurrent transactions T and T', here called *opponents*, both believing that the associated old leaders of T and T' respectively had fast decisions. So after f failures and ignoring the reply from the other opponent, i.e., -1, two opponents cannot both collect a sufficient number of replies, i.e., $\frac{N-f}{2}$, that summed up f form a fast quorum. Equation 2 follows.

$$N - FQ - 1 < \left\lfloor \frac{CQ}{2} \right\rfloor + 1 \quad (1) \qquad\qquad \frac{N-f}{2} + f - 1 < FQ \qquad (2)$$

If we minimize the ratio $\frac{N}{f}$ by still considering $f < \lceil \frac{N}{2} \rceil$, e.g., $N = 2f+1$, we obtain the following sizes for the classic and fast quorums, respectively:

$$CQ = f + 1 \qquad\qquad (3) \qquad\qquad FQ = f + \left\lfloor \frac{f+1}{2} \right\rfloor \qquad (4)$$

Note that CQ and FQ in Equations 3 and 4 have the same values adopted by EPaxos. The new recovery phase that applies under this optimization is a trivial extension of the recovery procedure as presented in EPaxos.

By using these new values for CQ and FQ, the *fast transaction decision* works as follows. After having collected FQ AckPropose in the *Proposal phase* for transaction T, T's leader can directly send the final decision via the Stable message if all the collected proposals are the same. Otherwise it proceeds in the classic way by entering the *Accept phase*.

4.2   Parallel Concurrency Control Layer

Each node in the system is equipped with a *Parallel Concurrency Control* layer (P-CC) that is responsible for executing transactions submitted by clients as well as processing the commit of transactions delivered by POB.

We can split P-CC's operations into two parts. The first part, the *execution phase*, is responsible for executing transactions optimistically. Following the classic multi-version concurrency control scheme implemented in state-of-the-art in-memory transactional systems [4], a transaction executes its read operations on the snapshot of memory present at the time of its beginning (i.e., which includes the set of commits applied before the transaction began), while its writes are buffered and can be applied atomically on all nodes only if the transaction can commit. At this stage, the transaction's read-set and write-set are also built.

The second part, the *commit phase*, is responsible for validating and committing the optimistic execution of transactions on all nodes. This is done by sending the commit message of a transaction T with T's read-set and write-set via the POB layer, and triggering the validation of T as soon as T is delivered by POB. A sufficient condition to guarantee that T appears as executed atomically on all nodes is to validate it by checking that no value read by T has been updated in between its beginning and its finalization.

The P-CC layer guarantees that *i)* every transaction, including aborted ones, observes a consistent state, and *ii)* the set of committed transactions satisfies Serializability (SR). Even if SR is one of the reference consistency criteria for transactional systems, it might be considered not necessary for several types of applications [20]. Such applications stand to benefit from requiring: *i)* that the transactional state never performs a transition to an incorrect state, and *ii)* that all operations always observe consistent states. While the former requires that only update transactions appear as executed sequentially (as demanded by SR), the latter allows to implement read-only transactions with lower guarantees.

In order to take advantage of the above considerations, ALVIN supports another strongly consistent criterion, besides SR, named Extended Update Serializability (EUS) [1], which can be considered as strong as Serializability for many common workloads [20,6]. Roughly speaking, EUS preserves Serializability of committed update transactions and disallows any transaction to observe incorrect states. However, with EUS, two read-only transactions might observe two different non-compatible histories of commits, caused by a different perceived commit order of non-conflicting update transactions. EUS gives the necessary flexibility to ALVIN for committing two update non-conflicting transactions T_h and T_k in an arbitrary order thanks to the POB layer, such that T_h completes before T_k on a node P_i, and vice-versa on another node P_j. At that point, transactions T_q and T_w that are executing on nodes P_i and P_j respectively, are allowed to observe two different serializations of T_h and T_k without providing any inconsistent view to the application. Then, in case T_q and T_w are read-only, they are also allowed to commit under EUS.

Therefore, in order to allow behaviors like the one described above, and to support EUS, P-CC can be configured to avoid the global certification of read-only transactions through POB at commit time, so that a read-only transaction can safely commit as soon as it has been processed locally. In fact, as described in [21], certification-less read-only transactions disallow Serializabiliy in case a total order on the commit of update transactions is not enforced.

Summarizing, since POB ensures a total order among commits of conflicting (both read-only and update) transactions and P-CC ensures that the read-set of committed transactions is not invalidated by concurrent transactions, then ALVIN enforces SR [21]. Moreover, P-CC guarantees that all read operations return the last value committed before the beginning of the transaction execution. This way, any transaction can never observe inconsistent states. Therefore, if read-only transactions are only processed locally without being submitted for a global certification, ALVIN guarantees SR restricted to committed update transactions, as demanded by EUS.

5   Evaluation

We evaluate ALVIN by comparing it against two certification-based transaction execution protocols [18] that rely on MultiPaxos [14] and EPaxos [17] for their ordering layer. MultiPaxos ensures serializability by total-ordering the commit

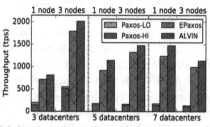
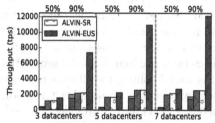

(a) Write-intensive workload for {3,5,7} sites and {1,3} nodes per site.

(b) 50% and 90% read-only transactions. One node per site.

Fig. 1. Throughput of TPC-C benchmark

requests for all write transactions, while serving read-only transactions locally leveraging multi-versioning. However, MultiPaxos is sequencer-based, thus the location of the node designated as the leader significantly affects its performance. In order to conduct a fair comparison, we used two versions of MultiPaxos: one with the leader located at a node with a point-to-point latency to other nodes that is higher than the average (*Paxos-HI*), and another where the connection latency is lower (*Paxos-LO*). We implemented ALVIN and competitors in the same transaction processing framework, using *Go* as the programming language.

We used two benchmarks in the evaluation: TPC-C [6] and Bank [11]. The former is a well known benchmark representative of on-line transaction processing workloads; the latter mimics operations of a monetary application where each transaction transfer amount of money among bank accounts. We ran our experiments on the Amazon EC2 infrastructure, using *r3.2xlarge* nodes in up to 7 geographically distributed sites (three in Asia, two in North America, one in South America and Europe). Each node has 8 CPU cores and 61GB RAM. Results are the average of 7 samples.

Figure 1 reports ALVIN's throughput of TPC-C benchmarks by varying the number of geographically distributed sites {3,5,7}. In Figure 1(a) we also changed the number of nodes per site as {1,3}, using a write intensive workload (<3% read-only). Results on read-dominated workloads are showed in Figure 1(b). Here we change the percentage of read-only transactions from 50% to 90% while using one node per datacenter. In this read dominated scenario we explore both versions of ALVIN, ensuring SR (ALVIN-SR) and EUS (ALVIN-EUS), with the purpose of assessing the effectiveness of EUS. In all depicted scenarios, we configured ALVIN to run with fast decisions enabled. We batch messages for all competitors, using a window of 20 to 50 msec, according to the nodes deployed.

TPC-C's transactions access several shared objects and have a non-negligible computation. This results in long transaction execution time and a complex dependency graph to be analyzed during the processing of commit requests in EPaxos. Rather, ALVIN is able to improve the parallelism thanks to the different delivery rules of POB, gaining up to 26% in throughput against EPaxos. Both EPaxos and ALVIN sustain their throughput while increasing the system's load

until 9 nodes (3 datacenters with 3 nodes each), then the system becomes over-
loaded and performance degrades due to increasing contention. MultiPaxos in
both its configurations performs worse than others due to the presence of single
remote leader that slows down the entire system's progress. In addition, here
transactions are long thus the sequential certification limits its performance.

Figure 1(b) shows the effectiveness of exploiting EUS in read-dominated workloads by avoiding to broadcast read-only transactions via the ordering layer.
Therefore ALVIN-EUS provides a speed up of up to 4.8× in throughput when
compared to ALVIN-SR and EPaxos. It is important to notice that in these
scenarios, MultiPaxos is also able to take advantage of local computation of
read-only transactions. In fact, its Paxos-LO configuration performs similar to
EPaxos and ALVIN-SR for the case of 90% of read-only transactions and 3 datacenters. In other scenarios, Paxos-LO saturates its leader's resources, slowing
down the ordering process. As before, Paxos-HI exposes poor performance due
to the high communication latency with the faraway designated leader. Regarding the comparison between EPaxos and ALVIN-SR, they follow about the same
trend observed in Figure 1(a) because they both process read-only transactions
in the same way.

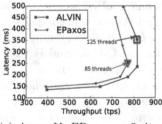
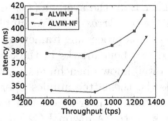

(a) ALVIN Vs EPaxos on 5 sites.   (b) Impact of fast decision.

Fig. 2. Throughput Vs Latency using TPC-C benchmark varying application threads

In Figure 2(a) we plot the latency increasing the system's load by adding
application threads per node from 15 to 125. Here, we used 5 sites and TPC-C
as the benchmark, adopting the same workload as in Figure 1(a). For increasing
the readability of the plot we excluded MultiPaxos because its results were 3×
slower than the other competitors. From the analysis of EPaxos's and ALVIN's
trends we observe that ALVIN has a lower transaction latency and it sustains its
throughput better than EPaxos. Specifically, with 85 threads per site EPaxos
stops scaling while ALVIN is still able to serve more requests. ALVIN reaches its
saturation point running 125 threads per site.

With the plot in Figure 2(b) we highlight the importance of configuring ALVIN
without the fast decision in high contention scenarios. In these situations, the
probability of taking a fast decision after having collected a fast quorum of
replies is low. Therefore the POB layer always pays the maximum number of
communication steps to reach a decision by contacting a fast quorum of nodes
in the *Proposal phase* and then falling back to the *Accept phase*. Disabling the

fast decision forces the leader to always collect replies from a classic quorum. We configured TPC-C as in Figure 1(a) with 7 sites and one node each, and we increased the load as before. ALVIN-NF (fast decisions disabled) improves the latency of ALVIN-F (fast decisions enabled) up to 30 msec, confirming that, in some scenarios, waiting for an unlikely fast decision does not pay off.

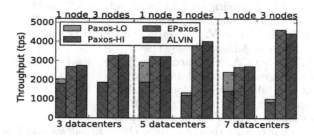

Fig. 3. Throughput under write-intensive workload for {3,5,7} sites and {1,3} nodes per site using Bank benchmark

The Bank benchmark has very small transactions (only few operations) and the amount of transactional work can be considered as negligible when compared to the coordination steps required for establishing the agreement on the global ordering. This makes the results of both ALVIN and EPaxos comparable in almost all configurations tested as we showed in Figure 3. Bank's accesses are uniformly distributed across all objects and we managed the total number of shared objects for having an average transaction's abort rate in the range of 10-20%.

EPaxos's dependency graph analysis does not slow down the transaction's critical path significantly because the strongly connected components with more than one node are only 1.7% of all, thus the main impacting factor on the performance is the number of communication delays used for delivering transactions and, with fast decisions enabled, both ALVIN and EPaxos use the same communication delays for delivering. However, it is worth noticing that all competitors relying on partial order instead of total order sustain their throughput when we increase the number of nodes until 7 datacenters, where they start degrading. MultiPaxos in both its configurations performs worse than others due to the presence of single remote leader that slows down the entire system's progress. The exception is Paxos-LO, which is the closest to others because it benefits from having a low latency leader when site count is limited.

6   Conclusion

At its core, the design of ALVIN shows that it is possible to achieve an effective tradeoff between performance and programmability in geo-replicated environments. An important insight of our work is that partial ordering of transactions can be significantly exploited to speed up local concurrency control through parallelism and that it can be determined without a unique leader, which increases scalability in a geo-replicated setting.

Acknowledgments. This work is supported in part by US National Science Foundation under grant CNS-1217385.

References

1. Adya, A.: Weak Consistency: A Generalized Theory and Optimistic Implementations for Distributed Transactions. PhD thesis AAI0800775. MIT (1999)
2. Almeida, S., Leitão, J., Rodrigues, L.: ChainReaction: A Causal+ Consistent Datastore Based on Chain Replication. In: 8th ACM EuroSys, pp. 85–98. ACM (2013)
3. Bernstein, P.A., Hadzilacos, V., Goodman, N.: Concurrency Control and Recovery in Database Systems. Addison-Wesley (1987)
4. Cachopo, J., Rito-Silva, A.: Versioned Boxes As the Basis for Memory Transactions. Sci. Comput. Program. 63(2), 172–185 (2006)
5. Corbett, J.C., et al.: Spanner: Google's Globally Distributed Database. ACM Trans. Comput. Syst. 31(3), 8:1–8:22 (2013)
6. TPC-C Benchmark, http://www.tpc.org/tpcc/
7. Défago, X., Schiper, A., Urbán, P.: Total Order Broadcast and Multicast Algorithms: Taxonomy and Survey. ACM Comput. Surv. 36(4), 372–421 (2004)
8. Guerraoui, R., Kapalka, M.: On the Correctness of Transactional Memory. In: 13th ACM SIGPLAN PPoPP, pp. 175–184. ACM (2008)
9. Guerraoui, R., Rodrigues, L.: Introduction to Reliable Distributed Programming. Springer (2006)
10. Guerraoui, R., Schiper, A.: Genuine Atomic Multicast in Asynchronous Distributed Systems. Theor. Comput. Sci. 254, 297–316 (2001)
11. Hirve, S., Palmieri, R., Ravindran, B.: Archie: A Speculative Replicated Transactional System. In: 15th ACM/IFIP/USENIX Middleware. ACM (2014)
12. Kraska, T., Pang, G., Franklin, M.J., Madden, S., Fekete, A.: MDCC: Multi-data Center Consistency. In: 8th ACM EuroSys, pp. 113–126. ACM (2013)
13. Lamport, L.: Generalized Consensus and Paxos. Technical report MSR-TR-2005-33, Microsoft Research (2005)
14. Lamport, L.: The Part-time Parliament. ACM Trans. Comput. Syst. 16(2), 133–169 (1998)
15. Lamport, L.: Fast Paxos. Distributed Computing 19(2), 79–103 (2006)
16. Mao, Y., Junqueira, F.P., Marzullo, K.: Mencius: Building Efficient Replicated State Machines for WANs. In: 8th USENIX OSDI, pp. 369–384. USENIX (2008)
17. Moraru, I., Andersen, D.G., Kaminsky, M.: There is More Consensus in Egalitarian Parliaments. In: 24th ACM SOSP, pp. 358–372. ACM (2013)
18. Pedone, F., Guerraoui, R., Schiper, A.: The Database State Machine Approach. Distrib. Parallel Databases 14(1), 71–98 (2003)
19. Pedone, F., Schiper, A.: Generic Broadcast. In: Jayanti, P. (ed.) DISC 1999. LNCS, vol. 1693, pp. 94–106. Springer, Heidelberg (1999)
20. Peluso, S., Ruivo, P., Romano, P., Quaglia, F., Rodrigues, L.: When Scalability Meets Consistency: Genuine Multiversion Update-Serializable Partial Data Replication. In: 32nd ICDCS, pp. 455–465. IEEE Computer Society (2012)
21. Schmidt, R., Pedone, F.: A Formal Analysis of the Deferred Update Technique. In: Tovar, E., Tsigas, P., Fouchal, H. (eds.) OPODIS 2007. LNCS, vol. 4878, pp. 16–30. Springer, Heidelberg (2007)
22. Sovran, Y., Power, R., Aguilera, M.K., Li, J.: Transactional Storage for Geo-replicated Systems. In: 23rd ACM SOSP, pp. 385–400. ACM (2011)
23. Zhang, Y., Power, R., Zhou, S., Sovran, Y., Aguilera, M.K., Li, J.: Transaction Chains: Achieving Serializability with Low Latency in Geo-distributed Storage Systems. In: 24th ACM SOSP, pp. 276–291. ACM (2013)

Distributed Local Approximation
of the Minimum k-Tuple Dominating Set
in Planar Graphs*

Andrzej Czygrinow[1], Michal Hanćkowiak[2], Edyta Szymańska[2],
Wojciech Wawrzyniak[2], and Marcin Witkowski[2]

[1] School of Mathematical and Statistical Sciences,
Arizona State University, Tempe, AZ,85287-1804, USA
aczygri@asu.edu
[2] Faculty of Mathematics and Computer Science,
Adam Mickiewicz University, Poznań, Poland
{mhanckow,edka,wwawrzy,mw}@amu.edu.pl

Abstract. In this paper we consider a generalization of the classical dominating set problem to the k-tuple dominating set problem (kMDS). For any positive integer k, we look for a smallest subset of vertices $D \subseteq V$ with the property that every vertex in $V \setminus D$ is adjacent to at least k vertices of D. We are interested in the distributed complexity of this problem in the model, where the nodes have no identifiers. The most challenging case is when $k = 2$, and for this case we propose a distributed local algorithm, which runs in a constant number of rounds, yielding a 7-approximation in the class of planar graphs. On the other hand, in the class of algorithms in which every vertex uses only its degree and the degree of its neighbors to make decisions, there is no algorithm providing a $(5 - \epsilon)$-approximation of the 2MDS problem. In addition, we show a lower bound of $(4 - \epsilon)$ for the 2MDS problem even if unique identifiers are allowed.

For $k \geq 3$, we show that for the problem kMDS in planar graphs, a trivial algorithm yields a $k/(k - 2)$-approximation. In the model with unique identifiers this, surprisingly, is optimal for $k = 3, 4, 5$, and 6, as we provide a matching lower bound.

1 Introduction

Let $G = (V, E)$ be a graph with $|V| = n$ and $|E| = m$. By $N_G(v) = \{u \in V : uv \in E\}$ we denote the neighborhood of a vertex $v \in V$ and by $N_G[v] = N_G(v) \cup \{v\}$ the closed neighborhood of v. A set $D \subseteq V$ is called a *dominating set* of G if every vertex in $V \setminus D$ is adjacent to a vertex in D. Equivalently, a subset $D \subseteq V$ is a dominating set of G if $|N_G(v) \cap D| \geq 1$ for every $v \in V \setminus D$. A *minimum dominating set* of a graph G, is a dominating set of G with the minimum cardinality and we refer to the problem of finding such a set as *the MDS problem*.

---

* The research is supported by grant N N206 565740.

M.K. Aguilera et al. (Eds.): OPODIS 2014, LNCS 8878, pp. 49–59, 2014.
© Springer International Publishing Switzerland 2014

One way of generalizing this notion, often desired in practice, is to look for a 'stronger' domination, where every vertex in G is dominated more than once. For any positive integer k, a *k-tuple dominating set* in G is a subset $D \subseteq V$ such that for every $v \in V \setminus D$, $|N(v) \cap D| \geq k$, that is, for every vertex $v \in V$, v is in D or v has at least k neighbors in D. In particular, a 1-tuple dominating set is simply a dominating set and the case when $k = 2$ is often called *double domination*. This paper is concerned with the problem of finding a minimum k-tuple dominating set called *the kMDS problem*.

Related notions of k-domination have been considered. For example, Harary and Haynes in [5] defined a generalization of domination by considering k-tuple total domination. In this, more restricted variant of the problem, for a positive integer k, a subset $D \subseteq V$ is called a *k-tuple total dominating set* of G if for every vertex $v \in V$, $|N[v] \cap D| \geq k$, that is, either v is in D and has at least $k-1$ neighbors in D, or v is in $V \setminus D$ and has at least k neighbors in D. Clearly, for a graph to have a k-tuple total dominating set its minimum degree must be at least $k - 1$.

The minimum dominating set problem has been in the main stream of interest for computer science due to its various applications and its complexity has been studied extensively in different classes of graphs. In general, the MDS problem is NP-hard [3], and moreover, it is also NP-hard to compute a $(C \log \Delta)$-approximation [11], for any constant $C > 0$, where Δ denotes the maximum degree of the graph. A simple sequential approximation algorithm matching this bound can be found in [7]. The complexity of the k-tuple total dominating problem was studied in [8], where the authors describe a $O(\ln |V| + 1)$-approximation algorithm in general graphs, and show that it cannot be approximated within the ratio of $(1 - \epsilon) \ln |V|$, for any $\epsilon > 0$ unless $NP \subseteq DTIME\left(|V|^{O(\log \log |V|)}\right)$. They also give a constant approximation algorithm for this version of the problem in the class of graphs of bounded degree and p-claw free graphs. Note that the algorithmic results carry over to the kMDS problem considered here.

The MDS and kMDS problems are particularly interesting in the distributed setting, where there is no central control. Then, to be able to perform efficient routing, communication or localization, one often looks for a partitioning of the nodes into clusters which in many situations can be obtained by means of a dominating set.

The same hardness result which holds for the sequential model, remains true in the distributed setting. In distributed systems a graph is an instance of the MDS problem and, at the same time, describes the communication network. In each synchronous round, each node may send a (different) message to each of its neighbors, receive messages from all of its neighbors and may perform arbitrary local computations. A distributed algorithm is called *local* if the number of its synchronous communication rounds is constant, independent of the size of the network. In addition, the following variations of the local model may be considered:

- ID model - each node in the networks is given a unique identifier ($O(\log n)$-bit label). This model is consistent with LOCAL model from [10] using constant time.
- PO model (also known as *anonymous network with a port numbering and orientation*) - each node of degree d has a linear order on the incident edges $1, 2, \ldots, d$, and for each edge, there is a linear order on the incident nodes and for oriented edges each endpoint knows which of them is the head and which is the tail.
- *fully anonymous model* - each node in the network makes a decision based on its degree and the degrees of its neighbors which are available to it.

The ID model is most general and in the case of local algorithms unique identifiers are often unnecessary [4]. Distributed approximation of the MDS problem has been considered for different classes of graphs (UDGs, bounded arboricity) both in the ID model and the PO model. In particular, the problem is much more tractable in the case of planar graphs. The current best deterministic approximation algorithm for the MDS problem in planar graphs was given in [9]. The algorithm there uses long messages (more than $O(\log n)$ bits) and its approximation ratio has been recently improved by Wawrzyniak in [13] to 52. At the same time, he proposed in [12] an algorithm which returns a 636-approximation of the minimum dominating set in planar graphs utilizing messages of length at most $O(\log n)$. In addition, the algorithm from [12] works in the PO model.

In this paper we study the distributed complexity of the kMDS problem in planar graphs. We extend the result from [12] and prove that there exists a constant approximation of the kMDS problem in planar graphs in the distributed local PO model. We are not aware of any prior work on the kMDS problem in the distributed model.

Although the case of the total domination is not addressed in the paper (and it would make sense only for small values of k), it would not be surprising if the techniques developed here applied to this variant of the problem mutatis mutandis.

In the remainder of the paper we call a k-tuple dominating set a k-*dominating set* for short.

First, we observe that for any $k \geq 3$, the set V of all vertices of a planar graph yields a constant approximation of the minimum k-dominating set. Recall that every planar graph on n vertices contains at most $3n - 6$ edges and if, in addition, it does not contain any triangles (i.e. is bipartite) then it is even more sparse and contains at most $2n - 4$ edges. We use these facts in our proofs.

Lemma 1. *Let* $G = (V, E)$ *be a planar graph and* D^* *be a minimum k-dominating set in* G. *Then, for every* $k \geq 3$,

$$|V \setminus D^*| < \frac{2}{k-2}|D^*|.$$

Proof. Let $H = (V, E(H))$ be a spanning subgraph of $G = (V, E)$, such that $E(H) = \{uv \in E : u \in D^* \wedge v \in V \setminus D^*\}$. Then H is a planar bipartite graph and thus, $|E(H)| \leq 2|V(H)| - 4 < 2|V \setminus D^*| + 2|D^*|$.

Furthermore, each vertex $v \in V \setminus D^*$ is k-dominated by the vertices from D^*. Hence, $|E(H)| \geq k|V \setminus D^*|$. Using both inequalities we obtain, that

$$k|V \setminus D^*| \leq |E(H)| < 2|V \setminus D^*| + 2|D^*|.$$

An immediate consequence of the above lemma is the following corollary which says that a trivial algorithm, i.e. an algorithm including all nodes into the kMDS, yields a good approximation of the optimum.

Corollary 1. *Let $k \geq 3$ be a fixed integer. Then a trivial distributed algorithm finds a $k/(k-2)$-approximation of the minimum k-dominating set problem in every planar graph.*

The above result is, in fact, best possible, as one can see in Lemma 3 given in Section 2.

The situation becomes more complicated when $k = 2$ and most part of our paper is devoted to this case. Our main result is a 7-approximation of the minimum double dominating set problem in planar graphs. Specifically, in Section 4 we give a distributed algorithm, called 2MDS, and prove the following.

Theorem 1. *The distributed algorithm 2MDS finds a 7-approximation of the minimum double dominating set problem in every planar graph in the port numbering model.*

Next, we show a lower bound which gives some information on the quality of the algorithm, though does not quite match it. Namely, we prove that in the fully anonymous computation model (as defined above), there is no algorithm providing a $(5 - \epsilon)$-approximation of the 2MDS in planar graphs.

All graphs considered in this paper are planar and we often identify a planar graph with its plane embedding.

The rest of the paper is organized as follows. In the next section we show the already mentioned lower bound. Also, we complement Cor.1 by providing a matching lower bounds for $k = 3, 4, 5, 6$ in the ID model and get $(4 - \epsilon)$ for $k = 2$. In Section 3, we introduce the notion of bunches and prove some facts about them useful in the analysis. In Section 4, we present our main algorithm and prove Theorem 1.

2   Lower Bounds

In this section we prove two lower bounds. The first result is in a very restrictive model in which every vertex makes the decision based on its degree and the degrees of its neighbors which we shall call a fully anonymous model. On the other hand our algorithm 2MDS has precisely this property and so the bound can be compared with Theorem 1. The second bound shows that the trivial algorithm from Corollary 1 cannot be improved even when vertices have unique identifiers.

Lemma 2. *For any $\epsilon > 0$ there is no deterministic algorithm in the fully anony-mous model, computing a $(5-\epsilon)$-approximation of a minimum double dominating set in planar graphs.*

Proof. Consider the following graph. Let H be obtained from the square of an even cycle $C = v_1v_2\ldots v_{2n}v_1$ by adding two new vertices v, w and edged vv_{2i-1}, wv_{2i}, for $i = 1,\ldots,n$. Note that H is planar (see Figure 1) and $|V(H)| = 2n+2$. Clearly, H has diameter three and for every constant d and v_i, the graph induced by $\{u \,|\, dist_H(u, v_i) \leq d\}$ is isomorphic to the same subgraph of H. As a result, every vertex v_i has the same information available to decide if it should be included in a double dominating set or not. Since any double dominating set in H contains vertices from C, a solution obtained by an algorithm will have at least $2n$ vertices. At the same time any optimal solution has $2 + \lceil 2n/5 \rceil$ vertices. Since $2n + 2 > (5 - \epsilon)(2 + \lceil 2n/5 \rceil)$, no local algorithm in the fully anonymous model can find a $(5 - \epsilon)$-approximation. $\quad\blacksquare$

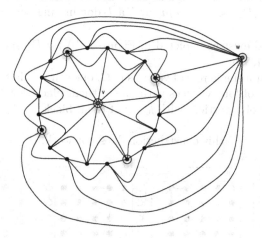

Fig. 1. Proof of Lemma 2

Corollary 2. *For every $\epsilon > 0$, there is a planar graph G such that algorithm 2MDS cannot find a $(5 - \epsilon)$-approximation of the 2MDS problem in G.*

We now turn our attention to the case when $k \geq 3$. In this case, surprisingly, the trivial approach from Corollary 1 is the best one can do in a constant number of rounds. Moreover, we cannot provide a better algorithm even for the LOCAL ([10]) model of computations, in which vertices have unique identifiers from $\{1, ..., n\}$, where n is the number of nodes. Notice that it carries over to the PO model automatically.

Lemma 3. *For $k = 3, 4, 5, 6$ and any $\epsilon > 0$, there is no deterministic local algorithm computing in $O(1)$ rounds a $(\frac{k}{k-2} - \epsilon)$-approximation of a minimum k-dominating set in planar graphs. Moreover, for $k = 2$ there is no deterministic local algorithm computing a $(4 - \epsilon)$-approximation of the 2MDS problem in a constant number of rounds.*

Proof. This result follows in a quite straightforward way from previous work. In particular, we use the strategy of repeated applications of Ramsey's theorem from [1] which was used in [6] to show a lower bound of $(7 - \epsilon)$ on the approximation ratio for the MDS problem in planar graphs. We sketch the idea of the proof indicating the places, where some adjustment for our case is needed. Let $\mathcal{A}_k$ be a deterministic distributed algorithm in LOCAL model, working in time $T = O(1)$ and finding a k-dominating set D_k in any planar graph G.

Let $G = G(r, m)$ be a planar graph given in [6], which is composed of r blocks of identical three-way grids of size $m \times m$, which are connected in a linear manner with additional 'parallel' edges as shown in Figure 2 a) ($r = 3$, $m = 7$). It is shown in [6] (see Lemma 1 therein) that for sufficiently large r and m there is an assignment of unique identifiers such that algorithm $\mathcal{A}_1$ appends almost all vertices of G to D_1. The same remains true for $\mathcal{A}_k$, that is, $|D_k| > (1 - \epsilon)n$, where $D_k = \mathcal{A}_k(G)$ and $n = |G| = rm^2$.

To prove our lemma it is enough to show that in G there exist: a 2-dominating set D_2 such that $|D_2| < (\frac{1}{4} + \epsilon)n$, and a k-dominating set D_k with $|D_k| < (\frac{k-2}{k} + \epsilon)n$.

Such sets indeed exist as depicted in Figure 2 (b) for D_2 and in Figure 3 for D_k, where $k = 3, 4, 5, 6$.

While looking at a single row of vertices in G, one can see that for D_2 there are repeating patterns of four vertices with one of them black, for D_k, $k = 3, 4, 5, 6$ there are patterns of k vertices with $k - 2$ of them black.

Notice that for $k \geq 7$ this construction does not work.

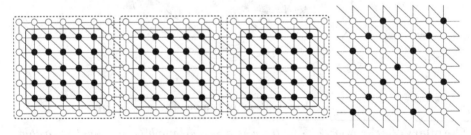

Fig. 2. a) A construction of $G(m, r)$ with $T = 1$, $r = 3$, $m = 7$. b) Black dots represent a 2-dominating set that contains $1/4$ fraction of vertices.

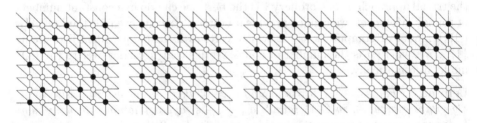

Fig. 3. Examples of $3, 4, 5$ and 6-dominating sets that contain $1/3$, $2/4$, $3/5$ and $4/6$ fraction of vertices, respectively

3   Notation and Tools

In the analysis of the algorithm from Section 4 we use some structural properties of planar graphs. In particular, the proof of its correctness relies on the concept of a bunch (introduced in [12]) which we discuss in this section. We follow the graph theoretic terminology of Diestel [2].

3.1   Bunches

Let $G = (V, E)$ be a plane embedding of a planar graph and let S, U be two disjoint subsets of V. A path vuw is called an S-U-S-path if $v, w \in S$ and $u \in U$. We use $\mathcal{P}$ to denote the set of all S-U-S-paths in G and $\mathcal{Q} \subseteq \mathcal{P}$ to denote a maximal subset of S-U-S paths in $\mathcal{P}$ such that every vertex $u \in U$ belongs to at most one path in $\mathcal{Q}$. In other words, if there is more than one S-U-S path in $\mathcal{P}$, which contains a vertex $u \in U$, then we discard all but one such path.

We say that G has a S-U-S-bunch of length m with poles at vertices v and w, where $v \neq w$, if G contains a sequence of S-U-S-paths $P_1, P_2, \ldots, P_m \in \mathcal{Q}$ with the following properties. Each P_i joins v with w. Furthermore, for each $i = 1, \ldots, m - 1$, the inner face of the cycle $P_i \cup P_{i+1}$ contains no vertex of S (it may, however, contain some vertices of $V \setminus S$). The inner face of the cycle $P_i \cup P_{i+2}$ contains $P_{i+1} - \{v, w\}$. Moreover, this sequence of paths is maximum in the sense that it is not possible to find a longer sequence. If B is an S-U-S-bunch of length m, then the vertices of $U \cap (V(P_1) \cup V(P_m))$ are called boundary vertices and the remaining vertices of U in B, that is, $U \cap (\bigcup_{i=2}^{m-1} V(P_i))$ are called internal vertices of B. A bunch of length $m = 1$ is a single path and it possesses only one boundary vertex and no internal vertices. An example of S-U-S-bunches is illustrated in Fig. 4.

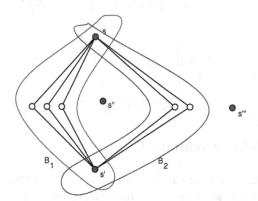

Fig. 4. An example of two S-U-S-bunches, B_1 of length 3 and B_2 of length 2; $s, s', s'', s''' \in S$

Let $\mathcal{B}$ be the set of all S-U-S-bunches in a plane graph G with respect to $\mathcal{Q}$. The lemma below provides a bound on the size of $\mathcal{B}$.

Lemma 4. $|\mathcal{B}| \leq 3|S| - 5$.

Proof. Let $b = |\mathcal{B}|$. We first define an auxiliary plane multigraph $H = H(G) = (V(H), E(H))$, where $V(H) = S$ and for each S-U-S-bunch B we pick exactly one path vuw in B and contract one of vu, uw preserving the topology of G. As a result, H is a plane multigraph with the property that each face of a cycle formed by two parallel edges contains a vertex from S. Moreover, $e := |E(H)|$ equals the number of S-U-S-bunches in G. Let $|S| = s$ and suppose that H has k components. Add $k - 1$ edges to H so that the new graph H' is plane and connected. Let $e' := |E(H')|$. Note that $s \geq 2$ and so $e' \geq 1$. We have

$$2 = s - e' + l, \tag{1}$$

where $e' = e + k - 1$ and l denotes the number of faces of H'.

Note that every face of H', except possibly the outer face, has at least three edges on its boundary. Indeed, for every two parallel edges e_1, e_2 of H', each of the two faces of $e_1 \cup e_2$ contains a vertex from S and so a face of H' cannot be bounded by just e_1, e_2. Therefore,

$$3(l - 1) \leq 2e' - 2,$$

as every edge of H' lies on the boundary of at most two faces and if $l > 1$ then the outer face is bounded by at least two edges. As a result,

$$l \leq \tfrac{2e'+1}{3}. \tag{2}$$

By (1) and (2),

$$s = 2 + e' - l \geq 2 + e' - \tfrac{2e'+1}{3}$$

and so,

$$3s \geq 6 + 3e' - 2e' - 1 = 5 + e' \geq 5 + e.$$

Thus,

$$e = |E(H)| = b \leq 3s - 5.$$

4   The 7-Approximation

In this section we give our main algorithm and show that it finds a double dominating set of size at most seven times the optimal. The algorithm is relatively simple and works as follows. In the first round all vertices of degree less than two or at least five are added to a 2-dominating set D. In the second round all vertices outside D which are not double dominated yet, are included in D.

Note that the only information required by a vertex to execute 2MDS is its degree, available in one communication round in the PO model and so, indeed, the algorithm works in the distributed setting where no identifiers are required.

Algorithm 1. 2MDS

1: $D := \emptyset$.
2: **for** $v \in V$ in parallel **do**
3:    **if** $d(v) < 2$ or $d(v) \geq 5$ **then**
4:        $D := D \cup \{v\}$
5:    **end if**
6: **end for**
7: **for** $v \in V \setminus D$ in parallel **do**
8:    **if** $|N(v) \cap D| < 2$ **then**
9:        $D := D \cup \{v\}$
10:    **end if**
11: **end for**

4.1   The Proof of Theorem 1

Now we are ready to prove Theorem 1. To this end, let D^* be a minimum double dominating set in G and let D be a double dominating set computed by 2MDS. In addition, let

$$D_1^* = \{v \in D^* : d(v) \geq 5\}, \ D_2^* = D^* \setminus D_1^*.$$

Let $\mathcal{B}_1$ be the set of all D_1^*-$(V \setminus D^*)$-D_1^* bunches and let $V_1 = \bigcup_{B \in \mathcal{B}_1} V(B)$, that is, the set of all vertices belonging to a bunch of $\mathcal{B}_1$. Note that every vertex of D is either in D^* or is double dominated by vertices from D^*. As a result, every vertex $x \in D \setminus D^*$ belongs to a bunch B in $\mathcal{B}_1$ or has at least one neighbor in D_2^*. Thus, to bound $|D|$, it is sufficient to bound the number of vertices in $D \setminus D^*$ coming from the bunches of $\mathcal{B}_1$, and the number of vertices of $D \setminus D^*$ adjacent to D_2^* and add the vertices from D_1^* that lie outside all bunches in $\mathcal{B}_1$.

First observe that, since the poles of every bunch $B \in \mathcal{B}_1$ have degree at least five, they are added to D in the first round of the algorithm (Step 4). In addition, in the same round, the algorithm can add at most two of the boundary vertices of B to $D \setminus D_1^*$ and possibly some of its internal vertices of degree at least five. Let v be an internal vertex of B with $d(v) \geq 5$. More precisely, say, let v be the internal vertex of a path $P \in B$, and P^-, P^+ be the paths in B such that no path of B goes through the two inner faces of $P^- \cup P \cup P^+$.

Since $d(v) \geq 5$, there is a vertex w in one of the two inner faces of $P^- \cup P \cup P^+$, say the inner face f of $P^- \cup P$. Next we show that the inner face f must contain a vertex belonging to D_2^*. The situation is depicted in Figure 5. Now recall that by definition of the bunches in $\mathcal{B}_1$, the inner face f contains no vertex of D_1^*. Moreover, if $w \notin D_2^*$, it must have at least two neighbors in D^*. However, since no path of B goes through the inner face f, there is a vertex w' inside f that belongs to D_2^*.

In view of the above, for every $v \in D$, which is an internal vertex of bunch B, we have that v is adjacent to at least one face that contains a vertex in D_2^*. It is possible, however, that two internal vertices of B are adjacent to the same face

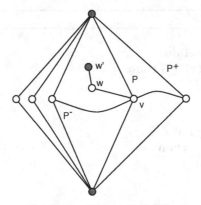

Fig. 5. Illustration of the proof of Theorem 1

f with $w' \in D_2^*$. As a result, the set D contains at most $2|D_2^*|$ internal vertices of bunches in $\mathcal{B}_1$.

Every pole of a bunch from $\mathcal{B}_1$ is in D_1^* and every other vertex in the bunch is double dominated by the poles. Thus, every vertex of D which belongs to a bunch from $\mathcal{B}_1$ has degree at least five in G. Consequently, we have

$$|D \cap V_1| \leq |D_1^* \cap V_1| + 2 \cdot 3|D_1^*| + 2|D_2^*|.$$

Indeed, every bunch has at most two boundary vertices, and the number of bunches in $\mathcal{B}_1$, by Lemma 4, is less than $3|D_1^*|$. In addition, as argued above, the number of internal vertices of degree at least five is at most $2|D_2^*|$.

Now, observe that every vertex $v \in V \setminus \bigcup_{B \in \mathcal{B}_1} V(B)$ belongs to D_2^* or has a neighbor in D_2^* or belongs to D_1^*. Also, every vertex in D_2^* has degree at most four. Hence, the number of vertices that do not belong to $\mathcal{B}_1$ and could be added to D by the algorithm is at most

$$|D_1^* \setminus V_1| + |D_2^*| + 4|D_2^*| = |D_1^* \setminus V_1| + 5|D_2^*|.$$

Therefore, in total, we have,

$$|D| \leq |D_1^* \cap V_1| + |D_1^* \setminus V_1| + 6|D_1^*| + 7|D_2^*| \leq 7(|D_1^*| + |D_2^*|) = 7|D^*|.$$

5   Summary

In this paper we present a simple distributed algorithm finding a 7-approximation of the minimum double dominating set problem in planar graphs. On the other hand, we show that in a planar graph, which is a subgraph of a three-way grid there is a lower bound of $5 - \epsilon$ for the approximation ratio in the class of algorithms in which every vertex makes its decision based on its degree and the degree of its neighbors. At the same time, a lower bound of $(4 - \epsilon)$ holds even if the vertices are equipped with unique identifiers.

Göös et al. asked a general question whether unique identifiers are needed for local approximation algorithms and in [4] they proved that port numbering is sufficient and equally efficient for a large class of problems in bounded degree graphs closed under lifts. It seems to us that the kMDS problem is a good candidate for a counterexample to this general hypothesis in planar graphs which may give different bounds in the ID model and the PO model.

References

1. Czygrinow, A., Hańćkowiak, M., Wawrzyniak, W.: Fast distributed approximations in planar graphs. In: Taubenfeld, G. (ed.) DISC 2008. LNCS, vol. 5218, pp. 78–92. Springer, Heidelberg (2008)
2. Diestel, R.: Graph Theory, 4th edn. Graduate texts in mathematics, vol. 173, pp. I–XVIII, 1–436. Springer (2012) ISBN 978-3-642-14278-9
3. Garey, M.R., Johnson, D.S.: Computers and Intractability. Freeman (1979)
4. Göös, M., Hirvonen, J., Suomela, J.: Lower bounds for local approximation. J. ACM 60(5), 39 (2013)
5. Harary, F., Haynes, T.W.: Double domination in graphs. Ars Combinatoria 55, 201–213 (2000)
6. Hilke, M., Lenzen, C., Suomela, J.: Brief announcement: Local approximability of minimum dominating set on planar graphs. In: 33rd ACM SIGACT-SIGOPS Symposium on Principles of Distributed Computing, PODC 2014, Paris, France (July 2014)
7. Johnson, D.: Approximation Algorithms for Combinatorial Problems. Journal of Computer and System Sciences 9(3), 256–278 (1974)
8. Klasing, R., Laforest, C.: Hardness results and approximation algorithms of k-tuple domination in graphs. Information Processing Letters 89(2), 75–83 (2004)
9. Lenzen, C., Pignolet, Y.A., Wattenhofer, R.: Distributed minimum dominating set approximations in restricted families of graphs. Distrib. Comput. 26(2), 119–137 (2013)
10. Peleg, D.: Distributed Computing: A Locality-Sensitive Approach. Society for Industrial and Applied Mathematics, Philadelphia (2000)
11. Raz, R., Safra, S.: A Sub-Constant Error-Probability Low-Degree Test, and a Sub-Constant Error-Probability PCP Characterization of NP. In: Proc. 29th Symposium on Theory of Computing (STOC), pp. 475–484 (1997)
12. Wawrzyniak, W.: Brief announcement: A local approximation algorithm for MDS problem in anonymous planar networks. In: PODC 2013, pp. 406–408 (2013)
13. Wawrzyniak, W.: A strengthened analysis of a local algorithm for the minimum dominating set problem in planar graphs. Inf. Process. Lett. 114(3), 94–98 (2014)

Time Lower Bounds
for Distributed Distance Oracles[*]

Taisuke Izumi[1] and Roger Wattenhofer[2]

[1] Nagoya Institute of Technology,
Gokiso-cho, Showa-ku, Nagoya, Aichi, 466-8555, Japan
t-izumi@nitech.ac.jp
[2] ETH Zurich, 8092 Zurich, Switzerland
wattenhofer@ethz.ch

Abstract. *Distributed distance oracles* consist of a labeling scheme which assigns a label to each node and a local data structure deployed to each node. When a node v wants to know the distance to a node u, it queries its local data structure with the label of u. The data structure returns an estimated distance to u, which must be larger than the actual distance but can be overestimated. The accuracy of the distance oracle is measured by *stretch*, which is defined as the maximum ratio between actual distances and estimated distances over all pairs (u, v).

In this paper, we focus on the time complexity of constructing distributed distance oracles with a given stretch. We show a number of time lower bounds depending on the stretch:

- Under the assumption that the popular combinatorial girth conjecture is true, any distributed algorithm constructing oracles with stretch $2t$ requires $\tilde{\Omega}(n^{1/(t+1)})$ rounds in unweighted graphs. This bound holds even if we only consider constant diameter graphs.
- For oracles with stretch $2t$ in weighted graphs, we have a lower bound of $\Omega(n^{\frac{1}{2}+\frac{1}{5t}})$ rounds, assuming the girth conjecture. This bound holds even if we only consider $O(\log n)$ diameter graphs.
- If we restrict the label size of oracles to $o(n^\epsilon)$ bits, where $\epsilon = 1/2t(t+1)$ in unweighted graphs and $\epsilon = (1/5t^2)$ in weighted graphs, the same lower bounds are obtained without assuming the girth conjecture.

To the best of our knowledge, this paper is the first that exhibits a non-trivial trade-off between time and stretch for distributed distance oracles.

1 Introduction

1.1 Background

The primary objective of routing protocols is to identify paths from sources to destinations, in order to route packets efficiently. While there are a number of

---

[*] This work is supported in part by KAKENHI No.25106507 and No.25289114.

M.K. Aguilera et al. (Eds.): OPODIS 2014, LNCS 8878, pp. 60–75, 2014.

criteria to measure efficiency (delay, bandwidth, reliability, and so on), the most popular choice for selecting a good path is the path's length (i.e., distance between two nodes). In other words, Internet routing is still often synonymous to shortest path routing. A well-known example is the distance-vector routing protocol BGP. Often, real-world routing protocols weigh the edges of the network to measure distance more precisely. Sometimes, this weight information is somewhat hidden. In the case of BGP, for example, a technique called AS-path prepending is used, where a node includes itself in the route several times in order to give an edge more weight, i.e. to discourage other nodes from routing through it.

Regarding distributed complexity, many distance problems are recognized as so-called *global* problems. That is, their distributed time-complexity is $\Omega(\Delta)$, where Δ is the unweighted (hop-count) diameter of the network. A common naive solution for such global problems is the centralized approach: A single node aggregates the whole topological information of the network (and the weights of all edges in the case of weighted graphs), and computes the solution locally. This solution gives a $O(\Delta)$-time matching upper bound for the model with unbounded communication on each edge. However, for large networks, this unbounded (often also called LOCAL) message passing model becomes unreasonable. Instead one should assume that communication messages are limited. An established model for distributed computation is the so-called CONGEST message passing model. It allows each message to have at most $O(\log n)$ bits, where n is the number of all nodes in the network. In this model, it is known that the conventional all-pairs distance computation or approximation requires $\tilde{\Omega}(n)$ rounds [5,11] even in unweighted graphs with constant diameters.[1] On the other hand, near tight upper bounds are also known for both weighted and unweighted graphs. In unweighted graphs, there is an algorithm constructing all-pair shortest paths in $O(n)$ rounds [9,6], and in weighted graphs, there is an algorithm computing an $(1 + o(1))$-approximation of all-pairs distances in $\tilde{O}(n)$ rounds [11]. These results imply that all-pairs distance computation is an expensive task, even in the approximation case.

1.2 Distance Oracles

The inherent difficulty behind the all-pairs distance computation is that each node must fill out its own distance table of $n-1$ entries (one of which corresponds to the distance to some other node). In other words, it is inherently necessary that each node must receive $\Omega(n)$ bits of information to fill out the table of size $\Theta(n)$. However, if we can have a more compact representation of distance tables, its construction can be achieved in sublinear time. This observation yields to the problem of *distributed distance oracles*. A *distance oracle* is a subquadratic-size data structure storing all-pairs approximated distances, which was originally introduced in the context of centralized algorithms [14]. A distributed distance

---

[1] The tilde complexity notation $\tilde{O}(f(n))$ hides a polylogarithmic factor in n, usually, in this line of work, caused by the $O(\log n)$ bits allowed in each message.

oracle consists of a labeling scheme giving a label to each node and a local data structure deployed to each node v in the network. When a node v wants to know the distance to another node u, v queries its local data structure with the label of u. The data structure returns an estimated distance to u, which must be larger than the actual distance but can be overestimated. The approximation factor of distance oracles is also called *stretch*, which is defined as the maximum ratio between actual distances and estimated distances over all pairs (u, v).

There have been two results about the construction time of distributed distance oracles so far. The first one is by Das Sarma et al. [1], which gives an algorithm guaranteeing stretch $2t - 1$, $\tilde{O}(n^{1/t})$-bit label size, and $\tilde{O}(n^{1/t}\Delta')$-round construction time, where t is a parameter trading time, space, and stretch, and Δ' is the shortest-path diameter of the graph (i.e., the maximum hop length over all-pairs shortest paths.[2] Since Δ' can become linear of n at the worst-case, the construction time of this algorithm can be superlinear. A second paper by Lenzen and Patt-Shamir [8] proposes an algorithm with stretch $2t(8t-3)$, $O(t(\log n))$-bit label size, and $\tilde{O}(n^{1/2+1/2k} + \Delta)$-round construction time. It also shows that any distributed distance oracle algorithm achieving an arbitrary non-trivial stretch must have $\tilde{\Omega}(\sqrt{n})$ construction time.

1.3   Our Contribution

In this paper, we present several time lower bounds for the construction of distributed distance oracles. The primary results of our paper are new improved lower bounds depending on stretch. More precisely, our contributions are as follows:

- Under the assumption that the popular combinatorial girth conjecture is true, any distributed algorithm constructing oracles with stretch $2t$ requires $\tilde{\Omega}(n^{1/(t+1)})$ rounds in unweighted graphs. This bound holds even if we only consider constant diameter graphs.
- For oracles with stretch $2t$ in weighted graphs, we have a lower bound of $\Omega(n^{\frac{1}{2}+\frac{1}{5t}})$ rounds, assuming the girth conjecture. This bound holds even if we only consider $O(\log n)$ diameter graphs.
- If we restrict the label size of oracles to $o(n^{\epsilon})$ bits, where $\epsilon = 1/2t(t + 1)$ in unweighted graphs and $\epsilon = (1/5t^2)$ in weighted graphs, the same lower bounds are obtained without assuming the girth conjecture.

To the best of our knowledge, these are the first results that exhibit a non-trivial trade-off between construction time and stretch.

1.4   Related Work

For unweighted graphs, there has been a lot of progress to understand the distributed complexity [9,5,6,12,2] of distance problems such as the single-source

---

[2] Note that $\Delta \leq \Delta'$ always holds because for the pair (u, v) giving the hop-count diameter path, the hop-length of the shortest path between u and v cannot be shorter than Δ.

shortest paths, all-pairs shortest paths, diameter, and distance oracles. A first paper by Frischknecht et al. [5] showed an $\tilde{\Omega}(n)$-time lower bound for the exact diameter computation in unweighted networks with constant diameters. A matching upper bound was shown by Holzer at al. [6], and Lenzen and Peleg [9]; they concurrently and independently proposed almost the same $O(n)$-time algorithm for all-pairs shortest paths. The hardness of the approximated diameter computation is also considered. An easy solution for a 2-approximation of the diameter is to construct a shortest path tree rooted at an arbitrary node u. Since shortest path trees and breadth-first search (BFS) are equivalent in unweighted networks, a 2-approximation is trivially achieved in $O(\Delta)$ time by running a simple BFS-tree construction. A result by [5] also showed that any $3/2$-approximation algorithm for the diameter problem requires $\tilde{\Omega}(\sqrt{n})$ time. This lower bound is improved to $\tilde{\Omega}(n)$ by [6]. Interestingly, regarding upper bounds, Peleg et al. showed that a $3/2$-approximated value of Δ is computable in $\tilde{O}(\sqrt{n}\Delta)$ time [12]. A recent paper [9] improves this time bound for a $3/2$-approximation to an additive $\tilde{O}(\sqrt{n} + \Delta)$ time. Holzer and et al. [6] show a more accurate approximation algorithm for the network diameter problem with $O(n/\Delta + \Delta)$ running time. Its approximation factor is $(1 + \epsilon)$ for an arbitrary small constant $\epsilon < 1$.

While a rich literature exists for unweighted networks, only a few papers consider distance problems in weighted networks. To the best of our knowledge, there are three papers directly related to weighted graphs. Das Sarma et al. [1] and Lenzen and Patt-Shamir [8] we already discussed in the introduction. The paper by Lenzen and Patt-Shamir [8] also considers several related problems, including (all-pairs) shortest paths or diameter. In addition there is a very recent result by Nanongkai [11]. It proposes faster distributed approximation algorithms for single-source shortest paths and all-pairs shortest paths.

1.5   Roadmap

The paper is organized as follows: We introduce fundamental definitions and notations in Section 2. In Section 3, we give our lower bound proof for unweighted graphs. It is extended to the weighted case in Section 4. The case for bounded label size oracles is considered in Section 5. Finally in Section 6, we conclude this paper.

2   Preliminaries

2.1   Round-Based Synchronous Systems

A distributed system consists of n nodes interconnected with communication links. We model it by a weighted undirected graph $G = (V, E, w)$, where $V = \{v_0, v_1, \cdots, v_{n-1}\}$ is the set of nodes, $E \subseteq V \times V$ is the set of links (edges), and $w : E \to \mathbb{N}$ is the edge-weight function. Since we consider undirected graphs, $w(u, v) = w(v, u)$ holds for any $u, v \in V$. We also consider the system modeled by unweighted graphs, which is a special case of weighted graphs where every edge has weight one.

Executions of the system proceed with a sequence of consecutive rounds. In each round, each process sends a (possibly different) message to each neighbor, and within the round, all messages are received. After receiving its messages, each process performs local computation. Throughout this paper, we restrict the number of bits transmittable through any communication link per one round to $O(\log n)$ bits. This is known as the CONGEST model. Note that in weighted networks the weight of each edge does not imply the delay of communication. It is guaranteed that messages transferred through weighted edges reach their destinations within one round.

A path P between u and v is a sequence $u = u_0, u_1, \cdots u_k = v$ such that $(u_{i-1}, u_i) \in E$ holds for any i ($1 \leq i \leq k$). The *distance* between u and v in graph G is the weighted length of the shortest path between them, which is denoted by $d_G(u, v)$.

2.2 Problem Definition

The *distributed distance oracle* is defined as the problem of constructing a labeling scheme $\lambda : V \to L$, where L is the domain of labels, and a local data structure $dest_v : L \to \mathbb{Z}$ deployed to each node $v \in V$, which locally computes the distance estimation from v by giving the label $\lambda(u)$ of any target node u. The value $dest_v(\lambda(u))$ returned by the local oracle at node v is always lower at least the actual distance $d_G(u, v)$. The *stretch* of a distributed distance oracle is defined as $\max_{u,v \in V} dest_v(\lambda(u))/d_G(v, u)$. The *label size* of a distributed distance oracle is defined as $\lceil \log |L| \rceil$.

3   Lower Bound for Unweighted Graphs

3.1   Two-Party Communication Complexity

Communication complexity, which was first introduced by Yao [15], reveals the amount of communication to compute a global function whose inputs are distributed in the network. The most successful scenario in communication complexity is *two-party* communication complexity, where two players, called Alice and Bob, have x-dimensional 0-1 vectors $\mathbf{a}$ and $\mathbf{b}$ respectively, and compute a global function $f : \{0,1\}^x \times \{0,1\}^x \to \{0,1\}$. The communication complexity of a two-party protocol is the number of one-bit messages exchanged by the protocol for the worst case input (if the protocol is randomized, it is defined as the expected number of bits exchanged for the worst-case input). One of the most useful problems in communication complexity theory is *set-disjointness*:

Definition 1. *The x-bit set-disjointness function* $disj_x : \{0,1\}^x \times \{0,1\}^x \to \{0,1\}$ *is defined as follows:*

$$disj_x(\mathbf{a}, \mathbf{b}) = \begin{cases} 1 \ if \ \exists i \in [0, x-1] : a_i = b_i = 1, \\ 0 \ otherwise \end{cases}$$

For this problem, the following theorem is known [13,7].

Theorem 1. *The communication complexity of the x-bit set-disjointness problem is $\Omega(x)$.*

In the following argument, we use a slightly different form of the set-disjointness problem: We first introduce a *base graph* $H = (W, F)$ such that $|W| = N, |F| = M$ for some value $N > 0$ and $M > 0$. Alice and Bob respectively have subsets F_a and F_b of F as their inputs. The goal of the two-party computation is to decide if $(W, F_a \cup F_b) = H$ holds or not. This problem is equivalent to the M-bit set-disjointness problem, i.e., each edge in G is one-bit entry of the set-disjointness, and $e \in F_a$ (resp. $e \in F_b$) implies that Alice's (resp. Bob's) corresponding bit is set to zero. Thus by Theorem 1, the communication complexity of this problem is $\Omega(M)$. In what follows, we refer to this form of the set-disjointness problem as the *graphic set-disjointness* over H. If an instance (F_a, F_b) satisfies $F_a \cup F_b = F$, we say that (F_a, F_b) is *disjoint*. Otherwise we say that (F_a, F_b) is *intersecting*. Two examples of the graphic set-disjointness are shown in Figure 1, where one instance is disjoint and another is intersecting. The black (resp. gray) lines represent the edges Alice and Bob have (resp. does not have), and the dotted line in the intersecting case is the the edge commonly lost by both players).

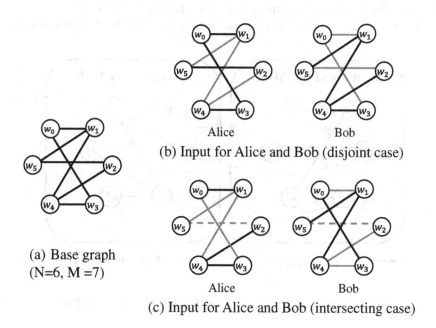

(a) Base graph
(N=6, M =7)

(b) Input for Alice and Bob (disjoint case)

(c) Input for Alice and Bob (intersecting case)

Fig. 1. Two examples of graphic set-disjointness instances

3.2   Gadget Construction

The core of the lower bound proof is a reduction from the graphic set-disjointness over some large-girth graph. The reduction scheme itself is similar to one introduced by [5]. This subsection shows the construction of the gadget for the reduction from the graphic set-disjointness over $H = (W, F)$. The constructed graph is denoted by $\Gamma_{H,\gamma}(F_a, F_b)$, where γ is a design parameter and (F_a, F_b) is any instance of the graphic set-disjointness over H. Letting $\Gamma_{H,\gamma}(F_a, F_b) = (V, E)$, V and E are constructed by the following steps:

1. The set of nodes V consists of two groups of N nodes $W^a = \{w_0^a, w_1^a, \cdots, w_{N-1}^a\}$ and $W^b = \{w_0^b, w_1^b, \cdots, w_{N-1}^b\}$.
2. For any i, $0 \leq i \leq N - 1$, each pair (w_i^a, w_i^b) is connected by an edge. The path (of length one) added in this step is called an *intra-cluster* path.
3. Each pair $(w_i^a, w_j^a) \in W^a$ (resp. $(w_i^b, w_j^b) \in W^b$) is connected by a path of length γ if and only if $(w_i, w_j) \in F_a$ (resp. $(w_i, w_j) \in F_b$) $(\gamma > 0)$. The path added in this step is called an *inter-cluster* path.

Informally, $\Gamma_{H,\gamma}(F_a, F_b)$ behaves as the weighted version of graph $H' = (W, F_a \cup F_b)$ (where each edge has weight γ). We can observe its behavior easily by clustering node pair w_i^a and w_i^b for each $i \in [0, N-1]$. Figure 2 gives an alternative drawing of $\Gamma_{H,\gamma}(F_a, F_b)$ for the instance shown in Figure 1. Each light-gray band corresponds to an edge e in H, which contains at least one actual path of length γ if and only if $e \in F_a$ or $e \in F_b$ holds. For this construction, we can show the following lemma:

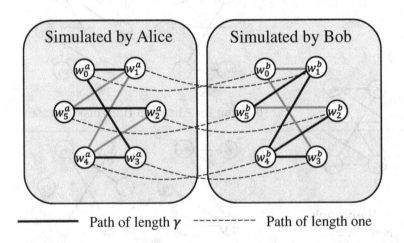

Path of length γ ---------- Path of length one

Fig. 2. Construction of $\Gamma_{H,\gamma}(E_a, E_b)$ for the disjoint instance in Figure 1

Lemma 1. *Let (F_a, F_b) be an instance of the graphic set-disjointness over $H = (W, F)$, $H' = (W, F_a \cup F_b)$, and $\Gamma = \Gamma_{H,\gamma}(F_a, F_b)$ for short. Then, for any integer $k > 0$, the following two properties hold:*

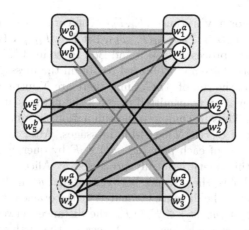

Fig. 3. An alternative view of $\Gamma_{H,\gamma}(E_a, E_b)$ shown in Figure 2

- If $d_{H'}(w_i, w_j) = 1$ $(i \neq j)$, $d_\Gamma(w_i^a, w_j^a) \leq \gamma + 2$.
- If $d_{H'}(w_i, w_j) = k$ $(k > 1,\ i \neq j)$, $d_\Gamma(w_i^a, w_j^a) \geq k\gamma$.

Proof. If $d_{H'}(w_i, w_j) = 1$, $(w_i, w_j) \in F_a$ or $(w_i, w_j) \in F_b$ holds. It implies that Γ contains an inter-cluster path between w_i^a and w_j^a or w_i^b and w_j^b. Since Γ always contains the edges (w_i^a, w_i^b) and (w_j^b, w_j^b), the first property obviously holds.

We look at the second property. Suppose for contradiction that there exists a simple path P between w_i^a and w_j^a whose length is less than $k\gamma$. Note that P is a concatenation of several inter-cluster paths and intra-cluster paths. It implies that P contains at most $k-1$ inter-cluster paths. Now let P' be the path obtained from P by contracting all the intra-cluster paths. Since P' is the concatenation of several inter-cluster paths, it can be represented by some sequence of the nodes where two inter-cluster paths are concatenated. Let $w_{\beta_0}^{\alpha_0}, w_{\beta_1}^{\alpha_1}, \cdots, w_{\beta_l}^{\alpha_l}$ be that sequence $(0 < l \leq k - 1)$. Then, for any $x \in [0, l-1]$, $w_{\beta_x}^{\alpha_x}$ and $w_{\beta_{x+1}}^{\alpha_{x+1}}$ must be connected by an inter-cluster path. That is, either $(w_{\beta_x}, w_{\beta_{x+1}}) \in F_a$ or $(w_{\beta_x}, w_{\beta_{x+1}}) \in F_b$ must hold. However, it implies that $H' = (W, F_a \cup F_b)$ contains a path from w_i to w_j with length $l(< k)$. It is a contradiction. $\square$

The main theorem utilizes the conjecture below:

Conjecture 1 (Girth conjecture). For any integers N and t, there exists a graph $H_{t,N}$ of N nodes and $\Theta(N^{1+1/t})$ edges whose girth is at least $2t + 2$.

Theorem 2. *Assume that the girth conjecture is true for some constant $t > 0$. Let ALG be a distributed algorithm constructing distance oracles with stretch $2t$. Then, its worst-case running time $\tau(n)$ must satisfy $\tau(n) \geq \Omega\left(n^{\frac{1}{t+1}}/\log n\right)$.*

Proof. The theorem is proved by the reduction from the graphic set-disjointness over $H_{t,N}$ claimed in Conjecture 1 (the value of N is determined later). That is, we construct from ALG a two-party protocol solving the graphic set-disjointness problem over $H_{t,N}$ for any instance (F_a, F_b). The core of the construction is to simulate the run of ALG in $\Gamma_{H_{t,N},8t}(F_a, F_b)$. Let $\Gamma = \Gamma_{H_{t,N},8t}(F_a, F_b)$ for short. Alice simulates all the processes in W^a and Bob those in W^b. To make the simulation proceed, both Alice and Bob need to obtain the messages exchanged on intra-cluster paths in the run of ALG. Since there are N intra-cluster paths, the amount of the information transmitted through the paths is at most $O(N \log n)$ bits per one round. Thus to complete the simulation, it suffices that Alice and Bob totally exchange $O(\tau(n)N \log n)$-bit messages. After the simulation, Alice checks the distance of each pair $(w_i^a, w_j^a) \in F$ by querying it to w_i^a's local oracle. Note that this query is locally processed at Alice. From Lemma 1, if $(w_i^a, w_j^a) \in F_a \cup F_b$ holds, the distance between w_i^a and w_j^a in Γ is at most $8t + 2$ (remind $\gamma = 8t$). Hence the distance estimated by the oracle is at most $2t(8t+2)$. On the other hand, if $(w_i, w_j) \notin F_a \cup F_b$, the distance between w_i and w_j in the graph $H_{t,N} \setminus (w_i, w_j)$ is at least $2t + 1$ because the girth of $H_{t,N}$ is at least $2t + 2$. Thus, by Lemma 1, the distance between w_i^a and w_j^b is at least $8t(2t + 1)$. These two facts imply that Alice can determine the disjointness of (F_a, F_b) from the query results: If all the queries return values at most $2t(8t + 2)$, (F_a, F_b) is disjoint. Otherwise, it is intersecting. Finally Alice sends one-bit information of the decision.

The two-party protocol explained above totally consumes $O(\tau(n)N \log n)$ bits in the worst case, which must be lower bounded by the communication complexity of the graphic set-disjointness over $H_{t,N}$, that is, $\Omega(N^{1+1/t})$ bits. Now we rewrite variable N by using only n and t. Since the number n of nodes in $\Gamma_{H_{t,N},8t}(F_a, F_b)$ is $2N + (8t - 1) \cdot \Theta(N^{1+1/t}) = \Theta(tN^{1+1/t})$, $N = \Theta((n/t)^{t/(t+1)})$ holds. Since t is a constant, we have $N = \Theta(n^{t/(t+1)})$. Thus the total amount of messages exchanged by the proposed two-party protocol is $\Theta((n^{t/(t+1)} \cdot (\tau(n) \log n))$. Since this is bounded by $\Omega(N^{1+1/t}) = \Omega(n)$. It follows that $\tau(n) = \Omega(n^{1/(t+1)}/ \log n)$. The theorem is proved. □

4   Lower Bound for Weighted Graphs

The lower bound in the previous section is extended to a stronger lower bound for weighted graphs. The fundamental idea of the extension is to utilize the framework by Das Sarma et al. [2]. Given values N and t, let $N^- = N^{\frac{1}{2} - \frac{1}{5t}}$ and $N^+ = N^{\frac{1}{2} + \frac{1}{5t}}$ for short. For simplicity, we assume that N^+ is a power of two. Note that this assumption is not essential and easily removed without affecting the asymptotic complexity we prove in this section. The gadget graph $\Gamma_H'(F_a, F_b)$ (say Γ' for short) is built by the following steps:

1. We first prepare N^- paths of length N^+, each of which is denoted by P_i ($0 \le i \le N^- - 1$). The nodes constituting P_i are identified by $v_{(i,0)}, v_{(i,1)}, \cdots, v_{(i,N^+-1)}$ from left to right. The weight of each edge constituting these paths is one. Furthermore, we give an alias to each endpoint node. We refer to nodes $v_{(i,0)}$ and $v_{(i,N^+-1)}$ as w_i^a and w_i^b respectively ($0 \le i \le N^- - 1$). We also define $W^a = \{w_0^a, w_1^a, \cdots w_{N^--1}^a\}$ and $W^b = \{w_0^b, w_1^b, \cdots w_{N^--1}^b\}$.

2. Construct a complete binary tree with N^+ leaves. The leaf nodes in the tree are labeled by $u_0, u_1, \cdots u_{N^+-1}$ from left to right. The weight of edges in the tree is $100N^+N^-t^2$.
3. Add edges $(u_i, v_{(j,i)})$ for any $i \in [0, N^+ - 1]$ and $j \in [0, N^- - 1]$. These edges also has weight $100N^+N^-t^2$.
4. Encode the instance (F_a, F_b) to the graph induced by W^a and W^b. That is, an edge (w_i^a, w_j^a) (resp. (w_i^b, w_j^b)) is connected by an edge of weight $8tN^+$ if and only if $(w_i, w_j) \in F_a$ (resp. $(w_i, w_j) \in F_b$).

The whole construction is illustrated in Figure 4. Note that the number n of nodes in $\Gamma_H'(F_a, F_b)$ is $\Theta(N)$, and its diameter is $D = O(\log n)$. This gadget has a structure similar to the unweighted case. We have the following lemma:

Lemma 2. *Let (F_a, F_b) be an instance of the graphic set-disjointness problem over $H = (W, F)$, and $H' = (W, F_a \cup F_b)$ for short. Then, for any integer $k > 0$, the following two properties hold:*

- *If $d_{H'}(w_i, w_j) = 1$ $(i \neq j)$, $d_{\Gamma'}(w_i^a, w_j^a) \leq (8t + 2)N^+$.*
- *If $d_{H'}(w_i, w_j) = k$ $(k > 1, i \neq j)$, $d_{\Gamma'}(w_i^a, w_j^a) \geq 8tkN^+$.*

Proof. Since all the edges augmented in Step 2 and 3 of the construction are too heavy, they are not contained in the shortest path between w_i^a and w_j^b for any i and j. Thus we can omit those edges in the proof (in Figure 4, they are grayed out). Then, the graph $\Gamma' = \Gamma_H'(F_a, F_b)$ can be seen as a weighted version of $\Gamma = \Gamma_{H,8t}(F_a, F_b)$: The length of the path between w_i^a and w_i^b $(0 \leq i \leq N^- - 1)$ is N^+ (which corresponds to intra-cluster paths in Γ) and each edge between $(w_i^a, w_j^a) \in F_a$ (resp. $(w_i^b, w_j^b) \in F_b$) has weight $8tN^+$. That is, we have $d_{\Gamma'}(w_i^a, w_j^a) = N^+ \cdot d_\Gamma(w_i^a, w_j^a)$. Consequently, the lemma is deduced from Lemma 1. □

The following theorem is the core of the reduction.

Theorem 3 (Das Sarma et al. [2]). *Let ALG be any algorithm running on the graph Γ', where H is an arbitrary graph of N^- nodes. Then there exists a two-party protocol satisfying the following three properties:*

- *At the beginning of the protocol, Alice (resp. Bob) knows the whole topological information of Γ' except for the subgraph induced by W^b (resp. W^a),*
- *after the run of the protocol, Alice and Bob output the internal states of the processes in W^a and W^b at round $N^+/2$ in the execution of ALG, respectively, and*
- *the protocol consumes at most $O(N^+(\log n)^2)$-bit communication.*

While the graph used in this paper is a slightly modified version of the original construction in [2], the theorem above is proved in the almost same way. So we just quote it without the proof.

The theorem above induces our lower bound via a reduction from two-party graphic set-disjointness:

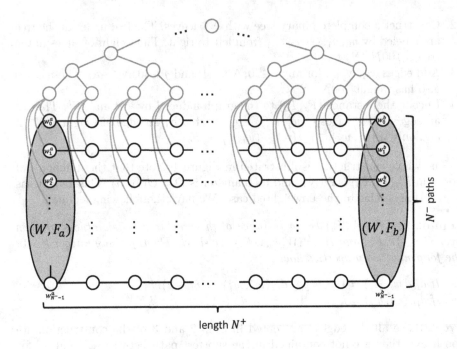

Fig. 4. Construction of $\Gamma'_H(F_a, F_b)$

Theorem 4. *Assume that the girth conjecture is true for some constant $t > 0$. Let ALG be an algorithm constructing distributed (weighted) distributed distance oracles with stretch 2t. Then, its worst-case running time $\tau(n)$ must satisfy*
$$\tau(n) = \Omega(N^+) = \Omega\left(n^{\frac{1}{2}+\frac{1}{5t}}\right).$$

Proof. The proof is almost the same as Theorem 2. Letting H_{t,N^-} be the graph claimed in Conjecture 1 and (F_a, F_b) be any instance of the graphic set-disjointness over H_{t,N^-}, we consider the run of ALG in the graph $\Gamma'_{H_{t,N^-}}(F_a, F_b)$. Suppose for contradiction that $\tau(n) < N^+/2$ holds. Then, by Theorem 3, we can have a two-party protocol where Alice and Bob simulate the run of ALG at the processes in W^a and W^b respectively. After the simulation, Alice queries the distance between w_i^a and w_j^b for each $(w_i, w_j) \in F$. Then, by Lemma 2 and the same argument as the proof of Theorem 2, Alice can determine the disjointness of (F_a, F_b). That is, if all the queries return values at most $2t(8t + 2)N^+$, (F_a, F_b) is disjoint. Otherwise, it is intersecting. Finally Alice sends the one-bit information of the decision. By Theorem 3, this protocol consumes only $O(N^+(\log n)^2)$-bit communication for deciding the disjointness of (F_a, F_b). However, the communication complexity of the graphic set-disjointness over H_{t,N^-} is bounded by the number of edges in the base graph H_{t,N^-}. That is, from Conjecture 1, it is lower bounded by $\Omega((N^-)^{(1+1/t)}) = \Omega((N^{\frac{1}{2}-\frac{1}{5t}})^{(1+1/t)}) = \Omega(N^{\frac{1}{2}+\frac{1}{5t}+\epsilon}) = \omega(N^+(\log n)^2)$, where ϵ is a small constant (depending on t). It is a contradiction. $\square$

5   Lower Bound for Bounded Label Size Oracles

In this section, we present an unconditional lower bound for the case of bounded label size oracles. For lack of space, we only focus on the bound for unweighted graphs, but its result is easily extended to the weighted case by combining the argument in Section 4.

The fundamental idea follows the proof in Section 3. We construct a reduction from the two-party graphic set-disjointness. The main difference is that we use a graph of $\Theta(N^{1+1/t-\epsilon})$ nodes with girth $(2t + 2)$ as the base graph (where ϵ is a small constant depending on t), but augment only N intra-cluster paths crossing Alice and Bob sides. The following lemma is an alternative to the girth conjecture.

Lemma 3. *Let $\epsilon \leq 1/2t^2$. For any sufficiently large integer N, there exists a bipartite graph $H = (U \cup W, F)$ such that $|U| = N^{1+1/t-\epsilon}$, $|W| = N$, and $|F| = N^{1+1/t}$ hold and the girth is at least $2t + 2$.*

Proof. The proof idea is based on the seminal one by Erdos's probabilistic method, which shows an existence of the graph with high chromatic number and girth [3,4]. We consider the random construction of a bipartite graph H^* whose node set is $U \cup W$. That is, fixing the vertex set U and W, for each pair $(u, w) \in U \times W$, we add an edge with probability $1/N^{1-\epsilon}$. Then the graph H^* satisfies the following two properties with a non-zero probability: (1) The number of edges is $\Omega(N^{1+1/t})$, and (2) there are only $o(N^{1+1/t})$ cycles with a length less than or equal to $2t$. Once we find a graph H^* with both properties, the desired graph H is obtained from H^* by removing $o(N^{1+1/t})$ edges from each short cycle, which still have $\Omega(N^{1+1/t})$ edges but there is no cycle with length less than $2t + 2$ (remind that the graph is bipartite and thus there is no cycle of length $2t + 1$). Thus the remaining part of the proof is to show that the properties (1) and (2) are simultaneously satisfied with a non-zero probability. More precisely, it suffices to show that each property is satisfied with a probability more than $1/2$. Then using the union bound, the probability that either property (1) or (2) fails becomes strictly smaller than one.

The first property is almost trivial. Let X be the number of edges in H'. Since the variable X is the sum of independent Poisson trials, we can apply Chernoff bounds [10]. Then it is not difficult to obtain $\Pr[X < E[X]/2] = o(1)$. That is, $X \geq E[X]/2$ holds with probability more than $1/2$. The expected number $E[X]$ of edges in H' is $|U||W| \cdot (1/N^{1-\epsilon}) = N^{1+1/t}$, and thus the first property holds.

We look at the second property. Let Y be the number of cycles with length less than $2t + 2$ in H'. The probability that a given sequence of $2k$ nodes ($k \leq t$) form a cycle is obviously bounded by $(1/N^{1-\epsilon})^{2k}$. Since we assume $\epsilon \leq 1/2t^2$, the expected number $E[Y]$ is bounded as follows:

$$E[Y] = \sum_{k=1}^{t} \binom{N^{1+\frac{1}{t}-\epsilon}}{k} \binom{N}{k} \cdot \left(\frac{1}{N^{1-\epsilon}}\right)^{2k}$$

$$\leq t N^{t(1+\frac{1}{t}-\epsilon)} N^t \cdot N^{-2t(1-\epsilon)}$$

$$\leq t N^{1+\epsilon t}$$

$$\leq O(N^{1+1/2t}),$$

Using Markov's inequality [10], we can have

$$\Pr[Y \geq 3E[Y]] \leq E[Y]/(3E[Y]) = 1/3.$$

Thus the property (2) is also satisfied with probability more than $1/2$. The lemma is proved.   □

Let $H = (U \cup W, F)$ be the graph proposed in Lemma 3 for $\epsilon = 1/2t^2$. The gadget graph $\hat{\Gamma}_{H,8t}(F_a, F_b)$ to encode the graphic set-disjointness (F_a, F_b) over H is constructed similarly to $\Gamma_{H,8t}(F_a, F_b)$ in Section 3. Only the difference is that we connect Alice and Bob sides only by edges (w_i^a, w_i^b) for any i ($0 \leq i \leq N-1$), but not connect u_i^a and u_i^b. The constructed gadget is presented in Figure 5. We define $U^a = \{u_0^a, u_1^a, \cdots, u_{N-1}^a\}$ and $U^b = \{u_0^b, u_1^b, \cdots, u_{N-1}^b\}$.

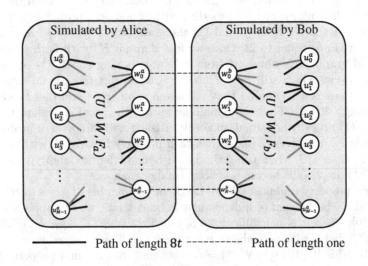

Fig. 5. Construction of $\hat{\Gamma}_{H,8t}(F_a, F_b)$

For this construction, we can have a lemma analogous to Lemma 1.

Lemma 4. *Let (F_a, F_b) be an instance of the graphic set-disjointness over H, $H' = (U, W, F_a \cup F_b)$, and $\hat{\Gamma} = \hat{\Gamma}_{H,8t}(F_a, F_b)$ for short. Then, for any integer $k > 0$, the following two properties hold:*

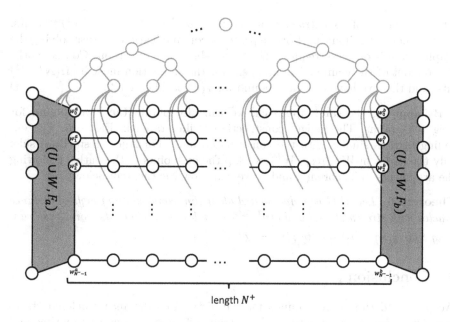

Fig. 6. Construction of the gadget for weighted and bounded label-size oracles

- *If* $(w_i, u_j) \in F_b$, $d_{\hat{\Gamma}}(w_i^a, u_j^b) \leq (8t + 1)$.
- *If* $(w_i, u_j) \notin F_a \cup F_b$, $d_{\hat{\Gamma}}(w_i^a, u_j^b) \geq 8t(2t + 1)$.

While the proof is omitted for lack of space, it is almost the same as that for Lemma 1. We show the main theorem:

Theorem 5. *Let ALG be an algorithm constructing distributed distance oracles with stretch $2t$ and $o(n^{1/2t(t+1)})$-bit label size. Then, its worst-case running time $\tau(n)$ must satisfy $\tau(n) \geq \Omega\left(n^{\frac{1}{t+1}}/\log n\right)$.*

Proof. The proof basically follows that for Theorem 2. To construct a two-party graphic set-disjointness protocol, Alice and Bob simulate the internal states of the processes W^a and U^a, and W^b and U^b in the run of ALG, respectively. After the simulation, since Bob knows all the labels assigned to the nodes in U^b, it sends them to Alice. This information allows Alice to estimate the distance between w_i^a and u_j^b for any i and j *locally*. Then, by Lemma 4, Alice can determine the existence of the edge (w_i^b, u_j^b) for i, j such that $(w_i^a, u_j^a) \notin F_a$ holds. That is, Alice first queries the distance between w_i^a and u_j^b, and then if the estimated distance is less than or equal to $2t(8t + 1)$, it decides $(w_i^a, u_j^b) \in F_b$. Repeating this kind of queries, Alice can determines the disjointness of (F_a, F_b).

Compared to the protocol proposed in the proof of Theorem 2, the extra communication incurred by this protocol is to send the labels of the nodes in U^b from Bob to Alice. Since the label size for one node is $o(n^{1/2t(t+1)}) = o(N^{1/2t^2})$

bits, the amount of the extra communication is $o(N^{1/2t^2}) \cdot |U^b| = o(N^{1+1/t})$ bits, which is not a dominant part of the protocol communication because solving the graphic set-disjointness requires $\Omega(N^{1+1/t})$-bit communication. Consequently, the amount of the communication spent for the simulation must be $\Omega(N^{1+1/t})$ bits, and thus we have the same bound for $\tau(n)$ as Theorem 2.   □

By applying the same approach, we can also obtain the lower bound for weighted graphs. The gadget construction is illustrated in Figure 6. The encoding of H is similar with the construction of $\hat{\Gamma}$. For Alice (resp. Bob) side, only the nodes in W^a (resp. W^b) overlap the endpoints of the paths. Following the arguments in Theorem 4 and 5, we can show the theorem below:

Theorem 6. *Let ALG be a distributed algorithm constructing weighted distance oracles with stretch 2t and and $o(n^{1/5t^2})$-bit label size. Then, its worst-case running time $\tau(n)$ must satisfy $\tau(n) = \Omega\left(n^{\frac{1}{2}+\frac{1}{5t}}\right)$.*

6 Conclusion

We presented time lower bounds for the distributed distance oracle construction. Our primary result is to exhibit a trade-off between construction time and stretch. More precisely, given stretch factor $2t$, our lower bounds have the form of $\tilde{\Omega}(n^{1/O(t)})$ rounds for unweighted graphs, and the form of $\tilde{\Omega}(n^{1/2+1/O(t)})$ rounds for weighted graphs. While we assume that the girth conjecture is true for proving the bounds, we can bypass it when we consider bounded label-size ofracles. Restricting the label size to n^ϵ for a small constant ϵ depending on t, the same lower bounds are unconditionally obtained. An open problem related to our results is to find algorithms whose running time gets close to our lower bounds. The currently best algorithm in weighted graphs takes $O(n^{\frac{1}{2}+\frac{1}{2t}})$ rounds for the construction and achieves $O(t^2)$ stretch. The algorithm whose stretch linearly depends on t but achieving $O(n^{\frac{1}{2}+\frac{1}{O(t)}})$-round construction time is still open. Faster solutions for unweighted graphs are also not known.

References

1. Das Sarma, A., Dinitz, M., Pandurangan, G.: Efficient computation of distance sketches in distributed networks. In: Proc. of the 24th ACM Symposium on Parallelism in Algorithms and Architectures (SPAA), pp. 318–326 (2012)
2. Das Sarma, A., Holzer, S., Kor, L., Korman, A., Nanongkai, D., Pandurangan, G., Peleg, D., Wattenhofer, R.: Distributed verification and hardness of distributed approximation. In: Proc. of the 43rd Annual ACM Symposium on Theory of Computing, pp. 363–372 (2011)
3. Diestel, R.: Graph Theory, 4th edn., vol. 173. Springer (2012)
4. Erdös, P.: Graph theory and probability. In: Classic Papers in Combinatorics, pp. 276–280 (1987)
5. Frischknecht, S., Holzer, S., Wattenhofer, R.: Networks cannot compute their diameter in sublinear time. In: Proc. of the Twenty-Third Annual ACM-SIAM Symposium on Discrete Algorithms (SODA), pp. 1150–1162 (2012)

6. Holzer, S., Wattenhofer, R.: Optimal distributed all pairs shortest paths and applications. In: Proc. of the 2012 ACM Symposium on Principles of Distributed Computing (PODC), pp. 355–364 (2012)
7. Kalyanasundaram, B., Schintger, G.: The probabilistic communication complexity of set intersection. SIAM Journal on Discrete Mathematics 5(4), 545–557 (1992)
8. Lenzen, C., Patt-Shamir, B.: Fast routing table construction using small messages: Extended abstract. In: Proc. of the 45th Annual ACM Symposium on Symposium on Theory of Computing (STOC), pp. 381–390 (2013)
9. Lenzen, C., Peleg, D.: Efficient distributed source detection with limited bandwidth. In: Proc. of the 2013 ACM Symposium on Principles of Distributed Computing (PODC), pp. 375–382 (2013)
10. Mitzenmacher, M., Upfal, E.: Probability and Computing: Randomized Algorithms and Probabilistic Analysis. Cambridge University Press (2005)
11. Nanongkai, D.: Distributed approximation algorithms for weighted shortest paths. In: Proc. of the 46th Annual ACM Symposium on Theory of Computing (STOC), pp. 565–573 (2014)
12. Peleg, D., Roditty, L., Tal, E.: Distributed algorithms for network diameter and girth. In: Czumaj, A., Mehlhorn, K., Pitts, A., Wattenhofer, R. (eds.) ICALP 2012, Part II. LNCS, vol. 7392, pp. 660–672. Springer, Heidelberg (2012)
13. Razborov, A.A.: On the distributional complexity of disjointness. Theoretical Computer Science 106(2), 385–390 (1992)
14. Thorup, M., Zwick, U.: Approximate distance oracles. Journal of the ACM 52(1), 1–24 (2005)
15. Yao, A.C.-C.: Some complexity questions related to distributive computing (preliminary report). In: Proc. of the 11th Annual ACM Symposium on Theory of Computing (STOC), pp. 209–213 (1979)

Erasure-Coded Byzantine Storage
with Separate Metadata

Elli Androulaki[1], Christian Cachin[1], Dan Dobre[2], and Marko Vukolić[3]

[1] IBM Research - Zurich, Rüschlikon, Switzerland
{lli,cca}@zurich.ibm.com
[2] Work Done at NEC Labs Europe, Germany
dan@dobre.net
[3] Department of Computer Science, ETH Zurich, Switzerland and Eurécom,
Sophia Antipolis, France
vukolic@eurecom.fr

Abstract. Although many distributed storage protocols have been introduced, a solution that combines the strongest properties in terms of availability, consistency, fault-tolerance, storage complexity, and concurrency has been elusive so far. Combining these properties is difficult, especially if the resulting solution is required to be efficient and incur low cost.

We present AWE, the first *erasure-coded* distributed implementation of a multi-writer multi-reader read/write register object that is, at the same time: (1) asynchronous, (2) wait-free, (3) atomic, (4) amnesic, (i.e., nodes store a bounded number of values), and (5) Byzantine fault-tolerant (BFT), using the optimal number of nodes. AWE maintains metadata separately from bulk data, which is encoded into fragments with a k-out-of-n erasure code and stored on dedicated *data nodes* that support only simple reads and writes. Furthermore, AWE is the first BFT storage protocol that uses only $n = 2t + k$ data nodes to tolerate t Byzantine faults, for any $k \geq 1$. Metadata, on the other hand, is stored using an atomic snapshot object, which may be realized from $3t + 1$ *metadata nodes* for tolerating t Byzantine faults.

AWE is efficient and uses only lightweight cryptographic hash functions. Moreover, we show that hash functions are needed by any BFT distributed storage protocol that stores the bulk data on $3t$ or fewer data nodes.

1 Introduction

Erasure coding is a key technique that saves space and retains robustness against faults in distributed storage systems. In short, an erasure code splits a large data value into n *fragments* such that from any k of them the input value can be reconstructed. Erasure coding is used by several large-scale storage systems [24, 28] that offer large capacity, high throughput, resilience to faults, and efficient use of storage space.

Whereas the storage systems in production use today only tolerate crashes or outages, storage systems in the *Byzantine failure model* survive also more severe faults, ranging from arbitrary state corruption to malicious attacks on processes. In this paper, we consider a model where multiple *clients* concurrently access a storage service provided by a distributed set of *nodes*, where t out of n nodes may be Byzantine. We model the storage service as an abstract read/write register object.

M.K. Aguilera et al. (Eds.): OPODIS 2014, LNCS 8878, pp. 76–90, 2014.

Although Byzantine-fault tolerant (BFT) erasure-coded distributed storage systems have received some attention in the literature [5, 9, 15, 18, 21], our understanding of their properties is not mature. The role of different quorums, the semantics of concurrent access, the latency of protocols, and the processing capabilities of the nodes have been investigated thoroughly for protocols based on *replication* [12, 27]; in contrast, our knowledge about *erasure-coded* distributed storage is far more limited. In fact, the existing BFT erasure-coded storage protocols suffer from multiple drawbacks: some require nodes to store an unbounded number of values [18] or rely on node-to-node communication [9], others need computationally expensive public-key cryptography [9, 21] or may block clients due to concurrent operations of other clients [21].

Contribution. This paper introduces AWE, the first erasure-coded distributed implementation of a multi-reader multi-writer (MRMW) register that is, at the same time, (1) asynchronous, (2) wait-free, (3) atomic, (4) amnesic, (5) tolerates the optimal number of Byzantine nodes, and (6) does not use public-key cryptography.

These properties are desirable, as *wait-freedom* [22] and *atomicity* (or *linearizability*) [23] are not only the most fundamental but also the strongest liveness and consistency properties (respectively) of distributed storage. Roughly, wait-free liveness means that any correct client operation terminates irrespective of the behavior of the faulty nodes and clients, whereas atomicity means that all operations appear to take effect instantaneously. Therefore, guaranteeing wait-freedom and atomicity under the weakest possible assumptions (asynchrony, Byzantine faults) is highly desirable. Furthermore, *amnesic* storage [11] in combination with erasure-coding minimizes the storage overhead, another important measure for distributed storage. Roughly speaking, in amnesic storage nodes store a bounded number of values and erase obsolete data. Finally, the absence of public-key cryptography contributes to an efficient implementation of AWE. Although different subsets of these robustness properties have been demonstrated so far, they have never been achieved together for erasure-coded storage. Combining these desirable properties, has been a longstanding open problem [18].

AWE distinguishes between *metadata* (short control information) and *bulk data* (the erasure-coded stored values) and introduces two separate *classes* of nodes that store metadata and bulk data. With this approach, AWE beats the lower bound of $n > 3t$ nodes needed for distributed BFT storage [26], for the class of *data nodes* (that store bulk data). This makes AWE novel, as all known erasure-coded BFT storage solutions comply with this bound for their bulk data storage.

More specifically, with a k-out-of-n erasure code, protocol AWE needs only $2t + k$ data nodes, for any $k \geq 1$. This approach saves resources in practice, as storage costs for the bulk data often dominate. The data nodes may be passive objects that support read and write operations but cannot execute code, as in Disk Paxos [1]. In practice, such services may be provided by the key-value stores (KVS) popular in cloud storage.

We formulate AWE in a modular way using an abstract *metadata service* that stores control information with an *atomic snapshot* object. A snapshot object may be realized in a distributed asynchronous system from simple read/write registers [3]. For making this implementation fault-tolerant, these registers must still be emulated from $n > 3t$ different *metadata nodes* , in order to tolerate t Byzantine nodes.

Finally, AWE uses simple cryptographic hash functions but no expensive public-key operations. To explain the use of cryptography in AWE, we show that separating data from metadata and reducing the number of data nodes to $3t$ or less implies the use cryptographic techniques. This result is interesting in its own right, as it implies that *any* distributed BFT storage protocol that uses $3t$ or fewer nodes for storing bulk data must involve cryptographic hash functions and place a bound on the computational power of the Byzantine nodes. As all existing BFT erasure-coded storage protocols (including AWE) rely on cryptography, this result does not pose a restriction on practical systems. However, it illustrates a fundamental limitation that is particularly relevant for $k = 1$, i.e., for replication-based BFT storage protocols.

Structure. The paper continues with the overview of related work in Section 2. The model is given in Section 3 and Protocol AWE is presented in Section 4. The communication and storage complexities of AWE are compared to those of existing protocols in Section 5. Section 6 establishes the necessity of cryptographic assumptions for BFT storage with less than $3t$ data nodes. Finally, Section 7 concludes the paper. Detailed proofs appear in a technical report [4].

2 Related Work

Table 1 summarizes this section that gives a brief overview of the relevant related work.

Earlier designs for erasure-coded distributed storage have suffered from potential aborts due to contention [16] or from the need to maintain an unbounded number of fragments at data nodes [18]. In the crash-failure model, ORCAS [15] and CASGC [10] achieve optimal resilience $n > 2t$ and low communication overhead, combined with wait-free (ORCAS) and FW-termination (CASGC), respectively. FW-termination ensures that read operations always progress only in executions with a finite number of writes.

In the model with Byzantine nodes, Cachin and Tessaro (CT) [9] introduced the first wait-free protocol with atomic semantics and optimal resilience $n > 3t$. CT uses a verifiable information dispersal protocol but needs node-to-node communication, which lies outside our model. Hendricks et al. (HGR) [21] present an optimally resilient protocol that comes closest to our protocol among the existing solutions. It offers many desirable features, that is, it has as low communication cost, works asynchronously, achieves optimal resilience, atomicity, and is amnesic. Compared to our work, it (1) uses public-key cryptography, (2) achieves only FW-termination instead of wait-freedom, and (3) requires *processing* by the nodes, i.e., the ability to execute complex operations beyond simple reads and writes.

To be fair, much of the (cryptographic) overhead inherent in the CT and HGR protocols defends against poisonous writes from Byzantine clients, i.e., malicious client behavior that leaves the nodes in an inconsistent state. We do not consider Byzantine clients in this work, since permitting arbitrary client behavior is problematic [20]. Such a client might write garbage to the storage system and wipe out the stored value at any time. However, even without the steps that protect against poisonous writes, HGR still requires processing by the nodes and is not wait-free.

Table 1. Comparison of erasure-coded distributed storage solutions. An asterisk (*) denotes optimal properties. The column labeled *Type* states the computation requirements on nodes: *Proc.* denotes processing; *Msg.* means sending messages to other nodes, in addition to processing; *R/W* denotes a read/write register.

Protocol	BFT	Liveness	Data nodes	Type	Amnesic	Cryptogr.
ORCAS [15]	—	Wait-free	$2t + 1$	Proc.	—	N/A
CASGC [10]	—	FW-term.	$2t + 1$	Proc.	✓*	N/A
CT [9]	✓*	Wait-free *	$3t + 1$	Msg.	—	Public-key
HGR [21]	✓*	FW-term.	$2t + k$, for $k > t$	Proc.	✓*	Public-key
M-PoWerStore [13]	✓*	Wait-free *	$3t + 1$	Proc.	—	Hash func. *
DepSky [5]	✓*	Obstr.-free	$3t + 1$	R/W *	—	Public-key
AWE (Sec. 4)	✓*	Wait-free *	$2t + k$, for $k \geq 1$ *	R/W *	✓*	Hash func. *

The M-PoWerStore protocol [13] employs a cryptographic "proof of writing" for wait-free atomic erasure-coded distributed storage without node-to-node communication. Similar to other protocols, M-PoWerStore uses $n > 3t$ nodes (with processing capabilities) and is not amnesic.

Several systems have recently addressed how to store erasure-coded data on multiple redundant cloud services but only few of them focus on wait-free concurrent access. HAIL [6], for instance, uses Byzantine-tolerant erasure coding and provides data integrity through proofs of retrievability; however, it does not address concurrent operations by different clients. DepSky [5] achieves regular semantics and uses lock-based concurrency control; therefore, one client may block operations of other clients.

A key aspect of AWE lies in the differentiation of (small) metadata from (large) bulk data: this enables a modular protocol design and an architectural separation for implementations. The concept also resembles the separation between agreement and execution used in the context of BFT replicated state machines in partially synchronous systems [29].

FARSITE [2] first introduced such a separation of metadata and data for replicated storage; their data nodes and their metadata abstractions require processing, however, in contrast to AWE. Non-explicit ways of separating metadata from data can already be found in several previous erasure coding-based protocols. For instance, the cross checksum, a vector with the hashes of all n fragments, has been replicated on the data nodes to ensure consistency [9, 18]. Separation of metadata has been also used in practical replicated crash-tolerant systems such as Hadoop Distributed File System.

Finally, Cachin et al. [7] have recently shown in a predecessor to this work that also with replication, separating metadata from bulk data has benefits. Their asynchronous wait-free BFT distributed storage protocol, called MDStore, reduces the number of data nodes to only $2t + 1$. When protocol AWE is reduced to use replication with the trivial erasure code ($k = 1$), it uses as few nodes as MDStore to achieve the same wait-free atomic semantics; unlike AWE, however, MDStore is not amnesic and uses processing nodes.

The connection between separating data from metadata, reducing the number of data nodes, and the necessity of cryptographic techniques appears novel. In a sense, this paper shows a novel connection between the resilience of a distributed BFT protocol and the existence of a cryptographic primitive.

3 Definitions

We use a standard asynchronous deterministic distributed system of processes that communicate with each other. Processes comprise a set C of m *clients*, and a set D of n *data nodes* $d_1, \ldots, d_n$. Clients can only crash and up to t data nodes can be Byzantine and exhibit NR-arbitrary faults.

Protocols are presented in a modular event-based notation [8]. Processes interact through events that are qualified by the process identifier to which the event belongs. An event *Sample* of a process m with a parameter x is denoted by $\langle\ m\text{-}Sample \mid x\ \rangle$. Processes execute *operations*, defined in terms of *invocation* and *response* events. We use the standard notions of operation precedence, histories, and linearizability [23].

A *read/write register* r is an object that stores a value from a domain V and supports exactly two operations: (1) a *Write* operation to r with invocation $\langle\ r\text{-}Write \mid v\ \rangle$, taking a value $v \in V$ that terminates with a response $\langle\ r\text{-}WriteAck\ \rangle$; and (2) a *Read* operation from r with invocation $\langle\ r\text{-}Read\ \rangle$ that terminates with a response $\langle\ r\text{-}ReadResp \mid v\ \rangle$, containing a parameter $v \in V$. The behavior of a register is given through its sequential specification, which requires that every *r-Read* operation returns the value written by the last preceding *r-Write* operation in the execution, or the special symbol $\perp \notin V$ if no such operation exists.

The goal of this work is to describe a protocol that emulates a linearizable register abstraction among the clients; such a register is also called *atomic*. Some of the clients may crash and some nodes may be Byzantine. A protocol is called *wait-free* [22] if every operation invoked by a correct client eventually completes, irrespective of how other clients and nodes behave.

We make use of cryptographic hash functions modeled by a distributed oracle accessible to all processes [8]. A hash function H maps a bit string x of arbitrary length to a short, unique representation of fixed length. We use a *collision-free* hash function; this property means that no process, not even a Byzantine process, can find two distinct values x and x' such that $H(x) = H(x')$.

4 Protocol AWE

Erasure code. An (n, k)-*erasure code (EC)* with domain V is given by an encoding algorithm, denoted *Encode*, and a reconstruction algorithm, called *Reconstruct*. We consider only *maximum-distance separable codes*, which achieve the Singleton bound in the following sense. Given a (large) value $v \in V$, algorithm $Encode_{k,n}(v)$ produces a vector $[f_1, \ldots, f_n]$ of n *fragments*, which are from a domain $\mathcal{F}$. A fragment is typically much smaller than the input, and any k fragments contain all information of v, that is, $|V| \approx k|\mathcal{F}|$.

For an n-vector $F \in \left(\mathcal{F} \cup \{\perp\}\right)^n$, whose entries are either fragments or the symbol $\perp$, algorithm $Reconstruct_{k,n}(F)$ outputs a value $v \in V$ or $\perp$. An output value of $\perp$ means that the reconstruction failed. The *completeness* property of an erasure code requires that an encoded value can be reconstructed from any k fragments. In other words, for every $v \in \mathcal{V}$, when one computes $F \leftarrow Encode_{k,n}(v)$ and then erases up to $n - k$ entries in F by setting them to $\perp$, algorithm $Reconstruct_{k,n}(F)$ outputs v.

Metadata service. The metadata service is implemented by a standard *atomic snapshot object* [3], called *dir*, that serves as a *directory*. A snapshot object extends the simple storage function of a register to a service that maintains one value for each client and allows for better coordination. Like an array of multi-reader single-writer (MRSW) registers, it allows every client to *update* its value individually; for reading it supports a *scan* operation that returns the vector of the stored values, one for every client. More precisely, the operations of *dir* are:

- An *Update* operation to *dir* is triggered by an invocation $\langle$ *dir-Update* $\mid c, v$ $\rangle$ by client c that takes a value $v \in \mathcal{V}$ as parameter and terminates by generating a response $\langle$ *r-UpdateAck* $\rangle$ with no parameter.
- A *Scan* operation on *dir* is triggered by an invocation $\langle$ *dir-Scan* $\rangle$ with no parameter; the snapshot object returns a vector V of $m = |\mathcal{C}|$ values to c as the parameter in the response $\langle$ *r-ScanResp* $\mid V$ $\rangle$, with $V[c] \in \mathcal{V}$ for $c \in \mathcal{C}$.

The sequential specification of the snapshot object follows directly from the specification of an array of m MRSW registers (hence, the snapshot initially stores the special symbol $\perp \notin \mathcal{V}$ in every entry). When accessed concurrently from multiple clients, its operations appear to take place atomically, i.e., they are linearizable. Snapshot objects are weak — they can be implemented from read/write registers [3], which, in turn, can be implemented from a set of a distributed processes subject to Byzantine faults. Wait-free amnesic implementations of registers with the optimal number of $n > 3t$ processes are possible using existing constructions [14, 19].

Data nodes. Data nodes provide a simple key-value store interface. We model the state of data nodes as an array $data[ts] \in \Sigma^*$, initially $\perp$, for $ts \in Timestamps$. Every value is associated to a timestamp, which consists of a sequence number sn and the identifier c of the writing client, i.e., $ts = (sn, c) \in Timestamps = N_0 \times (\mathcal{C} \cup \{\perp\})$; timestamps are initialized to $T_0 = (0, \perp)$. Data node d_i exports three operations:

- $\langle$ d_i-*Write* $\mid ts, v$ $\rangle$, which assigns $data[ts] \leftarrow v$ and returns $\langle$ d_i-*WriteAck* $\mid ts$ $\rangle$;
- $\langle$ d_i-*Read* $\mid ts$ $\rangle$, which returns $\langle$ d_i-*ReadResp* $\mid ts, data[ts]$ $\rangle$; and
- $\langle$ d_i-*Free* $\mid TS$ $\rangle$, which assigns $data[ts] \leftarrow \perp$ for all $ts \in TS$, and returns $\langle$ d_i-*FreeAck* $\mid TS$ $\rangle$.

4.1 Protocol Overview

AWE uses the metadata directory *dir* to maintain pointers to the fragments stored at the data nodes. The directory stores an entry for every writer; it contains the timestamp of its most recently written value, the identities of those nodes that have acknowledged to store a fragment of it, a vector with the hashes of the fragments for ensuring data integrity, and additional metadata to support concurrent reads and writes. The linearizable semantics of protocol AWE are obtained from the atomicity of the metadata directory.

At a high level, the writer first invokes *dir-Scan* on the metadata to read the highest stored timestamp, increments it, and uses this as the timestamp of the value to be written. Then it encodes the value to n fragments and sends one fragment to each data node. The data nodes store it and acknowledge the write. After the writer has received acknowledgments from $t + k$ data nodes, it writes their identities (together with the timestamp and the hashes of the fragments) to the metadata through *dir-Update*. The reader proceeds accordingly: it first invokes *dir-Scan* to obtain the entries of all writers; it determines the highest timestamp among them and extracts the fragment hashes and the identities of the data nodes; finally, it contacts the data nodes and reconstructs the value after obtaining k fragments that match the hashes in the metadata.

Although this simplified algorithm achieves atomic semantics, it does not address timely garbage-collection of obsolete fragments, the main problem to be solved for amnesic erasure-code distributed storage. If a writer would simply replace the fragments with those of the value written next, it is easy to see a concurrent reader may stall.

Protocol AWE uses two mechanisms to address this: first, the writer *retains* those values that may be accessed concurrently and exempts them from garbage collection so that their fragments remain intact for concurrent readers, which gives the reader enough time to retrieve its fragments. Secondly, some of the retained values may also be *frozen* in response to concurrent reads; this forces a concurrent read to retrieve a value that is guaranteed to exist at the data nodes rather than simply the newest value, thereby effectively limiting the amount of stored values. A similar freezing method has been used for wait-free atomic storage with replicated data [14, 19], but it must be changed for erasure-coded storage with separated metadata. The retention technique together with the separation of metadata appears novel. More specifically, metadata separation prevents straightforward applications of existing "freezing" techniques, whereas storage that is simultaneously wait-free and amnesic requires garbage collection method that we show here for the first time.

For the two mechanisms, i.e., retention and freezing, every reader maintains a *reader index*, both in its local variable *readindex* and in its metadata. The reader index serves for coordination between the reader and the writers. The reader increments its index whenever it starts a new *r-Read* and immediately writes it to *dir*, thereby announcing its intent to read. Writers access the reader indices after updating the metadata for a write and before (potentially) erasing obsolete fragments. Every writer w maintains a table *frozenindex* with its most recent recollection of all reader indices. When the newly obtained index of a reader c has changed, then w detects that c has started a new operation at some time after the last write of w.

When w detects a new operation of c, it does not know whether c has retrieved the timestamp from *dir* before or after the *dir-Update* of the current write. The reader may access either value; the writer therefore *retains* both the current and the preceding value for c by storing a pointer to them in *frozenptrlist* and in *reservedptrlist*. Clearly, both values have to be excluded from garbage collection by w in order to guarantee that the reader completes.

However, the operation of the reader c may access *dir* after the *dir-Update* of one or more subsequent write operation by w, which means that the nodes would have to retain every value subsequently written by w as well. To prevent this from happening and to limit the number of stored values, w *freezes* the currently written timestamp (as well as the value) and forces c to read this timestamp when it accesses *dir* within the same operation. In particular, the writer stores the current timestamp in *frozenptrlist* at index c and updates the reader index of c in *frozenindex*; then, the writer pushes both tables, *frozenindex* and *frozenptrlist*, to the metadata service during its next r-*Write*. The values designated by *frozenptrlist* (they are called *frozen*) and *reservedptrlist* (they are called *reserved*) are retained and excluded from garbage collection until w detects the next read of c, i.e., the reader index of c increases. Thus, the current read may span many concurrent writes of w and the fragments remain available until c finishes reading.

On the other hand, a reader must consider frozen values. When a slow read operation spans multiple concurrent writes, the reader c learns that it should retrieve the frozen value through its entry in the *frozenindex* table of the writer.

The protocol is amnesic because each writer retains at most two values per reader, a frozen value and a reserved value. Every data node therefore stores at most two fragments for every reader-writer pair plus the fragment from the currently written value. The combination of freezing and retentions ensures wait-freedom.

4.2 Details

Data structures. We use abstract data structures for compactness. In particular, given a timestamp $ts = (sn, c)$, its two fields can be accessed as $ts.sn$ and $ts.c$. A data type *Pointers* denotes a set of tuples of the form $(ts, set, hash)$ with $ts \in Timestamps$, $set \subseteq [1, n]$, and $hash[i] \in \Sigma^*$ for $i \in [1, n]$. Their initialization value is $Nullptr = ((0, \perp), \emptyset, [\perp, \dots, \perp])$.

A *Pointers* structure contains the relevant information about one stored value. For example, the writer locally maintains *writeptr* $\in$ *Pointers* designating to the most recently written value. More specifically, *writeptr.ts* contains the timestamp of the written value, *writeptr.set* contains the identities of the nodes that have confirmed to have stored the written value, and *writeptr.hash* contains the cross checksum, the list of hash values of the data fragments, of the written value.

The metadata directory *dir* contains a vector M with a tuple for every client $p \in C$ of the form

$$M[p] = \big(writeptr, frozenptrlist, frozenindex, readindex\big),$$

where the field *writeptr* $\in$ *Pointers* represents the *written value*, the field *frozenptrlist* is an array indexed by $c \in C$ such that *frozenptrlist*$[c] \in$ *Pointers* denotes a value *frozen by p for reader c*, and the integer *readindex* denotes the reader-index of p.

For preventing that concurrently accessed fragments are cleaned up too early, the writer maintains two tables, *frozenptrlist*, and *reservedptrlist*, each containing one *Pointers* entry for every reader in C. The second one, *reservedptrlist*, is stored only locally, together with the *frozenindex* table, which denotes the writer's most recently obtained copy of the reader indices. For the operations of the reader, only the local *readindex* counter is needed.

Every client maintains the following variables between operations: *writeptr*, *frozenptrlist*, *frozenindex*, and *reservedptrlist* implement freezing, reservations, and retentions for writers as mentioned, and *readindex* counts the reader operations. When clients access *dir*, they may not be interested to retrieve all fields or to update all fields; for clarity, we replace the fields to be ignored by $*$ in *dir-Scan* and *dir-Update* operations.

Algorithm 1. Protocol AWE, atomic register instance r for client c (part 1).

State

    // State maintained across write and read operations

    writeptr $\in$ *Pointers*, initially *Nullptr*   // Metadata of the currently written value

    frozenptrlist[p] $\in$ *Pointers*, initially *Nullptr*, for $p \in C$ // Frozen and retained for p

    reservedptrlist[p] $\in$ *Pointers*, initially *Nullptr*, for $p \in C$ // Reserved and retained for p

    frozenindex[p] $\in N_0$, initially 0, for $p \in C$ // Last known reader index of p

    readindex $\in N_0$, initially 0   // Reader index of c

    // Temporary state during operations

    prevptr $\in$ *Pointers*, initially *Nullptr*   // Metadata of the value written by c before

    readptr $\in$ *Pointers*, initially *Nullptr*   // Metadata of the value to be read by c

    readlist[i] $\in \Sigma^*$, initially $\perp$, for $i \in [1, n]$ // List of nodes that have responded during read

upon $\langle$ *r-Write* | v $\rangle$ **do**

    prevptr $\leftarrow$ *writeptr*

    invoke $\langle$ *dir-Scan* $\rangle$; **wait for** $\langle$ *dir-ScanResp* | M $\rangle$

    $(wsn, *) \leftarrow \max\{M[p].writeptr.ts \mid p \in C\}$ // Highest *ts* field in a *writeptr* in M

    writeptr.ts $\leftarrow (wsn + 1, c)$ // Construct metadata of the currently written value

    writeptr.set $\leftarrow \emptyset$

    $[v_1, \ldots, v_n] \leftarrow Encode_{k,n}(v)$

    forall $i \in [1, n]$ **do**

        writeptr.hash[i] $\leftarrow H(v_i)$

        invoke $\langle$ d_i-*Write* | *writeptr.ts*, v_i $\rangle$

upon $\langle$ d_i-*WriteAck* | *ats* $\rangle$ **such that** *ats* = *writeptr.ts* $\wedge$ |*writeptr.set*| $< t + k$ **do**

    writeptr.set $\leftarrow$ *writeptr.set* $\cup \{i\}$

    if |*writeptr.set*| $= t + k$ **then**

        // Update metadata at *dir* with currently written value and with frozen values

        invoke $\langle$ *dir-Update* | c, (*writeptr*, *frozenptrlist*, *frozenindex*, $*$) $\rangle$

        wait for $\langle$ *dir-UpdateAck* $\rangle$

        // Obtain current reader indices

        invoke $\langle$ *dir-Scan* $\rangle$; **wait for** $\langle$ *dir-ScanResp* | M $\rangle$

        freets $\leftarrow \{prevptr.ts\}$

        forall $p \in C \setminus \{c\}$ **do**

            $(*, *, *, index) \leftarrow M[p]$

            if *index* $>$ *frozenindex*[p] **then**

                // Client p may be concurrently reading *prevptr* or *writeptr*

                freets $\leftarrow$ *freets* $\cup \{frozenptrlist[p].ts, reservedptrlist[p].ts\}$

                frozenptrlist[p] $\leftarrow$ *writeptr*; *frozenindex*[p] $\leftarrow$ *index*

                reservedptrlist[p] $\leftarrow$ *prevptr*

        freets $\leftarrow$ *freets* $\setminus \bigcup_{p \in C}\{frozenptrlist[p].ts, reservedptrlist[p].ts\}$

        forall $j \in [1, n]$ **do**   // Clean up fragments except for current, frozen, and reserved

            invoke $\langle$ d_j-*Free* | *freets* $\rangle$

        invoke $\langle$ *r-WriteAck* $\rangle$

Operations. At the start of a write operation, the writer w saves the current value of *writeptr* in *prevptr*, to be used later during its operation, if w should reserve and retain

---

Algorithm 2. Protocol AWE, atomic register instance r for client c (part 2).

upon $\langle$ r-Read $\rangle$ **do**
    forall $i \in [1, n]$ **do** $readlist[i] \leftarrow \perp$
    $readindex \leftarrow readindex + 1$
    invoke $\langle$ dir-Update $\mid c, (*, *, *, readindex)$ $\rangle$; **wait for** $\langle$ dir-UpdateAck $\rangle$
    // Parse the content of dir and extract the highest timestamp, potentially frozen for c
    invoke $\langle$ dir-Scan $\rangle$; **wait for** $\langle$ dir-ScanResp $\mid M$ $\rangle$
    $readptr \leftarrow highestread(M, c, readindex)$
    if $readptr.ts = (0, \perp)$ **then**
        invoke $\langle$ r-ReadResp $\mid \perp$ $\rangle$
    else // Contact the data nodes to obtain the data fragments
        forall $i \in readptr.set$ **do**
            invoke $\langle$ d_i-Read $\mid readptr.ts$ $\rangle$

upon $\langle$ d_i-ReadResp $\mid vts, v$ $\rangle$ **such that** $vts = readptr.ts \wedge readlist[i] = \perp$ **do**
    if $v \neq \perp \wedge H(v) = readptr.hash[i]$ **then**
        $readlist[i] \leftarrow v$
        if $\left| \{j \mid readlist[j] \neq \perp\} \right| = k$ **then**
            $readptr \leftarrow Nullptr$
            $retval \leftarrow Reconstruct_{k,n}(readlist)$
            invoke $\langle$ r-ReadResp $\mid retval$ $\rangle$

---

that value. Then w determines the timestamp of the current operation, which is stored in $writeptr.ts$. After computing the fragments of v, sending them to the data nodes, and obtaining $t + k$ acknowledgements, the writer updates its metadata entry. It writes $writeptr$, pointing to v, together with $frozenptrlist$ and $frozenindex$, as they resulted after the previous write to dir. Then w invokes dir-$Scan$ and acquires the current metadata M, which it uses to determine values to freeze and to retain. It compares the acquired reader indices with the ones obtained during its last write (as stored in $frozenindex$). When w detects a read operation by c because $M[c].readindex > frozenindex[c]$, it freezes the current value (by setting $frozenptrlist[p]$ to $writeptr$) and reserves the previously written value (by setting $reservedptrlist[p]$ to $prevptr$). Finally, the writer deletes all fragments at the data nodes except for those of the currently written and the retained values.

To determine the timestamps for retrieving fragments, the reader uses the following two functions:

function $readfrom(M, c, p, index)$ **is**
    if $index > M[p].frozenindex[c]$ **then**
        return $M[p].writeptr$
    else   // $index = M[p].frozenindex[c]$
        return $M[p].frozenptrlist[c]$

function $highestread(M, c, index)$ **is**
    $max \leftarrow Nullptr$
    forall $p \in C$ **do**
        $ptr \leftarrow readfrom(M, c, p, index)$
        if $ptr.ts > max.ts$ **then**
            $max \leftarrow ptr$
    return max

Upon retrieving the array M from dir, the reader sets

$$readptr \leftarrow highestread(M, c, readindex),$$

which implements the logic of accessing frozen timestamps. The details of AWE appear in Algorithms 1–2.

Remarks. AWE does not rely on a majority of correct data nodes for correctness, as this is encapsulated in the directory service. For liveness, though, the protocol needs responses from $t + k$ data nodes during write operations, which is only possible if $n \geq 2t + k$. Furthermore, several optimizations may reduce the storage overhead in practice, e.g., readers can clean up values that are no longer needed by anyone.

5   Complexity Comparison

This section compares the communication and storage complexities of AWE to existing erasure-coded distributed storage solutions, in a setting with n data nodes and m clients. We denote the size of each stored value $v \in \mathcal{V}$ by $\ell = \lceil \log_2 |\mathcal{V}| \rceil$. In line with the intended deployment scenarios, we assume that ℓ is much larger (by several orders of magnitude) than n^2 and m^2, i.e., $\ell \gg n^2$ and $\ell \gg m^2$.

We examine the worst-case communication and storage costs incurred by a client in protocol AWE and distinguish metadata operations (on dir) from operations on the data nodes. The metadata of one value written to dir consists of a pointer, containing the cross checksum with n hash values, the $t + k$ identities of the data nodes that store a data fragment, and a timestamp. Moreover, the metadata entry of one writer contains also the list of m pointers to frozen values, the m indices relating to the frozen values, and the writer's reader index. Assuming a collision-resistant hash function with output size λ bits and timestamps no larger than λ bits, the total size of the metadata is $O(m^2 n \lambda)$. In the remainder of this section, the size of the metadata is considered to be negligible and is ignored, though it would incur in practice.

According to the above assumption, the complexity of AWE is dominated by the data itself. When writing a value $v \in \mathcal{V}$, the writer sends a fragment of size ℓ / k and a timestamp of size λ to each of the n data nodes. Assuming further that $\ell \gg \lambda$, the total storage space occupied by v at the data nodes amounts to $n\ell / k$ bits. Similarly, a read operation incurs a communication cost of $(t + k)k / \ell$ bits.

With respect to storage complexity, protocol AWE freezes and reserves two timestamps and their fragments for each writer-reader pair, and additionally stores the fragments of the last written value for each writer. This means that the storage cost is at most $2m^2 n\ell / k$ bits in total. The improvement described in a remark of Section 4.2 reduces this to $2mn\ell / k$ in the best case.

Table 2 shows the communication and storage costs of protocol AWE and the related protocols. Observe that in CASGC [10] and HGR [21], a read operation concurrent with an unbounded number of writes may not terminate, hence we state their cost as ∞. Moreover, in contrast to AWE, DepSky [5] is neither wait-free nor amnesic and M-PoWerStore [13] is not amnesic. It is easy to see that the communication complexity of AWE is lower than that of most storage solutions.

Table 2. Comparison of the communication and space complexities of erasure-coded distributed storage solutions. There are m clients, n data nodes, the erasure code parameter is $k = n - 2t$, and the data values are of size ℓ bits. An asterisk (*) denotes optimal properties.

Protocol	Communication cost		Storage cost
	Write	Read	
ORCAS-A [15]	$(1 + m)n\ell$	$2n\ell$	$n\ell$
ORCAS-B [15]	$(1 + m)n\ell/k$	$2n\ell/k$	$mn\ell/k$
CASGC [10]	$n\ell/k$ *	∞	$mn\ell/k$
CT [9]	$(n + m)n\ell/(k + t)$	ℓ *	$n\ell/(k + t)$ *
HGR [21]	$n\ell/k$ *	∞	$mn\ell/k$
M-PoWerStore [13]	$n\ell/k$ *	$n\ell/k$	∞
DepSky [5]	$n\ell/k$ *	$n\ell/k$	∞
AWE (Sec. 4)	$n\ell/k$ *	$(t + k)\ell/k$	$2m^2 n\ell/k$

6 Necessity of Cryptography

In this section, we show that every BFT storage protocol that maintains bulk data (as opposed to short metadata) on $3t$ or fewer nodes while tolerating t Byzantine faults implies the existence of cryptographic hash functions. We strengthen this result by considering single-writer single-reader implementations of a register object with value domain $\mathcal{V}$ where n data nodes are aided by one *metadata service (MDS)* process; intuitively, the role of the MDS in an implementation is to store coordination data, but not values. Up to t data nodes may exhibit Byzantine faults, yet the MDS is a correct process. We do not rely on self-verifying data [25] — the processes have no way to check to tell apart "valid" from "invalid" values.

We consider a *computational* model and adopt a cryptographic security notion [17]. Let κ be a security parameter. Suppose every process is implemented by an *efficient* algorithm, that is, an algorithm whose running time is bounded by some polynomial in κ; the length of the input values and the internal state of every process are also bounded by this polynomial. We assume the storage emulation takes inputs of length $\ell(\kappa)$, a polynomial in κ, i.e., $|\mathcal{V}| \leq 2^{\ell(\kappa)}$. Suppose that any MDS implementation has *small state* in the sense that its internal memory is restricted to $\phi(\kappa)$ bits such that there exists a constant $c > 1$ such that for all $\kappa > 0$, $\ell(\kappa) > \phi(\kappa)^c$. This ensures that a register emulation cannot simply store the written at the MDS.

We abstract the hash function as follows.

Definition 1 (Digest oracle). *A digest oracle D is a distributed atomic object accessible to all processes. It supports only one operation that takes a bit string x of arbitrary length as input and outputs a bit string d (denoted $D(x)$) of fixed length $\lambda(\kappa)$, where λ is a polynomial in κ.*

The operation of D may be probabilistic but it implements a mathematical function in the sense that when queried with an input that has already been queried before, it returns again the same output. Furthermore, D satisfies the following collision-resistance property. Consider any efficient adversarial process $\mathcal{A}$ with access to D that attempts to find a collision in D. The probability that $\mathcal{A}$ outputs two values x and x' such that

$D(x) = D(x')$ is negligible in κ. (A function μ is called negligible when for every integer $c > 0$ there exists an integer κ_c such that for all $\kappa > \kappa_c$, it holds $|\mu(\kappa)| < \kappa^{-c}$.)

The principal result of this section, stated next in Theorem 1, combines a standard indistinguishability argument about a concurrent system with a cryptographic reduction.

Theorem 1. *Consider a deterministic emulation Π of a safe register, which uses a meta-data service MDS and $n \leq 3t$ data nodes such that up to t of the data nodes may be Byzantine and controlled by an adversary. If MDS has small state, then a collision-free digest oracle D can be implemented.*

Proof. We first define D, which is implemented from a simulation of the storage protocol Π that uses MDS. More precisely, to compute the digest of a value x, a simulator executes Π by simulating one writer process w that executes $write(x)$, the n data nodes, and MDS. Then the simulator outputs the internal state md of process MDS as the return value of D. Whenever D is invoked, the simulator starts from the initial state and uses the same schedule; this ensures that two invocations of D with the same input give the same output.

We now show that D constructed from Π is collision-free. Towards a *contradiction*, assume there exist two distinct values a and b in $\mathcal{V}$ such that $D(a) = D(b)$. We now argue that Π is not a safe register emulation by describing multiple executions of Π. For simplicity, assume that $n = 3t$ and divide the n data nodes into three groups of t each, called A, B, and F.

Consider first an execution α of Π where initially w writes a using the schedule of the emulation of D. Suppose the nodes in A and F participate in this emulation and let t_α denote the time when the simulation of D returns md_a, the state of M. No messages from the writer are delivered to nodes in B.

Second, in execution β of Π, the value b is written. The execution is the same as α, except that the nodes in B participate instead of those in A and no messages from the writer are delivered to nodes in A. Note that $md_b = D(b) = D(a) = md_a$ by the assumption on a and b — the state of MDS is the same after $write(a)$ in α as after $write(b)$ in β.

Consider now an execution $\bar{\alpha}$ that extends α beyond t_α. At time t_α, the processes in A are being delayed indefinitely and do not take any further steps; as in α, no messages from w to nodes in B are ever delivered before the execution ends and the nodes in B continue operating from their initial state. Next, a reader r invokes *read*, interacts with the nodes in $B \cup F$ and with MDS, and returns a according to the safety property of the storage emulation.

Finally, consider an execution $\bar{\beta}$ that extends β beyond t_α. Here, the processes in B are delayed indefinitely from time t_α onward. Again, the nodes in A have still their initial state and continue now to participate in the execution. Furthermore, all nodes in F exhibit a Byzantine fault and *replace their state* with their state at time t_α in α; after that they again follow Π. Next, a reader r invokes *read* and only interacts with the nodes in $A \cup F$ and with MDS. Recall the state of MDS in β is the same as in α at time t_α. Since the nodes in A have the initial state and those in F and process MDS have the same state as in α at time t_α, execution $\bar{\beta}$ resumes from the same state as in $\bar{\alpha}$ except that the roles of the nodes in A and B are exchanged. However, as the emulation is deterministic, the reader cannot distinguish $\bar{\beta}$ from $\bar{\alpha}$ and returns a. This violates the safety of the storage emulation as $write(b)$ precedes $read$ in $\bar{\beta}$ but $read$ returns a. A contradiction.

7 Conclusion

This paper has presented AWE, the first *erasure-coded* distributed implementation of a multi-writer multi-reader read/write register object that is, at the same time, (1) asynchronous, (2) wait-free, (3) atomic, (4) amnesic, (i.e., with data nodes storing a bounded number of values) and (5) Byzantine fault-tolerant (BFT) using the optimal number of nodes. AWE is efficient since it does not use public-key cryptography and requires data nodes that support only reads and writes, further reducing the cost of deployment and ownership of a distributed storage solution. Notably, AWE stores metadata separately from k-out-of-n erasure-coded fragments. This enables AWE to be the first BFT protocol that uses as few as $2t + k$ data nodes to tolerate t Byzantine nodes, for any $k \geq 1$.

Future work should address how to optimize protocol AWE and to reduce the storage consumption for practical systems; this could be done at the cost of increasing its conceptual complexity and losing some of its ideal properties. For instance, when the metadata service is moved from a storage abstraction to a service with processing, it is conceivable that fewer values have to be retained at the nodes.

Acknowledgment. We thank Radu Banabic, Nikola Knežević, and Alessandro Sorniotti for inspiring discussions during the early stages of this work. This work is supported in part by the EU CLOUDSPACES (FP7-317555) and SECCRIT (FP7-312758) projects.

References

[1] Abraham, I., Chockler, G., Keidar, I., Malkhi, D.: Byzantine disk Paxos: Optimal resilience with Byzantine shared memory. Distributed Computing 18(5), 387–408 (2006)

[2] Adya, A., Bolosky, W.J., Castro, M., Cermak, G., Chaiken, R., Douceur, J.R., Howell, J., Lorch, J.R., Theimer, M., Wattenhofer, R.P.: FARSITE: Federated, available, and reliable storage for an incompletely trusted environment. In: Proc. Symp. Operating Systems Design and Implementation (2002)

[3] Afek, Y., Attiya, H., Dolev, D., Gafni, E., Merritt, M., Shavit, N.: Atomic snapshots of shared memory. Journal of the ACM 40(4), 873–890 (1993)

[4] Androulaki, E., Cachin, C., Dobre, D., Vukolić, M.: Erasure-coded Byzantine storage with separate metadata. Report arXiv:1402.4958, CoRR (2014)

[5] Bessani, A., Correia, M., Quaresma, B., André, F., Sousa, P.: DepSky: Dependable and secure storage in a cloud-of-clouds. In: Proc. European Conference on Computer Systems, pp. 31–46 (2011)

[6] Bowers, K.D., Juels, A., Oprea, A.: HAIL: A high-availability and integrity layer for cloud storage. In: Proc. ACM Conference on Computer and Communications Security, pp. 187–198 (2009)

[7] Cachin, C., Dobre, D., Vukolić, M.: Separating data and control: Asynchronous BFT storage with $2t + 1$ data replicas. In: Felber, P., Garg, V. (eds.) SSS 2014. LNCS, vol. 8756, pp. 1–17. Springer, Heidelberg (2014)

[8] Cachin, C., Guerraoui, R., Rodrigues, L.: Introduction to Reliable and Secure Distributed Programming, 2nd edn. Springer (2011)

[9] Cachin, C., Tessaro, S.: Optimal resilience for erasure-coded Byzantine distributed storage. In: Proc. Dependable Systems and Networks, pp. 115–124 (2006)

[10] Cadambe, V.R., Lynch, N., Medard, M., Musial, P.: Coded atomic shared memory emulation for message passing architectures. CSAIL Technical Report MIT-CSAIL-TR-2013-016. MIT (2013)

[11] Chockler, G., Guerraoui, R., Keidar, I.: Amnesic distributed storage. In: Pelc, A. (ed.) DISC 2007. LNCS, vol. 4731, pp. 139–151. Springer, Heidelberg (2007)

[12] Chockler, G., Guerraoui, R., Keidar, I., Vukolić, M.: Reliable distributed storage. IEEE Computer 42(4), 60–67 (2009)

[13] Dobre, D., Karame, G., Li, W., Majuntke, M., Suri, N., Vukolić, M.: PoWerStore: Proofs of writing for efficient and robust storage. In: Proc. ACM Conference on Computer and Communications Security (2013)

[14] Dobre, D., Majuntke, M., Suri, N.: On the time-complexity of robust and amnesic storage. In: Baker, T.P., Bui, A., Tixeuil, S. (eds.) OPODIS 2008. LNCS, vol. 5401, pp. 197–216. Springer, Heidelberg (2008)

[15] Dutta, P.S., Guerraoui, R., Levy, R.R.: Optimistic erasure-coded distributed storage. In: Taubenfeld, G. (ed.) DISC 2008. LNCS, vol. 5218, pp. 182–196. Springer, Heidelberg (2008)

[16] Frølund, S., Merchant, A., Saito, Y., Spence, S., Veitch, A.: A decentralized algorithm for erasure-coded virtual disks. In: Proc. Dependable Systems and Networks, pp. 125–134 (2004)

[17] Goldreich, O.: Foundations of Cryptography, vol. I & II. Cambridge University Press (2001–2004)

[18] Goodson, G.R., Wylie, J.J., Ganger, G.R., Reiter, M.K.: Efficient Byzantine-tolerant erasure-coded storage. In: Proc. Dependable Systems and Networks, pp. 135–144 (2004)

[19] Guerraoui, R., Levy, R.R., Vukolić, M.: Lucky read/write access to robust atomic storage. In: Proc. Dependable Systems and Networks, pp. 125–136 (2006)

[20] Hendricks, J.: Efficient Byzantine Fault Tolerance for Scalable Storage and Services. Ph.D. thesis, School of Computer Science, Carnegie Mellon University (2009)

[21] Hendricks, J., Ganger, G.R., Reiter, M.K.: Low-overhead Byzantine fault-tolerant storage. In: Proc. ACM Symposium on Operating Systems Principles (2007)

[22] Herlihy, M.: Wait-free synchronization. ACM Transactions on Programming Languages and Systems 11(1), 124–149 (1991)

[23] Herlihy, M.P., Wing, J.M.: Linearizability: A correctness condition for concurrent objects. ACM Transactions on Programming Languages and Systems 12(3), 463–492 (1990)

[24] Huang, C., Simitci, H., Xu, Y., Ogus, A., Calder, B., Gopalan, P., et al.: Erasure coding in Windows Azure Storage. In: Proc. USENIX Annual Technical Conference (2012)

[25] Malkhi, D., Reiter, M.K.: Byzantine quorum systems. Distributed Computing 11(4), 203–213 (1998)

[26] Martin, J.P., Alvisi, L., Dahlin, M.: Minimal Byzantine storage. In: Malkhi, D. (ed.) DISC 2002. LNCS, vol. 2508, pp. 311–325. Springer, Heidelberg (2002)

[27] Vukolić, M.: Quorum Systems: With Applications to Storage and Consensus. Synthesis Lectures on Distributed Computing Theory. Morgan & Claypool (2012)

[28] Wong, W.: Cleversafe grows along with customers' data storage needs. Chicago Tribune (2013)

[29] Yin, J., Martin, J.P., Alvisi, A.V.L., Dahlin, M.: Separating agreement from execution in Byzantine fault-tolerant services. In: Proc. ACM Symposium on Operating Systems Principles, pp. 253–268 (2003)

BChain: Byzantine Replication with High Throughput and Embedded Reconfiguration

Sisi Duan[1], Hein Meling[2], Sean Peisert[1], and Haibin Zhang[1]

[1] University of California, Davis, USA
{sduan,speisert,hbzhang}@ucdavis.edu
[2] University of Stavanger, Norway
hein.meling@uis.no

Abstract. In this paper, we describe the design and implementation of BChain, a Byzantine fault-tolerant state machine replication protocol, which performs comparably to other modern protocols in fault-free cases, but in the face of failures can also quickly recover its steady state performance. Building on chain replication, BChain achieves high throughput and low latency under high client load. At the core of BChain is an efficient Byzantine failure detection mechanism called *re-chaining*, where faulty replicas are placed out of harm's way at the end of the chain, until they can be replaced. Our experimental evaluation confirms our performance expectations for both fault-free and failure scenarios. We also use BChain to implement an NFS service, and show that its performance overhead, with and without failures, is low, both compared to unreplicated NFS and other BFT implementations.

1 Introduction

Building online services that are both highly available and correct is challenging. Byzantine fault tolerance (BFT), a technique based on state machine replication [25,31], is the only known *general* technique that can mask *arbitrary* failures, including crashes, malicious attacks, and software errors. Thus, the behavior of a service employing BFT is indistinguishable from a service running on a correct server.

There are two broad classes of BFT protocols that have evolved in the past decade: broadcast-based [5,24,1,12] and chain-based protocols [18,34]. The main difference between these two classes is their performance characteristics. Chain-based protocols aim at achieving high throughput, at the expense of higher latency. However, as the number of concurrent client requests grows, it turns out that chain-based protocols can actually achieve lower latency than broadcast-based protocols. The downside however, is that chain-based protocols are less resilient to failures, and typically relegate to broadcasting when failures are present. This results in a significant performance degradation.

In this paper we propose *BChain*, a fully-fledged BFT protocol addressing the performance issues observed when a BFT service experiences failures. Our evaluation shows that BChain can quickly recover its steady-state performance, while Aliph-Chain [18] and Zyzzyva [24] experience significantly reduced performance, when subjected to a simple crash failure. At the same time, the steady-state performance of BChain is comparable to Aliph-Chain, the state-of-the-art, chain-based BFT protocol. BChain also

M.K. Aguilera et al. (Eds.): OPODIS 2014, LNCS 8878, pp. 91–106, 2014.
© Springer International Publishing Switzerland 2014

Table 1. Characteristics of state-of-the-art BFT protocols tolerating f failures with batch size b. Bold entries mark the protocol with the lowest cost. The critical path denotes the number of one-way message delays. *Two message delays is only achievable with no concurrency.

	PBFT	Q/U	HQ	Zyzzyva	Aliph	Shuttle	BChain-3	BChain-5
Total replicas	$3f+1$	$5f+1$	$3f+1$	$3f+1$	$3f+1$	**$2f+1$**	$3f+1$	$5f+1$
Crypto ops	$2+\frac{8f+1}{b}$	$2+8f$	$4+4f$	$2+\frac{3f}{b}$	$1+\frac{f+1}{b}$	$2+\frac{2f}{b}$	$1+\frac{3f+2}{b}$	$1+\frac{4f+2}{b}$
Critical path	4	**2^***	4	3	$3f+2$	$2f+2$	$2f+2$	$3f+2$
Additional Requirements	None	None	None	Correct Clients	Protocol Switch	Olympus; Reconfig.	Reconfig.	None

outperforms broadcast-based protocols PBFT [5] and Zyzzyva with a throughput improvement of up to 50% and 25%, respectively. We have used BChain to implement a BFT-based NFS service, and our evaluation shows that it is only marginally slower (1%) than a standard NFS implementation.

BChain in a Nutshell. BChain is a self-recovering, chain-based BFT protocol, where the replicas are organized in a chain. In common case executions, clients send their requests to the head of the chain, which orders the requests. The ordered requests are forwarded along the chain and executed by the replicas. Once a request reaches a replica that we call the *proxy tail*, a reply is sent to the client.

When a BFT service experiences failures or asynchrony, BChain employs a novel approach that we call *re-chaining*. In this approach, the head reorders the chain when a replica is suspected to be faulty, so that a fault cannot affect the critical path.

To facilitate re-chaining, BChain makes use of a novel failure detection mechanism, where any replica can suspect its successor and only its successor. A replica does this by sending a signed suspicion message up the chain. No proof that the suspected replica has misbehaved is required. Upon receiving a suspicion, the head issues a new chain ordering where the accused replica is moved out of the critical path, and the accuser is moved to a position in which it cannot continue to accuse others. In this way, correct replicas help BChain make progress by suspecting faulty replicas, yet malicious replicas cannot *constantly* accuse correct replicas of being faulty.

Our re-chaining approach is inexpensive; a single re-chaining request corresponds to processing a single client request. Thus, the steady-state performance of BChain has minimal disruption. The latency reduction caused by re-chaining is dominated by the failure detection timeout.

Our Contributions in Context. We consider two variants of BChain—BChain-3 and BChain-5, both tolerating f failures. BChain-3 requires $3f+1$ replicas and a reconfiguration mechanism coupled with our detection and re-chaining algorithms, while BChain-5 requires $5f+1$ replicas, but can operate without the reconfiguration mechanism. We compare BChain-3 and BChain-5 with state-of-the-art BFT protocols in Table 1. All protocols use MACs for authentication and request batching with batch size b. The number of MAC operations for BChain at the bottleneck server tends to one for gracious executions. While this is also the case for Aliph-Chain [18], Aliph requires that clients take responsibility for switching to another slower BFT protocol in the presence of failures, to ensure safety and liveness. Thus, a single dedicated adversary might render the system much slower. Shuttle [34] can tolerate f faulty replicas using only $2f + 1$ replicas. However, it relies on a trusted auxiliary server. BChain does not require an auxiliary service, yet its critical path of $2f + 2$ is identical to that of Shuttle.

Our contributions can be summarized as follows:

1. We present BChain-3 and its sub-protocols for re-chaining, reconfiguration, and view change (§3). Re-chaining is a novel technique to ensure liveness in BChain. Together with re-chaining, the reconfiguration protocol can replace failed replicas with new ones, outside the critical path. The view change protocol deals with a faulty head.
2. We present BChain-5 and how it can operate without reconfiguration (§4).
3. In §5 we evaluate the performance of BChain for both gracious and uncivil executions under different workloads, and compare it with other BFT protocols. We also ran experiments with a BFT-NFS application and assessed its performance compared to the other relevant BFT protocols.

2   System Model

We assume a Byzantine fault tolerant system, where replicas communicate over pairwise channels and may behave arbitrarily. Our system can mask up to f faulty replicas, using n replicas. We write t, where $t \leq f$, to denote the number of faulty replicas that the system currently has. A computationally bounded adversary can coordinate faulty replicas to compromise safety only if more than f replicas are compromised.

Safety of our system holds in any asynchronous environment, where messages may be delayed, dropped, or delivered out of order. Liveness is ensured assuming *partial synchrony* [15]: synchrony holds only after some unknown global stabilization time, but the bounds on communication and processing delays are themselves unknown.

We use non-keyed *message digests*. The digest of a message m is denoted $D(m)$. We also use *digital signatures*. The signature of a message m signed by replica p_i is denoted $\langle m \rangle_{p_i}$. We say that a signature is *valid* on message m, if it passes the verification w.r.t. the public-key of the signer and the message. A vector of signatures of message m signed by a set of replicas $\mathcal{U} = \{p_i, \ldots, p_j\}$ is denoted $\langle m \rangle_{\mathcal{U}}$.

We classify the replica failures according to their behaviors. Weak semantics levy fewer restrictions on the possible behaviors than strong semantics. Apart from the weakest failure semantics (i.e., Byzantine failure), we are also interested in various other stronger failure semantics. *Crash failures*, occur when the replicas might halt permanently and no longer produce any output. By *timing failures*, we mean any replica failures that produce correct results but deliver them outside of a specified time window.

3   BChain-3

We now describe the main protocols and principles of BChain. Our description here uses digital signatures; later we show how they can be replaced with MACs, along with other optimizations. BChain-3 has five sub-protocols: (1) chaining, (2) re-chaining, (3) view change, (4) checkpoint, and (5) reconfiguration. The *chaining* protocol orders clients requests, while *re-chaining* reorganizes the chain in response to failure suspicions. Faulty replicas are moved to the end of the chain. The *view change* protocol selects a new head when the current head is faulty, or the system is slow. Our *checkpoint* protocol is similar to that of PBFT [5]. It is used to bound the growth of message logs and reduce the cost of view changes. We do not describe it in this paper. The *reconfiguration* protocol is responsible for reconfiguring faulty replicas.

To tolerate f failures, BChain-3 needs n replicas such that $f \leq \lfloor \frac{n-1}{3} \rfloor$. In the following, we assume $n = 3f + 1$ for simplicity.

3.1 Conventions and Notations

In BChain, the replicas are organized in a metaphorical *chain*, as shown in Figure 1. Each replica is uniquely identified from a set $\Pi = \{p_1, p_2, \cdots, p_n\}$. Initially, we assume that replica IDs are numbered in ascending order. The first replica is called the *head*, denoted p_h, the last replica is called the *tail*, and the $(2f+1)^{\text{th}}$ replica is called the *proxy tail,* denoted p_p. We divide the replicas into two subsets. Given a specific chain order, $\mathcal{A}$ contains the first $2f + 1$ replicas, initially p_1 to p_{2f+1}. $\mathcal{B}$ contains the last f replicas in the chain, initially p_{2f+2} to p_{3f+1}. For convenience, we also define $\mathcal{A}^{p\!\!/} = \{\mathcal{A} \setminus p_p\}$, excluding the proxy tail, and $\mathcal{A}^{h\!\!/} = \{\mathcal{A} \setminus p_h\}$, excluding the head.

The chain order is maintained by every replica and can be changed by the head and is communicated to replicas through message transmissions. (This is in

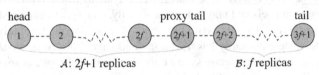

A: 2f+1 replicas   B: f replicas

Fig. 1. BChain-3. Replicas are organized in a chain

contrast to Aliph-Chain, where the chain order is fixed and known to all replicas and clients beforehand.) For any replica except the head, $p_i \in \mathcal{A}^{h\!\!/}$, we define its *predecessor* $\overleftarrow{p}_i$, initially p_{i-1}, as its preceding replica in the current chain order. For any replica except the proxy tail, $p_i \in \mathcal{A}^{p\!\!/}$, we define its *successor* $\overrightarrow{p}_i$, initially p_{i+1}, as its subsequent replica in the current chain order.

For each $p_i \in \mathcal{A}$, we define its *predecessor set* $\mathcal{P}(p_i)$ and *successor set* $\mathcal{S}(p_i)$, whose elements depend on their individual positions in the chain. If a replica $p_i \neq p_h$ is one of the first $f + 1$ replicas, its predecessor set $\mathcal{P}(p_i)$ consists of all the preceding replicas in the chain. For every other replica in $\mathcal{A}$, the predecessor set $\mathcal{P}(p_i)$ consists of the preceding $f + 1$ replicas in the chain. If p_i is one of the last $f + 1$ replicas in $\mathcal{A}$, the successor set $\mathcal{S}(p_i)$ consists of all the subsequent replicas in $\mathcal{A}$. For every other replica in $\mathcal{A}$, the successor set $\mathcal{S}(p_i)$ consists of the subsequent $f + 1$ replicas. Note that the cardinality of any replica's predecessor set or successor set is at most $f + 1$.

3.2 Protocol Overview

In a gracious execution, as shown in Figure 2, the first $2f + 1$ replicas (set $\mathcal{A}$) reach an agreement while the last f replicas (set $\mathcal{B}$) correspondingly update their states based on the agreed-upon requests from set $\mathcal{A}$. BChain transmits two types of messages along the chain: $\langle\text{CHAIN}\rangle$ messages transmitted from the head to the proxy tail, and $\langle\text{ACK}\rangle$ messages transmitted in reverse from the proxy tail to the head. A request is *executed* after a replica accepts the $\langle\text{CHAIN}\rangle$ message; a request *commits* at a replica if it accepts the $\langle\text{ACK}\rangle$ message.

Upon receiving a client request, the head sends a $\langle\text{CHAIN}\rangle$ message representing the request to its successor. As soon as the proxy tail accepts the $\langle\text{CHAIN}\rangle$ message, it sends a reply to the client and generates an $\langle\text{ACK}\rangle$ message, which is sent backwards along the chain until it reaches the head. Once a replica in $\mathcal{A}$ accepts the $\langle\text{ACK}\rangle$ message, it completes the request and forwards its $\langle\text{CHAIN}\rangle$ message to replicas in $\mathcal{B}$ to ensure that the message is committed at all the replicas.

To handle failures and ensure liveness, BChain incorporates failure detection and re-chaining protocol that works as follows: Every replica in $\mathcal{A}^y$ starts a timer after sending a $\langle\text{CHAIN}\rangle$ message. Unless an $\langle\text{ACK}\rangle$ is received before the timer expires, it sends a $\langle\text{SUSPECT}\rangle$ message to the head and also along the chain towards the head. Upon seeing $\langle\text{SUSPECT}\rangle$ messages, the head starts the re-chaining, by moving faulty replicas to set $\mathcal{B}$ where, if needed, replicas may be replaced in the reconfiguration protocol. In this way, BChain remains robust until new failures occur.

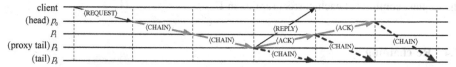

Fig. 2. BChain-3 common case communication pattern. All the signatures can be replaced with MACs. All the $\langle\text{CHAIN}\rangle$ and $\langle\text{ACK}\rangle$ messages can be batched. The $\langle\text{CHAIN}\rangle$ messages with dotted, blue lines are the forwarded messages that are stored in logs. *No* conventional broadcast is used at any point in our protocol. For a given batch size b.

3.3 Chaining

We now describe the sequence of steps of the chaining protocol, used to order requests, when there are no failures.

Step 0: *Client sends a request to the head.* A client c requests the execution of state machine operation o by sending a request $m = \langle\text{REQUEST}, o, T, c\rangle_c$ to the replica that it believes to be the head, where T is the timestamp.

Step 1: *Assign sequence number and send chain message.* When the head p_h receives a valid $\langle\text{REQUEST}, o, T, c\rangle_c$ message, it assigns a sequence number and sends message $\langle\text{CHAIN}, v, ch, N, m, c, \mathcal{H}, R, \Lambda\rangle_{p_h}$ to its successor, where v is the view number, ch is the number of re-chainings that took place during view v, $\mathcal{H}$ is the hash of its execution history, R is the hash of the reply r to the client containing the execution result, and Λ is the current chain order. Both of $\mathcal{H}$ and R are empty in this step.

Step 2: *Execute request and send chain message.* A replica p_j receives from its predecessor a valid $\langle\text{CHAIN}, v, ch, N, m, c, \mathcal{H}, R, \Lambda\rangle_{\mathcal{P}(p_j)}$ message, which contains valid signatures by replicas in $\mathcal{P}(p_j)$. The replica p_j updates $\mathcal{H}$ and R fields if necessary, appends its signature to the $\langle\text{CHAIN}\rangle$ message, and sends to its successor. Note that the $\mathcal{H}$ and R fields are empty if p_j is among the first f replicas, and both $\mathcal{H}$ and R must be verified before proceeding.

Each time a replica $p_j \in \mathcal{A}^{\mathscr{V}}$ sends a $\langle$CHAIN$\rangle$ message, it sets a timer, expecting an $\langle$ACK$\rangle$ message, or a $\langle$SUSPECT$\rangle$ message signaling some replica failures.

Step 3: *Proxy tail sends reply to the client and commits the request.* If the proxy tail p_j accepts a $\langle$CHAIN$\rangle$ message, it computes its own signature and sends the client the reply r, along with the $\langle$CHAIN$\rangle$ message it accepts. It also sends to its predecessor an $\langle$ACK$, v, ch, N, D(m), c\rangle_{p_j}$ message. In addition, it *forwards* to all replicas in $\mathcal{B}$ the corresponding $\langle$CHAIN$, v, ch, N, m, c, \mathcal{H}, R, \Lambda\rangle_{p_j}$ message . The request commits at the proxy tail.

Step 4: *Client completes the request or retransmits.* The client completes the request if it receives a $\langle$REPLY$\rangle$ message from the proxy tail with signatures by the last $f + 1$ replicas in the chain. Otherwise, it retransmits the request to all replicas.

Step 5: *Other replicas in $\mathcal{A}$ commit the request.* A valid $\langle$ACK$, v, ch, N, D(m), c\rangle_{\mathcal{S}(p_j)}$ message is sent to replica p_j by its successor, which contains valid signatures by replicas in $\mathcal{S}(p_j)$. The replica appends its own signature and sends to its predecessor.

Step 6: *Replicas in $\mathcal{B}$ execute and commit request.* The replicas in $\mathcal{B}$ collects $f + 1$ matching $\langle$CHAIN$\rangle$ messages, and executes the operation, completing the request. Thus, the request commits at each correct replica in $\mathcal{B}$.

3.4   Re-chaining

To facilitate failure detection and ensure that BChain remains live, we introduce a protocol we call *re-chaining*. With re-chaining, we can make progress with a bounded number of failures, despite incorrect suspicions. The algorithm ensures that, eventually all faulty replicas are identified and appropriately dealt with. The strategy of the re-chaining algorithm is to move replicas that are *suspected* to set $\mathcal{B}$, where if deemed necessary, they are rejuvenated.

Algorithm 1. Failure detector at replica p_i

1: **upon** $\langle$CHAIN$\rangle$ sent by p_i
2:     $starttimer(\Delta_{1,p_i})$

3: **upon** $\langle$Timeout$, \Delta_{1,p_i}\rangle$ {Accuser p_i}
4:     send $\langle$SUSPECT$, \vec{p}_i, m, ch, v\rangle_{p_i}$ to $\vec{p}_i$ and p_h

5: **upon** $\langle$ACK$\rangle$ from $\vec{p}_i$
6:     $canceltimer(\Delta_{1,p_i})$

7: **upon** $\langle$SUSPECT$, p_y, m, ch, v\rangle$ from $\vec{p}_i$
8:     forward $\langle$SUSPECT$, p_y, m, ch, v\rangle$ to $\overleftarrow{p}_i$
9:     $canceltimer(\Delta_{1,p_i})$

BChain Failure Detector. The objective of the BChain failure detector is to identify faulty replicas, and issue a new chain configuration and to ensure that progress can be made. It is implemented as a timer on $\langle$CHAIN$\rangle$ messages, as shown in Algorithm 1. On sending a $\langle$CHAIN$\rangle$ message m, replica p_i starts a timer, Δ_{1,p_i}. If the replica receives an $\langle$ACK$\rangle$ for the message before the timer expires, it cancels the timer and starts a new one for the next request in the queue, if any. Otherwise, it sends a $\langle$SUSPECT$, \vec{p}_i, m, ch, v\rangle$ to both the head and its predecessor to signal the failure of its successor. Moreover, if p_i receives a $\langle$SUSPECT$\rangle$ message from its successor, the message is forwarded to p_i's predecessor, along the chain until it reaches the head. To prevent that a faulty replica fails to forward the ⟨SUSPECT⟩ message, it is also sent directly to the head. Passing it along the chain allows us to cancel timers and reduce the number of suspect messages.

Let p_i be the *accuser*; then the *accused* can only be its successor, $\vec{p}_i$. This is ensured by having the accuser sign the ⟨SUSPECT⟩ message, just as an ⟨ACK⟩ message.

On receiving a ⟨SUSPECT⟩, the head starts re-chaining via a new ⟨CHAIN⟩ message. If the head receives multiple ⟨SUSPECT⟩ messages, only the one *closest* to the proxy tail is handled. Handling a ⟨SUSPECT⟩ message is done by increasing ch, selecting a new

---

Algorithm 2. BChain-3 Re-chaining-I (At head, p_h)

1: **upon** ⟨SUSPECT, p_y, m, ch, v⟩ from p_x
2:     **if** $p_x \neq p_h$ **then**   {p_x is not the head}
3:         p_z is put to the 2nd position   {$p_z = \mathcal{B}[1]$}
4:         p_x is put to the $(2f+1)^{\text{th}}$ position
5:         p_y is put to the end

---

chain order Λ, and sending a ⟨CHAIN⟩ message to order the same request again.

Re-Chaining Algorithms. We provide two re-chaining algorithms for BChain-3 as shown in Algorithm 2 and 3. To explain these algorithms, assume that the head, p_h, has received a ⟨SUSPECT⟩ message from a replica p_x suspecting is successor p_y. Let p_z be the first replica in set $\mathcal{B}$. Both algorithms show how the head selects a new chain order. Both are *efficient* in the sense that the number of re-chainings needed is proportional to the number of existing failures t instead of the maximum number f. We levy no assumptions on how failures are distributed in the chain.

Re-chaining-I—crash failures handled first. Algorithm 2 is reasonably efficient; in the worst case, t faulty replicas can be removed with at most $3t$ re-chainings. More specifically, if the head is correct and $3t \leq f$, the faulty replicas are moved to the end of chain after at most $3t$ re-chainings; if $3t > f$, at most $3t$ re-chainings are necessary and at most $3t - f$ replicas are replaced in the reconfiguration protocol (§3.6), assuming that any individual replica

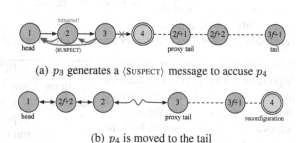

(a) p_3 generates a ⟨SUSPECT⟩ message to accuse p_4

(b) p_4 is moved to the tail

Fig. 3. Example (1). A faulty replica is denoted by a double circle. After the timer expires, replica p_3 issues a ⟨SUSPECT⟩ message to accuse p_4 (which is faulty). The head moves p_3 to the proxy tail position and the faulty replica p_4 to the tail.

can be reconfigured within f re-chainings. Algorithm 2 is even more efficient when handling timing and omission failures, with one such replica being removed using only one re-chaining. Despite the succinct algorithm, the proof of correctness for the general case is complicated. We omit the details due to lack of space. To help grasp the underlying idea, consider the following *simple* examples.

▷ Example (1): In Figure 3, replica p_4 has a timing failure. This causes p_3 to send a ⟨SUSPECT⟩ message up the chain to accuse p_4. According to our re-chaining algorithm, p_3 is moved to the $(2f+1)^{\text{th}}$ position and becomes the proxy tail, and p_4 is moved to the end of the chain and becomes the tail. Our fundamental design principle is that timing failures should be given top priority.

▷ Example (2): In Figure 4, p_3 is the only faulty replica. We consider the circumstance where p_3 sends the head a ⟨SUSPECT⟩ message to frame its successor p_4 even if p_4 follows the protocol. According to our re-chaining algorithm, replica p_4 will be moved to the tail, while p_3 becomes the new proxy tail. However, from then on, p_3 can no longer accuse any replicas. It either follows the specification of the protocol, or chooses not to participate in the agreement, in which case p_3 will be moved to the tail. The example illustrates another important designing rationale that an adversarial replica cannot constantly accuse correct replicas.

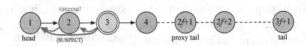

(a) p_3 generates a ⟨SUSPECT⟩ message to *maliciously* accuse p_4

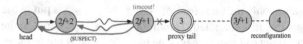

(b) p_{2f+1} generates a ⟨SUSPECT⟩ message to accuse p_3

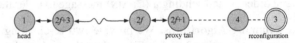

(c) p_3 is moved to the tail and reconfigured

Fig. 4. Example (2). Replica p_3 maliciously accuse p_4 by sending a ⟨SUSPECT⟩ message. The head moves p_3 to the proxy tail and p_4 to the tail. If p_3 does not behave, it will be accused by its predecessor p_{2f+1} such that in another round of re-chaining p_3 is moved to the tail.

Re-chaining-II—improved efficiency. Algorithm 3 can improve efficiency for the *worst* case. The underlying idea is simple: every time the head receives a ⟨SUSPECT⟩ message, both the accuser and the accused are moved to

Algorithm 3. BChain-3 Re-chaining-II
___
1: **upon** ⟨SUSPECT, p_y, m, ch, v⟩ from p_x
2:   **if** $p_x \neq p_h$ **then**   {p_x is not the head}
3:       p_x is put to the $(3f)^{\text{th}}$ position
4:       p_y is put to the end

the end of the chain. Algorithm 3 does not prioritize crash failures, and relies on a stronger reconfiguration assumption. If the head is correct and $2t \leq f$, the faulty replicas are moved to the end of chain after at most $2t$ re-chainings; if $2t > f$, at most $2t$ re-chainings are necessary and at most $2t - f$ replica reconfigurations (§3.6) are needed, assuming that any individual replica can be reconfigured within $\lfloor f/2 \rfloor$ re-chainings. When an accused replica is moved to the end of chain, the reconfiguration process is initialized, either offline or online. The replicas moved to the end of the chain are all "tainted" and reconfigured, as we discuss in §3.6.

Timer Setup and Preventing Timer-Based Performance Attacks. Existing BFT protocols typically only keep timers for view changes, while BChain also requires timers for ⟨ACK⟩ and ⟨CHAIN⟩ messages. To achieve accurate failure detection, we need different values for each timer in each replica in the chain.

The timeout for each replica $p_i \in \mathcal{A}$ is defined as $\Delta_{1,i} = \mathcal{F}(\Delta_1, l_i)$, where $\mathcal{F}$ is a fixed and efficiently computable function, Δ_1 is the base timeout, and l_i is p_i's position in the chain order. Note that for p_h, we have that $l_h = 1$ and thus $\mathcal{F}(\Delta_1, 1) = \Delta_1$. Correspondingly, for p_p we have that $l_p = 2f + 1$ and $\mathcal{F}(\Delta_1, 2f + 1) = 0$. It is reasonable to adopt a *linear function* with respect to the position of each replica as the timer function, e.g., $\mathcal{F}(\Delta_1, l_i) = \frac{2f+1-l_i}{2f} \Delta_1$. As an example with $n = 4$ and $f = 1$, we may set $\Delta_{1,p_1} = \mathcal{F}(\Delta_1, 1) = \Delta_1$, $\Delta_{1,p_2} = \mathcal{F}(\Delta_1, 2) = \Delta_1/2$, and $\Delta_{1,p_3} = \mathcal{F}(\Delta_1, 3) = 0$.

To detect and deter misbehaving replicas that always delay requests to the upper bound timeout value to increase system latency, we also verify the processing delays for the average case and allow replicas to suspect other replicas who frequently do so. Concretely, each replica p_i maintains an additional performance threshold timer Δ'_{1,p_i} such that $\Delta'_{1,p_i} < \Delta_{1,p_i}$, which is used to detect slow or faulty replicas as mentioned above. That is, we ask the replica to further suspect its successor if their average delay exceeds Δ'_{1,p_i}. This will allow us to thwart dedicated performance attacks on messages delays while preventing temporarily slow replicas from being accused prematurely. We will show in §5.1 how to efficiently set up and maintain the timers in actual experiments.

3.5 View Change

The view change protocol has two functions: (1) to select a new head when the current head is deemed faulty, and (2) to adjust the timers to ensure eventual progress, despite deficient initial timer configuration.

A correct replica p_i votes for view change if either (1) it suspects the head to be faulty, or (2) it receives $f + 1$ ⟨VIEWCHANGE⟩ messages. The replica votes for view change and moves to a new view by sending all replicas a ⟨VIEWCHANGE⟩ message that includes the new view number, the current chain order, a set of valid checkpoint messages, and a set of requests that commit locally with proof of execution. For each request that commits locally, if $p_i \in \mathcal{A}$, then a proof of execution for a request contains a ⟨CHAIN⟩ message with signatures from $\mathcal{P}(p_i)$ and an ⟨ACK⟩ message with signatures from $\mathcal{S}(p_i)$. Otherwise, a proof of execution contains $f + 1$ ⟨CHAIN⟩ messages. Upon sending a ⟨VIEWCHANGE⟩ message, p_i stops receiving messages except ⟨CHECKPOINT⟩, ⟨NEWVIEW⟩, or other ⟨VIEWCHANGE⟩ messages. When the new head collects $2f + 1$ ⟨VIEWCHANGE⟩ messages, it sends all replicas a ⟨NEWVIEW⟩ message which includes the new chain order, in which the head of the old view has been moved to the end of the chain, a set of valid ⟨VIEWCHANGE⟩ messages, and a set of ⟨CHAIN⟩ messages.

The other function of view change is to adjust the timers. In addition to the timer Δ_1 maintained for re-chaining, BChain has two timers for view changes, Δ_2 and Δ_3. Δ_2 is a timer maintained for the current view v when a replica is waiting for a request to be committed, while Δ_3 is a timer for ⟨NEWVIEW⟩, when a replica votes for a view change and waits for the ⟨NEWVIEW⟩. Algorithm 4 describes how to initialize, maintain, and adjust these timers.

The view change timer Δ_2 at a replica is set up for the first request in the queue. A replica sends a ⟨VIEWCHANGE⟩ message to all replicas and votes for view change if Δ_2 expires or it receives $f + 1$ ⟨VIEWCHANGE⟩ messages. In either case, when a replica votes for view change, it cancels its timer Δ_2. After a replica collects $2f + 1$ ⟨VIEWCHANGE⟩ messages (including its own), it starts a timer Δ_3 and waits for the ⟨NEWVIEW⟩ message. If the replica does not receive ⟨NEWVIEW⟩ message before Δ_3 expires, it starts a *new* ⟨VIEWCHANGE⟩ and updates Δ_3 with a new value $g_3(\Delta_3)$.

Algorithm 4. View Change Handling and Timers at p_i

1: $\Delta_2 \leftarrow init_{\Delta_2}$; $\Delta_3 \leftarrow init_{\Delta_3}$	10: **upon** $2f + 1$ ⟨VIEWCHANGE⟩
2: *voted* ← **false**	11: *starttimer*(Δ_3)
3: **upon** ⟨Timeout, Δ_2⟩	12: **upon** ⟨Timeout, Δ_3⟩
4: send ⟨VIEWCHANGE⟩	13: $\Delta_3 \leftarrow g_3(\Delta_3)$
5: *voted* ← **true**	14: send *new* ⟨VIEWCHANGE⟩
6: **upon** $f + 1$ ⟨VIEWCHANGE⟩ ∧ ¬*voted*	15: **upon** ⟨NEWVIEW⟩
7: send ⟨VIEWCHANGE⟩	16: *canceltimer*(Δ_3)
8: *voted* ← **true**	17: $\Delta_1 \leftarrow g_1(\Delta_1)$
9: *canceltimer*(Δ_2)	18: $\Delta_2 \leftarrow g_2(\Delta_2)$

When a replica receives the ⟨NEWVIEW⟩ message, it sets Δ_1 and Δ_2 using $g_1(\Delta_1)$ and $g_2(\Delta_2)$, respectively. In practice, the functions $g_1(\cdot)$, $g_2(\cdot)$, and $g_3(\cdot)$ could simply double the current timeouts. However, to avoid the circumstance that the timeouts for Δ_1 and Δ_2 increase without bound, we introduce upper bounds for both of them. Once either timer exceeds the prescribed bound, the system starts reconfiguration.

3.6 Reconfiguration

Reconfiguration [26] is a general technique, often abstracted as stopping the current state machine and restarting it with a new set of replicas, usually reusing non-faulty replicas in the new configuration. In BChain we use reconfiguration in concert with re-chaining to replace faulty replicas with new ones. The reconfiguration operates *out-of-band* in the $\mathcal{B}$ replica set, and imposes only negligible overhead on client request processing being done by replicas in $\mathcal{A}$. We omit the details due to lack of space.

3.7 Optimizations

In general, signatures for ⟨CHAIN⟩ and ⟨ACK⟩ cannot be replaced with MACs. However, we can replace other signatures with MACs. Moreover, we can combine all-MAC-based and signature-based BChain approaches such that the failure-free case uses MACs only and re-chaining uses signatures. We also developed a highly efficient *purely* MAC-based variant of BChain for $n = 4$ and $f = 1$, which does not rely on reconfiguration.

4 BChain without Reconfiguration

We now discuss BChain-5, which uses $n = 5f + 1$ replicas to tolerate f Byzantine failures, just as Q/U [1] and Zyzzyva5 [24]. With $5f + 1$ replicas at our disposal, we design an efficient re-chaining algorithm, which allows the faulty replicas to be identified easily without relying on reconfiguration. Meanwhile, a Byzantine quorum of replicas can reach agreement. BChain-5 relies on the concept of Byzantine quorum protocols [28]. Set $\mathcal{A}$ is a Byzantine quorum which consists of $\lceil \frac{n+f+1}{2} \rceil = 3f + 1$ replicas, while set $\mathcal{B}$ consists of the remaining of $2f$ replicas.

BChain-5 has four sub-protocols: chaining, re-chaining, view change, and checkpoint. In contrast, BChain-3 additionally requires a reconfiguration protocol. The protocols for BChain-3 and BChain-5 are identical with respect to message flow. The main difference lies in the size of the $\mathcal{A}$ set, which now consists of $3f + 1$ replicas. BChain-5 also uses Algorithm 3, modifying only Line 3 to put p_x to the $(5f)^{\text{th}}$ position.

Assuming the timers are accurately configured and that the head is non-faulty, it takes at most f re-chainings to move f failures to the tail set $\mathcal{B}$. The proofs for safety and liveness of BChain-5 are easier than those of BChain-3 due to a different re-chaining algorithm and the absence of the reconfiguration procedure.

To Reconfigure or Not to Reconfigure? The primary benefit of BChain-5 over BChain-3 is that it eliminates the need for reconfiguration to achieve liveness. This is beneficial, since reconfiguration needs additional resources, such as machines to host reconfigured replicas. However, since BChain-5 can identify and move faulty replicas to the tail set $\mathcal{B}$, we can still leverage the reconfiguration procedure on the replicas in $\mathcal{B}$, to provide long-term system safety and liveness. This does not contradict the claim that BChain-5 does not need reconfiguration; rather, it just makes the system more robust. Furthermore, BChain-5 provides flexibility with respect to when the system should be reconfigured. Specifically, reconfiguration can happen any time after the system achieves a stable state or simply has run for a "long enough" period of time.

5   Evaluation

This section studies the performance of BChain-3 and BChain-5 and compares them with three well-known BFT protocols—PBFT [5], Zyzzyva [24], and Aliph [18]. Aliph uses Chain for gracious execution under high concurrency. Aliph-Chain enjoys the highest throughput when there are no failures, however, as we will see, it cannot sustain its performance during failure scenarios by itself, where BChain is superior.

We study the performance using two types of benchmarks: the micro-benchmarks by Castro and Liskov [5] and the Bonnie++ benchmark [10]. We use micro-benchmarks to assess throughput, latency, scalability, and performance during failures of all the five protocols. In the x/y micro-benchmarks, clients send x kB requests and receive y kB replies. Clients invoke requests in a *closed-loop*, where a client does not start a new request before receiving a reply for a previous one. All the protocols implement batching of concurrent requests to reduce cryptographic and communication overheads.

All experiments were carried out on DeterLab [4], utilizing a cluster of up to 65 identical machines equipped with a 2.13 GHz Xeon processor and 4GB of RAM. They are connected through a 100Mbps switched LAN.

We have assessed the performance of all protocols under gracious execution, and find that both BChain-3 and BChain-5 achieve higher throughput and lower latency than PBFT and Zyzzyva especially when the number of concurrent client requests is large, while BChain-3 has performance similar to the Aliph-Chain protocol. Our experiment bolsters the point of view of Guerraoui *et al.* [18] that (authenticated) chaining replication can increase throughput and reduce latency under high concurrency. We omit the detailed evaluation for gracious execution.

In addition to micro-benchmarks, we have also evaluated a BFT-NFS service implemented using PBFT [5], Zyzzyva [24], and BChain-3. We show that performance overhead of BChain-3, with and without failure, is low, both compared to unreplicated NFS and other BFT implementations.

In case of failures, both BChain-3 and BChain-5 outperform all the other protocols by a wide margin, due to BChain's unique re-chaining protocol. Through the timeout adjustment scheme, we show that a faulty replica cannot reduce the performance of the system by manipulating the timeouts.

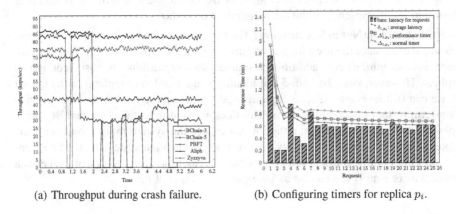

(a) Throughput during crash failure.   (b) Configuring timers for replica p_i.

Fig. 5. Performance under failure

5.1 Performance under Failures

We compare the performance of BChain with the other BFT protocols under two scenarios: a simple crash failure scenario and a performance attack scenario. As the results in Figure 5(a) show, BChain has superior reaction to failures. When BChain detects a failure, it will start re-chaining. At the moment when re-chaining starts, the throughput of BChain temporarily drops to zero. After the chain has been re-ordered, BChain quickly recovers its steady state throughput. The dominant factor deciding the duration of this throughput drop (i.e. increased latency) is the failure detection timeout, not the re-chaining. We also show that BChain can resist a timer-based performance attack, i.e., a faulty replica cannot intentionally manipulate timeouts to slow down the system.

Crash Failure. We compare the throughput during crash failure for BChain-3, BChain-5, PBFT, Zyzzyva, and Aliph. The results are shown in Figure 5(a). We use $f = 1$, message batching, and 40 clients. To avoid clutter in the plot, we used different failure inject times for the protocols: BChain-3, BChain-5, and PBFT all experience a failure at 1s, while Zyzzyva and Aliph experience a failure at 1.5s and 2s, respectively.

We note that Aliph [18,36] generally switches between three protocols: Quorum, Chain, and a backup, e.g., PBFT. For our experiments, we adopt the same setting as in the Aliph paper [18], i.e., it uses a combination of Chain and PBFT as backup and a configuration parameter k, denoting the number of requests to be executed when running with the backup protocol. We use both $k = 1$ and $k = 2^i$.

Even though Aliph exhibits slightly higher throughput than BChain-3 prior to the failure, its throughput takes a significant beating upon failure, dropping well below that of the PBFT baseline. The overall performance depends on how often failures occur and how often Aliph switches between main and backup protocols, i.e., parameter k. On the other hand, the throughput of PBFT does not change in any obvious way after failure injection, showing its stability during failure scenarios. Zyzzyva, in comparison, in the presence of failures, uses its slower backup mode (i.e., clients collects and sends certificate) which exhibits even lower throughput than PBFT.

We configured BChain with a fairly high timeout value (100ms). In fact, BChain can use much smaller timeouts, since one re-chaining only takes about the same time as it takes for BChain to process a single request. On the other hand, Aliph's signature-based, view-change like switching approach introduces a significant delay.

We claim that even in presence of a Byzantine failure, the throughput of BChain-3 and BChain-5 would not change significantly, except that there might be two (instead of one) short periods where the throughput drops to zero. That is, BChain-3 uses at most two re-chainings to handle a Byzantine faulty replica, while BChain-5 uses only one.

Timer Setup and Performance Attack Evaluation. We now show how to set up the timers for replicas in the chain as discussed in §3.4. Initially, there are no faulty replicas and we set the timers based on the average latency of the first 1000 requests. Figure 5(b) illustrates the timer setup procedure for a correct replica p_i, where each bar represents the actual latency of a request, the lowest line is the average latency δ_{1,p_i}, the middle line is the performance threshold timer Δ'_{1,p_i} used to deter performance attacks, and the upper line is the normal timer Δ_{1,p_i}. In our experiment, we set $\Delta'_{1,p_i} = 1.1\delta_{1,p_i}$ and $\Delta_{1,p_i} = 1.3\delta_{1,p_i}$. That is, we expect the performance reduction to be bounded to 10% of the actual latency during a performance attack by a dedicated adversary.

To evaluate the robustness against a timer-based performance attack, we ran 10 experiments using the 0/0 benchmark, each with a sequence of 10000 requests. We assume there are no faulty replicas initially and we use the first 1000 request to train the timers. For each experiment, starting from the 1001th request, we let a replica mount a performance attack by intentionally delaying messages sent to its predecessor. To simulate different attacks, we simply let the faulty replica sleep for an "appropriate" period of time following different strategies. As expected, our findings show that the possible actions of a faulty replica is very limited: it either needs to be very careful not to be accused, thus imposing only a marginal performance reduction, or it will be suspected which will lead to a re-chaining and then a reconfiguration.

5.2   A BFT Network File System

We now evaluate a BFT-NFS service implemented using PBFT [5], Zyzzyva [24], and BChain-3. The BFT-NFS service exports a file system, which can then be mounted on a client machine. Upon receiving client requests, the replication library and the NFS daemon is called to reach agreement on the order in which to process client requests. Once processing is done, replies are sent to clients. The NFS daemon is implemented using a fixed-size memory-mapped file.

We use the Bonnie++ benchmark [10] to compare our three implementations with NFS-std, an unreplicated NFS V3 implementation, using an I/O intensive workload. We evaluate the Bonnie++ benchmark with the following directory operations (DirOps): (1) create files in numeric order; (2) stat() files in the same order; (3) delete them in the same order; (4) create files in an order that will appear random to the file system; (5) stat() random files; (6) delete the files in random order. We measure the average latency achieved by the clients while up to 20 clients run the benchmark concurrently.

Table 2. NFS DirOps evaluation in fault-free cases

BChain-3	Zyzzyva	BFS	NFS-std
$41.66s(1.10\%)$	$42.47s(2.99\%)$	$43.04s(4.27\%)$	$41.20s$

As shown in Table 2, the latency achieved by BChain-3 is 1.10% lower than NFS-std, in contrast to BFS and Zyzzyva.

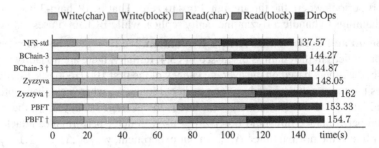

Fig. 6. NFS evaluation with the Bonnie++ benchmark. The † symbol marks experiments with one failure.

We also evaluate the performance using the Bonnie++ benchmark when a failure occurs at time zero, as detailed in Figure 6. The bar chart also includes data points for the non-faulty case. The results shows that BChain can perform well even with failures, and is better than the other protocols for this benchmark.

6   Related Work

Chandra and Toueg [7] introduced the notion of unreliable failure detectors, which could be used to solve consensus in the presence of crash failures. In their design, the failure detector outputs the identity of processes suspected to have crashed. In contrast to crash failures, Byzantine failures are not context-free, and thus it is impossible to define a general failure detector in a Byzantine environment, independently of the algorithm using the failure detector [13]. Some previous work [27,13,23,3] has extended the failure detector notion to cover a wider range of failures. For example, the muteness failure detector [13] interacts with the algorithm of a remote process to detect if the remote process has turned mute. BChain can prevent performance attacks, including those invoked by a mute process, without algorithmic help due to its chain structure.

Fault diagnosis [29,2,30,32,37,33,22] aims to identify faulty replicas. The basic idea is that a *proof of misbehavior* for a replica is collected by executing a modified BFT protocol. However, it usually requires several protocol rounds to collect the necessary information to provide such a proof. An adversary can render the system even less practical by intermittently following and violating the protocol specification. Similarly, PeerReview [19] can detect and deter failures by exploiting accountability. It also uses a "sufficient" number of witnesses to discover faulty replicas. BChain fault diagnosis, though *not* perfectly accurate, does not have the above-mentioned properties. There is no need to regularly collect evidence, and no additional latency is induced by intermittent adversaries. We note that Hirt, Maurer, and Przydatek [21] used the idea of the "imperfect fault detection" to achieve general multi-party computation in synchronous environments, but their techniques are very different from ours.

7 Conclusion

We have presented BChain, a new chain-based BFT protocol that outperforms prior protocols during failure scenarios, while offering comparable performance for the failure-free case. In the presence of failures, instead of switching to a slower, backup protocol, BChain leverages a novel technique—re-chaining—to efficiently detect and deal with the failures such that it can quickly recover its steady-state performance. BChain does not rely on any trusted components or unproven assumptions.

Acknowledgement. This research was supported in part by the National Science Foundation under grants CCF-1018871 and CNS-1228828. Hein Meling was supported by the Tidal News project under grant number 201406 from the Research Council of Norway. The authors thank Tiancheng Chang, Matt Franklin, Leander Jehl, Karl Levitt, Keith Marzullo, Phil Rogaway, Marko Vukolic, and anonymous reviewers for their helpful comments.

References

1. Abd-El-Malek, M., Ganger, G., Goodson, G., Reiter, M., Wylie, J.: Fault-scalable Byzantine fault-tolerant services. In: SOSP, pp. 59–74. ACM Press (2005)
2. Adams, J., Ramarao, K.: Distributed diagnosis of Byzantine processors and links. In: ICDCS, pp. 562–569. IEEE Computer Society (1989)
3. Baldoni, R., Helary, J., Raynal, M.: From crash fault-tolerance to arbitrary-fault tolerance: Towards a modular approach. In: DSN, pp. 273–282 (2000)
4. Benzel, T.: The science of cyber security experimentation: The DETER project. In: ACSAC (2011)
5. Castro, M., Liskov, B.: Practical Byzantine fault tolerance. In: OSDI, pp. 173–186. USENIX Association (1999)
6. Chandra, T., Hadzilacos, V., Toueg, S.: The weakest failure detector for solving consensus. J. ACM 43(4), 685–722 (1996)
7. Chandra, T., Toueg, S.: Unreliable failure detectors for reliable distributed systems. Journal of the ACM 43(2), 225–267 (1996)
8. Chiang, M., Wang, S., Tseng, L.: An early fault diagnosis agreement under hybrid fault model. Expert Syst. Appl. 36(3), 5039–5050 (2009)
9. Clement, A., Wong, E., Alvisi, L., Dahlin, M., Marchetti, M.: Making Byzantine fault tolerant systems tolerate Byzantine faults. In: NSDI, pp. 153–168. USENIX Association (2009)
10. Coker, R.: http://www.coker.com.au/bonnie++
11. Clement, A., Kapritsos, M., Lee, S., Wang, Y., Alvisi, L., Dahlin, M., Riche, T.: UpRight cluster services. In: SOSP, pp. 277–290. ACM Press (2009)
12. Cowling, J., Myers, D., Liskov, B., Rodrigues, R., Shrira, L.: HQ replication: A hybrid quorum protocol for Byzantine fault tolerance. In: OSDI, pp. 177–190. USENIX Association (2006)

13. Doudou, A., Garbinato, B., Guerraoui, R., Schiper, A.: Muteness failure detectors: Specification and implementation. In: Hlavicka, J., Maehle, E., Pataricza, A. (eds.) EDDC 1999. LNCS, vol. 1667, pp. 71–87. Springer, Heidelberg (1999)

14. Doudou, A., Garbinato, B., Guerraoui, R.: Encapsulating Failure Detection: From Crash to Byzantine Failures. In: Blieberger, J., Strohmeier, A. (eds.) Ada-Europe 2002. LNCS, vol. 2361, pp. 24–50. Springer, Heidelberg (2002)

15. Dwork, C., Lynch, N., Stockmeyer, L.: Consensus in the presence of partial synchrony. J. ACM 35(2), 288–323 (1988)

16. Fischer, M., Lynch, N., Paterson, M.: Impossibility of distributed consensus with one faulty process. J. ACM 32(2), 374–382 (1985)

17. Ghemawat, S., Gobioff, H., Leung, S.: The Google file system. In: SOSP, pp. 29–43 (2003)

18. Guerraoui, R., Knezevic, N., Quema, V., Vukolic, M.: The next 700 BFT protocols. In: EuroSys, pp. 363–376. ACM (2010)

19. Haeberlen, A., Kouznetsov, P., Druschel, P.: PeerReview: practical accountability for distributed systems. In: SOSP, pp. 175–188. ACM (2007)

20. Hendricks, J., Sinnamohideen, S., Ganger, G., Reiter, M.: Zzyzx: Scalable fault tolerance through Byzantine locking. In: DSN, pp. 363–372. IEEE Computer Society (2010)

21. Hirt, M., Maurer, U.M., Przydatek, B.: Efficient secure multi-party computation (Extended Abstract). In: Okamoto, T. (ed.) ASIACRYPT 2000. LNCS, vol. 1976, pp. 143–161. Springer, Heidelberg (2000)

22. Hsiao, H., Chin, Y., Yang, W.: Reaching fault diagnosis agreement under a hybrid fault model. IEEE Transactions on Computers 49(9) (September 2000)

23. Kihlstrom, K.P., Moser, L.E., Melliar-Smith, P.M.: Byzantine Fault Detectors for Solving Consensus. Comput. J. 46(1), 16–35 (2003)

24. Kotla, R., Alvisi, L., Dahlin, M., Clement, A., Wong, E.: Zyzzyva: Speculative Byzantine fault tolerance. In: SOSP, pp. 45–58. ACM (2007)

25. Lamport, L.: Using time instead of timeout for fault-tolerant distributed systems. Trans. on Programming Languages and Systems 6(2), 254–280 (1984)

26. Lamport, L., Malkhi, D., Zhou, L.: Reconfiguring a state machine. SIGACT News 41(1), 63–73 (2010)

27. Malkhi, D., Reiter, M.: Unreliable intrusion detection in distributed computations. In: CSFW, pp. 116–125 (1997)

28. Malkhi, D., Reiter, M.: Byzantine quorum systems. Distributed Computing 11(4) (1998)

29. Preperata, F., Metze, G., Chien, R.: On the connection asssignment problem of diagnosable systems. IEEE Transactions on Electronic Computers EC-16(6), 848–854 (1967)

30. Ramarao, K., Adams, J.: On the diagnosis of Byzantine faults. In: Proc. Symp. Reliable Distributed Systems, pp. 144–153 (1988)

31. Schneider, F.: Implementing fault-tolerant services using the state machine approach: A tutorial. ACM Computing Surveys 22(4), 299–319 (1990)

32. Serafini, M., Bondavalli, A., Suri, N.: Online diagnosis and recovery: On the choice and impact of tuning parameters. IEEE Trans. Dependable Sec. Comput. 4(4), 295–312 (2007)

33. Shin, K., Ramanathan, P.: Diagnosis of processors with Byzantine faults in a distributed computing system. In: Proc. Symp. Fault-Tolerant Computing, pp. 55–60 (July 1987)

34. van Renesse, R., Ho, C., Schiper, N.: Byzantine chain replication. In: Baldoni, R., Flocchini, P., Binoy, R. (eds.) OPODIS 2012. LNCS, vol. 7702, pp. 345–359. Springer, Heidelberg (2012)

35. van Renesse, R., Schneider, F.B.: Chain replication for supporting high throughput and availability. In: OSDI, pp. 91–104. USENIX Association (2004)

36. Vukolic, M.: Abstractions for asynchronous distributed computing with malicious players. PhD thesis. EPFL, Lausanne, Switzerland (2008)

37. Walter, C., Lincoln, P., Suri, N.: Formally verified on-line diagnosis. IEEE Trans. Software Eng. 23(11), 684–721 (1997)

RoBuSt: A Crash-Failure-Resistant Distributed Storage System*

Martina Eikel, Christian Scheideler, and Alexander Setzer

University of Paderborn, Germany
{martinah,scheideler,asetzer}@mail.upb.de

Abstract. In this work we present the first distributed storage system that is provably robust against crash failures issued by an adaptive adversary, i.e., for each batch of requests the adversary can decide based on the entire system state which servers will be unavailable for that batch of requests. Despite up to $\gamma n^{1/\log\log n}$ crashed servers, with $\gamma > 0$ constant and n denoting the number of servers, our system can correctly process any batch of lookup and write requests (with at most a polylogarithmic number of requests issued at each non-crashed server) in at most a polylogarithmic number of communication rounds, with at most polylogarithmic time and work at each server and only a logarithmic storage overhead.

Our system is based on previous work by Eikel and Scheideler (SPAA 2013), who presented IRIS, a distributed information system that is provably robust against the same kind of crash failures. However, IRIS is only able to serve lookup requests. Handling both lookup and write requests has turned out to require major changes in the design of IRIS.

Keywords: Theory of Distributed Systems, DHT, Crash Failures, Denial-of-Service Attacks.

1 Introduction

One of the main challenges of a distributed system is that it is able to work correctly even if parts of the system fail to work. If a server experiences a *crash failure* it becomes unavailable to the other servers, i.e., it does not issue or respond to requests any more. Crash failures can be temporary or permanent, and if it is temporary, a server may either be back to its state when it crashed, or it may have lost all of its state. We will focus on crash failures where, whenever a server becomes available again, it is back to its state when it crashed. This is a reasonable assumption since for commercial servers it is extremely rare that their state cannot be recovered. However, a temporary unavailability is not that uncommon and can have many causes such as maintenance work, hardware or software glitches, or denial-of-service attacks. Especially denial-of-service attacks

---

* This work was partially supported by the German Research Foundation (DFG) within the Collaborative Research Center "On-The-Fly Computing" (SFB 901) and by the EU within FET project MULTIPLEX under contract no. 317532

M.K. Aguilera et al. (Eds.): OPODIS 2014, LNCS 8878, pp. 107–122, 2014.

can be a serious threat because they are normally unpredictable, hard to prevent and they can cause the unavailability of a server for an extended period of time.

Predominant approaches in information and storage systems to deal with the threat of crash failures are to use redundancy: information that is replicated among multiple machines is likely to remain accessible even if some servers are unavailable. Unfortunately, in systems that consist of thousands of servers a complete replication of the data over all servers is not feasible. Hence, one needs to find an appropriate tradeoff between the amount of redundancy and the number of crashed servers the system can handle. One can easily show that if $\Theta(\log n)$ copies of a data item are placed randomly among n servers, and these random positions are not known to the adversary, then any strategy of the adversary that blocks half of the servers will not block all of the copies, with high probability[1]. The situation is completely different, however, when considering an adaptive adversary, i.e., someone who has complete knowledge about the system.

In a previous work, Eikel and Scheideler [7] presented a distributed information system, called IRIS, that just needs a constant storage redundancy in order to be robust against an adaptive adversary that can crash up to $\Theta(n^{1/\log\log n})$ servers. Unfortunately, the system lacks the important ability to handle write requests, i.e., to add, remove and update data items. This work solves this problem.

1.1   Model and Preliminaries

We assume that the storage system consists of a static set $\mathcal{S} = \{s_1, \ldots, s_n\}$ of n reliable servers of identical type. The servers are responsible for storing the data as well as handling the user requests. We assume that all data items are of the same size, and that any data item d is uniquely identified by a key $key(d)$. The universe of all possible keys is denoted by U, and $m := |U|$ is assumed to be polynomial in n. Furthermore, we assume that the size of the data items is at least $\Omega(\log n \log m)$. There are two types of user requests: lookup(k) for $k \in U$, and write(k, d) for $k \in U$ and a data item d. The user can issue a request by sending it to one of the servers in $\mathcal{S}$. Given a lookup(k) request, the system is supposed to either return the data item d with $key(d) = k$, or to return NULL if no such data item exists. Given a write(k, d) request, the system is supposed to store data item d with key k such that subsequent lookup(k) requests can be answered correctly. Note that with a write($\cdot$) request the user can also update or remove data.

Every server knows about all other servers and can therefore directly communicate with any one of them. This does not endanger scalability since millions of IP addresses can easily be stored in main memory in any reasonable computer today and we assume the set of servers to be static. We use the standard synchronous message passing model for the communication between the servers. That is, time proceeds in synchronized *communication rounds*, or simply *rounds*,

---

[1] "With high probability", or short, "w.h.p.", means a probability of at least $1 - 1/n^c$ where the constant c can be made arbitrarily large.

and in each round each server first receives all messages sent to it in the previous round, processes all of them, and then sends out all messages that it wants to send out in this round. Note that assuming the synchronous model is just a simplification and that our protocols only require the message delays to be bounded. In addition, we use the synchronous model because describing all protocols in an asynchronous setting would significantly blow up the construction and would hide the main innovations behind our system. We assume that the time needed for internal computations is negligible, which is reasonable as the operations in the protocols we describe are simple enough to satisfy this property.

For the crash failures, we assume a *batch-based* adaptive adversary. This means the following: We assume that time is divided into *periods* consisting of a polylogarithmic number of rounds. The adversary has complete knowledge of the current system, but cannot predict the (future) random choices of the system. Based on his knowledge, he can select an arbitrary set of $O(n^{1/\log\log n})$ servers to be crashed. A server that is crashed will not send any message nor react to messages sent from other servers. We assume that the servers have a failure detector that allows them to determine whether a server is crashed so that statements like "if server i is crashed then ..." are allowed in the protocol. Note that assuming bounded message delays, failure detection can simply be implemented using timeouts. After that, the adversary may issue an arbitrary collection of requests to the system by sending up to $\omega \in \mathbb{N}$ lookup($\cdot$) requests and up to ω write($\cdot$) requests to each server. In order to keep the presentation of RoBuSt as clear as possible, throughout this work we assume $\omega = 1$. RoBuSt can in principle handle arbitrary values of ω, but in that case the bound on the work required by each server for serving all requests must be multiplied with ω.[2] There are no further limitations, i.e., the keys selected by the adversary may or may not be associated with data items stored in the system, and the adversary is also allowed to issue multiple lookup requests for the same key. The task of the system is to correctly handle *all* of these requests. We assume that any period is long enough such that the system has enough time to perform all necessary computations and to answer all requests. After any period, the adversary may select a different set of $\Theta(n^{1/\log\log n})$ servers to be crashed. We assume that the set of crashed servers does not change during a fixed period, which is why we use the notion of a batch-based adaptive adversary. Of course, allowing crash failures at arbitrary times would make the model much stronger, yet it would significantly complicate the system design, which is why we leave this to future research. Note that we assume links between intact (i.e., non-crashed) servers to be reliable. Unreliable links can be dealt with using, for example, at-least-once delivery or error correction strategies, which are out of scope for our design since it is already complex enough.

In order to measure the quality of the storage system, we introduce the following notation. A storage strategy is said to have a *redundancy* of r if r times

---

[2] Note that our system would not be able to answer all requests with at most polylogarithmic work if $\omega > polylog(n)$, but this would trivially hold for any storage system.

more storage (including any control storage) is used for the data than storing the plain data. We call a storage system *scalable* if its redundancy is at most $\text{polylog}(n)$, *efficient* if any collection of lookup and write requests specified by the adversary can be processed correctly in at most $\text{polylog}(n)$ many communication rounds in which every server sends and receives at most $\text{polylog}(n)$ many messages of at most $\text{polylog}(n)$ size, and *robust* if any collection of lookup and write requests specified by the adversary can be processed correctly even if a set of up to $\Theta(n^{1/\log\log n})$ servers specified by the adversary crash.

1.2 Related Work

Over the past years, distributed storage systems have gained a lot of importance. Popular examples include the storage solutions offered by Google, Apple, or Amazon. Since availability and retrievability of the stored data is a key aspect of distributed storage systems, these systems should be able to work correctly despite common failures. Often failures in distributed systems are divided into the following types [4]: crash failures, omission failures, timing failures, and Byzantine failures. In crash failures the affected component (for instance a server) completely stops working. In receive (send) omission failures the affected component cannot receive (send) any further messages. A timing failure leads a component to not respond within a specified time interval. In case of a Byzantine failure, the affected component may react in an arbitrary, even malicious manner.

This work focuses on crash failures. Many works dealing with crash failures in distributed systems focus on crash failure recovery and crash failure detection [15,12,8]. But to the best of our knowledge, no previous work has considered how to secure a distributed storage system against many (e.g., more than a polylogarithmic number) simultaneous crash failures controlled by an adaptive adversary while using only polylogarithmic work, time and redundancy. That is, we do not seek to prevent failures or attacks, but rather focus on how to maintain a good availability and performance even in spite of them. Our system is based on the distributed hash table (DHT) paradigm (e.g., [3,5,9,14,16]), with the additional twist of using coding and arranging the used DHTs in an appropriate structure. Various systems based on DHTs that are resistant against Denial-of-Service (DoS) attacks (which represent a special type of crash failures) have already been proposed [10,11,13]. But these do not work for adaptive adversaries. The first DHTs that are robust against past insider crash failures were proposed in [1,2], where a past insider only has complete knowledge of the information system up to some *past* time point t_0. For this kind of insider, it is possible to design an information system so that any information that was inserted or last updated *after* t_0 is safe against crash failures [1,2]. But the constructions proposed in these papers would not work at all for a current insider because they are heavily based on randomization to ensure unpredictability. Eikel and Scheideler were the first to present a distributed information system, called IRIS, that is provably robust even against a current insider that crashes up to $\Theta(n^{1/\log\log n})$ servers. The authors showed that IRIS can correctly answer any set of lookup requests (with one request per server that is not crashed) with polylogarithmic time and work at each server and only a constant redundancy. Still it remained open whether it is possible to design a distributed storage system that can efficiently handle lookup and write requests under the presence of crash failures. We answer this question positively by proposing such a system.

1.3   Our Contribution

We present the first scalable distributed storage system, called *Robust Bucket Storage (in short RoBuSt)*, that is provably robust against adaptive crash failures and that supports both lookup and write requests. Concretely, we allow the adversary to have complete knowledge about the storage system and to have the power to crash any set of $\gamma n^{1/\log\log n}$ servers, for $\gamma > 0$ constant. The task of the system is to serve any collection of lookup and write requests in an efficient way despite the crash failures.

RoBuSt expands some of the ideas in IRIS, a distributed storage system that we proposed in SPAA 2013 [7]. The system presented in this work tolerates a number of crashed servers that is similar to the number of servers blocked by a DoS attack that the Basic IRIS version can tolerate and achieves comparable efficiency bounds (up to a logarithmic factor). In contrast to IRIS, which can only handle lookup requests, RoBuSt is able to additionally handle write requests. Although in the lookup protocol we are able to adapt some of the underlying ideas of IRIS, adding the write functionality required significant changes in the whole structure. To simplify the description for readers who are familiar with IRIS, we try to re-use terminology whenever there are similarities (e.g., Probing Stage, Decoding Stage).

One aspect is that IRIS organizes data into layers of n data items each, and each layer is encoded separately using distributed coding that involves all n servers. This means that whenever a data item needs an update, all n servers have to update their information for the corresponding layer. Since we allow any set of write requests, it may happen that every write request involves a different layer, which would create an enormous update work. To solve this issue, in RoBuSt we store the data items in so-called buckets that are organized in a binary tree. For each data item, there are a logarithmic number of buckets that are a potential storage location for the data item. For a data item there may exist different versions of it in different buckets. But our system ensures that the highest bucket (i.e., the bucket with minimum distance to the root in the underlying binary tree over the buckets) that contains a version of the data item always holds the most recent version.

Furthermore, IRIS uses a fixed set of hash functions to specify anchor locations for the data so that afterwards lookup requests can be served efficiently despite an adversarial DoS attack. However, using fixed hash functions in RoBust would enable the adversary to annul the fair distribution of data in a bucket. Therefore, RoBuSt chooses new, random hash functions whenever write requests have to be served.

Another complication is the fact that a server may not know whether its information is up-to-date. This is because at the time when write requests were executed that required an update in that server, the server might have been crashed. Our organization of the data and our protocols ensure that any server that answers a request always returns the most recent version of a data item.

Nevertheless, RoBuSt makes sure that all data can still be efficiently found while the storage overhead is at most a logarithmic factor.

Theorem 1. *RoBuSt is a scalable and efficient distributed storage system that only needs a logarithmic redundancy to protect itself against batch-based adaptive crash failures in which up to $\gamma \cdot n^{1/\log\log n}$ servers can crash for a constant $\gamma > 0$, w.h.p.*

2   Underlying Datastructure

In the following, we assume keys are potentially from an address space of size at most n^p, i.e., we need $\Lambda := p \log n$ bits for each address. We introduce the following definitions: For a data item d, denote the *address* of d by $\mathsf{key}(d) = d_{p\log n - 1} \ldots d_1 d_0 \in \{0,1\}^{p\log n}$ and let $\mathsf{bit}_d(i) := d_i$.

Our data structure is based on a binary tree with $\Lambda + 1$ levels, so-called *zones*. We denote the nodes of each zone as *buckets* where each bucket will hold a set of data items. The internal storage strategy of the buckets is described in Section 2.1. Zone 0 consists of a single bucket, bucket B_ε. Each bucket B that is not in zone Λ has two children, denoted by 0-$child(B)$ and 1-$child(B)$. For each data item d there is not only a single possible bucket in which to store d but there are $\Lambda + 1$ possible buckets for d, one in each zone. Bucket B_ε may hold any data item. Any data item d that may belong to bucket B in zone ℓ, may also belong to 0-$child(B)$ if $\mathsf{bit}_d(\ell) = 0$ or to 1-$child(B)$ if $\mathsf{bit}_d(\ell) = 1$. In the following, let $\mathcal{B}$ be the set of all buckets and let $bucket(z,d) : \{0,\ldots,\Lambda\} \times U \to \mathcal{B}$ be a function that returns the unique possible bucket of a data item d at zone z. Initially, a bucket does not contain any data. During the runtime of the system the following invariant is satisfied: Each bucket, excluding bucket B_ε, stores either 0 or between n and $2n$ data items. Bucket B_ε stores at most $2n$ data items.

2.1   Internal Storage Strategy of the Buckets

The idea of storing a set D of data items into a bucket B is to reuse the basic concepts of the storage strategy for individual layers from IRIS [7]. Roughly speaking this strategy works as follows: In order to achieve the desired robustness, we first create $c \geq 18 \log m$ *pieces* $d_1, \ldots, d_c$ for each data item $d \in D$ using Reed Solomon coding. Using c hash functions chosen uniformly and independently at random, these pieces are then mapped to servers. Finally, all these pieces are encoded with each other, such that at the end each intact server holds for each piece some parity information resulting from the encoding process. Besides encoded data pieces each bucket B additionally stores c hash functions and a timestamp $t(B)$. The timestamp is used to handle out-dated information a server might hold if it has crashed in a previous period in which write requests were served.

In the following we roughly describe the coding strategy presented in [7]. The coding strategy is a block-based distributed strategy that follows the topology of a k-ary butterfly as described in the following. For $k \in \mathbb{N}$ we use the notation $[k] = \{0, \ldots, k-1\}$.

Definition 1. *For any $d, k \in \mathbb{N}$, the d-dimensional k-ary butterfly $BF(k, d)$ is a graph $G = (V_k, E)$ with node set $V_k = [d+1] \times [k]^d$ and edge set E with*

$$E = \{\{(i, x), (i+1, (x_1, \ldots, x_i, b, x_{i+2}, \ldots, x_d))\}$$
$$\mid x = (x_1, \ldots, x_d) \in [k]^d, \ i \in [d], \ and \ b \in [k]\}.$$

A node u of the form (ℓ, x) is said to be on butterfly level ℓ of G. Furthermore, $LT(u)$ is the unique k-ary tree of nodes reached from u when going downwards the butterfly (i.e., to nodes on butterfly levels $\ell' > \ell$) and $UT(u)$ is the unique k-ary tree of nodes reached from u when going upwards the butterfly. Moreover, for a node u at level ℓ, let $BF(u)$ be the unique k-ary sub-butterfly of dimension ℓ ranging from butterfly level 0 to ℓ in $BF(k, d)$ that contains u.

A visualization of a k-ary butterfly is given in Figure 1.

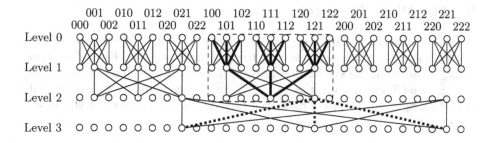

Fig. 1. Visualization of a k-ary butterfly $BF(k, d)$ for $k = d = 3$. For a better readability most of the edges from level two and three are omitted. The dashed box denotes the sub-butterfly $BF((2, 111))$. The thick solid lines in the dashed box denote the edges of $UT((2, 111))$. The thick dotted lines denote the edges of $LT((2, 121))$.

In the following let $BF(k, d)$ be a k-ary butterfly with $n = k^d$ and with server s_i, $i \in \{0, \ldots, n-1\}$, emulating the butterfly nodes $(0, i), \ldots, (d, i)$. That is, whenever a butterfly node (j, i), $j \in \{0, \ldots, d\}$ is supposed to perform an action or store data, this is done by server s_i. We say a server s is *connected via the k-ary butterfly* to another server s', if there is an edge (u, v) in the butterfly such that u is emulated by s and v is emulated by s'.

While in IRIS each server holds $O(1)$ data pieces per layer, in our system each server holds $O(\log n)$ data pieces per bucket. This is due to the fact that each bucket contains $O(n)$ data items and for each data item $c = \Theta(\log m)$ pieces are created and distributed evenly among the servers. Hence, we simply concatenate the data pieces a server s_i holds in a bucket B and denote the resulting data *block* as b_i.

In order to encode the data blocks $b_0, \ldots, b_{n-1}$ assigned to the servers $s_0, \ldots, s_{n-1}$ in bucket B, initially, b_i is placed in node $(0, i)$ for every $i \in \{0, \ldots, n-1\}$. Given that in butterfly level ℓ we have already assigned data blocks $d(\ell, x)$ to the nodes (ℓ, x) we use the coding strategy presented in [7] to assign data blocks $d(\ell+1, x)$ to the nodes at butterfly level $\ell+1$. The used coding strategy is based on some simple parity computations and ensures the following property: If at most one butterfly node $(\ell + 1, x_j)$ from the set of nodes $\{(\ell + 1, x_1), \ldots, (\ell + 1, x_k)\}$ is crashed, then the information in the remaining nodes $(\ell + 1, x_i)$, $i \in \{1, \ldots, k\} \backslash \{j\}$, suffices to recover $d(\ell, x_1), \ldots, d(\ell, x_k)$. Furthermore, with Lemma 2.4 in [7] the storage amount of each server s_i, $i \in \{0, \ldots, n-1\}$, required for the encoding of a single bucket is upper bounded by $(1+e)z$, where z denotes the maximum size of the data blocks stored at any server s_j, $j \in \{0, \ldots, n-1\}$. Since there may exist outdated data items in the system, but for each level at most one, i.e. in total at most $\Lambda + 1 = O(\log n)$ many for each data item, the redundancy of our system increases to $O(\log n)$.

Corollary 1. *RoBuSt has a redundancy of $O(\log n)$.*

3   The Write Protocol

In the following let D with $|D| \leq (1 - \delta)n$ and $\delta < 1/72 \cdot n^{1/\log\log n}$, be the set of data items for which intact servers received write requests. For a data item d that is stored in the system denote the c pieces that have been created from d using Reed Solomon coding as $d_1, \ldots, d_c$. Furthermore, denote the server that is holding d_1 (after the pieces have been spread over the n servers) as the server *maintaining* d.

3.1   Preprocessing Stage

In this stage, for each crashed server s_i, a unique intact server is determined, denoted as the *representative* of s_i, such that at the end of this stage each crashed server is the representative of at most two other servers. The idea of the representatives is to let them take over the roles of the according crashed servers in actions (e.g. routing, computations) the crashed servers were supposed to perform. For this, we additionally need to ensure that each intact server knows the representatives of all crashed servers it is connected to in the underlying k-ary butterfly.

The determination of the representatives and the introduction of the representatives to the appropriate servers can be done in the same manner as in the *butterfly completion stage* of [7], which can be carried out in $(2 + o(1)) \log n$ rounds with a congestion of at most $O(\log n)$ (see Lemma 2.11 in [7]). In contrast to [7], we do not need to compute a so-called *decoding depth* here that gives information about the minimum level of the butterfly that the decoding must be initiated from, which would take $O(\log n)$ rounds. In the following, we denote by $s(i)$ the representative of s_i if s_i is crashed or s_i itself otherwise.

3.2   Writing Stage Overview

In order to keep the specification of our system simple, we first give a high-level overview of how a set of write requests is handled. Further details are given in the following subsection.

The Writing Stage consists of up to $\Lambda + 1$ phases. Each phase $z \in \{0, \dots, \Lambda\}$ deals with a single bucket B_z from zone z only and receives a set of data items D_z to be inserted into B_z. At the beginning, $D_0 := D$ is the set of all data items for which there are write requests. In the following, $\mathcal{D}(B_z)$ denotes the set of data items that are stored in bucket B_z (at the beginning of phase z). Phase $z \in \{0, \dots, \Lambda\}$ consists of the following steps.

1. Completely decode B_z and send all decoded pieces of a data item $d \in \mathcal{D}(B_z)$ to the server maintaining d (for details, see the full version [6]).
2. If $|\mathcal{D}(B_z) \cup D_z| \leq 2n$: Add the data items from D_z to $\mathcal{D}(B_z)$, choose c new hash functions $h_1, \dots, h_c : U \rightarrow V$ uniformly at random for B_z, and reencode B_z (see [6] and below).
3. Else ($|\mathcal{D}(B_z) \cup D_z| > 2n$):
   (a) The intact servers agree on a subset $D_{z+1} \subseteq \mathcal{D}(B_z) \cup D$ of size n with the property that for all $d, d' \in D_{z+1}$, $\text{bit}_d(z) = \text{bit}_{d'}(z) = b \in \{0, 1\}$ (for details, see the full version [6]).
   (b) Reencode the data items in $(D_z \cup \mathcal{D}(B)) \setminus D_{z+1}$ in bucket B_z and choose c new hash functions $h_1, \dots, h_c : U \rightarrow V$ uniformly at random for B_z. (see below)
   (c) Set $B_{z+1} :=0\text{-}child(B_{z+1})$ if $b = 0$ and $B_{z+1} :=1\text{-}child(B_{z+1})$ if $b = 1$ and propagate the data items in D_{z+1} to the next phase (for details, see the full version [6]).

Each phase of the Writing Stage can be performed in $O(\log n)$ rounds with a congestion of $O(\log n)$ at each server in each round (see [6]). Since there are at most $O(\log n)$ phases in the Writing Stage, the overall runtime is $O(\log^2 n)$ rounds.

Encoding of a Bucket. In the following we describe how a set of data items is reencoded into a bucket, as required in step 2 and step 3b. Note that the reencoding of a bucket does not only consist of the simple encoding of the data items belonging to that bucket but it consists of some additional steps, as described in the following.

First, in contrast to IRIS, $s(1)$ chooses c hash functions $h_1, \dots, h_c : U \rightarrow V$ uniformly at random that will be used to map data pieces of this bucket to servers. While in IRIS the hash functions that map data pieces to servers are never changed, we need to choose new hash functions for a bucket B whenever

B is (re)encoded. The reason for this is that otherwise the adversary would be able to generate write requests that overload certain servers.

Note that the hash functions need to satisfy certain expansion properties, but if c is chosen sufficiently large ($c \geq 18 \log m$) they do so, w.h.p. (more information is provided in [6]). After that, $s(1)$ distributes the c hash functions to all other intact servers $s(i)$. This distribution can be realized by simply broadcasting the hash functions in the k-ary butterfly from level $\log_k n$ to level 0. In addition, s_1 distributes a current timestamp $t(B_z)$ to all other intact servers and each intact server $s(i)$ sets its current timestamp for bucket B_z to that value. Each server $s(i)$ now creates for each data item which it maintains or which it has received write requests for and which are not propagated to the next phase the c pieces $d_1, \ldots, d_c$ of d using Reed Solomon coding (Section 2). Here, d_j, $j \in \{1, \ldots, c\}$, is supposed to be sent to the server s' responsible for $h_j(d)$ or to its representative if s' is crashed. Unfortunately, a server $s(i)$ does not necessarily know the representative of the server s' if that server is crashed. Thus, instead of sending the data pieces directly, the servers initiate a bottom-up routing in the underlying k-ary butterfly in order to determine the representative of $h_j(d)$ for each $1 \leq j \leq c$. Obviously, this takes only $\log_k n$ rounds and can be performed with a congestion of $O(k)$ per node. Once $s(i)$ knows the representative of $h_j(d)$, it directly sends d_j to $h_j(d)$ for all $1 \leq j \leq c$.

After the pieces of data items have been distributed, the servers encode the data items in $(\mathcal{D}(B_z) \cup D_z) \setminus D_{z+1}$ in a distributed fashion. Note that the set of data blocks for server i in zone z is completely overwritten for each server $s(i)$ in this process. This can be done by a simple top-down approach using the coding strategy for IRIS (see Section 2.1 in [7]). In addition, we also store the timestamp of the bucket along with the data block by appending it to the composed data block.

The following lemma holds during the encoding step, regardless of the current phase.

Lemma 1. *Assume the adversary blocks less than $(\gamma/2) \cdot 2^{\log_k n}$ servers, with $\gamma = 1/36$. Then, for any data item d that is (re-)written during the current period, and any level $0 \leq \ell \leq \log_k n$, there are at most $c/6$ pieces of d that are mapped to sub-butterflies $BF(v)$ (for some v at level ℓ) with at least $\lceil 2^{\ell-1} \rceil$ crashed servers in $BF(v)$, w.h.p.*

The lemma plays an important role in the proof of the correctness of the Lookup Protocol. A proof of this lemma can be found in the full version [6].

4   The Lookup Protocol

In order to keep the specification of our system simple, we provide the description of the lookup protocol as a separate protocol that is executed after the execution of the Write Protocol. The lookup protocol is divided into two stages: the Preprocessing Stage (Section 4.1) and the Zone Examination Stage (Section 4.2). The former is similar to the Preprocessing Stage of the Write Protocol (Section 3.1).

The latter is performed for each zone individually and split into two further stages: the Probing Stage and the Decoding Stage. The basic idea of the Probing Stage is to answer a request by directly collecting a sufficient number of data pieces. If this is not possible, either because too many of the servers holding a piece are crashed or because of congestion, the Decoding Stage tries to recover a data item by utilizing the distributed coding described in Section 2.1. Note that both a Probing Stage as well as a Decoding Stage can be found in IRIS ([7]), too. While they match in their general structure, there are important differences that are caused by the differences in the underlying structure and the implications of the write functionality. For example, servers may now store obsolete data items without being aware of that.

4.1 The Preprocessing Stage

The Preprocessing Stage is exactly the same as in Section 3.1. If at least one write request has been handled in the current period, we can thus skip this part and re-use the established k-ary butterfly and the unique representatives.

4.2 The Zone Examination Stage

In the following let $\mathcal{D}$ be the set of data items for which a lookup request arrived at an intact server. The idea of this stage is to successively perform a lookup for each $d \in \mathcal{D}$ in each zone until a copy of d has been found and returned to the appropriate server. The zone examination stage is performed for at most $\Lambda + 1$ zones starting with zone 0.

In each phase $z \in \{0, \ldots, \Lambda\}$, beginning with $z = 0$, each server with an unserved lookup request for some data item d initiates a lookup request for d in bucket $bucket(z, d)$. Any server that receives a copy of the data item it requested during the lookup in zone z, as described in the following, returns that copy and is finished. All remaining lookup requests are handled in the next phase, phase $z := z + 1$. This procedure is repeated until each lookup request is served.

Handling a set of lookup requests in one phase z is done by performing the Probing Stage and the Decoding Stage as described in the following.

Probing Stage. In the following let s be an intact server that has an unserved lookup request for a data item d at the beginning of phase z. The idea of the Probing Stage is to either achieve $c/3$ up-to-date pieces such that d can be recovered. Or to assign the request for d to a level $\{1, \ldots, \log_k n\}$ (as defined later) in order to further handle the request in the next stage, the Decoding Stage. In the following, for a server s', an index $i \in \{1, \ldots, c\}$, and a data piece d' we denote by $P_i(s', d')$ the unique path of length $\log_k n$ in the k-ary butterfly from the butterfly node on level $\log_k n$ emulated by s' to the butterfly node on level 0 emulated by the server that is responsible for $h_i(d')$.

On a high level view, in phase z, server s performs the following steps.

1. Acquire current hash functions and timestamp t_d for bucket $bucket(z, d)$.

2. Choose c intact servers $s(d_1), \ldots, s(d_c)$ uniformly and independently at random.
3. Send a $\mathsf{probe}(d, i, t_d)$ message to $s(d_i)$, $i \in \{1, \ldots, c\}$, in order to initiate the forwarding of the $\mathsf{probe}(\cdot)$ message along the c paths $P_i(s(d_i), d_i)$.

Note that acquiring the hash functions in step 1 is necessary since s may have been crashed in the last period in which a write occured in bucket $bucket(Z, d)$ (at which the hash functions were replaced). Acquiring the current hash functions and the timestamp works as follows: First of all, s randomly chooses $\kappa := \Theta(\log n)$ intact servers and asks them for their timestamp in bucket $bucket(Z, d)$. The intact servers can be found in $O(1)$ communication rounds, w.h.p., by selecting κ random servers in each round until κ intact servers have been found. Let t_d be the maximum timestamp s received. If t_d is greater than the timestamp s stores for $bucket(Z, d)$, s knows that it does not have the current hash functions and asks one server from which it received t_d for the c hash functions for bucket $bucket(Z, d)$. Note that during this process each server only receives $O(\log n)$ requests throughout this process, w.h.p.

Once s knows the correct hash functions, its goal is to retrieve at least $c/3$ pieces of d. Since contacting the servers holding the c pieces of d directly may cause a too high congestion at these servers, we use the method of forwarding c probes from uniformly chosen intact servers $s(d_1), \ldots, s(d_c)$ to the servers responsible for the c pieces of d along the c paths $P_1(s(d_1), d_1), \ldots, P_c(s(d_c), d_c)$ (step 2, step 3). Analogously to step 1 choosing the c intact servers in step 2 takes $O(1)$ communication rounds, w.h.p.

In the following we describe how the nodes from the paths $P_1(s_1, d_1), \ldots,$ $P_c(s_c, d_c)$ react on incoming messages during this phase. Let u be a butterfly node on level $\ell \in \{0, \ldots, \log_k n\}$ that has received a $\mathsf{probe}(d, i, t_d)$ message. In order to reduce redundancy u combines probes for the same piece of d (and thus the same target) and u marks itself as the new origin of the probe (technique of *splitting and combining* [7]). In the following we denote a butterfly node u as *congested* if it has received more than $\alpha \cdot c$ $\mathsf{probe}(\cdot)$ messages for different probes, for a sufficiently large constant $\alpha > 0$. Whenever u receives a $\mathsf{probe}(d, i, t_d)$ message, u performs the following steps.

1. If u is congested:
2.   Stop forwarding the probe and send a $\mathsf{fail}(d, i, \ell)$ message to the origin of the probe message.
3. Else:
4.   If $\ell \neq 0$: Forward $\mathsf{probe}(d, i, t_d)$ message to the butterfly node on level $\ell - 1$ on the path $P_i(s(d_i), d_i)$.
5.   If $\ell = 0$: (probe has reached its destination)
6.   If u's current version of bucket $bucket(Z, d)$ has timestamp t_d:
7.   If u holds piece d_i of d: Send requested piece d_i to the origin of the probe message.
8.   Else: Send $\mathsf{notexists}(d)$ message to the origin of the probe message.
9.   Else: Send $\mathsf{fail}(d, i, 0)$ to the origin of the probe message.

If a butterfly node on level $\ell \in \{0, \ldots, \log_k n - 1\}$ receives a data item, a fail($\cdot$), or a notexists($\cdot$) message, it forwards this answer to the origin of the request to which this message was an answer to (along the same path that the request was routed). A butterfly node on level $\log_k n$ emulated by $s(d_i)$, $i \in \{1, \ldots, c\}$, that received an answer for a probe for data piece d_i simply forwards this answer to the server that initiated the forwarding of that probe. These answers ensure that after $O(\log_k n)$ rounds the server s that received a lookup request for a data item d has received for all initially sent probe($\cdot$) messages a piece of d, or a notexists(d) message, or the level at which the probing failed. Depending on which kinds of answers s has received, it reacts as follows:

- If s received at least $c/3$ up-to-date pieces of d, s recovers d using Reed Solomon coding and answers the request.
- Else if s receives a notexists(d) message, s answers that the requested data item does not exist in the system.
- Else if s receives more than $c/3$ fail($d, i, 0$) messages, s declares the request for d to *belong to level* ℓ, where $\ell \in \{0, \ldots, \log_k n\}$ is the smallest level that contains at least $5c/6$ active probes for d_i, i.e., probes for d_i that were not aborted at level $\ell - 1$ or earlier.

It is easy to see that the Probing Stage takes at most $O(\log n)$ communication rounds per phase with at most $O(\log^2 n)$ congestion at every server in each round. Since the only reason for a piece of a data item to be deactivated at level 0 is that it is outdated, the following is a direct corollary from Lemma 1.

Corollary 2. *No data item can ever belong to level* 0.

For the analysis of the runtime of the lookup protocol, the following lemma plays an important role.

Lemma 2. *If the adversary can only block less than* $(\gamma/2) \cdot 2^{\log_k n}$ *servers, then for every* $\ell \in \{1, \ldots, \log_k n\}$, *the number of data items belonging to level ℓ is at most* $2\gamma n/k^\ell$ *with* $\gamma = 1/36$.

The general idea and structure of the proof of Lemma 2 is based on the proof of Lemma 2.16 in [7]. In contrast to [7], no requests are aborted due to crashed nodes here. Furthermore, we have a different definition of when a node belongs to level ℓ here (we require at least $5c/6$ active probes instead of $c/2$) and a different value of γ. A proof of Lemma 2 can be found in the full version [6].

Decoding Stage. The Decoding stage proceeds in $\log_k n$ sub-phases. In the following, for a server s that holds a lookup request for some data item d that has not been answered before this sub-phase, we define $s_i^{(\ell)}(d)$ as the node at level ℓ on the unique path of length $\log_k n$ from the butterfly node on level $\log_k n$ emulated by $s_i(d)$ to the butterfly node on level 0 responsible for $h_i(d)$.

On a high level view, the Decoding Stage works as follows: During each sub-phase $1 \leq \ell \leq \log_k n$, starting with level 1, we try to recover the data items belonging to level ℓ. In order to recover a data item d, we need to collect at least $c/3$ pieces of d. To do so, we randomly choose $5c/6$ requests for pieces of d that were active at level ℓ in the Probing Stage and for each of these pieces d_i we determine whether $BF(s_i^{(\ell)}(d))$ can be decoded without congestion (as described later). If $BF(s_i^{(\ell)}(d))$ can be decoded without congestion, the decoding is initiated and the result of this is sent back to the origin. (Throughout the whole process, we use the same combining/splitting approach of messages as in the Probing Stage.) Otherwise, the origin is informed that the according piece of d could not be decoded. If for a data item d not sufficiently many (i.e., less than $c/3$) pieces could be recovered, the request for d is declared to belong to level $\ell + 1$ and will be considered again in the next sub-phase. Note that requests for non-existing data items may be handled in the Decoding Stage. However, these can be treated as existing items (with the only difference being that one intact server taking part in the decoding is sufficient to tell that the data item does not exist).

In the following, we describe the operation of any sub-phase ℓ in more detail. First of all, each server s that is responsible for a lookup request of a data item d that belongs to level ℓ chooses $5c/6$ among the at least $5c/6$ indices of pieces of d that were active at level ℓ in the Probing Stage. For such a piece d_i of d with current timestamp t, s sends a $\mathsf{decode}(d,i,t)$ message from $s_i^{(\log_k n)}(d)$ to $v := s_i^{(\ell)}(d)$ (which is done by simply routing through the k-ary butterfly into the direction of $h_i(d)$ for ℓ rounds). In order to determine whether $BF(v)$ can be decoded without congestion, v first checks whether it is congested, i.e., it received more than βck $\mathsf{decode}(\cdot)$ messages for a sufficiently large constant β and, if not, then issues a $\mathsf{decodeCHECK}(d,i)$ message, which is spread to all nodes in $UT(v)$. During this spreading, whenever a further forwarding of all messages received by a node u at a level $\ell - \kappa$, $1 \le \kappa < \ell$, could lead to congestion (i.e., u received more than βck $\mathsf{decodeCHECK}(d',i')$ messages for distinct (d',i') pairs), u stops the forwarding of all messages and instead spreads a $\mathsf{cong}(\cdot)$ message in $BF(u)$. In addition, it sends a $\mathsf{fail}(\cdot)$ message to all neighbors at level $\ell - \kappa + 1$. Each node on a level ℓ', $\ell - \kappa + 1 \le \ell' < \ell$, that receives such a $\mathsf{fail}(\cdot)$ message forwards this message to all neighbors at level $\ell' + 1$ from which it received a $\mathsf{decodeCHECK}(\cdot)$ mesage. By this it is ensured that whenever a node in $BF(u)$ is congested each node v' at level ℓ with $v' \in BF(u)$ receives a $\mathsf{fail}(\cdot)$ message after at most 2ℓ rounds. Each node u' at level $\ell - \kappa$, $1 \le \kappa < \ell$, that received a $\mathsf{cong}()$ message initiates the same spreading of $\mathsf{cong}()$ messages in $UT(u')$. If v had not been congested before the spreading and v has not received any $\mathsf{fail}(\cdot)$ message after 2ℓ rounds, it knows that any piece of a data item for which v received a $\mathsf{decode}(\cdot)$ message can be decoded if not outdated nodes in $BF(v)$ forbid this. Thus, it initiates the decoding for each of the pieces, which may fail due to outdated nodes. If the decoding is possible, it recovers all of these pieces within $O(\ell)$ communication rounds with a congestion of at most βck^2 per node (using the distributed decoding described in [7]). These are then forwarded to the origins of the requests. If, however the decoding fails, or if v was congested or received a $\mathsf{fail}(\cdot)$ message, it sends a $\mathsf{fail}(\cdot)$ message to the origins of the $\mathsf{decode}(\cdot)$

messages it received (which, again, are forwarded up to the initiator of that decode($\cdot$) message). Finally, if a server s that is responsible for a lookup request of a data item d receives at least $c/3$ successfully decoded pieces, it determines d and answers the request. Otherwise, it changes the request to belong to level $\ell + 1$ such that it will be processed again in the next sub-phase.

It is easy to see that the Decoding Stage satisfies the following property:

Lemma 3. *The Decoding Stage takes at most $O(\log n)$ communication rounds per sub-phase with at most $O(\log^3 n)$ congestion in every node at each round, w.h.p.*

Similarly to Lemma 2 of the Probing Stage, for the Decoding Stage the following lemma holds:

Lemma 4. *At the beginning of each sub-phase $\ell \in \{1, \ldots, \log_k n\}$, the number of data items with requests belonging to level ℓ is at most $\varphi n/k^\ell$ with $\varphi = \Theta(k)$.*

A proof of Lemma 4 can be found in the full version [6]. The previous lemmas and results imply Corollary 3, which proves Theorem 1.

Corollary 3. *RoBuSt correctly serves any set of lookup and write requests (with one request per intact server) in at most $O(\log^4 n)$ communications rounds, with a congestion of at most $O(\log^3 n)$ at every server in each round and a redundancy of $O(\log n)$ even if up to $1/72 \cdot n^{1/\log\log n}$ servers are crashed.*

5 Conclusion and Future Work

We presented the first scalable distributed storage system that is provably robust against batch-based crash failures with up to $\gamma n^{1/\log\log n}$ crashes allowed ($\gamma > 0$ constant). An interesting question that has not been investigated in this work is whether the techniques that enabled the Enhanced IRIS system [7] to tolerate a larger number of failed servers could be adapted for RoBuSt in order to increase the number of crashed servers allowed up to ϱn (for some constant $\varrho > 0$) while (as a minor drawback) also increasing the redundancy to $O(\log n)$, such as it is the case in Enhanced IRIS.

Moreover, while we assume batch-based failures, it would be interesting to see whether a scalable distributed storage system can be designed that can tolerate failures occuring at arbitrary points in time. Dealing with a similar issue, it would also be interesting to enhance our system to allow dynamics (i.e. joins and leaves of servers) in our system in order to model P2P networks.

A further interesting challenge is to enhance our distributed storage system such that additional types of attacks can be handled, for example Byzantine attacks.

References

1. Awerbuch, B., Scheideler, C.: A Denial-of-Service Resistant DHT. In: Pelc, A. (ed.) DISC 2007. LNCS, vol. 4731, pp. 33–47. Springer, Heidelberg (2007)

2. Baumgart, M., Scheideler, C., Schmid, S.: A dos-resilient information system for dynamic data management. In: Proc. SPAA, pp. 300–309 (2009)
3. Bhargava, A., Kothapalli, K., Riley, C., Scheideler, C., Thober, M.: Pagoda: A Dynamic Overlay Network for Routing, Data Management, and Multicasting. In: Proc. of SPAA, pp. 170–179 (2004)
4. Cristian, F.: Understanding fault-tolerant distributed systems. Commun. ACM 34(2), 56–78 (1991)
5. Rowstron, A., Druschel, P.: Pastry: Scalable, decentralized object location, and routing for large-scale peer-to-peer systems. In: Guerraoui, R. (ed.) Middleware 2001. LNCS, vol. 2218, pp. 329–350. Springer, Heidelberg (2001)
6. Eikel, M., Scheideler, C., Setzer, A.: RoBuSt: A Crash-Failure-Resistant Distributed Storage System (full version). ArXiv e-prints (September 2014)
7. Eikel, M., Scheideler, C.: IRIS: A Robust Information System Against Insider DoS-Attacks. In: Proceedings of the 25th ACM Symposium on Parallelism in Algorithms and Architectures, SPAA 2013, pp. 119–129. ACM (2013)
8. Gupta, I., Chandra, T.D., Goldszmidt, G.S.: On scalable and efficient distributed failure detectors. In: Proceedings of the Twentieth Annual ACM Symposium on Principles of Distributed Computing, PODC 2001, pp. 170–179. ACM, New York (2001)
9. Harvey, N.J.A., Jones, M.B., Saroiu, S., Theimer, M., Wolman, A.: SkipNet: A Scalable Overlay Network with Practical Locality Properties. In: Proc. of USITS, p. 9 (2003)
10. Kargl, F., Maier, J., Weber, M.: Protecting Web Servers from Distributed Denial of Service Attacks. In: Proc. of WWW, pp. 514–524 (2001)
11. Keromytis, A.D., Misra, V., Rubenstein, D.: SOS: Secure Overlay Services. In: Proc. of SIGCOMM, pp. 61–72 (2002)
12. Leners, J.B., Wu, H., Hung, W.-L., Aguilera, M.K., Walfish, M.: Detecting failures in distributed systems with the falcon spy network. In: Proceedings of the Twenty-Third ACM Symposium on Operating Systems Principles, SOSP 2011, pp. 279–294. ACM, New York (2011)
13. Morein, W.G., Stavrou, A., Cook, D.L., Keromytis, A.D., Misra, V., Rubenstein, D.: Using Graphic Turing Tests to Counter Automated DDoS Attacks Against Web Servers. In: Proc. of CCS, pp. 8–19 (2003)
14. Ratnasamy, S., Francis, P., Handley, M., Karp, R., Shenker, S.: A Scalable Content-Addressable Network. In: Proc. of SIGCOMM, pp. 161–172 (2001)
15. Sistla, A.P., Welch, J.L.: Efficient distributed recovery using message logging. In: Proceedings of the Eighth Annual ACM Symposium on Principles of Distributed Computing, PODC 1989, pp. 223–238. ACM, New York (1989)
16. Stoica, I., Morris, R., Liben-Nowell, D., Karger, D., Kaashoek, M.F., Dabek, F., Kalakrishnan, H.: Chord: A Scalable Peer-to-Peer Lookup Service for Internet Applications, Technical Report MIT (2002)

Checkpointing in Parallel
State-Machine Replication

Odorico M. Mendizabal[1,2], Parisa Jalili Marandi[3],
Fernando Luís Dotti[1], and Fernando Pedone[3]

[1] Pontifícia Universidade Católica do Rio Grande do Sul – PUCRS
Porto Alegre, Brazil
odoricomendizabal@furg.br, fernando.dotti@pucrs.br
[2] Universidade Federal do Rio Grande – FURG
Rio Grande, Brazil
[3] University of Lugano – USI
Lugano, Switzerland
{fernando.pedone,parisa.jalili.marandi}@usi.ch

Abstract. State-machine replication is a popular approach to building
fault-tolerant systems, which relies on the sequential execution of commands to guarantee strong consistency. Sequential execution, however,
threatens performance. Recently, several proposals have suggested par-
allelizing the execution model of the replicas to enhance state-machine
replication's performance. Despite their success in accomplishing high
performance, the implications of these models on checkpointing and recovery is mostly left unaddressed. In this paper, we focus on the checkpointing problem in the context of Parallel State-Machine Replication.
We propose two novel algorithms and assess them through simulation
and a real implementation.

Keywords: State-machine replication, checkpointing, fault tolerance.

1 Introduction

State-machine replication (SMR) is a well-established approach to implementing fault-tolerant services. Replicas in state-machine replication start in the same
initial state and execute an identical and ordered set of client commands sequentially and deterministically [10,16]. Therefore, all the replicas traverse the same
sequence of internal states and produce the same outputs. Consensus is often
used to ensure that commands are totally ordered across replicas [11].

Sequential execution of commands can be a performance bottleneck and a
waste of resources, in particular when replicas have access to multicore processors. To overcome this limitation, techniques that allow concurrent execution of
commands in state-machine replication have been proposed [8,9,12,13]. These
techniques are based on the observation that some commands are *independent*,
that is, they access disjoint portions of the replica's state or do not modify
shared parts of the state. Therefore, independent commands can be executed

M.K. Aguilera et al. (Eds.): OPODIS 2014, LNCS 8878, pp. 123–138, 2014.
© Springer International Publishing Switzerland 2014

concurrently without compromising the service's consistency. Dependent commands, however, those that modify shared parts of the state, must be executed sequentially, in the same order across replicas.

This paper focuses on checkpointing in Parallel State-Machine Replication (P-SMR) [12], a scalable multithreaded replication model, whose scalability stems from the absence of a centralized component in the execution path of independent commands (e.g., no local scheduler [9]). In P-SMR, replicas alternate between the execution of concurrent commands (i.e., those mutually independent) and the execution of sequential commands. Recovering a failed replica in classic SMR requires retrieving the commands the replica executed but "forgot" due to the failure and the commands the replica missed while it was down. To speed up recovery, replicas can periodically checkpoint their state against stable storage so that upon recovering, a replica can start with a state not too far behind the other replicas, after reading its local checkpoint from stable storage or retrieving a checkpoint from a remote operational replica. Performing checkpoints efficiently in P-SMR is more challenging than in classic SMR because the checkpoint operation must account for the execution of concurrent commands.

We propose two checkpoint techniques for P-SMR: coordinated and uncoordinated. The coordinated algorithm executes checkpoints when replicas are in sequential mode. The uncoordinated algorithm is more complex but can checkpoint a replica's state during both sequential and concurrent execution modes. The fundamental differences between the two approaches are three-fold: (a) With the coordinated mechanism, any two replicas save the same sequence of checkpoints throughout the execution; with uncoordinated checkpoints, replicas may save different states. Saving the same sequence of checkpoints has performance implications during recovery, as we explain in the paper. (b) Since an uncoordinated checkpoint can be started while a replica is executing commands concurrently, faster threads will be idle for shorter periods when waiting for slow threads in the uncoordinated technique than in the coordinated approach. (c) Coordinated checkpoints incur system-wide synchronization, while uncoordinated checkpoints are local to a replica. We discuss in the paper the implications of each technique using simulation models and an in-memory database service.

This paper makes the following contributions: (a) it discusses recovery of failed replicas in the context of parallel state-machine replication, a topic that has received little attention until now; (b) it proposes two checkpoint techniques for P-SMR, coordinated and uncoordinated, and compares their pros and cons; and (c) it assesses the performance of the two techniques using simulation models and an in-memory database service.

The remainder of this paper is organized as follows. In Section 2, we present the system model and assumptions. In Section 3, we recall parallel state-machine replication and provide a consensus-based algorithm that implements P-SMR. In Section 4, we discuss recovery in classic state-machine replication and introduce two checkpoint algorithms for P-SMR. We assess the performance of our proposed algorithms in Section 5 and relate them to the state of the art in Section 6. We conclude the paper in Section 7.

2   System Model and Assumptions

We assume a distributed system composed of interconnected processes. There is an unbounded set $C = \{c_1, c_2, \ldots\}$ of client processes and a bounded set $R = \{r_1, r_2, \ldots\}$ of replica processes. We do not make any assumptions about the relative speed of processes or message delays, i.e., the system is asynchronous.

We assume the *crash-recovery* model and exclude malicious or arbitrary behavior. A process can be either *up* or *down*, and it switches between these two modes when it fails (i.e., from up to down) and when it recovers (i.e., from down to up). Replicas are equipped with volatile memory and stable storage. Upon a crash, a replica loses the content of its volatile memory, but the content of its stable storage survives crashes.

Processes communicate by message passing, using either one-to-one or one-to-many communication. One-to-one communication is performed through primitives *send(m)* and *receive(m)*, where m is a message. If m's sender transmits m "enough times" and m's destination does not fail, then m is eventually received. One-to-many communication relies on the consensus abstraction, defined next.

The consensus problem can be described in terms of processes that propose values and processes that must agree upon a decided value. Consensus is defined by the primitives *propose(v)* and *decide(v)*, where v is an arbitrary value. A consensus protocol ensures the following safety requirements: (i) any value decided must have been proposed; (ii) a process can decide at most one value; and (iii) two different processes cannot decide different values. Solving consensus requires additional assumptions about the system model [6]. In this paper, we simply assume consensus can be solved without explicitly extending the model with these assumptions (e.g., [3,4]).

State-machine replication can be implemented with a sequence of consensus *rounds*, where the i-th consensus round decides on the i-th command to be executed by the replicas. We identify the decision of the i-th consensus round as *decide(i,v)*. In order to simplify our algorithms, we modify the propose primitive above such that a value proposed by a non-faulty process is eventually decided in some consensus round.

Our consistency criterion is *linearizability*: a system is linearizable if there is a way to reorder the client commands in a sequence that (i) respects the semantics of the commands, as defined in their sequential specifications, and (ii) respects the real-time ordering of commands across all clients [1].

3   Parallel State-Machine Replication

In contrast to classic state-machine replication (SMR), where the execution of commands is sequential, in parallel state-machine replication (P-SMR) independent commands can be executed concurrently. To understand the interdependencies between commands, assume commands C_i and C_j, where W_i and W_j indicate the commands' writeset and R_i and R_j indicate their readset. According to [9], C_i and C_j are *dependent* if any of the following conditions hold:

(i) $W_i \cap W_j \neq \emptyset$, (ii) $W_i \cap R_j \neq \emptyset$, or (iii) $R_i \cap W_j \neq \emptyset$. In other words, if the writeset of a command intersects with the readset or the writeset of another command, the two commands are dependent. Two commands are independent if they are not dependent.

P-SMR parallelizes the *agreement* and the *execution* of commands. Instead of using a single sequence of consensus rounds to order commands as in SMR, P-SMR uses multiple sequences of consensus. More precisely, if there are $n + 1$ threads at each replica, $t_0, ..., t_n$, P-SMR requires $n + 1$ consensus sequences, $\gamma_0, ..., \gamma_n$, where thread t_0 (at each replica) participates in consensus sequence γ_0 only, and thread t_i, $0 < i \leq n$, participates in consensus sequences γ_0 and γ_i. To ensure that t_i handles commands in the same order across replicas, despite participating in two consensus sequences, t_i orders messages from its two consensus sequences using a deterministic merge procedure (e.g., handling decisions for the sequences in round-robin fashion). To ensure progress, every consensus sequence must have a never-ending stream of consensus rounds, which can be achieved by having one or more processes proposing *nil* values if no value is proposed in a consensus sequence after some time [14]. Obviously, replicas discard *nil* values decided in a consensus round.

P-SMR ensures two important invariants. First, commands decided in consensus sequence γ_0 are serialized with any other commands at a replica and executed by thread t_0 in the same order across replicas (*sequential execution mode*). Second, commands decided in the same round in consensus sequences $\gamma_1, ..., \gamma_n$ are executed by threads $t_1, ..., t_n$ concurrently at a replica (*concurrent execution mode*).

Clients propose a command by choosing the consensus sequence that guarantees ordered execution of dependent commands while maximizing parallelism of independent commands. The mapping of commands onto consensus sequences is application dependent. In the following, we illustrate two such mappings.

- (Concurrent reads and sequential writes.) Commands that read the replica's state are proposed in any arbitrary consensus sequence γ_i, $0 < i \leq n$; commands that modify the replica's state are proposed in sequence γ_0.

- (Concurrent reads and writes.) Divide the service's state into disjoint partitions $P_1, ..., P_n$ so that commands that access partition P_i only are proposed in γ_i and commands that access multiple partitions are proposed in γ_0.

Clients must be aware of the mapping of commands onto consensus sequences and must be able to identify commands that read the service's state only or modify the state, in the first case above, or to identify commands that access a single partition (and which partition) or multiple partitions, in the second case. Algorithm 1 presents P-SMR in detail. For each thread t_i, $round[i]$ (line 3) indicates the number of the next consensus round to be handled (or being handled) by t_i, for all consensus sequences involving t_i. Threads use semaphores $S[0..n]$ (line 4) to alternate between sequential and concurrent modes and, as shown in the next section, to create a checkpoint. Variable $next[i]$ (line 5) determines whether t_i is in *sequential* or *concurrent* mode.

Thread t_0 tracks decisions in consensus sequence γ_0 only (line 8). The "decided $[\gamma_0](r, \langle - \rangle)$ and $r = round[0]$" condition holds when there is a decision in consensus sequence γ_0 that matches $round[0]$. If the value decided in $round[0]$ is a command (line 9), t_0 waits for every other thread t_i (line 10) and then handles the request (line 11). After the command is executed, t_0 signals the other threads to continue their execution (line 12). Whatever value is decided in the round, $round[0]$ is incremented (line 13). Note that a *nil* decision in consensus sequence γ_0 does not cause threads to synchronize.

Algorithm 1: P-SMR

1: *Initialization:*
2:   **for** $i : 0..n$ **do**   {*for each thread t_i:*}
3:     $round[i] \leftarrow 1$ {*all threads start in the same round*}
4:     $S[i] \leftarrow 0$ {*semaphore used to implement barriers*}
5:     $next[i] \leftarrow$ SQ   {*start in sequential mode*}
6:   start threads $t_0, ..., t_n$

7: *Thread t_0 at a replica executes as follows:*
8:   **upon** decided $[\gamma_0](r, \langle cid, cmd \rangle)$ **and** $r = round[0]$
9:     **if** $cmd \neq nil$ **then**   {*if cmd is a real command...*}
10:      **for** $i : 1..n$ **do** wait($S[0]$)   {*barrier: wait for threads $t_1, .., t_n$*}
11:      execute cmd and reply to cid {*execute command and reply to client*}
12:      **for** $i : 1..n$ **do** signal($S[i]$)   {*let threads $t_1, .., t_n$ continue*}
13:      $round[0] \leftarrow round[0] + 1$ {*pass to the next round*}

14: *Thread t_i in $t_1, ..., t_n$ at a replica executes as follows:*
15:   **upon** decided $[\gamma_0](r, \langle cid, cmd \rangle)$ **and** $r = round[i]$ **and** $next[i] =$ SQ
16:     **if** $cmd \neq nil$ **then**   {*if decided on a real command...*}
17:       signal($S[0]$)   {*barrier: signal semaphore $S[0]$ (see line 10)*}
18:       wait($S[i]$)   {*...and wait to continue (see line 12)*}
19:       $next[i] \leftarrow$ CC   {*set execution mode as concurrent*}

20:   **upon** decided $[\gamma_i](r, \langle cid, cmd \rangle)$ **and** $r = round[i]$ **and** $next[i] =$ CC
21:     **if** $cmd \neq nil$ **then**   {*if decided on a command...*}
22:       execute cmd and reply to cid {*execute command and reply to client*}
23:       $next[i] \leftarrow$ SQ   {*set execution mode as sequential*}
24:       $round[i] \leftarrow round[i] + 1$ {*pass to the next round*}

Each thread t_i, $0 < i \leq n$, alternates between executing in sequential and concurrent modes (lines 15 and 20). If t_i decides a value in consensus sequence γ_0 for its current round and the current execution mode is sequential (line 15), t_i checks whether the command is not *nil* (line 16) and in such a case t_i signals thread t_0 (line 17) and waits for t_0 to continue (line 18). Thread t_i then sets $next[i]$ to CC (line 19), meaning that it is in concurrent mode now. When t_i decides a value in consensus sequence γ_i for round $round[i]$ and $next[i] =$ CC (line 20), t_i executes the command if it is not *nil* (lines 21–22), sets the execution mode as sequential (line 23), and passes to the next round (line 24).

4   Checkpointing in P-SMR

Recovery in classic SMR is conceptually simple: Replicas log commands before executing them (e.g., as part of consensus) and periodically (e.g., after k commands) save the application state or the changes made since the last recorded checkpoint in stable storage. When a replica recovers from a failure, it retrieves a checkpoint from its local storage or from a remote replica and resumes operation after installing this checkpoint. The recovering replica also needs to recover the value decided in consensus rounds not included in the installed checkpoint.

Checkpoints speed up recovery and save storage space. Checkpoints shorten recovery time since a recovering replica does not need to start with an empty state and (re-)execute every decided command to catch up with the other replicas. Checkpoints save storage space since commands decided in "old" consensus rounds can be garbage collected. A sufficient condition to remove data related to the i-th consensus round is that all replicas have recorded a checkpoint containing the effects of the command decided in the i-th round.

We propose next two novel checkpointing algorithms for P-SMR. In the first algorithm, coordinated checkpointing, replicas must converge to a common state before taking a checkpoint; in the second algorithm, uncoordinated checkpointing, replicas take checkpoints independently and may not be in an identical state when the checkpoint takes place. We conclude the section with a comparison between the two algorithms.

4.1   Coordinated Checkpointing

The idea behind our coordinated checkpointing algorithm is to force replicas to undergo the same sequence of checkpointed states. To this end, we define a checkpoint command CHK that depends on all other commands. Therefore, CHK is executed in sequential mode in P-SMR and ordered by consensus sequence γ_0. Since replicas implement a deterministic strategy to merge consensus sequences, command CHK is guaranteed to be executed after each replica reaches a certain common state.

Algorithm 2 presents the coordinated checkpoint algorithm in detail. When a replica recovers from a failure (line 1), it first retrieves the latest checkpoint stored at the replica or requests one from a remote replica (line 2). Tuple $\langle last\_rnd[0] \rangle$ identifies the retrieved checkpoint. (Every replica stores an initialization checkpoint, empty and identified by $\langle 1 \rangle$.) The replica then initializes variables S, $round$ and $next$ (lines 3–10) and starts all threads (line 11).

Thread t_0's only difference with respect to Algorithm 1 is that it must check whether a decided command is a checkpoint request (line 16), in which case t_0 stores the replica's state on stable storage and identifies the checkpoint as $\langle round[0] \rangle$ (line 17). Threads $t_1, ..., t_n$ execute the same pseudocode in Algorithms 1 and 2.

Algorithm 2: Coordinated checkpoint

1: **upon** starting **or** recovering from a failure
2:   retrieve latest/remote checkpoint, which has id $\langle last\_rnd[0]\rangle$
3:   **for** $i : 0..n$ **do**   {*for each thread t_i...*}
4:     $S[i] \leftarrow 0$ {*semaphore used to implement barriers*}
5:     **if** $i = 0$ **then**   {*thread t_0...*}
6:       $round[i] \leftarrow last\_rnd[0] + 1$ {*goes to the next round in...*}
7:       $next[i] \leftarrow$ SQ   {*...sequential mode*}
8:     **else**   {*threads $t_1, ..., t_n$...*}
9:       $round[i] \leftarrow last\_rnd[0]$ {*stay in this round in...*}
10:      $next[i] \leftarrow$ CC   {*...concurrent mode*}
11: start threads $t_0, ..., t_n$

12: *Thread t_0 at a replica executes as follows:*
13: **upon** decided $[\gamma_0](r, \langle cid, cmd\rangle)$ **and** $r = round[0]$
14:    **if** $cmd \neq nil$ **then**   {*if cmd is a command/checkpoint request...*}
15:      **for** $i : 1..n$ **do** wait($S[0]$)   {*barrier: wait n times on semaphore*}
16:      **if** $cmd = CHK$ **then**   {*if cmd is a checkpoint request...*}
17:        store checkpoint with id $\langle round[0]\rangle$ {*take checkpoint*}
18:      **else**   {*else...*}
19:        execute cmd and reply to cid {*execute command and reply to client*}
20:      **for** $i : 1..n$ **do** signal($S[i]$)   {*let each thread t_i continue*}
21:      $round[0] \leftarrow round[0] + 1$ {*one more handled decision*}

22: **each** Δ time units **do**   {*ideally done by a single replica only:*}
23:    propose$[\gamma_0](\langle t_0, CHK\rangle)$ {*request a system-wide checkpoint*}

24: *Thread t_i in $t_1, ..., t_n$ at a replica executes as follows:*
25: **upon** decided $[\gamma_0](r, \langle cid, cmd\rangle)$ **and** $r = round[i]$ **and** $next[i] =$ SQ
26:    **if** $cmd \neq nil$ **then**   {*if cmd is a command/checkpoint request...*}
27:      signal($S[0]$)   {*implement barrier (see line 15)*}
28:      wait($S[i]$)   {*...and wait to continue (see line 20)*}
29:      $next[i] \leftarrow$ CC   {*set execution mode as concurrent*}

30: **upon** decided $[\gamma_i](r, \langle cid, cmd\rangle)$ **and** $r = round[i]$ **and** $next[i] =$ CC
31:    **if** $cmd \neq nil$ **then**   {*if cmd is an actual command...*}
32:      execute cmd and reply to cid {*execute command and reply to client*}
33:      $next[i] \leftarrow$ SQ   {*set execution mode as sequential*}
34:      $round[i] \leftarrow round[i] + 1$ {*one more handled decision*}

4.2   Uncoordinated Checkpointing

We now present an alternative algorithm that does not coordinate checkpoints across replicas: each replica decides locally when checkpoints will happen. Unlike the coordinated checkpointing algorithm, where all replicas record identical checkpoints, with the uncoordinated algorithm the checkpoints vary across the replicas.

The main difficulty with uncoordinated checkpoints is that a checkpoint request may be received any time during a thread's execution. Thus, one thread may receive a checkpoint request when in sequential execution mode while another thread receives the same request when in concurrent execution mode. Essentially, this happens because we do not order checkpoint requests with consensus decisions, as in the coordinated version of the algorithm.

In brief, our algorithm works as follows. First, thread t_0 requests a checkpoint by sending a local message to the other threads. Second, the handling of a checkpoint request at a replica does not change the sequence of commands executed by threads t_i, $0 < i \leq n$, which still alternate between sequential and concurrent execution modes in each round. To guarantee this property, when t_0 requests a checkpoint it tracks the signal it receives from t_i: If t_i signals t_0 upon receiving the checkpoint request, then after the checkpoint, t_0 releases t_i so that t_i can proceed with the next command. If t_i signals t_0 because it started the sequential execution mode, after the checkpoint t_0 keeps t_i waiting until t_0 also goes through the sequential execution of commands. In this case, when t_i later receives the checkpoint request, it simply discards it.

Algorithm 3 presents the uncoordinated checkpointing algorithm in detail. When a replica recovers from a failure, it retrieves the last saved checkpoint from its local storage or from a remote replica (line 2). This checkpoint identifies the round and the execution mode the thread must be in, after the checkpoint is installed (lines 4–5). (A replica is initialized with an empty checkpoint, identified as $\langle 2, \text{SQ}[, 1, \text{CC}]_{\times n} \rangle$.) Variable $last\_sync[i]$ contains the last round when t_i started in sequential mode and signaled t_0 (line 8); $waiting[i]$ tells whether upon executing a command t_0 must wait for t_i (line 9).

The execution of a sequential command by t_0 is similar in both the coordinated and uncoordinated algorithms, with the exception that t_0 only waits for t_i if it is not already in waiting mode (line 14); this happens if t_i signals t_0 because it started sequential execution mode but t_0 started a checkpoint. After the execution of the sequential command, all threads are released (lines 17–18). To execute a checkpoint, t_0 sends a message to all threads and waits for them (lines 21–23). If t_i signaled t_0 because it entered sequential mode in t_0's current round or some round ahead (line 26), which happens if the value decided in t_0's current round is nil, t_0 keeps track that t_i is waiting (line 27); otherwise t_0 signals t_i to continue (line 29).

The execution of commands for threads $t_1, ..., t_n$ is similar in both checkpoint algorithms, with the exception that before signaling the start of sequential execution mode, t_i sets $last\_sync[i]$ with its round number (line 33). Upon receiving a checkpoint request $\langle r, CHK \rangle$ that satisfies condition $last\_sync[i] < r \leq round[i]$ (line 42), t_i signals t_0 and waits for t_0's signal (lines 43–44). If $last\_sync[i] \geq r$, then it means that t_i has already signaled t_0 when entering sequential execution mode; thus, it does not do it again. If $r > round[i]$, then the checkpoint request is for a round ahead of t_i's current round. This request will be considered when t_i reaches round r.

Algorithm 3: Uncoordinated checkpoint

1: **upon** recovering from a failure
2:    retrieve checkpoint, which has id $\langle rnd[0], nxt[0], ..., rnd[n], nxt[n] \rangle$
3:    **for** $i : 0..n$ **do**   *{for each thread t_i, $0 \leq i \leq n$:}*
4:       $round[i] \leftarrow rnd[i]$ *{t_i's round and...}*
5:       $next[i] \leftarrow nxt[i]$ *{... execution mode when checkpoint taken}*
6:       $S[i] \leftarrow 0$ *{semaphore used to implement barriers}*
7:    **for** $i : 1..n$ **do**   *{for each thread t_i, $1 \leq i \leq n$:}*
8:       $last\_sync[i] \leftarrow 0$ *{last round t_i entered sequential mode}*
9:       $waiting[i] \leftarrow$ false   *{initially t_i isn't waiting}*
10:    start threads $t_0, ..., t_n$

11: *Thread t_0 at a replica executes as follows:*
12:    **upon** decided $[\gamma_0](r, \langle cid, cmd \rangle)$ **and** $r = round[0]$
13:       **if** $cmd \neq nil$ **then**   *{if decided on a command...}*
14:          **for** $i : 1..n$ **do if** $\neg waiting[i]$ **then** wait($S[0]$)   *{wait for each active t_i}*
15:          execute cmd and reply cid *{execute command and reply to client}*
16:          **for** $i : 1..n$ **do**
17:             $waiting[i] \leftarrow$ false   *{after sequential mode no thread waits}*
18:             signal($S[i]$)   *{ditto!}*
19:       $round[0] \leftarrow round[0] + 1$ *{t_0 passes to the next round}*

20:    **each** Δ time units **do**   *{t_0 periodically triggers a local checkpoint}*
21:       **for** $i : 1..n$ **do**
22:          send $\langle round[0], CHK \rangle$ to t_i *{send checkpoint request to t_i}*
23:          **if** $\neg waiting[i]$ **then** wait($S[0]$)   *{wait for each active thread t_i}*
24:       store checkpoint with id $\langle round[0], next[0], round[1], ... \rangle$ *{take checkpoint}*
25:       **for** $i : 1..n$ **do**   *{for each t_i}*
26:          **if** $last\_sync[i] \geq round[0]$ **then**   *{if t_i entered sequential mode...}*
27:             $waiting[i] \leftarrow$ true   *{keep t_i waiting until t_0 catches up}*
28:          **else**   *{else...}*
29:             signal($S[i]$)   *{let t_i proceed}*

30: *Thread t_i in $t_1, ..., t_n$ at a server executes as follows:*
31:    **upon** decided $[\gamma_0](r, \langle cid, cmd \rangle)$ **and** $r = round[i]$ **and** $next[i] = $ SQ
32:       **if** $cmd \neq nil$ **then**   *{if decided on a real command...}*
33:          $last\_sync[i] \leftarrow round[i]$ *{take note that entered sequential mode}*
34:          signal($S[0]$)   *{implement barrier}*
35:          wait($S[i]$)   *{...and wait to continue}*
36:       $next[i] \leftarrow $ CC   *{set execution mode as concurrent}*

37:    **upon** decided $[\gamma_i](r, \langle cid, cmd \rangle)$ **and** $r = round[i]$ **and** $next[i] = $ CC
38:       **if** $cmd \neq nil$ **then**   *{if cmd is an actual command...}*
39:          execute cmd and reply to cid *{execute command and reply to client}*
40:       $next[i] \leftarrow $ SQ   *{set execution mode as sequential}*
41:       $round[i] \leftarrow round[i] + 1$ *{pass to the next round}*

42:    **upon** receive $\langle r, CHK \rangle$ from t_0 **and** $last\_sync[i] < r \leq round[i]$
43:       signal($S[0]$)   *{checkpoints are done in mutual exclusion}*
44:       wait($S[i]$)   *{ditto!}*

4.3 Coordinated versus Uncoordinated Checkpointing

With coordinated checkpoints, a checkpoint only happens after each thread receives a *CHK* request and finishes executing all the commands decided before the request. With uncoordinated checkpoints, a checkpoint is triggered within a replica and is not ordered with commands. These mechanisms have important differences, as we discuss next.

First, with coordinated checkpoints every replica saves the same state upon taking the k-th checkpoint. Saving the same state across replicas is important for *collaborative state transfer* [2], a technique that improves performance by involving multiple operational replicas in the transferring of a saved checkpoint to the recovering replica, each replica sending part of the checkpointed state. Collaborative state transfer is not possible with uncoordinated checkpoints.

Second, coordinated checkpoints take place when replicas are in sequential execution mode; hence, no checkpoint contains a subset of commands executed concurrently. Uncoordinated checkpoints, however, can save states of a replica during concurrent execution mode. The implication on performance is that threads that execute commands more quickly when in concurrent mode do not have to wait for slower threads to catch up so that a checkpoint can be taken.

Third, the interval between the time when a checkpoint is triggered at a replica and the time when it takes place in the replica in the uncoordinated technique is lower than in the coordinated technique. In addition to requiring a consensus execution, which introduces some latency, a checkpoint request in the coordinated technique can only be handled after previously decided commands are executed at the replicas.

5 Performance Analysis

In this section, we assess the impact of the proposed approaches on the system performance by means of a simulation model and a prototype. Our simulations focus mostly on the cost of synchronization due to checkpointing. Aspects inherent to recovery (e.g., state transferring) are highly dependent on the application and sensitive to the data structures used by the service, the workload, and the size of checkpoints. We consider such aspects with our prototype, which implements an in-memory database with operations to read and write database entries. In our experiments, we generate sustained workloads with independent commands only. With this strategy we maximize the use of threads to execute commands, removing the possibility of thread idleness due to the synchronization needed by dependent commands.

5.1 Simulations

We implemented a discrete-event simulation model in C++ and configured each experiment to run until the 98% confidence interval of the command response time was a small fraction of the average value. We evaluated replicas without checkpointing enabled and with the two proposed checkpoint algorithms, and considered different classes of workload in terms of requests execution time: (i) fixed-duration commands (i.e., all commands take the same time to execute), (ii) uniformly distributed command duration, and (iii) exponentially distributed command duration. In the last case, a majority of commands have low execution times, while a small number of commands take long to execute.

We start by evaluating the scalability of both techniques. Figure 1 shows the maximum throughput achieved by a replica according to the number of threads, where each thread is associated with a processing unit (i.e., core). In these experiments, we used workloads (i), (ii), and (iii), described above, with average command execution time of 0.5 units. Checkpoints are taken every 200 time units, and the checkpoint duration is 0. By not considering the time taken to create a checkpoint, the results reveal the overhead caused exclusively by checkpoint synchronization. The throughput of P-SMR without checkpoints scales proportionally to the number of threads. The overhead of uncoordinated checkpointing is lower than the overhead of the coordinated technique and the difference between the two increases with the number of threads.

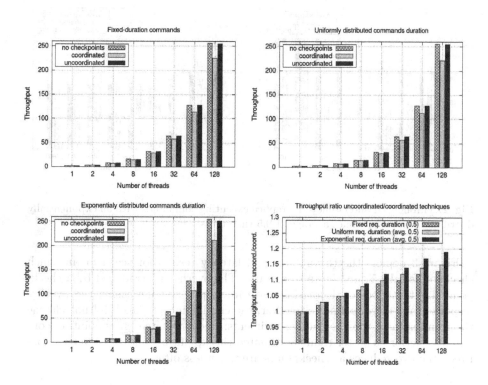

Fig. 1. Throughput of a replica with the number of threads for different commands execution duration workloads and the ratio of the two techniques with the number of threads

The bottom right graph of Figure 1 depicts the throughput ratio between the uncoordinated and the coordinated techniques under different workloads, as we increase the number of threads. Two facts stand out: First, uncoordinated checkpointing outperforms coordinated checkpointing in all scenarios and the difference increases with the number of threads. Second, the difference between the two techniques is more important when there is more variation in the command execution time. This happens because "faster threads" (i.e., those executing shorter commands) wait longer for "slow threads" during a checkpoint in the coordinated technique than in the uncoordinated approach.

Next, we evaluate the impact caused by the checkpoint frequency. Figure 2 shows the throughput and latency of replicas with 16 threads. In this experiment, the command duration follows the exponential distribution. The checkpointing interval varies from 12 to 1600 time units and the checkpointing duration is 0. The workload generated for this experiment reaches a throughput equivalent to 75% of the maximum. Although the uncoordinated checkpointing algorithm outperforms the coordinated algorithm in all configurations, the difference between the two decreases as checkpoints become more infrequent.

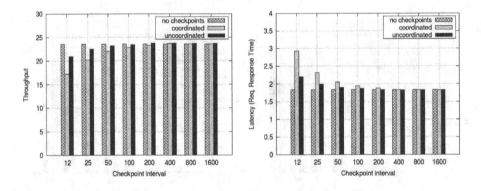

Fig. 2. Throughput and latency of a replica executing commands with an exponentially distributed execution time (average of 0.5 time units)

Figure 3 depicts the throughput and latency results for scenarios in which checkpoints take 5 time units to execute. The overhead introduced by a checkpoint has the effect of decreasing the throughput and increasing the average response time of commands. However, the checkpoint overhead did not change the trend seen in the previous experiments: uncoordinated checkpointing consistently performs better than coordinated checkpointing, and the difference between the two reduces as checkpoints are taken less often.

5.2   Implementation

We implemented consensus using Multi-Ring Paxos [14], where each consensus sequence is mapped to one Paxos instance. To achieve high performance, each

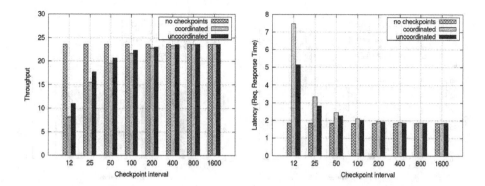

Fig. 3. Throughput and latency of a replica executing commands with an exponentially distributed execution time (average of 0.5 time units) and checkpoint duration of 5 time units

thread t_i decides several times on consensus sequence γ_i before deciding on sequence γ_0. Moreover, multiple commands proposed to a consensus sequence are batched by the group's coordinator (i.e., the coordinator in the corresponding Paxos instance) and order is established on batches of commands. Each batch has a maximum size of 8 Kbytes. The system was configured so that each Paxos instance uses 3 acceptors and can tolerate the failure of one acceptor.

The service is a simple in-memory database, implemented as a hash table, with operations to create, read, write, and remove entries. Each entry has an 8-byte key and an 8-byte value. A checkpoint duplicates the hash table in memory (using copy-on-write) and writes the duplicated structure to disk, either synchronously or asynchronously. We ran our experiment on a cluster with Dell PowerEdge R815 nodes equipped with four octa-core AMD Opteron processors and 128 GB of main memory (replicas), and Dell SC1435 nodes equipped with two dual-core AMD Opteron processors and 4 GB of main memory (Paxos's acceptors and clients). Each node is equipped with one 1Gb network interface. The nodes ran CentOS Linux 6.2 64-bit with kernel 2.6.32.

Figure 4 shows the throughput and the corresponding 90% percentile of the response time of both techniques. Checkpoints are taken once every 5 seconds and each one takes approximately 3.2 seconds to complete. When a checkpoint happens the database has approximately 10 million entries. The results show that uncoordinated checkpointing has a slight advantage over coordinated checkpoint in some of the configurations. Given the high rate of commands executed per second and the frequency of checkpoints, these results corroborate those presented in the previous section.

6   Related Work

In this section, we briefly review other models for parallel state-machine replication and compare their checkpointing mechanisms. A comprehensive survey of

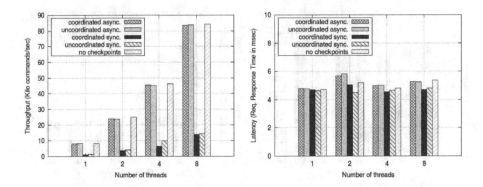

Fig. 4. Throughput and response time of coordinated and uncoordinated checkpointing with asynchronous and synchronous disk writes

checkpointing and recovery techniques for message-passing environments can be found in [5].

In [9], a parallelizer on each replica delivers commands in total order across replicas and distributes them among a set of threads for parallel execution. The parallelizer serializes the execution of dependent requests and ensures that their execution order follows the order decided by the agreement layer. This work also proposes a synchronization primitive executed on the replicas, but invoked by the agreement layer, to select a sequence number for checkpoints. Each replica blocks the execution of all the requests delivered after this sequence number until the checkpoint is completed. Since the selected sequence numbers may vary across the replicas, the recorded checkpoints are not identical across the replicas, similarly to our uncoordinated algorithm.

Eve [8] is a parallel replication model in which replicas first execute commands and then verify the equality of their states through a verification stage. Eve distinguishes one of the replicas as the primary to which clients send their requests. The primary groups commands into batches, assigns to each batch a unique sequence number, and transmits the batched commands to the other replicas. All the replicas, including the primary, are equipped with a deterministic mixer that converts a batch of requests into a set of parallel batches such that all the requests in a parallel batch can be executed in parallel. Once the execution of a batch terminates, replicas calculate a token based on their current state and send it to the verification stage. The verification stage checks the equality of the tokens. If the tokens are equal, replicas commit the executed batch, update their most recent stable sequence number, and respond to the clients. Otherwise, replicas must roll back the execution and re-execute the commands in the order determined by the primary. In Eve, checkpointing happens after the execution of each batch of commands and thus is more frequent than in traditional state-machine replication approach. Similar to a diverging replica, a recovering replica can request state changes from other replicas to build a consistent state.

In [7], the authors present Rex, a replication approach that benefits from multicore architectures. In Rex, one of the replicas, the primary, serves incoming client requests concurrently. There is no ordering requirements on delivering and executing client requests. The relative order among conflicting requests is given by synchronization primitives that coordinate access to shared data. The primary records causal dependencies among synchronization events during the execution and periodically proposes up-to-date traces to the secondary replicas, which reach a consensus on a sequence of traces to be executed. Secondary replicas faithfully follow the traces to replay the primary's execution. Checkpoints are used by Rex to allow a replica to recover from failures, to implement rollback on a downgrading replica (i.e., a leader replica that switches to a secondary role), and to facilitate garbage collection [7]. When a checkpoint is created, the primary pauses all threads before taking on any new requests and marks this particular point in the trace. A secondary receiving such a trace waits until the replay of the marked checkpoint and creates a snapshot. Once the checkpoint is created, the secondary replica resumes its execution and copies the checkpoint to other replicas in background. Since the checkpoint mark proposed by the primary establishes a consistent cut among replicas, the checkpoint generated is identical to each replica, similarly to our coordinated algorithm.

Another parallel replication technique is proposed in [15], where consensus is implemented to benefit from the multicore processors. Since the execution of commands is not parallelized on the replicas and follows the order decided by consensus, recovery can be implemented as in classic state-machine replication.

7 Conclusion

In this paper, we proposed two novel algorithms to address checkpointing in parallel state-machine replication [12]. The difference between our algorithms lies in the way checkpoint requests are synchronized with service commands. In coordinated checkpointing, checkpoints happen either before or after the execution of a batch of concurrent commands decided in a round. In uncoordinated checkpointing, a checkpoint can contain states of a replica in between two serialized commands. These two techniques have implications on performance, which increase in importance as the number of threads augments and checkpoints become more frequent.

Acknowledgments. We would like to thank the anonymous reviewers for their suggestions to improve the paper. This work was supported in part by CAPES PVE - Special Visiting Researcher - Grant No. 88881.062190/2014-01.

References

1. Attiya, H., Welch, J.: Distributed Computing: Fundamentals, Simulations, and Advanced Topics. Wiley-Interscience (2004)

2. Bessani, A., Santos, M., Felix, J., Neves, N., Correia, M.: On the efficiency of durable state machine replication. In: ATC (2001)
3. Chandra, T.D., Toueg, S.: Unreliable failure detectors for reliable distributed systems. Journal of the ACM 43(2), 225–267 (1996)
4. Dwork, C., Lynch, N., Stockmeyer, L.: Consensus in the presence of partial synchrony. Journal of the ACM (JACM) 35(2), 288–323 (1988)
5. Elnozahy, E.N.M., Alvisi, L., Wang, Y.M., Johnson, D.B.: A survey of rollback-recovery protocols in message-passing systems. ACM Comput. Surv. 34(3), 375–408 (2002), http://doi.acm.org/10.1145/568522.568525
6. Fischer, M.J., Lynch, N.A., Paterson, M.S.: Impossibility of distributed consensus with one faulty process. Journal of the ACM (JACM) 32(2), 374–382 (1985)
7. Guo, Z., Hong, C., Yang, M., Zhou, D., Zhou, L., Zhuang, L.: Rex: Replication at the speed of multi-core. In: Proceedings of the Ninth European Conference on Computer Systems, p. 11. ACM (2014)
8. Kapritsos, M., Wang, Y., Quema, V., Clement, A., Alvisi, L., Dahlin, M.: All about eve: execute-verify replication for multi-core servers. In: OSDI, pp. 237–250. USENIX Association (2012)
9. Kotla, R., Dahlin, M.: High throughput byzantine fault tolerance. In: DSN (2004)
10. Lamport, L.: Time, clocks, and the ordering of events in a distributed system. Communications of the ACM 21(7), 558–565 (1978)
11. Lamport, L.: The part-time parliament. ACM Transactions on Computer Systems (TOCS) 16(2), 133–169 (1998)
12. Marandi, P.J., Bezerra, C.E.B., Pedone, F.: Rethinking state-machine replication for parallelism. In: ICDCS (2013)
13. Marandi, P.J., Primi, M., Pedone, F.: High performance state-machine replication. In: DSN (2011)
14. Marandi, P.J., Primi, M., Pedone, F.: Multi-Ring Paxos. In: DSN (2012)
15. Santos, N., Schiper, A.: Achieving high-throughput state machine replication in multi-core systems. In: ICDCS (2013)
16. Schneider, F.B.: Implementing fault-tolerant services using the state machine approach: A tutorial. ACM Computing Surveys (CSUR) 22(4), 299–319 (1990)

Strong Equivalence Relations
for Iterated Models

Zohir Bouzid[1,*], Eli Gafni[2], and Petr Kuznetsov[1]

[1] Télécom ParisTech
[2] UCLA

Abstract. The Iterated Immediate Snapshot model (IIS), due to its elegant geometrical representation, has become standard for applying topological reasoning to distributed computing. Its modular structure makes it easier to analyze than the more realistic (non-iterated) read-write Atomic-Snapshot memory model (AS). It is known that AS and IIS are equivalent with respect to *wait-free task* computability: a distributed task is solvable in AS if and only if it is solvable in IIS. We observe, however, that this equivalence is not sufficient in order to explore solvability of tasks in *sub-AS* models (i.e. proper subsets of AS runs) or computability of *long-lived* objects, and a stronger equivalence relation is needed.

In this paper, we consider *adversarial* sub-AS and sub-IIS models specified by the sets of processes that can be *correct* in a model run. We show that AS and IIS are equivalent in a strong way: a (possibly long-lived) object is implementable in AS under a given adversary if and only if it is implementable in IIS under the same adversary. Therefore, the computability of any object in shared memory under an adversarial AS scheduler can be equivalently investigated in IIS.

1 Introduction

Iterated memory models (see a survey in [25]) proved to be a convenient tool to investigate and understand distributed computing. In an iterated model, every process passes, one by one, through a series of disjoint communication-closed memories M_1, M_2, ... Each memory M_i is a distinct set of shared memory locations that can only be accessed a bounded number of times. The most popular iterated model is the *Iterated Immediate Snapshot* model (IIS) [4]. Here a process accesses each memory with the *immediate snapshot* operation [3] that writes to the memory and returns a snapshot of the memory contents. Once memory M_k is accessed, a process never comes back to it. IIS has many advantages over the more realistic (non-iterated) read-write Atomic-Snapshot memory model (AS) [1]. Moreover, nice geometrical representation of IIS [21,24] makes it suitable for topological reasoning in analyzing algorithms and proving

---

* The work of this author was partially supported by the French ANR project Displexity.

M.K. Aguilera et al. (Eds.): OPODIS 2014, LNCS 8878, pp. 139–154, 2014.
© Springer International Publishing Switzerland 2014

their correctness. It is therefore natural to look for a generic trasformation that would map any problem in AS to an *equivalent* problem in IIS.

It has been shown by Borowski and Gafni [4] that the complete sets of runs of IIS and AS are, in a strict sense, *equivalent*: a distributed task is (wait-free) solvable in AS if and only if it is (wait-free) solvable in IIS. They established the result by presenting a *forward simulation* that, in every AS run, simulates an IIS run [3], and a *backward simulation* that, in every IIS run, simulates an AS run [4]. The equivalence turned out to be instrumental, e.g., in deriving the impossibility of wait-free set agreement [2,19]. More generally, the equivalence enables the topological characterization of task solvability in AS [19,15].

However, tasks can be seen as *one-shot* abstraction: a process is invoking a task at most once. As a result, any nonblocking [20] task solution is also wait-free, so the equivalence established in [3,4] allows for equating AS and IIS in terms of (wait-free) task solvability. However, for *long-lived* objects, this equivalence turns out to be insufficient. The goal of this paper is to establish a stronger one using elaborate model simulations.

Another motivation for a stronger equivalence between AS and IIS is the question of task solvability in *sub-AS models* (i.e., proper subsets of AS runs). We focus on *adversarial* sub-AS models [5,22], specified by sets of processes that can be *correct* in a model run. Note that the original AS model is described by the adversary consisting of *all* non-empty sets of processes. Since the introduction of adversaries in [5], the models have become popular for investigating task computability [10,11,17]. But how to define an IIS "equivalent" for an adversarial sub-AS model?

In IIS, a correct yet "slow" process may be never noticed by other processes: a process may go through infinitely many memories $M_1, M_2, \ldots$ without appearing in the snapshots of any other process. Instead, we specify adversarial *sub-IIS* models using the sets of *strongly correct* processes [27] (originally referred to as *fast* processes [7]). Informally, a process is strongly correct in an IIS run if it belongs to the largest set of processes that "see" each other infinitely often in the run. More precisely, we match the run with a directed graph whose vertices are processes, and there is an edge from i to j if and only if j sees i infinitely often in the run. Now the processes that constitute the largest strongly connected component in the graph are called strongly correct (we show that there is exactly one such component in the graph).

A topological characterization of task computability in sub-IIS models has been recently derived [12]: given a task T and an sub-IIS model M, solvability of T in M is equated with the existence of a specific continuous map between geometrical structures modelling inputs and outputs of the task. The characterization of [12] extends the celebrated asynchronous computability theorem [19] to sub-IIS models, and may provide deeper insights about task (in)solvability in sub-IIS than conventional *operational* reasoning can give. But is sub-IIS characterization of [12] relevant for (more realistic) adversarial sub-AS models?

In this paper, we show that the answer is "yes". We prove that sub-models of IIS and AS that are governed by the same adversary are equivalent in a strong sense: An object is implementable in AS under a given adversary if and only if it is implementable in IIS under the same adversary. This holds regardless of whether the object is one-shot, like a distributed task, or long-lived, like a queue or a counter. To achieve this result, we present a two-way simulation protocol that provides an equivalent sub-IIS model for any sub-AS model and which guarantees that the set of correct processes in an AS run coincides with the set of strongly correct processes in the simulated IIS run, and vice versa:

- We propose an "AS to IIS" simulation which ensures that a process appears strongly correct in the simulated IIS run if and only if it is correct in the orignal AS run. Every correct (in AS) process is "noticed" infinitely often by other correct (in AS) processes in the simulated IIS run, even if the process is much slower than the others. To this goal, we simulate IIS steps with the RAP (Resolver Agreement Protocol) [11] and employ a "fair" simulation strategy—at each point, we first try to promote the most "left behind" process in the currently simulated run. Even if the RAP-based simulation "blocks" because of a disagreement between the simulators (unavoidable in asynchronous fault-prone systems [6]), we guarantee that the blocked process is eventually noticed by more advanced simulated processes.
- To obtain our "IIS to AS" simulation, we extend the multiple-shot IS simulations [3] with a "helping" mechanism, reminiscent to the one employed in the atomic-snapshot simulation [1]. Here even if a process i is not able to complete its simulated read, it may adopt the snapshot published by a concurrent process j, under the condition that j has seen the most recent write of i. Since every move by a strongly correct process is eventually seen by every other strongly correct process, we derive the desired property that every strongly correct process makes progress in the simulated run.

Equating the set of correct processes with the set of strongly correct processes in the iterated simulated run is illuminating, because it equips any adversarial model [5,22] with an iterated equivalent. Our simulations also preserve the set of processes considered to be participating in the original run, which implies that the recent topological characterization of task computability in sub-IIS models [12] can be applied to sub-AS models too.

An important property of our simulation algorithms is that they are model-independent, i.e., they deliver the promised guarantees without making any assumptions on the model runs. The algorithms are therefore *wait-free* in the sense that they do not involve any form of waiting based using the assumptions about the sub-AS (or sub-IIS) model they are used in.

Roadmap. Section 2 relates our results to earlier work. Our model definitions, including the discussion of the AS and IIS models, the definition of strongly connected processes in IIS, and the definition of a simulation, are given in Section 3. Sections 4 and 5 present our two-way simulation.

2 Related Work

The IIS model introduced by Borowsky and Gafni [4] has become standard in topological reasoning about distributed computing [18,2,4,19,15]. The IIS model is precisely captured by the standard chromatic subdivision of the input complex [24,21], and thus enables intuitive and elegant reasoning about its computability power, in particular, distinguishing solvable and unsolvable. The IIS model is equivalent to the classical read-write model with respect to (wait-free) task solvability [3,4,13,27].

On the one hand, Borowsky and Gafni [3] have shown that one round of IIS can be implemented wait-free in AS, thus establishing a wait-free simulation of multi-round IIS. But the simulation only ensures that *one* correct process appears as strongly correct in the IIS run. Our algorithm ensures that *every* correct processes appear as strongly correct in the simulation.

On the other hand, IIS can simulate AS in the non-blocking manner, i.e., making sure that at least one process that participates in infinitely many rounds of IIS manages to simulate infinitely many steps of AS [4]. Later, Gafni and Rajsbaum [13] generalized the simulation of [4] to superset-closed adversaries (aslo called $\mathcal{L}$-resilient adversaries [11]). The simulation in [13] guarantees that at least one set in $\mathcal{L}$ will appear correct in the simulated execution. Raynal and Stainer [27] presented an extension of the simulation in [4] and sketched a proof that the set of correct processes in the simulated AS run is equal to the set of strongly correct processes in the "simulating" IIS run. In this paper, we propose an algorithm that achieves this property using the idea of the original atomic-snapshot implementation by Afek et al. [1], which we believe to be more intuitive and simpler to understand.

The relations between different simulation protocols are summarized in the following table (here SC denotes the set of strongly correct processes in an IIS run):

	correct(AS) $\subseteq$ SC(IIS)?	SC(IIS) $\subseteq$ correct(AS)?
From AS to IIS		
Borowsky and Gafni [3]	$\exists p \in$ correct(AS): $p \in$ SC(IIS)	✓
This paper	✓	✓
From IIS to AS		
Borowsky and Gafni [4]	✓	$\exists p \in$ SC(IIS): $p \in$ correct(AS)
Gafni and Rajsbaum [13]	✓	$\exists X \subseteq$ SC(IIS): $X \subseteq$ correct(AS), $X \neq \emptyset$
Raynal and Stainer [27]	✓	✓
This paper	✓	✓

The notion of a strongly correct process in IIS was introduced by Gafni in [7] (under the name of a *fast* process) and formally treated by Raynal and Stainer in [27,28]. The equivalence between adversarial restrictions of AS and IIS we establish in this paper motivated formulating a generalized topological characterization of task computability in sub-IIS [12].

Our AS-to-IIS simulation presented in Section 4 offers a novel use of the Resolver Agreement Protocol (RAP) proposed in [11], where a set of simulators try to maintain the balance between the simulated processes by promoting the "most behind" process that is not "blocked." Our IIS-to-AS simulation presented in Section 5 is based on the non-blocking simulation of [4], with the helping mechanism similar to the one used in the original atomic snapshot construction [1].

Herlihy and Rajsbaum [16] considered the problem of simulating task solutions in a variety of models, but their results only concern colorless tasks, which boils down to a very restricted notion of simulation. Rajsbaum et al. [26] introduced the Iterated *Restricted* Immediate Snapshot (*IRIS*) framework, where the restriction is defined via a specific failure detector on the per-round basis.

3   Definitions

In this section, we recall how the standard read-write and IIS models are defined, discuss the notion of a strongly correct process in the IIS model, and explain what we mean by simulating one model in another.

Standard Shared-Memory Model. We consider a standard atomic-snapshot model (AS) in which a collection $\Pi = \{1, \ldots, n\}$ of processes communicate via atomically updating their distinct registers in the memory and taking atomic snapshots of the memory contents. AS is equivalent to the standard read-write shared-memory model [1]. Without loss of generality, we assume that every process writes its input value in the first step and then alternates taking snapshots with updating its register with the result of it latest snapshot. This is known as a *full-information* protocol. We say that a process *participates* in a run E if it performs at least one update operation. Let $part(E)$ denote the set of participating processes in E. A process i is *correct* in E if i takes infinitely many steps in E. Let $correct(E)$ denote the set of correct processes in E.

IIS Model. In the IIS memory model, each process is supposed to go through a series of independent memories M_1, M_2, ... Each memory is accessed by a process with a single *immediate snapshot* (IS) operation [3]. Informally, in IS, processes write their values and then take atomic snapshots, so that the execution can be represented as a sequence of blocks where, in each block, a distinct set of processes first write and then take identical snapshots.

A run E in IIS is a sequence of non-empty sets of processes $S_1 \supseteq S_2 \supseteq \ldots$, with each $S_r \subseteq \{1, \ldots, n\}$ consisting of the processes that participate in the rth iteration of immediate snapshot (IS). Furthermore, each S_r is equipped with an ordered partition: $S_r = S_r^1 \cup \cdots \cup S_r^{n_r}$ (for some $n_r \leq n$), corresponding to the order in which processes are invoked in the respective IS.

Fix a run $E = S_1, S_2, \ldots$. The processes $i \in S_1$ are called *participating*. If j appears in all the sets S_k, we say that j is *infinitely participating* in E. The sets of participating and infinitely participating processes in a run E are denoted $part(E)$ and $\infty\text{-}part(E)$, respectively.

If $i \in S_r$ (i *participates in round* r), let V_{ir} denote the set of processes appearing in i's r-th *snapshot* in E, defined as the union of all sets in the partition of S_r preceding and including $S_r^m \subseteq S_r$ such that $i \in S_r^m$: $V_{ir} = S_r^1 \cup \cdots \cup S_r^m$. It is immediate that for all processes i, j and rounds r, such that i and j participate in r, the following properties are satisfied [3]: (self-inclusion) $i \in V_{ir}$; (containment) $V_{ir} \subseteq V_{jr} \vee V_{jr} \subseteq V_{ir}$; and (immediacy) $i \in V_{jr} \Rightarrow V_{ir} \subseteq V_{jr}$.

Our definitions can be interpreted operationally as follows. S_r is the set of processes accessing memory M_r, and each S_r^j is the set of processes obtaining the same *snapshot* after accessing M_r. Recall that in IS, the view of a process $i \in S_r^j$ is defined by the values written by the processes in $S_r^1 \cup \cdots \cup S_r^j$.

Strongly Correct Processes. It is convenient then to define, for each round r of E, a directed graph G_E^r with processes that participate in r as nodes and a directed edge from i to j if $j \in V_{ir}$. $G_E^{(r)}$ is then the union of the graphs $G_E^r, G_E^{r+1}, \ldots$

We say that *process i is aware of round r of process j* in an IIS execution E if there exists a path from i to j in $G_E^{(r)}$. The *participating set of a process i in a run E*, denoted by $part(E, i)$, is the set of processes j, such that i is aware of the first round of j in E.

A process is *strongly correct* in E if every process in $\infty\text{-}part(E)$ is aware of each of i's round in E. Let $SC(E)$ denote the set of strongly correct processes in E. Intuitively, $SC(E)$ is the largest set of processes that "see" each other (appear in each other's views) infinitely often in E.

Formally, denote by G_E^* the graph limit $\lim_{r \to \infty} G_E^{(r)}$. That is, i is a vertex of G_E^* if it is in $\infty\text{-}part(E)$ and (i, j) is an edge of G_E^* if E contains infinitely many rounds r such that $j \in V_{ir}$, i.e., i is aware of infinitely many rounds of j. By the containment property of IIS snapshots, in every round r, either $i \in V_{jr}$ or $j \in V_{ir}$. Hence, for all $i, j \in \infty\text{-}part(E)$, we are guaranteed that G_E^* contains at least one of the edges (i, j) and (j, i). Therefore, G_E^* has a single strongly connected component. It is then immediate that $SC(E)$ is exactly the set of processes in the strongly connected component of G_E^*.

In the following, we prove a property about strongly correct processes:

Proposition 1. *For all E in IIS, $i \in SC(E)$ iff there exists r_0, such that for all $r \geq r_0$, $G_E^{(r)}$ contains a path between every process in V_{ir} and i.*

Proof. $\Rightarrow$ Let $i \in SC(E)$. Since $SC(E) \subseteq \infty\text{-}part(E)$, i belongs also to $\infty\text{-}part(E)$. Take r_0 as the first round such that for all $r \geq r_0$: $V_{ir} \subseteq \infty\text{-}part(E)$. r_0 is well defined since the processes not belonging to $\infty\text{-}part(E)$ can appear only finitely often in the snapshots of i.

Since i is strongly correct, for all $r \geq r_0$, $G_E^{(r)}$ contains a path between every process $j \in \infty\text{-}part(E)$ and i. But as $r \geq r_0$, $V_{ir} \subseteq \infty\text{-}part(E)$. Hence, $G_E^{(r)}$ contains a path between every process of V_{ir} and i.

$\Leftarrow$ Let i be a process and r_0 a round such that for all $r \geq r_0$, $G_E^{(r)}$ contains a path between every process in V_{ir} and i. We need to prove $i \in SC(E)$.

Note that the containment property of IS snapshots guarantees that every process $j \in \infty\text{-}part(E) \setminus V_{ir}$ obtains a snapshot that contains V_{ir}. That is, $(j, i) \in G_E^r$ and hence $(j, i) \in G_E^{(r)}$.

Thus, we conclude that $G_E^{(r)}$ contains a path between every process in $\infty\text{-}part(E)$ and i.

Since there are infinitely many such rounds r and the number of possible paths is bounded, it follows that G_E^* contains a path between every process in $\infty\text{-}part(E)$ and i. Consequently, $i \in SC(E)$.

Model Simulations. WLOG, we consider runs of AS or IIS in which every process alternates writes with taking snapshots of (iterated or non-iterated) memory, using the result of its latest snapshot (or its input value initially) as the value to write. A set of such runs is called a *model*. Notice that the writes do not return any meaningful response, just an indication that the operation is complete. Thus, the state evolution of a process i in such a run E is characterized by the sequence $V_{i,1}^E, V_{i,2}^E, \ldots\ldots$ of the snapshots i takes in E.

By simulation of a run of a model B in another model A, we mean a distributed algorithm that in every run of A outputs at every process a sequence of snapshots so that all these sequences are consistent with some run of B and, moreover, reflect the inputs and the participating set of A. The latter intuitively filters out any "fake" simulation that produces a run of B that has nothing to do with the original run of A.

Formally, in every run E of A, a simulation $Sim_{A,B}$ outputs, at every simulator $i \in \{1, \ldots, n\}$ a (finite or infinite) sequence of snapshot values $U_{i,1}, U_{i,2}, \ldots\ldots$ There exists a run E' of B such that:

- For all i, $V_{i,1}^{E'}, V_{i,2}^{E'}, \ldots$ is exactly $U_{i,1}, U_{i,2}, \ldots\ldots$;
- for every $i \in correct(E)$ (resp., $SC(E)$ if A is an IIS model), $part(E, i) = part(E', i)$.

For the sake of brevity, we assume that in the simulated algorithm, as its local state, each process i simply maintains a vector storing the numbers of snapshots taken by other processes i is aware of so far. The process writes the vector as its current state in write operation. Each time a new snapshot is taken, the process updates its vector and simply increments its number of steps in it. Initially, the vector of process i stores 1 in position i and 0 at every other position. The reader can easily convince herself that this simplification does not bring a loss of generality, i.e., provided a simulation for such an algorithm, we can derive a simulation for the full-information algorithm.

4   From AS to IIS: Resolving and Bringing to the Front

The goal of this section is to provide an algorithm $AS \rightarrow IIS$ that simulates an execution of an IIS model where the set of processes that appear strongly correct coincides with the set of correct processes in the original AS (Algorithm 1).

Overview. To simulate a round of IIS, we use the original implementation of (one-shot) IS [3] using AS. To ensure fairness of the simulation, each process tries to advance the process that is currently the most behind.

Recall that the original IS construction [3] involves n "levels" of recursion. Processes start from level n and proceed down to level 1. In level ℓ, a process registers its participation and then takes an atomic snapshot. If the size of the snapshot is less than ℓ, then the process recursively proceeds to level $\ell - 1$, otherwise it returns the snapshot as its output in the IS simulation. Since at most n processes start at level n and at least one process (the one that writes the last) drops the simulation at each level, at most ℓ processes can reach any level ℓ. Thus, in the worst case, a process returns after reaching level 1.

In our $AS \rightarrow IIS$ algorithm, in order to promote the next step of a given process, the simulators use an *agreement protocol* [2,23] for each level of the IS simulation [3]. More precisely, to simulate the atomic snapshot obtained by the process in level ℓ, the simulator takes an atomic snapshot itself and computes the set of other simulated processes that also reached level ℓ. If the cardinality of the set is exactly ℓ, then the simulator proposes 1 to the agreement algorithm. Otherwise, it proposes 0. If the agreement protocol returns 1, then the simulated process completes the IS iteration by outputting the set of ℓ processes in level ℓ. If the agreement protocol returns 0, the process gets down to level $\ell - 1$ in the current IS iteration.

To make sure that the simulation is safe, we need to guarantee that the simulators agree on the outcome of each simulated step. For this we use the recently proposed *Resolver Agreement Protocol (RAP)* [11]. This protocol guarantees agreement (no two processes output different values) and validity (every output value was previously proposed). Moreover, if all proposed values are the same, then the algorithm terminates (this feature can be implemented using the *commit-adopt* (*CA*) algorithm [8]). Otherwise, if two different values are proposed, RAP may enter in the *blocked* state. The blocked state can be *resolved* by the process whose step is being simulated (i.e., the simulator with the same process id): the simulated process writes the value it adopted from CA in a dedicated register so that every correct process would eventually read the value and terminate.

Formally, RAP exports one operation $propose(v)$, $v \in \{0, 1\}$ that returns a value in $\{0, 1, \bot\}$ and is associated with a unique *resolver* process. Then the following guarantees are provided: (i) Every returned non-$\bot$ value is a proposed value; (ii) If all processes propose the same input value, then no process returns $\bot$; (iii) The resolver never returns $\bot$; (iv) No two different non-$\bot$ values are returned. Additionally, RAP guarantees that every process returns in a finite number of its steps.

```
1  Shared: R[1], ..., R[n] := [⊥, ..., ⊥], ..., [⊥, ..., ⊥];
2  Shared: Counter₁, ..., Counterₙ := 0, ..., 0;
3  Local: countf[1, ..., n] := [0, ..., 0] ;          // counters for ''frozen'' processes
4  Local: lastf[1, ..., n] := [0, ..., 0] ;      // last rounds in which processes were ''frozen''

5  R[i][i] := (run, 0, n) ;                     // start with highest level of the first iteration
6  while true do
7  │   Counterᵢ + +;
8  │   S := snapshot of R[1], ..., R[n] ;
9  │   if i is blocked in S then
10 │   │   p := i ;
11 │   else
12 │   │   for each j ∈ Π do
13 │   │   │   x := the largest round such that Vⱼₓ are aware of round x of j (in S) ;
14 │   │   │   if x > lastf[j] then
15 │   │   │   │   lastf[j] := x ;
16 │   │   │   │   countf[j] := Counterⱼ ;                                          // freeze j
17 │   │   │   end
18 │   │   end
19 │   │   repeat
20 │   │   │   cands := {j | j is not blocked and Counterⱼ > countf[j]} ;          // ignoring
       │   │       non-participants
21 │   │   │   Counterᵢ + +;
22 │   │   until cands ≠ ∅;
23 │   │   p := argmin_{j∈cands}(round-level(j, S), (j + round(j, S)) mod n) ;      // choose the
       │           ''most-behind'' process
24 │   end
25 │   (r, ℓ) := round-level(p, S) ;             // compute current round and level of p
26 │   U := {j | (*, r, ℓ) ∈ S[*][j]} ;          // all processes that reached (r, ℓ)
27 │   v := RAP_{p,r,ℓ}(|U| = ℓ) ;               // the result of next step of p
28 │   if v = 1 then
29 │   │   R[i][p] := S[i][p] · (run, r + 1, n) ;                    // p completes round r
30 │   │   V_{pr} := U ;                         // output the snapshot of p in round r
31 │   else
32 │   │   if v = 0 then
33 │   │   │   R[i][p] := S[i][p] · (run, r, ℓ - 1) ;                // p proceeds to (r, ℓ - 1)
34 │   │   else
35 │   │   │   R[i][p] := S[i][p] · (blocked, r, ℓ) ;                // p blocks in (r, ℓ)
36 │   │   end
37 │   end
38 end
```

Algorithm 1. The $AS \to IIS$ simulation algorithm: code for process i.

Operation. Algorithm 1 operates as follows. For every process i, the algorithm maintains a shared array $R[i][]$, written by i and read by all, that stores i's perspective on the current simulation. In particular, the sequence of iterations r and levels ℓ that a process j has passed through, *as witnessed by* i, is stored in $R[i][j]$.

The r-th view of process p in the simulated run is stored at a local variable V_{pr}. As we show below, in every execution of Algorithm 1, these views are evaluated identically by different simulators.

After taking a snapshot S in line 8 of the current simulated state, the *simulator* i first checks if the *simulated process* i is *blocked* (line 10). A process p is considered blocked if for every $S[j][p]$ that contains (d, r, ℓ) with $(r, \ell) = $ *round-level*(p, S), we have $d = $ *blocked*. (Here *round-level*(p, S) denotes the maximal round-level reached by p in S, i.e., the maximal value of the last element in $S[*][p]$, computed lexicographically.)

If simulated process i is blocked, simulator i retrieves the round-level (r, ℓ) at which it is blocked (line 25) and participates in $RAP_{i,r,\ell}$. We assign i to be the *resolver* of each RAP instance $RAP_{i,r,\ell}$, and thus the instance returns a non-$\bot$ which "unblocks" simulated process i.

If simulated process i is not blocked, simulator i checks if some process j has completed a *new* (not considered by i in previous rounds of the simulation) round r_j, such that all processes in V_{jr_j} are aware of round r_j of j (line 13). Every such process j is then *frozen* by i, i.e., j is put on hold and not simulated until simulator j performs a "physical" step (in lines 7 or 21).

In the set of remaining processes, the simulator chooses the "slowest" *non-blocked* and *non-frozen* process (line 23). To make sure that the notion of the slowest process is well-defined, we introduce a total order on the tuples (i, r, ℓ), $i \in \Pi$, $r \in \mathbb{N}$, $\ell \in \mathbb{N}_n$ as follows. We say that $(i, r_i, \ell_i) < (j, r_j, \ell_j)$ if $(r_i < r_j) \vee ((r_i = r_j) \wedge (\ell_i > \ell_j)) \vee ((r_i = r_j) \wedge (\ell_i = \ell_j) \wedge ((i + r_i) \bmod n < (j + r_i) \bmod n))$. This way *argmin* in line 23 returns a single process, ties are broken by choosing the process associated with the current iteration (the association is done in round-robin).

The slowest process p currently observed (by i) in round-level (r, ℓ) is then simulated using p's next instance of RAP, $RAP_{p,r,\ell}$, which accepts either 1 (exactly ℓ processes have appeared on round-level (r, ℓ) in S) or 0 (otherwise). If $RAP_{p,r,\ell}$ returns 1, p outputs the set of ℓ processes in (r, ℓ) as its snapshot in round r, denoted V_{pr}, and then p is promoted to round $r + 1$ (lines 29 and 30). If $RAP_{p,r,\ell}$ returns 0, p is promoted to level $\ell - 1$ of the same round r (line 33). Otherwise, if $RAP_{p,r,\ell}$ is blocked, we mark the status of i as *blocked* in (r, ℓ) (line 35).

Correctness Intuition. Our algorithm tries to always promote the process that is the "most left behind" process (that is not *blocked* or *frozen*) to the front of the simulation.

Observe that a simulated process i can only get blocked if two simulators proposed two different values to some $RAP_{i,r,\ell}$: one simulator finds exactly ℓ processes in (r, ℓ) and, thus, believes that i should complete round r by outputting the ℓ processes, and the other found strictly less processes in (r, ℓ) and thus believes that i should go one level down in round r and output a smaller snapshot. Therefore, a process i is blocked because another process appeared at its round-level (r, ℓ) and two simulators disagreed whether the other process was there or not: one simulator finds exactly ℓ processes at the level and the other— less than n. The last such "missed" process p will now be considered the slowest process in the simulation and, thus, will be chosen to be promoted in line 23 by any other simulator. Note that p cannot be blocked in (r, ℓ), because every simulator that found p in (r, ℓ) will also find exactly ℓ processes in (r, ℓ). This is because p is the last process to reach (r, ℓ). Moreover, p completes iteration r having i in its snapshot: since p completes r in level ℓ reached by i, p sees i in round r in the simulated run. By repeating this reasoning inductively, even though i is blocked, another process p carries this information to the "front" of the simulation, thus making sure that every other simulated process will eventually be aware of round r of i. Process i unblocks itself by completing its own $RAP_{i,r,\ell}$ an thus providing it with a non-$\perp$ output.

Now observe that a process can only get frozen if it produced a new snapshot in a round r and all the processes appearing in this snapshot became aware of it. By Proposition 1, strongly correct processes in the simulated run are frozen infinitely often. Therefore, only a correct simulator i may appear as strongly correct in the simulated run: otherwise the corresponding simulated process i would get frozen after simulator i crashes and stay frozen forever (only simulator i can "unfreeze" simulated process i in the simulation).

Thus, intuitively, a correct process i either gets blocked infinitely often or gets frozen infinitely often. In both cases, i is "seen" infinitely often by other correct processes. Moreover, a faulty process eventually either (i) gets faulty or frozen forever, or (ii) becomes invisible to the remaining processes in the simulated run. In both cases, the faulty process does not appear strongly correct in the simulation. Since process i starts the algorithm by registering its participation in round-level $(0, n)$ (line 5), the set of participants as witnessed by a strongly correct process i in the simulated IIS run is the set of participants in the original AS run. Thus:

Theorem 1. *Algorithm 1 provides a simulation of the IIS model in the AS model such that, for each AS run E, the simulated IIS run E' satisfies (1) $correct(E) = SC(E')$, and (2) $\forall i \in correct(E)$: $part(E) = part(E', i)$.*

Proof. Take any run E of Algorithm 1. The simulated run E' is defined as a collection of all sets V_{ir}, $i \in \{1, \ldots, n\}$, $r \in \mathbb{N}$ produced in E. By the correctness of the IS simulation [3] and the use of the RAP agreement protocol [11] for each atomic snapshot taken in the simulation of [3], we conclude that for all r, all sets V_{ir} satisfy containment, self-inclusion and immediacy (defined in Section 3). Notice that by the algorithm, every correct process i produces a snapshot V_{ir} in every iteration r.

Every Strongly Correct Process Is Correct. Assume, by contradiction, that $i \in SC(E')$ but $i \notin correct(E)$. Define r_0 to be the first simulated round of E' such that in all $r \geq r_0$, (i) only processes of $\infty\text{-}part(E')$ are simulated and (ii) V_{ir} contain only strongly correct processes. r_0 is well defined since the processes that appear infinitely often in the snapshots of strongly correct processes are necessarily also strongly correct.

Take a round $r \geq r_0$ where V_{ir} is simulated after the crash of i in E (recall that $i \notin correct(E)$). Since we assumed that i is strongly correct in E', all the processes in $\infty\text{-}part(E')$ (including V_{ir}) will eventually be aware of round r of i. But the fact that the processes in V_{ir} are strongly correct means that the processes of $\infty\text{-}part(E')$ are aware of infinitely many of their rounds. Therefore, every process in $\infty\text{-}part(E')$ eventually knows that the processes in V_{ir} were aware of a round of i. Hence, i will be frozen by all of them. Since i has crashed in E before its round r was simulated, it will never be unfrozen and, thus, takes no simulated steps after round r—a contradiction. Consequently, $i \in correct(E)$ and $SC(E') \subseteq correct(E)$.

Every Correct Process Is Strongly Correct. By Proposition 1, i is strongly correct in E' if and only if there exists a round such that for all later rounds r, the processes of V_{ir} are aware of round r of i. Hence, if i is not strongly correct in E', the condition in line 13 can apply to i only in a finite number of rounds. Thus, after a certain round r_0, a process can be frozen (line 15) only if it is strongly correct.

Now we show that $correct(E) \subseteq SC(E')$. Suppose not, i.e., there are processes $i, j \in correct(E)$ and a round $r \geq r_0$ such that i is never aware of round r of j in E'. Since $r \geq r_0$, j cannot be frozen by i. Let r_i be the round of process i at the moment when j completes round r, i.e., outputs V_{rj} (line 30). Let $|V_{rj}| = \ell$, i.e., j sees exactly ℓ process in round r.

Take r' to be the first round greater than r_i, such that $(i + r') \mod n+1 = n$, i.e., i has the lowest priority in round r'. Thus, before i is simulated at some level ℓ' of r', any other process that is not frozen or blocked must have completed its simulation of round r' or reached a level that is lower than ℓ'. We choose ℓ' to be the level at which i obtains its snapshot in r', and let m be some simulator that simulated $V_{ir'}$.

We observe first that $(r, \ell) < (r', \ell')$: otherwise, i will eventually reach level $\ell'' \geq \ell'$ of round r, find exactly ℓ'' processes (including j) at that level, and output its snapshot V_{ir} such that $j \in V_{ir}$—a contradiction with the assumption that i is never aware of round r of j.

Consider the time after i reaches (r', ℓ') and before it obtains the snapshot $V_{ir'}$. By the algorithm, the simulator m must choose the slowest non-blocked and non-frozen process to simulate. Suppose that j is never observed blocked by m after i reaches (r', ℓ'). Since j cannot be frozen by m after r_0, the algorithm guarantees that eventually, m would bring j to level (r', ℓ') and, thus, simulates a snapshot $V_{ir'}$ such that $j \in V_{ir'}$—a contradiction.

Now suppose that m observes j as blocked in round r or later. Without loss of generality, suppose that j is observed as blocked by m in round r. Indeed, if i is never aware of round r of j in, it is never aware of any later round of j.

We claim that at the moment when the first simulator took its snapshot on behalf of j for round r (in line 8), there was another blocked process k that reached (r, ℓ) and later was observed as resolved by another simulator. Indeed, the only reason for j to block in $RAP_{j,r,\ell}$ is that two simulators proposed conflicting sets of processes that have been observed to reach (r, ℓ). Moreover, by the algorithm, one of these sets contains exactly ℓ processes and the other contains strictly less. Consider any process in the difference between these two atomic snapshots. Every such process was considered blocked by one of the simulators at the moment it took its snapshot in line 8, otherwise it would appear in all obtained snapshots or would be chosen to be simulated as the slowest process. For the last such process s to reach level (r, ℓ), $RAP_{s,r,\ell}$ cannot get blocked, because all simulators will propose exactly ℓ processes that reached (r, ℓ). Thus, s obtains V_{sr} with $j \in V_{sr}$ and enters level n of round $r + 1$. Hence, by our assumption, i is never aware of round $r + 1$ of s (otherwise it would transitively get aware of round r of j). Moreover, by the arguments above, we have $(r+1, n) < (r', \ell')$. Therefore, s is not strongly correct and cannot be frozen as $r + 1 \geq r_0$. Moreover, s does not block in round r, thus m should eventually try simulating s in round $r + 1$.

By repeating the argument inductively, we locate a process k that reaches round r' and is aware of round r of i. Moreover, since i has the lowest priority in round r', every not yet blocked process should be simulated by m ahead of it. Thus, eventually some process that is aware of j in round r reaches (r', ℓ') and, hence, gets in $V_{ir'}$. Therefore, i must be aware of round r of j—a contradiction.

Finally, since every process starts the algorithm by registering its participation at level $(0, n)$ (line 5), the set of participating processes in E is automatically the participating set for every correct process in E'.

5   From IIS to AS: Identical Snapshots and Helping

We now describe our $IIS \rightarrow AS$ algorithm that, in any run of the IIS model, simulates a run of the AS model in which every process alternates updates with atomic snapshots [1].

As a basis, we take the non-blocking simulation proposed by Borowsky and Gafni [4]. In this algorithm, each process i maintains a local *counter* vector $C_i[1, \ldots, n]$ where each $C_i[j]$ stores the number of simulated snapshots of j *as currently witnessed by i*. To simulate a snapshot operation, process i accesses the iterated memories, writing its counter vector C_i, taking a snapshot of counter vectors of other processes, and updating each position $C_i[k]$ with the maximal value of $C_j[k]$ across all counter vectors read in the iteration. In each iteration r of the IIS memory, this is expressed as a single $WriteRead_r(C_i)$ operation the outcome of which satisfies the self-inclusion, containment, and immediacy properties specified in Section 3. If all these vectors are identical, i outputs the vector as the result of its next snapshot operation. Initially and each time a process i completes its next snapshot operation, it simulates an update operation by incrementing $C_i[i]$.

We first observe that the original simulation of the AS model proposed in [4] is, in the worst case, only non-blocking. Indeed, it admits runs in which some strongly correct process is never able to complete its snapshot operation, even though it is "noticed" infinitely often. Consider, for example, the following IIS run: $[\{1\}\{2,3\}]$, $[\{3\}, \{1,2\}]$, $[\{1\}\{2,3\}]$, $\ldots$, i.e., all the three processes are strongly correct and in every iteration, one of the processes in $\{1,3\}$ only sees itself and, thus, completes its new snapshot. Thus, in every round one of the processes in $\{1,3\}$ outputs a new snapshot, while the remaining process 2 sees two different vectors and thus does not complete its simulated snapshot. As a result, process 2 never manages to completes its first snapshot in the simulated AS run, even though it is strongly correct!

To fix this issue, we equip the algorithm of [4] with a helping mechanism, similar to the helping mechanism proposed in the atomic snapshot simulation in [1]. In addition to its counter vector, in each iteration of our Algorithm 2, a process also writes the result of its last snapshot: $WriteRead_r(C_i)$ (line 4).

```
1  C_i[1,...,n] := [0,...,0]; C_i[i] := 1; r := 0; SI_i := [0,...,0];
2  while true do
3  │   r ++;
4  │   S := WriteRead_r(C_i, SI_i);
5  │   if ∃SI such that (∀(C_j, SI_j) ∈ S : C_j = SI) or (∃(C_j, SI) ∈ S : SI[i] = C_i[i]) then
6  │   │   SI_i := SI;
7  │   │   output SI; // Output the next atomic snapshot
8  │   │   C_i[i] ++;
9  │   end
10 │   C_i := max(C_1,...,C_n) ; // Adopt the maximal counter value for each process j
11 end
```

Algorithm 2. The $IIS \rightarrow AS$ simulation algorithm: code for process i.

Now a process i outputs a new snapshot not only if it sees that everybody agrees on the clock vector, but also if another process produces a snapshot containing i's latest counter value.

Theorem 2. *Algorithm 2 provides a simulation of the AS model in the IIS model such that, for each run E in IIS, the simulated run E' satisfies (1) $SC(E) = correct(E')$ and (2)$\forall i \in SC(E)$: $part(E, i) = part(E')$.*

Proof. Consider any run E of Algorithm 2. First we observe that all atomic snapshots of the simulated processes output in E are all related by containment, i.e., for every two snapshot U and U' output in the algorithm in line 7, we have $U \leq U'$ or $U' \leq U$, when the two vectors are compared position-wise. Indeed, for every output snapshot U, there is a round r and a process i, such that all processes that appear in i's immediate snapshot in round r have put U as their clock vectors. Since in the algorithm the clock vector C_i is maintained to have the maximal value seen so far for every process j and by the containment property of immediate snapshot, every process that took the immediate snapshot in round r or later will compute a clock vector $U' \geq U$.

Therefore, we order all atomic snapshots output in E based on the containment order, let $U_1, U_2, \ldots$ be the resulting sequence (here $U_\ell \leq U_{\ell+1}$ for each $\ell = 1, 2, \ldots$). Then for each $\ell = 1, 2, \ldots$ and for each process i, $U_{\ell+1}[i] \neq U_\ell[i]$, we add an update operation in which i increments its counter (initially 1) and writes the result to position i in the memory just before $U_{\ell+1}$. Notice that since a process only increments its counter after it has output a snapshot, $U_{\ell+1}[i] \neq U_\ell[i]$ implies that $U_{\ell+1}[i] = U_\ell[i] + 1$.

We call the resulting sequence E' and observe that it is a run of the AS model. Indeed, the snapshots taken in E' are related by containment and, by construction, each snapshot returns the latest written value for each process. By construction, E and E' agree on the sequence of snapshots taken by every given process. Moreover, since the clock vector of process i contains the most up-to-date value for every other process and in the first step each process simply writes its initial clock vector in the memory, the set of participating processes as observed by i in E is the same as the set of participating processes observed by i in E'. Thus, E is an AS run, and Algorithm 2 simulates AS in IIS.

Every update of the counter of a strongly correct process i eventually appears in the snapshot of every other strongly correct process. Thus, every simulated snapshot of a strongly correct process eventually completes and $part(E, i) = part(E')$. If a process is not strongly correct, it eventually blocks in trying to complete its snapshot. Thus, $SC(E) = correct(E')$.

6 Conclusion

This paper presents two simulation algorithms that, taken together, maintain the equality between the set of *correct* processes in AS and the set of *strongly correct* processes in IIS. This equality enables a strong equivalence relation between AS and IIS *sub-models*: an object is implementable in an adversarial sub-AS model if and only if it is implementable in the corresponding adversarial sub-IIS model. The result holds regardless of whether the object is one-shot, like a distributed task, or long-lived, like a queue or a counter. (Naturally, in IIS, we guarantee liveness of object operations to the strongly correct processes only.) The equivalence presented in this paper motivates the recent topological characterization of task computability in sub-IIS models [12] and suggests further exploration of iterated models that capture, besides adversaries [5], the use of generic tasks like the Möbius task [14] or of a task from the family of 0-1 exclusion [9].

References

1. Afek, Y., Attiya, H., Dolev, D., Gafni, E., Merritt, M., Shavit, N.: Atomic snapshots of shared memory. J. ACM 40(4), 873–890 (1993)
2. Borowsky, E., Gafni, E.: Generalized FLP impossibility result for t-resilient asynchronous computations. In: STOC, pp. 91–100. ACM Press (May 1993)
3. Borowsky, E., Gafni, E.: Immediate atomic snapshots and fast renaming. In: PODC, pp. 41–51. ACM Press, New York (1993)
4. Borowsky, E., Gafni, E.: A simple algorithmically reasoned characterization of wait-free computation (extended abstract). In: PODC, pp. 189–198 (1997)
5. Delporte-Gallet, C., Fauconnier, H., Guerraoui, R., Tielmann, A.: The disagreement power of an adversary. Distributed Computing 24(3-4), 137–147 (2011)
6. Fischer, M.J., Lynch, N.A., Paterson, M.S.: Impossibility of distributed consensus with one faulty process. J. ACM 32(2), 374–382 (1985)
7. Gafni, E.: On the wait-free power of iterated-immediate-snapshots (1998), http://www.cs.ucla.edu/~eli/eli/wfiis.ps (unpublished manuscript)
8. Gafni, E.: Round-by-round fault detectors (extended abstract): Unifying synchrony and asynchrony. In: PODC, pp. 143–152 (1998)
9. Gafni, E.: The 0–1-exclusion families of tasks. In: Baker, T.P., Bui, A., Tixeuil, S. (eds.) OPODIS 2008. LNCS, vol. 5401, pp. 246–258. Springer, Heidelberg (2008)
10. Gafni, E., Kuznetsov, P.: Turning adversaries into friends: Simplified, made constructive, and extended. In: Lu, C., Masuzawa, T., Mosbah, M. (eds.) OPODIS 2010. LNCS, vol. 6490, pp. 380–394. Springer, Heidelberg (2010)
11. Gafni, E., Kuznetsov, P.: Relating $\mathcal{L}$-resilience and wait-freedom via hitting sets. In: Aguilera, M.K., Yu, H., Vaidya, N.H., Srinivasan, V., Choudhury, R.R. (eds.) ICDCN 2011. LNCS, vol. 6522, pp. 191–202. Springer, Heidelberg (2011)

12. Gafni, E., Kuznetsov, P., Manolescu, C.: A generalized asynchronous computability theorem. In: PODC (2014)
13. Gafni, E., Rajsbaum, S.: Distributed programming with tasks. In: Lu, C., Masuzawa, T., Mosbah, M. (eds.) OPODIS 2010. LNCS, vol. 6490, pp. 205–218. Springer, Heidelberg (2010)
14. Gafni, E., Rajsbaum, S., Herlihy, M.P.: Subconsensus tasks: Renaming is weaker than set agreement. In: Dolev, S. (ed.) DISC 2006. LNCS, vol. 4167, pp. 329–338. Springer, Heidelberg (2006)
15. Herlihy, M., Kozlov, D.N., Rajsbaum, S.: Distributed Computing Through Combinatorial Topology. Morgan Kaufmann (2014)
16. Herlihy, M., Rajsbaum, S.: Simulations and reductions for colorless tasks. In: PODC, pp. 253–260 (2012)
17. Herlihy, M., Rajsbaum, S.: The topology of distributed adversaries. Distributed Computing 26(3), 173–192 (2013)
18. Herlihy, M., Shavit, N.: The asynchronous computability theorem for t-resilient tasks. In: STOC, pp. 111–120 (May 1993)
19. Herlihy, M., Shavit, N.: The topological structure of asynchronous computability. J. ACM 46(2), 858–923 (1999)
20. Herlihy, M., Shavit, N.: On the nature of progress. In: Fernàndez Anta, A., Lipari, G., Roy, M. (eds.) OPODIS 2011. LNCS, vol. 7109, pp. 313–328. Springer, Heidelberg (2011)
21. Kozlov, D.N.: Chromatic subdivision of a simplicial complex. Homology, Homotopy and Applications 14(1), 1–13 (2012)
22. Kuznetsov, P.: Understanding non-uniform failure models. Bulletin of the EATCS 106, 53–77 (2012)
23. Kuznetsov, P.: Universal model simulation: BG and extended BG as examples. In: Higashino, T., Katayama, Y., Masuzawa, T., Potop-Butucaru, M., Yamashita, M. (eds.) SSS 2013. LNCS, vol. 8255, pp. 17–31. Springer, Heidelberg (2013)
24. Linial, N.: Doing the IIS (2010) (unpublished manuscript)
25. Rajsbaum, S.: Iterated shared memory models. In: López-Ortiz, A. (ed.) LATIN 2010. LNCS, vol. 6034, pp. 407–416. Springer, Heidelberg (2010)
26. Rajsbaum, S., Raynal, M., Travers, C.: The iterated restricted immediate snapshot model. In: Hu, X., Wang, J. (eds.) COCOON 2008. LNCS, vol. 5092, pp. 487–497. Springer, Heidelberg (2008)
27. Raynal, M., Stainer, J.: Increasing the power of the iterated immediate snapshot model with failure detectors. In: Even, G., Halldórsson, M.M. (eds.) SIROCCO 2012. LNCS, vol. 7355, pp. 231–242. Springer, Heidelberg (2012)
28. Raynal, M., Stainer, J.: Synchrony weakened by message adversaries vs asynchrony restricted by failure detectors. In: PODC (2013)

The Opinion Number of Set-Agreement[*]

Pierre Fraigniaud[1,**], Sergio Rajsbaum[2,***],
Matthieu Roy[3,†], and Corentin Travers[4,‡]

[1] CNRS and University Paris Diderot, France
[2] Instituto de Matemáticas, UNAM, Mexico
[3] CNRS, LAAS, Univ. Toulouse, France
[4] CNRS and U. of Bordeaux, France

Abstract. This paper carries on the effort to bridging runtime verification with distributed computability, studying necessary conditions for monitoring failure prone asynchronous distributed systems. It has been recently proved that there are correctness properties that require a large number of opinions to be monitored, an opinion being of the form true, false, perhaps, probably true, probably no, etc. The main outcome of this paper is to show that this large number of opinions is not an artifact induced by the existence of artificial constructions. Instead, monitoring an important class of properties, requiring processes to produce at most k different values does require such a large number of opinions. Specifically, our main result is a proof that it is impossible to monitor k-set-agreement in an n-process system with fewer than $\min\{2k, n\} + 1$ opinions. We also provide an algorithm to monitor k-set-agreement with $\min\{2k, n\} + 1$ opinions, showing that the lower bound is tight.

1 Introduction

Monitoring correctness properties at runtime, is a well established research domain. The essential objective of runtime verification is to determine, at any point in time, whether a system is in a legal or illegal state, with respect to a given specification.

In runtime verification, *monitors* are hardware or software components in charge of supervising the state of the system. In particular, the case of a distributed system whose execution is observed by several monitors has been considered in [4,6,18]. As soon as a violation of the legality of the execution is revealed by any of these monitors at runtime, recovery code can be executed for bringing the system back to a legal state. The monitors may communicate with each other, and every monitor produces a local *opinion*. The simplest case considers that opinions are in {true, false}, and a recovery code is fired as soon as one of these opinions is false.

---

[*] This work was supported in part by ECOS-NORD project #M12M01.
[**] Additional supports from ANR project DISPLEXITY, and INRIA project GANG.
[***] Additional support from UNAM-PAPIIT project and CONACYT LAISLA.
[†] Additional support from CNRS PICS.
[‡] Additional support from ANR project DISPLEXITY.

M.K. Aguilera et al. (Eds.): OPODIS 2014, LNCS 8878, pp. 155–170, 2014.
© Springer International Publishing Switzerland 2014

When processes may fail and the distributed system is asynchronous, it is however no longer sufficient for the monitors to produce only binary opinions. It has indeed been shown in [12] that there are correctness properties that cannot be monitored by interpreting the opinions produced by the monitors using the logical conjunction of all the opinions, where the system is in legal state if and only if all opinions are true. Furthermore, it has been later shown in [13] that there are correctness properties for which no decentralized monitoring exists if the number of possible opinions is constant, whatever the interpretation of the opinions. These results hold in distributed systems composed of n asynchronous processes communicating by reading and writing to a shared memory, and in which any number of processes may fail by crashing.

In this work, we consider a wait-free shared memory model, and assume as in [12,13] that each process has a variable that contains the output of a computation that we want to monitor. More precisely, the correctness property to be monitored describes the set of values that are allowed to appear in these variables simultaneously. Such a set of values is called a *instance*. The set of correct instances associated to the correctness property defines a *distributed language*. The following illustrative example was detailed in [13], as it is arising often in practice [5]. Let us consider a system where *requests* are sent by clients, and *acknowledged* by servers. The execution is correct when (1) all requests have been acknowledged, and (2) every received acknowledgement corresponds to a previously sent request. Each monitor i observes a variable reporting a pair (R_i, A_i), namely, the subset R_i of requests that have been received by the servers, and the subset A_i of acknowledgements that have been sent by the servers. The corresponding distributed language is the set of instances whose variables satisfy conditions (1) and (2).

To monitor the system, processes observe the values in their variables, and communicate with each other as long as needed. Eventually each process must produce its opinion about the legality of the instance with respect to the given language. It is required that the multisets of opinions produced in case the instance is legal differs from those produced for illegal instances. This partition of multisets of opinion is captured by an *interpretation*. In the most common setting, each opinion is in the set {true, false}, and the (global) interpretation of opinions is the logical conjunction of these opinions. However, as mentioned above, there are languages that require more opinions and more complex interpretations than logical conjunction [12,13].

In [13], it was shown that, for any k, $1 \leq k \leq n$, there exists a distributed language requiring monitors to produce at least k distinct opinions in a system with n monitors. To establish this result, an ad hoc language was constructed. The existence of a real distributed task that would be arbitrarily difficult in terms of the number of opinions needed to monitor it was left open.

In this paper, we study the number of opinions needed to monitor the family of *set agreement* tasks, widely studied in the distributed computing literature since its introduction in [8]. Recall that, for $n \geq 1$ processes, and $1 \leq k \leq n$, the (n, k)-set-agreement task is specified as follows. Each processes proposes a *value*, from some ground set. After communicating with the other processes, each process has to decide on one of the proposed values, such that at most k different proposed values are decided. Hence, if $k = 1$ then we get the classic *consensus* task [9], and, as we increment the value of k, we get easier agreement tasks, until we get the trivial task for $k = n$.

More precisely, we consider the language $\mathcal{L}_{n,k}$ corresponding to (n, k)-set-agreement. The value to be monitored at each process i, $1 \leq i \leq n$, consists of a pair (s_i, t_i). Consider the set of pairs $u = \{(s_{i_1}, t_{i_1}), (s_{i_2}, t_{i_2}), \ldots, (s_{i_\ell}, t_{i_\ell})\}$ present at some time in the system, where the pair (s_{i_j}, t_{i_j}) is the variable of process i_j, $j = 1, \ldots, \ell$. Note that we may have $\ell < n$ since some processes may crash or be arbitrarily slow. Then u is in the language $\mathcal{L}_{n,k}$ (i.e., represents a correct execution of an algorithm $\mathcal{A}$ pretending to solve (n, k)-set-agreement) if and only if (1) there are at most k different t_{i_j}, $j = 1, \ldots, \ell$ (i.e., at most k different values have been decided by $\mathcal{A}$), and (2) for each t_{i_j}, there must exist a pair $(s_{i_{j'}}, t_{i_{j'}})$ in u such that $s_{i_{j'}} = t_{i_j}$ (i.e., t_i has been proposed by at least one participating process). Again, a monitor for the set agreement language $\mathcal{L}_{n,k}$ must produce opinions that enable to distinguish legal instances from illegal ones, for some interpretation of those opinions. To establish an impossibility result regarding monitoring, we must prove that there is no monitor enabling to decide $\mathcal{L}_{n,k}$ whatever the interpretation of the opinions is.

Our Results. We first prove that the language $\mathcal{L}_{n,k}$ can be monitored using $\min\{2k, n\} + 1$ opinions. Then, our main result is a proof that it is impossible to monitor $\mathcal{L}_{n,k}$ with fewer opinions. Thus, in particular, consensus can be monitored with 3 opinions, and cannot be monitored with less than 3 opinions, even using interpretations different from the logical conjunction of boolean opinions.

The upper bound is a simple adaptation of the universal algorithm presented in [13]. Our proof technique for the lower bound is also inspired from the lower bound in [13], and, as such, is also based on combinatorial topology arguments using Sperner's lemma. However, a careful analysis of the *alternation number* of the set agreement language had to be achieved in this paper. This parameter captures the number of times a sequence of instances can alternate between legal and illegal over an execution of the system.

Hence, the main outcome of this paper is the perhaps surprising result, that the difficulty of monitoring set agreement (in terms of number of opinions) is captured by the formula $\min\{2k, n\} + 1$. The difficulty grows linearly as k is increased, at double the rate, and only until $2k = n$. For larger values of k, the number of opinions stays at its maximum, equal to $n + 1$ (no task requires more opinions). Indeed, monitoring the most important case in distributed computing, which is $k = n - 1$, does not require more opinions than when $2k = n$. We conjecture that the number of opinions has a direct relation with the number of logic values needed to design a temporal logic framework, and it is well known that more logic values dramatically increase the difficulty of reasoning in the logic. Hence this paper continues motivating further research at the border between runtime verification and distributed computability, in the context of studying necessary conditions for monitoring asynchronous distributed systems susceptible to failures.

Related Work. Most work on decentralized monitoring, is based on logical frameworks, using LTL or variants of it, see e.g. [4], where a formula ϕ is decomposed into local formulas, so monitor i evaluates locally ϕ_i, and emits a boolean-valued opinion. In our terminology, a logical conjunction interpretation is used. That is, it is assumed a global violation can always be detected locally by a process. See [13] for a more detailed discussion logic-based runtime verification. To the best of our knowledge, the effects of asynchrony and failures in a decentralized monitoring setting were considered for the first time in [12], and subsequently in [13]. In distributed computing, monitoring has been investigated for stable property detection in a failure-free message-passing environment [7], and distributed program checking in the context of self-stabilization [3].

2   Preliminaries

We consider the standard *wait-free shared memory model* [17]. The system consists of n processes denoted $\{p_1, \ldots, p_n\}$ that communicate via a shared memory. Processes are asynchronous and any number of them may fail by crashing (i.e., halt and never recover.). The shared memory is made of n shared registers, one per process, that support atomic read and write operations. To simplify the design of protocol, we assume that *atomic snapshots* [1] are available. An atomic snapshot consists in an n-components array R and supports two operations *update(v)* and *snapshot()*. *update(v)* by process p_i sets the value of $R[i]$ to v and *snapshot()* reads atomically all the components of R [1].

A *task* specifies for each process possible inputs and for each possible assignment of inputs to the processes, which are the valid outputs. More precisely, an *input* or *output set* is a non-empty set $s = \{(id_1, v_1), \ldots, (id_\ell, v_\ell)\}$ where $id_1, \ldots, id_k$ are distinct processes identities[1] and $v_1, \ldots, v_k$ are values from some set $\mathcal{V}$. s specifies an assignment of input or output values to each processes in $\mathrm{ID}(s) = \{id_1, \ldots, id_k\}$. The set of values in s is denoted by $val(s)$, or $mval(s)$ when considering the multiset of values. Two input or output sets s, t *match* if $\mathrm{ID}(s) = \mathrm{ID}(t)$. A task is specified by a triplet $(\mathcal{I}, \mathcal{O}, \Delta)$ where $\mathcal{I}$ and $\mathcal{O}$ are inclusion-closed sets of input and output sets respectively, and Δ maps each input set $s \in \mathcal{I}$ to a non-empty set of matching output sets.

In the *(n, k)-set agreement* task [8], each process input is a value taken from some set $\mathcal{V}$ of cardinality larger than k. Each process is required to output a value which is the input value of a participating process such that no more than k distinct values are output. More precisely, $\mathcal{I}$ is the set of the input sets s with $val(s) \subseteq \mathcal{V}$ and $t = \{(id_1, v_1), \ldots, (id_\ell, v_\ell)\} \in \mathcal{O}$ if and only if $|val(t)| \leq k$. For any input set $s \in \mathcal{I}$, $\Delta(s) = \{t \in \mathcal{O} : \mathrm{ID}(s) = \mathrm{ID}(t) \text{ and } val(t) \subseteq val(s)\}$.

---

[1] Identities are equal to processes indexes. That is, the identity of process p_i is i.

Distributed Languages. Distributed languages have been introduced in [13] as a simple way to represent correctness properties of distributed systems. Given an alphabet of symbols A, a *distributed language* over A is a set of non-empty input sets with, for each $s \in \mathcal{L}$, $val(s) \subseteq A$. In the context of distributed languages, input sets s with $val(s) \subseteq A$ are called *instances*. An instance s is *legal* if $s \in \mathcal{L}$ and *illegal* otherwise. A distributed language might specify a global property that a distributed system should satisfy. In this case, each element of A encodes a local state and the sets in $\mathcal{L}$ represent the global states of the system that satisfy the property.

We focus on checking that the result of a distributed computation satisfies the specification of a given task. More precisely, given a task $T = (\mathcal{I}, \mathcal{O}, \Delta)$, we define the *language* $\mathcal{L}_T$ *induced by* T as follows. Let $V_{\mathcal{I}}$ and $V_{\mathcal{O}}$ be respectively the sets of possible input and output values of T. Language $\mathcal{L}_T$ is defined over the alphabet $V_{\mathcal{I}} \times V_{\mathcal{O}}$. That is, each process input is a pair $(s_i, t_i) \in V_{\mathcal{I}} \times V_{\mathcal{O}}$. Given an input set $u = \{(id_1, (s_1, t_1)), \ldots, (id_\ell, (s_\ell, t_\ell))\}$, let $s_u = \{(id_1, s_1), \ldots, (id_\ell, s_\ell)\}$ and $t_u = \{(id_1, t_1), \ldots, (id_\ell, t_\ell)\}$. Then, $u \in \mathcal{L} \iff t_u \in \Delta(s_u)$. In other words, an input set u belongs to $\mathcal{L}_T$ if and only if each process i starting with input value s_i is allowed to decide t_i, according to the specification of T.

The language $\mathcal{L}_{n,k}$ induced by (n, k)-set agreement is defined as follows. The alphabet over which $\mathcal{L}_{n,k}$ is defined is $\mathcal{V} \times \mathcal{V}$ where $\mathcal{V}$ is the set from which values are proposed in the (n, k)-set agreement task. Then, for any instance $s = \{(id_1, (s_1, t_1)), \ldots, (id_\ell, (s_\ell))\}$, $s \in \mathcal{L}_{n,k}$ if and only if $|\{t_1, \ldots, t_\ell\}| \leq k$ and $\{t_1, \ldots, t_\ell\} \subseteq \{s_1, \ldots, s_\ell\}$.

3   Wait-Free Languages Monitoring

As defined in [13], monitoring the correctness specified by a distributed language $\mathcal{L}$ over an alphabet A involves two components: an *opinion-maker* M, and an *interpretation* μ. The opinion-maker is a distributed protocol. For each process p_i, its input is a pair (id_i, a_i) where $a_i \in A$ and its output is an *opinion* about the legality of the input set. The processes running this algorithm are called *monitors*, and the (finite) set of possible individual opinions U, the *opinion set*.

The interpretation μ specifies how the collection of individual opinions produced by the monitors should be interpreted. It is required that the opinions of the monitors should be able to distinguish legal input sets from illegal ones according to $\mathcal{L}$. Thus, $\mu = (\mathbf{Y}, \mathbf{N})$ is a partition of all multi-sets of at most n elements over U. $\mathbf{Y}$ is called the "yes" set, and $\mathbf{N}$ is called the "no" set.

A pair (M, μ) is *a monitor for* language $\mathcal{L}$ over alphabet A if the following holds:

- For the opinion-maker protocol M, input of of each process i is any element a_i of A and the output an opinion u_i. M is required to be wait-free: each participating process is required to decide an opinion in a finite number of its own steps, regardless of the behavior of the other processes.

– For any input set s, in any execution of M with input s where all participating processes decide, the multiset S of opinions that are decided satisfies:

$$s \in \mathcal{L} \iff S \in \mathbf{Y}.$$

Thus, if the input set s is illegal, i.e., $s \notin \mathcal{L}$, then the processes must produce a multiset of opinions in $\mathbf{N}$.

By extension, a pair (M, μ) is a *monitor for a task* T if (M, μ) is a monitor for the induced language $\mathcal{L}_T$.

Example: AND-Interpretation. When the set of opinions consists only in two values, i.e., $U = \{0, 1\}$, a natural interpretation μ is induced by the AND-operator [10,12] as follows. For every multi-set of opinions S, $S \in \mathbf{Y}$ if every opinion in S is 1, otherwise, $S \in \mathbf{N}$. Intuitively, a process outputs 1 if according to its view the instance is legal and 0 otherwise. Note that the interpretation is id-oblivious: for each opinion, the identities of the processes that produce that opinion is not taken into account by the interpretation.

Not every language has a wait-free monitor with an AND-interpretation. Indeed, the languages that can be monitored with the AND-interpretation are exactly those that are *projection-closed*[2] [12]. In particular, the language induced by the consensus task is not projection-closed and therefore does not have a wait-free monitor with the AND-interpretation. In fact, our main result implies that three opinions are necessary and sufficient to wait-free monitor consensus.

Opinion Number and Alternation Number. The *opinion number* #opinion($\mathcal{L}$) of a language $\mathcal{L}$, is the smallest size of the opinion set U for which there exists a monitor with opinion U for $\mathcal{L}$. The *alternation number* #altern($\mathcal{L}$) of a language $\mathcal{L}$ is the longest sequence of increasing instances that alternates between legal and illegal instances. More precisely, #altern($\mathcal{L}$) is the largest integer ℓ for which there exists instances $s_1 \subset s_2 \subset \ldots \subset s_\ell$ such that for every i, $1 \leq i < \ell$, $s_i \subset s_{i+1}$, and either $s_i \in \mathcal{L} \wedge s_{i+1} \notin \mathcal{L}$ or $s_i \notin \mathcal{L} \wedge s_{i+1} \in \mathcal{L}$.

A strong relationship between opinion and alternation numbers of languages is exposed in [13]. First, it is established that the alternation number gives an upper bound on the opinion number: For every language $\mathcal{L}$, #opinion($\mathcal{L}$) $\leq$ #altern($\mathcal{L}$) + 1. Second, it is shown that for any $k, 1 \leq k \leq n$, there exists a n-processes language $\mathcal{L}$ with #altern($\mathcal{L}$) = k that requires at least k opinions to be monitored, i.e., #opinion($\mathcal{L}$) $\geq k$. For (n, k)-set agreement, the alternation number of the induced language $\mathcal{L}_{n,k}$ is:

Lemma 1. #altern($\mathcal{L}_{n,k}$) = $\min\{2k + 1, n\}$.

Proof. Consider the following sequence of instances: $s_1 = \{(1, (0, 1))\}$, $s_2 = \{(1, (0, 1)), (2, (1, 1))\}$, $\ldots$, $s_{2i} = \{(1, (0, 1)), (2, (1, 1)), \ldots, (2i, (i, i))\}$, $s_{2i+1} = \{(1, (0, 1)), \ldots, (2i, (i, i)), (2i + 1, (i, i + 1))\}, \ldots$ for $i : 2i, 2i + 1 \leq n$. Note that

---

[2] Language $\mathcal{L}$ is projection-closed if and only if for each $s \in \mathcal{L}$ and each $s' \subset s$, $s' \in \mathcal{L}$.

$s_{2i+1} \notin \mathcal{L}_{n,k}$ as this instance corresponds to the case in which a value, namely $i + 1$ is decided whereas it has not been proposed. In s_{2i}, however, i values are decided and each of them is also a proposed value. Thus, if the number of decided values is less than or equal to k, that is, $i \leq k$, $s_{2i} \in \mathcal{L}_{n,k}$. Hence, the increasing sequence $s_1, s_2, \ldots, s_{\min\{2k+1,n\}}$ alternates between legal and illegal instances. Thus, $\#\text{altern}(\mathcal{L}_{n,k}) \geq \min\{2k+1, n\}$.

Let $s_1 \subset \ldots \subset s_x$ be a sequence of increasing instances such that for every $i, 1 \leq i < x$, $s_i \in \mathcal{L}_{n,k} \wedge s_{i+1} \notin \mathcal{L}_{n,k}$ or $s_i \notin \mathcal{L}_{n,k} \wedge s_{i+1} \in \mathcal{L}_{n,k}$. Let $D(i)$ denote the size of the set of values decided in s_i. That is, if $s_i = \{(id_1, (a_1, b_1)), \ldots, (id_\ell, (a_\ell, b_\ell))\}$, $D(i) = |\{b_1, \ldots, b_\ell\}|$. Note that $D(1) \geq 1$. Consider instances s_i, s_{i+1} with i such that with $s_i \in \mathcal{L}_{n,k}$. As $s_i \subset s_{i+1}$, and since in s_i no more that k values are decided and every decided value is valid, it follows that a value not decided in s_i is decided in s_{i+1}, i.e., $D(i) < D(i+1)$.

Let $s_{x'}$ be the largest legal instance in the sequence. Note that $x' = x$ or $x' = x - 1$. On one hand, we have $D(x') \leq k$. On the other hand, $1 + \lfloor \frac{x'-1}{2} \rfloor \leq D(x')$, since the number of alternations from a legal instance to an illegal one in the sequence $s_1, \ldots, s_{x'}$ is at least $\lfloor \frac{x'-1}{2} \rfloor$ and $D(1) \geq 1$. Hence, $x' \leq 2k$, and as $x \leq x' + 1$, we obtain $x \leq 2k + 1$. Therefore, any sequence of increasing and alternating instances is of length at most $2k + 1$. Hence, $\#\text{altern}(\mathcal{L}_{n,k}) \leq \min\{2k+1, n\}$.

$\square$

The two next sections focus on determining how many opinions are needed to monitor (n, k)-set agreement. We show that $\#\text{opinion}(\mathcal{L}_{n,k}) = \#\text{altern}(\mathcal{L}_{n,k})$ if $\#\text{altern}(\mathcal{L}_{n,k}) < n$ and $\#\text{opinion}(\mathcal{L}_{n,k}) = n+1$ otherwise, that is, $\#\text{opinion}(\mathcal{L}_{n,k}) = \min\{2k, n\} + 1$.

4   Monitoring k-Set Agreement

We prove in this section that (n, k)-set-agreement can be wait-free monitored using $\min\{2k, n\} + 1$ opinions.

In [13], an *universal monitor* is presented that uses at most $n + 1$ opinions for any n-processes language. The opinion-maker of this monitor depends on the language being checked whereas the interpretation depends solely on the opinions output by the processes. Independently of the language, the opinions produced by each process belongs to a set of size $n+1$. If $k > \lfloor \frac{n}{2} \rfloor$, $\min\{2k, n\}+1 = n+1$ and thus in this case the language $\mathcal{L}_{n,k}$ induced by (n, k)-set-agreement can be checked by the universal monitor described in [13].

For the case $k \leq \lfloor \frac{n}{2} \rfloor$, we present a monitor (M, μ) for (n, k)-set agreement that uses $2k + 1$ opinions. The set of opinions is:

$$U = \{\text{red}\} \cup \{(\text{green}, \ell) : 1 \leq \ell \leq k)\} \cup \{(\text{orange}, \ell) : 1 \leq \ell \leq k)\}.$$

Note that $|U| = 2k + 1$. The partition $(\mathbf{Y}, \mathbf{N})$ is defined as follows. For any multiset S with values in U, let $\text{maxlevel}(S) = \max\{\ell : (\text{green}, \ell) \in S \vee (\text{orange}, \ell) \in S\}$. Then,

$$S \in \mathbf{Y} \iff (\text{green}, \text{maxlevel}(S)) \in S \wedge \text{red} \notin S \qquad (1)$$

Equivalently, any multiset S that contains the value red or a pair (orange, ℓ) and no pair (green, ℓ') with $\ell \leq \ell'$ is in the "no" set $\mathbf{N}$.

Recall that each process p_i starts with an input-output pair (s_i, t_i), runs the opinion-maker protocol and produces an opinion based on what it sees during the execution of the protocol. Since we consider wait-free protocol and processes run asynchronously, it may be the case that a process p_i does not see the input-output pair of some participating processes.

Intuitively, a process outputs red if in the collection (s', t') of input-output pairs it sees does not fit the specification of (n, k)-set-agreement, and this cannot be fixed by completing (s', t') with other input-output pairs. This occurs if agreement is broken, i.e., more than k values are decided in t'. In the case where agreement is satisfied (t' contains at most k distinct values) as well as validity (every decided values is proposed, that is, appears in s'), the process outputs (green, d) where $d \leq k$ is the number of decided values. In the last case, namely agreement is satisfied but validity is not (a decided value in t' does not appear in s'), (orange, d) is output. Note that in this case, the global input (s, t) consisting in the collection of input-output pairs of each participating process may still fit the specification of (n, k)-set agreement. The pseudo-code of the opinion-maker appears in Fig. 1.

Opinion-maker M at process i with input $(i, (s_i, t_i))$:
  $R.\text{update}(i, (s_i, t_i))$;
  $r \leftarrow R.\text{snapshot}()$;
  $(s', t') \leftarrow (\{(j, s_j) : r[j] \neq \bot\}, \{(j, t_j) : r[j] \neq \bot\})$;
  let $val(s')$ and $val(t')$ denote the set of values in s' and t' respectively;
  if $|val(t')| > k$ then $decide$ red
  else if $val(t') \subseteq val(s')$ then $decide$ (green, $|val(t')|$)
        else $decide$ (orange, $|val(t')|$)

Fig. 1. Opinion-maker for (n, k)-set-agreement, $k \leq \lfloor \frac{n}{2} \rfloor$

Correctness of the (n, k)-Set Agreement Monitor. We prove that, for $k \leq \lfloor \frac{n}{2} \rfloor$, the opinion-maker M of Fig. 1 used with the interpretation μ described in Equation 1 produce a wait-free monitor for (n, k)-set agreement task $(\mathcal{I}, \mathcal{O}, \Delta)$. Note that the opinion set of M is of size $2k + 1$, as desired.

Let V denote the set of possible input values for the (n, k)-set agreement task. Let us consider an execution e of the opinion-maker with input set $\{(id_1, (s_1, t_1)), \ldots, (id_\ell, (s_\ell, t_\ell))\}$ where $(s_i, t_i) \in V \times V$ for each $i, 1 \leq i \leq \ell$. Assume that every participating process decides, and let S denote the multiset of opinions that are decided in this execution. In addition, let $s = \{(id_1, s_1), \ldots, (id_\ell, s_\ell)\}$ and $t = \{(id_1, t_1), \ldots, (id_\ell, t_\ell)\}$.

A process p_i decides red if more than k values are decided in the output set t' it sees. Even if i has only a partial view of the input, i.e., $(s', t') \subsetneq (s, t)$, the lack of agreement cannot be fixed in (s, t) since every value decided in t' is also decided in t. Therefore, $t \notin \Delta(s)$, and consequently, it is correct that $S \in \mathbf{N}$.

Suppose that no processes decides red. Let p_i be the last process that writes its input (s_i, t_i) into shared memory. The invocation of R.snapshot() by p_i thus starts after every participating process p_j has updated the shared object with its input (s_j, t_j). It thus follows that p_i observes the full input (s, t). Denote by d the number of decided values in t. Since no process decides red, $d \leq k$. We consider two cases, depending on whether $t \in \Delta(s)$ or not.

- Assume first that $t \in \Delta(s)$. Since p_i sees the full output, it decides (green, d). As in every partial input $(s', t') \subseteq (s, t)$, the number of decided values is at most d, each process that sees a partial input decides either (green, d') or (orange, d') with $d' \leq d$. Therefore, maxlevel(S) = d, and thus $S \in \mathbf{Y}$.
- Assume now that $t \notin \Delta(s)$. Since no process decides red, the number of decided values in t is at most k. As $t \notin \Delta(s)$, then at least one decided value is not valid, i.e., there exists a value $u \in val(t)$ that is not contained in $val(s)$. Therefore p_i decides (orange, d). Assume for contradiction that a process p_j decides (green, ℓ) with $\ell \geq d$. Observe that, at process j, the collection of input-output pairs (s', t') seen by j is such that $(s', t') \subseteq (s, t)$. Hence, every value decided in t' is also decided in t. Thus $\ell = |val(t')|$, the number of decided values seen by p_j, is smaller than or equal to d.
  Therefore $\ell = d$ and processes p_i and p_j see the same output set $W = val(t) = val(t')$ of decided values. Moreover, as p_j decides (green, ℓ), validity holds for the pair (s', t'), that is $W = val(t') \subseteq val(s')$. Consequently, as $s' \subseteq s$, each value $v \in W$ is also proposed in s. Therefore, agreement and validity are verified in (s, t) and p_i should decide (green, d): a contradiction. Hence, for every (green, ℓ) $\in S$, $\ell < d$ from which we conclude that $S \in \mathbf{N}$.   □

The following Lemma summarizes the result of this section:

Lemma 2. (n, k)-set-agreement has a monitor that uses $\min\{2k, n\} + 1$ opinions, i.e., #opinion($\mathcal{L}_{n,k}$) $\leq \min\{2k, n\} + 1$.

5 Opinion Number of (n, k)-Set Agreement

We now establish a lower bound on the number of opinions required to monitor $\mathcal{L}_{n,k}$. The proof is by contradiction.

Assume for contradiction that $\mathcal{L}_{n,k}$ has a wait-free monitor (M, μ) with opinion set U, $|U| < \min\{2k, n\} + 1$. The existence of this monitor implies that a specific task $T_{U,\mu}$ has a wait-free protocol. When solving a task, each participating process starts with a private input value and has to eventually decide irrevocably on an output value. Let $\mathcal{V}, \mathcal{V} \supseteq \{0, \ldots, k\}$, the alphabet of the language $\mathcal{L}_{n,k}$. In task $T_{U,\mu}$, process p_i's input is any pair $(s_i, t_i) \in \mathcal{V} \times \mathcal{V}$ and it is required to output a value $u_i \in U$. When the input set s is a legal instance, i.e., $s \in \mathcal{L}_{n,k}$, the multiset of output value must belong to the "yes" set $\mathbf{Y}$ and in the "no" set $\mathbf{N}$ otherwise. We show that as long as the set of opinions U is of cardinality less than $\min\{2k, n\} + 1$, whatever the interpretation μ, task $T_{U,\mu}$ is not wait-free solvable. To that end, we rely on the representation of wait-free computations by topological objects, as in, e.g., [2,15,16]. Our main tool is a variant of Sperner's Lemma.

5.1   Preliminaries

Basic Notions of Topology. We first review some basic definitions. A *complex* $\mathcal{K}$ is a set of vertices $V(\mathcal{K})$, and a family of finite, nonempty subsets of $V(\mathcal{K})$, called *simplexes*, satisfying: (1) if $v \in V(\mathcal{K})$ then $\{v\}$ is a simplex, and (2) if s is a simplex, so is every nonempty subset of s. The *dimension* of a simplex s is $|s| - 1$, the dimension of $\mathcal{K}$ is the largest dimension of its simplexes, and $\mathcal{K}$ is *pure* of dimension k if every simplex belongs to a k-dimensional simplex. A simplex τ is a *face* of a simplex σ if τ is a subset of σ. If τ is not equal to σ then τ is a *proper face* of σ. The *complex induced by a simplex* σ consists in σ and all its faces.

In distributed computing, a vertex represents a local state, a simplex a global state and a complex a collection of global states. Hence, one of the labels of each vertex is an identity in $[n]$. We denote by $\mathrm{ID}(\sigma)$ the identities of the vertexes of σ. A simplex is *chromatic* if it is properly colored with ids in $[n]$. A complex is *chromatic* if each of its simplex is chromatic.

A *simplicial map* f from complex $\mathcal{K}$ to complex $\mathcal{L}$ is a function from $V(\mathcal{K})$ to $V(\mathcal{K})$ that preserves simplexes. That is, if $\tau = \{v_1, \ldots, v_\ell\}$ is a simplex of $\mathcal{K}$, then $\{f(v_1), \ldots, f(v_\ell)\}$ is a simplex of $\mathcal{L}$. In addition, if $\mathcal{K}$ and $\mathcal{L}$ are chromatic complexes, f is said to be *id-preserving* if for any simplex $\tau = \{v_1, \ldots, v_\ell\} \in \mathcal{K}$, $\mathrm{ID}(\tau) = \mathrm{ID}(\{f(v_1), \ldots, f(v_\ell)\})$.

Pseudomanifold and Divided Images. A complex $\mathcal{K}$ of dimension n is a *pseudomanifold with boundary* if it is strongly connected, and each $(n-1)$-simplex in $\mathcal{K}$ is a face of precisely one or two n-simplexes. For simplicity, a pseudomanifold with boundary will simply be called a *pseudomanifold*. We sometimes write n-pseudomanifold as a shorthand for a n-dimensional pseudomanifold. Let $\mathcal{K}$ be a n-pseudomanifold. A $(n-1)$-simplex σ is said to be *internal* if it is a face of exactly two n-simplexes of $\mathcal{K}$ and *external* otherwise. The boundary of $\mathcal{K}$, denoted $\partial\mathcal{K}$, is the sub-complex of $\mathcal{K}$ induced by its external simplexes. More precisely, $\partial\mathcal{K}$ consists in each $(n-1)$-simplex of $\mathcal{K}$ that is a face of exactly one n-simplex together with all its faces.

Divided images of complexes have been introduced in [2] as a combinatorial tool to represent certain classes of executions of read/write wait-free protocols. A subdivision of a complex is a divided image, but subdivided images are not always subdivisions. Divided images capture the essential properties to study wait-free computability.

Definition 1 ([2], Definition 4.1). *Let $\mathcal{K}$ and $\mathcal{L}$ be finite n-dimensional complexes and ψ a function that maps every simplex of $\mathcal{K}$ to a subcomplex of $\mathcal{L}$. The complex $\mathcal{K}$ is a divided image of $\mathcal{L}$ under ψ if and only if*

1. *$\psi(\emptyset) = \emptyset$,*
2. *for every 0-simplex $\sigma \in \mathcal{L}$, $\psi(\sigma)$ is a single vertex,*
3. *for every $\sigma, \sigma' \in \mathcal{L}$, $\psi(\sigma \cap \sigma') = \psi(\sigma) \cap \psi(\sigma')$, and*
4. *for every $\sigma \in \mathcal{L}$, $\psi(\sigma)$ is a $\dim(\sigma)$-pseudomanifold with $\partial\psi(\sigma) = \psi(\partial\sigma)$.*

When ψ is clear from the context or not relevant, we simply say that $\mathcal{K}$ *is a subdivided image of* $\mathcal{L}$. Next Lemma state some properties of divided:

Lemma 3 ([2], Lemma 4.2). *Let* $\mathcal{K}, \mathcal{L}$ *be n-dimensional complexes such that* $\mathcal{K}$ *is a divided image of* $\mathcal{L}$ *under* ψ.

1. *For every* $\sigma, \sigma' \in \mathcal{L}$, *if* $\sigma \subseteq \sigma'$, *then* $\psi(\sigma) \subseteq \psi(\sigma)$.
2. *For every pair of j simplexes* $\sigma, \sigma' \in \mathcal{K}$, *if* $\sigma \neq \sigma'$ *and* $\sigma \cap \sigma' \neq \emptyset$, *then* $\psi(\sigma \cap \sigma')$ *is a pseudomanifold of dimension strictly smaller than j.*
3. *A $(n-1)$-dimensional simplex* $\tau \in \mathcal{K}$ *is external if and only if for some external $(n-1)$-dimensionnal simplex* $\sigma \in \mathcal{L}$, $\tau \in \psi(\sigma)$.

The *carrier* of simplex $\tau \in \mathcal{K}$, denoted $\mathrm{carrier}(\tau)$ is the simplex $\sigma \in \mathcal{L}$ of smallest dimension such that $\tau \in \psi(\sigma)$. Note that, by Definition 1(3), $\mathrm{carrier}(\tau)$ is well defined and by Lemma 3(2), it is unique. If $\mathcal{L}$ is a chromatic complex, $\mathcal{K}$ is a *chromatic divided image* [2] of $\mathcal{L}$ if $\mathcal{K}$ is a divided image of $\mathcal{L}$, $\mathcal{K}$ is a *chromatic* complex, and for any $\tau \in \mathcal{K}$, $\mathrm{ID}(\tau) \subseteq \mathrm{ID}(\mathrm{carrier}(\tau))$. Note in particular that if $dim(\tau) = dim(\mathrm{carrier}(\tau))$, τ is properly colored with the ids in $\mathrm{ID}(\tau)$.

Combinatorial Implication of Wait-Free Computability. Simplexes and complexes are a convenient way to represent tasks and distributed protocol. A tasks $T = (\mathcal{I}, \mathcal{O}, \Delta)$ can be equivalently described by an input complex $\widetilde{\mathcal{I}}$, an output complex $\widetilde{\mathcal{O}}$ and a function $\widetilde{\Delta}$ that maps each simplex of $\widetilde{\mathcal{I}}$ to a subcomplex of $\widetilde{\mathcal{O}}$. Vertexes of $\widetilde{\mathcal{I}}$ and $\widetilde{\mathcal{O}}$ are labeled with an identity and a value and there is a simplex $s = \{(id_1, v_1), \dots, (v_\ell, id_\ell)\}$ in $\widetilde{\mathcal{I}}$ (respectively, $\widetilde{\mathcal{O}}$) if and only if s is an input set in $\mathcal{I}$ (respectively, an output set in $\mathcal{O}$). Similarly, $t \in \widetilde{\Delta}(s)$ if and only if the corresponding output set t is in $\Delta(s)$. In the following, we consider the topological representation for tasks and drop the $\sim$ notation.

Without loss of generality, a read/write wait-free protocol consists in a certain number B of (asynchronous) rounds. In each round, process p_i writes its state in its cell $R[i]$, takes a snapshot of the memory and updates its state. The process initial state is its input. At the end of the B rounds, a final state is reached on which depends the process decision. A *protocol complex* $\mathcal{P}$ represents all possible final states for some execution. Each vertex is labeled with an id and a possible final state. $\sigma = \{(id_1, v_1), \dots, (id_\ell, v_\ell)\}$ is a simplex in $\mathcal{P}$ if there is an execution at the end of which process p_i with identity id_i is in state v_i, for $1 \leq i \leq \ell$.

An *immediate snapshot execution* can be divided into blocks. In each block, a subset of the participating processes are active. They first simultaneously write before taking a snapshot. One important result of the topological approach is a characterization of the structural properties of the protocol complex of immediate snapshot executions, namely the immediate snapshot protocol complex is a chromatic divided image of the input complex [2]. If a protocol solves a task $T = (\mathcal{I}, \mathcal{O}, \Delta)$, in any execution, the final states can be mapped to decision values in such a way that the output set is allowed for the input set of the execution according to Δ. As considering only a subset of all possible executions might be sufficient to derive an impossibility result, we obtain the following necessary condition for read/write wait-free solvability of tasks:

Theorem 1 ([2], Theorem 5.10). *Let* $T = (\mathcal{I}, \mathcal{O}, \Delta)$ *a task. If there is a read/write wait-free protocol which solves* T, *then there is a chromatic divided image* $\mathcal{I}^*$ *of* $\mathcal{I}$ *and a id-preserving simplicial map* δ *from* $\mathcal{I}^*$ *to* $\mathcal{O}$ *that agrees with* Δ.

A Variant of Sperner's Lemma Given a function $f : V(\mathcal{K}) \longrightarrow U$, for each $\sigma = \{v_0, \ldots, v_\ell\}$ simplex of $\mathcal{K}$, $f(\sigma)$ denote the set of labels of the vertexes of σ by f, i.e., $f(\sigma) = \{f(v_0), \ldots, f(v_\ell)\}$.

Lemma 4. *Let* $\mathcal{K}$ *be a n-pseudomanifold, let* U *be a set of at least* $n+1$ *elements and let* $f : V(\mathcal{K}) \to U$. *If for some subset* $B \subset U$ *of size* n, *there is an odd number of B-labeled* $(n-1)$*-simplexes in the boundary of* $\mathcal{K}$, *then there is an odd number of C-labeled n-simplexes in* $\mathcal{K}$, *for some set* C *of* $n+1$ *elements,* $B \subset C \subseteq U$:

$$\left| \{\sigma \in \partial \mathcal{K} : dim(\sigma) = n - 1 \text{ and } f(\sigma) = B\} \right| \text{ is odd} \implies$$

$$\exists\, C, |C| = n+1, \left| \{\sigma \in \mathcal{K} : dim(\sigma) = n \text{ and } f(\sigma) = C\} \right| \text{ is odd}.$$

The proof is omitted and can be found in [11].

5.2   The Lower Bound

Lemma 5. *Let* M *be a wait-free opinion-maker with opinion set* U *and* μ *be an interpretation over* U. *If* (M, μ) *is a monitor for* (n, k)*-set-agreement,* $|U| \geq \min\{2k, n\} + 1$.

Proof. Recall that in the (n, k)-set agreement task, each process starts with a value from some set $\mathcal{V}$ and is required to decide an initial value of some process such that no more than k distinct values are decided. For the proof, we assume without loss of generality that $\mathcal{V} = \{0, \ldots, k\}$.

The proof is by contradiction. We assume that there exists a monitor (M, μ) with opinion set $U, |U| < \min\{n, 2k\} + 1$. The opinion-maker M is a wait-free protocol. In any of its executions, each participating process i starts with a pair $(s_i, t_i) \in \mathcal{V} \times \mathcal{V}$ and is directed to decided a value $u_i \in U$. The interpretation μ defines a partition $(\mathbf{Y}, \mathbf{N})$ of all the multisets with values in U; the multiset of decided values belongs to $\mathbf{Y}$ if and only if the set of initial pairs (s_i, t_i) verify the specification of (n, k) set agreement.

M is therefore a wait-free protocol for the task $T = (\mathcal{I}, \mathcal{U}, \Delta_\mu)$ where $\mathcal{I} = complex(\mathcal{V} \times \mathcal{V}, n)$, $\mathcal{U} = complex(U, n)$. Given any finite set X, $complex(X, n)$ is the $(n-1)$-dimensional *pseudosphere* [15] complex induced by X: for each $i \in [n]$ and each $x \in X$, there is a vertex labeled (i, x) in the vertex set of $complex(X, n)$ and $s = \{(id_1, x_1), \ldots, (id_\ell, x_\ell)\}$ is a simplex of $complex(X, n)$ if and only if $\{x_1, \ldots, x_\ell\} \subseteq X$ and s is properly colored with identities. The relation Δ_μ is defined next.

Each simplex $s = \{v_1, \ldots, v_\ell\}$ of $\mathcal{I}$ represents the initial and decided values of the participating processes in some execution of a (n, k)-set agreement protocol.

In more details, each vertex v_j has two labels: an id id_j, a pair $(s_i, t_i) \in V \times V$ representing the process input and output value for the (n, k)-set agreement tasks. Let us define two predicates, $agree_k(s)$ and $valid(s)$, that are verified if the set of input-output pairs (s_i, t_i) in s satisfy the agreement property and validity property, respectively, of (n, k)-set agreement. That is,

$$agree_k(s) \iff |\{t_j : (id_j, (s_j, t_j)) \in s\}| \leq k$$
$$valid(s) \iff \{t_j : (id_j, (s_j, t_j)) \in s\} \subseteq \{s_j : (id_j, (s_j, t_j)) \in s\}$$

A vertex of a simplex $t \in \mathcal{U}$ has two labels: and identity id and an opinion $u \in U$. Recall that $ID(t)$ and $mval(t)$ then denote the set of ids and the multiset of opinions, respectively, of the vertexes of t. Δ_μ is then defined as follows. For any $s \in \mathcal{I}, t \in \mathcal{U}$,

$$t \in \Delta_\mu(s) \iff ID(t) = ID(s) \text{ and } \begin{cases} mval(t) \in \mathbf{Y} \text{ if } agree_k(s) \text{ and } valid(s) \\ mval(t) \in \mathbf{N} \text{ otherwise.} \end{cases}$$

We associate with each simplex $s \in \mathcal{I}$ a value in $\{+1, -1\}$ depending on whether agreement and validity hold in s:

$$\forall s \in \mathcal{I} : sign(s) = \begin{cases} +1 \text{ if } agree_k(s) \text{ and } valid(s) \\ -1 \text{ otherwise.} \end{cases}$$

We focus on the following simplexes of $\mathcal{I}$ (See also Figure 2):

$$\sigma_{2i} = \{(1, (0, 1)), (2, (1, 1)), (3, (1, 2)), (4, (2, 2)), \ldots, (2i - 1, (i - 1, i)), (2i, (i, i))\}, \ (i \geq 1),$$

$$\sigma_{2i+1} = \{(1, (0, 1)), (2, (1, 1)), (3, (1, 2)), (4, (2, 2)), \ldots, (2i, (i, i)), (2i + 1, (i, i + 1))\}, \ (i \geq 0).$$

Note that $\sigma_1 \subset \sigma_2 \subset \ldots \subset \sigma_n$ and that for each simplex σ_i, the complex induced is a $(i - 1)$-pseudomanifold. Furthermore, we have

$$\begin{cases} valid(\sigma_i) \text{ and } agree_k(\sigma_i) & \text{if } 1 \leq i \leq 2k \text{ and } i \text{ is even} \\ \neg valid(\sigma_i) \text{ and } agree_k(\sigma_i) & \text{if } 1 \leq i \leq 2k \text{ and } i \text{ is odd} \\ \neg agree_k(\sigma_i) & \text{if } 2k < i \end{cases}$$

Hence, $sign(\sigma_i) = (-1)^i$ if $1 \leq i \leq 2k + 1$ and $sign(\sigma_i) = -1$ for $2k + 1 \leq i$. Another noteworthy property is the fact that every $(dim(\sigma_i) - 1)$-dimensional face of σ_i except σ_{i-1} has the same sign as σ_i:

$$\forall i \in [2, 2k + 1], \forall \sigma \subset \sigma_i, dim(\sigma) = dim(\sigma_i) - 1 \wedge \sigma \neq \sigma_{i-1} \implies sign(\sigma) = sign(\sigma_i) \tag{2}$$

Since there is a wait-free protocol for the task $T = (\mathcal{I}, \mathcal{U}, \Delta_\mu)$, namely the opinion-maker M, it follows from Theorem 1 that there is a chromatic divided image $\mathcal{I}^*$ of $\mathcal{I}$ under a function ψ and a id-preserving simplicial map $\delta : \mathcal{I}^* \to \mathcal{U}$ that agrees with Δ_μ. Since δ is a simplicial map, it maps each vertex $v \in V(\mathcal{I}^*)$ to a vertex $\delta(v)$ in $\mathcal{U}$. Recall that each vertex of $\mathcal{U}$ has two labels: a process id $\in [n]$ and an opinion $u \in U$. δ thus implies a (not necessarily proper) coloring $c : V(\mathcal{I}^*) \to U$ on the vertexes of $\mathcal{I}^*$: For each vertex v of $\mathcal{I}^*$, $c(v) = val(\delta(v))$. Given a simplex $s = \{v_1, \ldots, v_\ell\} \in \mathcal{I}^*$ we denote by slight abuse of notation $c(s) = \{c(v_1), \ldots, c(v_\ell)\} \subseteq U$ the *multiset* of colors induced by δ of the vertex of s. Figure 3 represents a chromatic divided image of the input simplex σ_3.

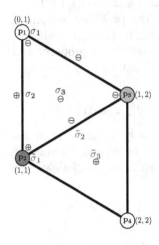

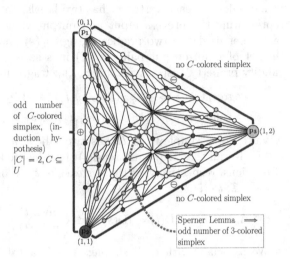

Fig. 2. The simplexes $\sigma_1 \subset \sigma_2 \subset \sigma_3$, and $\tilde{\sigma}_1 \subset \tilde{\sigma}_2 \subset \tilde{\sigma}_3$ and their sign

Fig. 3. Proof strategy

Claim. Let σ, σ' be j-simplexes in $\mathcal{I}$ with $sign(\sigma) = -sign(\sigma')$ and let s, s' be j-simplexes in $\mathcal{I}^*$. If $s \in \psi(\sigma)$ and $s' \in \psi(\sigma')$, $c(s) \neq c(s')$.

Proof of Claim 5.2. $c(s)$ is the multiset of opinions output by the participating processes (those processes with ids $ID(s) = ID(\sigma)$) in some execution e of the opinion-maker M in which the initial configuration is σ. Similarly, in some execution e' of M with initial configuration σ', the opinions collectively output by the participating processes is $c(s')$. As $sign(\sigma) = -sign(\sigma')$, the validity and agreement properties of (n, k)-set agreement in one initial configuration are satisfied but this is not the case in the other execution. Therefore the same multiset of opinion cannot be output in both executions. For example, in Figure 3, for any 1-dimensional simplex s in the subdivided left edge $\{(p_1, (0, 1)), (p_2, (1, 1))\}$ and any s' in the subdivided top right edge $\{(p_1, (0, 1)), (p_3, (1, 2))\}$, $c(s) \neq c(s')$. □

Next, we show that for each $j, 1 \leq j \leq \min\{2k+1, n\}$, there exists at least one execution of the opinion-maker where the input configuration is σ_j in which the j participating processes output j distinct opinions. More precisely, we establish by induction on j the following claim (See Figure 3 for an illustration of the proof strategy).

Claim. For each $j, 1 \leq j \leq \min\{2k+1, n\}$, there exists a set $C_j \subseteq U$ of size j such that $\big|\{s \in \psi(\sigma_j) : c(s) = C_j\}\big|$ is odd

Proof of Claim 5.2.

- Base case $j = 1$. By definition, σ_1 is a vertex. By Definition 1(2), $\psi(\sigma_1)$ is also a vertex v of $\mathcal{I}^*$. Since δ is a simplicial map, $\delta(\psi(\sigma_1))$ is a vertex in $\mathcal{U}$.

Let $C_0 = val(\delta(\psi(\sigma_1)))$, i.e., $C_0 = \{c(v)\}$. That is, C_0 is the singleton consisting in the opinion output by process 1 when it runs alone with input/output pair $(0, 1)$.

– Induction step. Let $j \geq 2$ and assume that the claim is true for $j - 1$.

By Definition 1(4), $\psi(\sigma_j)$ is a $(j - 1)$-pseudomanifold. In order to apply our variant of Sperner's Lemma, we establish that the number of $(j - 2)$-simplexes in the boundary of $\psi(\sigma_j)$ that are colored with C_{j-1} is odd. To that end, let $s \in \partial\psi(\sigma_j)$ be an $(dim(\sigma_j) - 1)$ dimensional simplex and assume that $c(s) = C_{j-1}$. We prove that $s \in \psi(\sigma_{j-1})$.

Note that $dim(s) = j - 2$. By Lemma 3(3), there exist a $(j - 2)$ dimensional face σ of σ_j such that $s \in \psi(\sigma)$. Suppose for contradiction that $\sigma \neq \sigma_j$. It then follows that $sign(\sigma) = sign(\sigma_j)$ (Equation (2) from which we have $sign(\sigma) = -sign(\sigma_{j-1})$. By the induction hypothesis, there is a least one $(j-2)$-simplex $s' \in \psi(\sigma_{j-1})$ with $c(s') = C_{j-1}$. Since $sign(\sigma) = -sign(\sigma_{j-1})$, it follows from Claim 5.2 that no $(j - 2)$-simplex in $\psi(\sigma)$ is C_{j-1}-colored. In particular, $c(s) \neq C_{j-1}$: a contradiction.

Therefore, the only simplexes $s \in \partial\psi(\sigma_j)$ that are C_{j-1}-colored are the simplexes s in $\psi(\sigma_{j-1})$ such that $c(s) = C_{j-1}$. By the induction hypothesis, the number of such simplex is odd. It thus follows from Lemma 4 that there exists a set $C = C_j \supset C_{j-1} \supseteq U$ of j elements such that the number of $(j - 1)$-simplexes $s \in \psi(\sigma_j)$ colored with C_j is odd.

If $\min\{2k + 1, n\} = 2k + 1$, we can stop here: we have just proved that $\psi(\sigma_j) \subseteq \mathcal{I}^*$ contains at least one $2k$-dimensional simplex colored with $2k + 1$ distinct colors. Hence $|U| \geq 2k + 1 = \min\{2k, n\} + 1$, as desired.

Suppose that $n < 2k+1$. A similar reasoning can be carried over the collection of simplexes $\tilde{\sigma}_1 \subset \ldots \subset \tilde{\sigma}_n$ defined as follows (See also Figure 2):

$$\tilde{\sigma}_{2i-1} = \{(2, (1, 1)), (3, (1, 2)), \ldots, (2i-1, ((i-1), i)), (2i, (i, i))\}, \text{ for } 1 \leq i \leq \left\lfloor \frac{n}{2} \right\rfloor,$$

$$\tilde{\sigma}_{2i} = \{(2, (1, 1)), (3, (1, 2)), \ldots, (2i, (i, i)), (2i+1, (i, i+1))\}, \text{ for } 1 \leq i \leq \left\lfloor \frac{n-1}{2} \right\rfloor,$$

$$\tilde{\sigma}_n = \begin{cases} \{(2, 1, 1), (3, 1, 2), (n, \frac{n}{2}, \frac{n}{2}), \ldots, (1, \frac{n}{2}, \frac{n}{2} + 1)\} & (n \text{ even}) \\ \{(2, 1, 1), (3, 1, 2), (n, \frac{n-1}{2}, \frac{n-1}{2} + 1), \ldots, (1, \frac{n-1}{2} + 1, \frac{n-1}{2} + 1)\} & (n \text{ odd}) \end{cases}$$

Note that $sign(\tilde{\sigma}_i) = (-1)^{i+1}$ and if $\tilde{\sigma}$ is a $(dim(\tilde{\sigma}_i) - 1)$ face of $\tilde{\sigma}_i$ different from $\tilde{\sigma}_{i-1}$, then $sign(\tilde{\sigma}) = sign(\tilde{\sigma}_i)$. We have established above (Claim 5.2) that $\psi(\sigma_n)$ contains at least one $(n - 1)$-dimensional simplex with $c(s) = C_n$, where C_n is a set of n opinions. By applying the same reasoning, considering simplexes $\tilde{\sigma}_1, \ldots, \tilde{\sigma}_n$ instead, one can show that $\psi(\tilde{\sigma}_n)$ contains a $(n - 1)$-dimensional simplex $\tilde{s}$ colored with a set $\tilde{C}_n \subseteq U$ of size n. Since $sign(\tilde{\sigma}_n) = (-1)^{n+1} = -sign(\sigma_n)$, $\tilde{C}_n \neq C_n$ (Claim 5.2), from which we conclude that $|U| \geq n + 1$.

References

1. Afek, Y., Attiya, H., Dolev, D., Gafni, E., Merritt, M., Shavit, N.: Atomic Snapshots of Shared Memory. J. ACM 40(4), 873–890 (1993)

2. Attiya, H., Rajsbaum, S.: The Combinatorial Structure of Wait-free Solvable Tasks. SIAM Journal of Computing 31(4), 1286–1313 (2002)
3. Awerbuch, B., Patt-Shamir, B., Varghese, G.: Self-stabilization by Local Checking and Correction. In: 32nd Annual IEEE Symposium on Foundations of Computer Science (FOCS), pp. 268–277 (1991)
4. Bauer, A., Falcone, Y.: Decentralised LTL monitoring. In: Giannakopoulou, D., Méry, D. (eds.) FM 2012. LNCS, vol. 7436, pp. 85–100. Springer, Heidelberg (2012)
5. Bauer, A., Leucker, M., Schallhart, C.: Comparing LTL Semantics for Runtime Verification. J. Log. and Comput. 20(3), 651–674 (2010)
6. Berkovich, S., Bonakdarpour, B., Fischmeister, S.: Gpu-based Runtime Verification. In: 27th IEEE International Parallel & Distributed Processing Symposium (IPDPS), pp. 1025–1036 (2013)
7. Chandy, M., Lamport, L.: Distributed Snapshots: Determining Global States of Distributed Systems. ACM Trans. Comput. Syst. 3(1), 63–75 (1985)
8. Chaudhuri, S.: More Choices Allow more Faults: Set Consensus Problems in Totally Asynchronous Systems. Information and Computation 105(1), 132–158 (1993)
9. Fischer, M., Lynch, N., Paterson, M.: Impossibility of Distributed Consensus with One Faulty Process. J. ACM 32(2), 374–382 (1985)
10. Fraigniaud, P., Korman, A., Peleg, D.: Local Distributed Decision. In: 52nd Annual IEEE Symposium on Foundations of Computer Science (FOCS), pp. 708–717 (2011)
11. Fraigniaud, P., Rajsbaum, S., Roy, M., Travers, C.: The Opinion Number of Set-Agreement. Technical report hal-01073578 (2014), http://hal.inria.fr/hal-01073578/PDF/
12. Fraigniaud, P., Rajsbaum, S., Travers, C.: Locality and Checkability in Wait-free Computing. Distributed Computing 26(4), 223–242 (2013)
13. Fraigniaud, P., Rajsbaum, S., Travers, C.: On the Number of Opinions Needed for Fault-Tolerant Run-Time Monitoring in Distributed Systems. In: Bonakdarpour, B., Smolka, S.A. (eds.) RV 2014. LNCS, vol. 8734, pp. 92–107. Springer, Heidelberg (2014)
14. Henle, M.: A Combinatorial Introduction to Topology. Dover (1994)
15. Herlihy, M., Kozlov, D., Rajsbaum, S.: Distributed Computing Through Combinatorial Topology. Morgan Kaufmann-Elsevier (2013)
16. Herlihy, M., Shavit, N.: The Topological Structure of Asynchronous Computability. J. ACM 46(6), 858–923 (1999)
17. Raynal, M.: Concurrent Programming - Algorithms, Principles, and Foundations. Springer (2013)
18. Sen, K., Vardhan, A., Agha, G., Rosu, G.: Decentralized Runtime Analysis of Multithreaded Applications. In: 20th International IEEE Parallel & Distributed Processing Symposium (IPDPS) (2006)

On the Importance of Registers
for Computability

Rati Gelashvili, Mohsen Ghaffari, Jerry Li, and Nir Shavit

Massachusetts Institute of Technology, Cambridge USA
{gelash,ghaffari,jerryzli}@mit.edu, shanir@csail.mit.edu

Abstract. All consensus hierarchies in the literature assume that we have, in addition to copies of a given object, an unbounded number of registers. But why do we really need these registers?

This paper considers what would happen if one attempts to solve consensus using various objects but without any registers. We show that under a reasonable assumption, objects like queues and stacks cannot emulate the missing registers. We also show that, perhaps surprisingly, initialization, shown to have no computational consequences when registers are readily available, is crucial in determining the synchronization power of objects when no registers are allowed. Finally, we show that without registers, the number of available objects affects the level of consensus that can be solved.

Our work thus raises the question of whether consensus hierarchies which assume an unbounded number of registers truly capture synchronization power, and begins a line of research aimed at better understanding the interaction between read-write memory and the powerful synchronization operations available on modern architectures.

1 Introduction

In a seminal paper [7], Herlihy introduced the *consensus hierarchy*, where the synchronization power of an object is measured by its *consensus number*, defined as the maximum number of processes for which wait-free consensus is solvable using instances of the object and as many *read-write registers* as needed. But do we really need these read-write registers? In this paper we consider what would happen if one attempts to solve consensus (henceforth we will use the term "solve" to mean a wait-free solution) using various objects without any registers.

Consider the following interesting example. It is well known [7] that a single queue initialized with two items and with two registers, can solve two process consensus. We show that this is possible even if the queue is in an arbitrary initial state, and that a queue can solve two process consensus even without registers if it is initialized properly. Moreover, two queues in arbitrary initial states are sufficient for solving two process consensus. On the other hand, we prove that it is impossible to solve two process consensus using a single empty queue. In other words, unless you have multiple queues or multiple registers, a

M.K. Aguilera et al. (Eds.): OPODIS 2014, LNCS 8878, pp. 171–185, 2014.
© Springer International Publishing Switzerland 2014

queue's ability to solve consensus is completely dependent on its initialization. This example motivates us to better understand the computational effects of the number of objects and their initialization when no registers are available.

We begin our investigation by considering a general class of objects we refer to as *consistent sets*, that includes natural objects such as queues, stacks and priority queues. Most of the above examples for queues are specific instances of our results for consistent set objects. We show that it is possible to solve two process consensus with a single consistent set object and two registers or with two consistent set objects, even when the objects are initialized in arbitrary states. We also show the corresponding generalization for the impossibility result mentioned above:

Theorem 1. *It is impossible to solve consensus for two processes using a single consistent set object initialized in an empty state.*

As far as we know this is the first result showing that initialization to a different natural state matters for reaching agreement. At its core, the proof involves inductively constructing an interleaving of two solitary executions, such that the processes cannot distinguish between running alone and running in this interleaved execution. However, obtaining the indistinguishability guarantees is rather involved. It requires a new technique to adapt the interleaving to the state of the consistent set object, and involves constructing successive pieces of the interleaved execution separately and then merging them. The challenge is to maintain indistinguishability, which we prove is possible because of the properties of a consistent set object.

We have so far focused on whether two processes can solve consensus using a limited number of objects. This question has practical value as typically small numbers of objects are used in most data structure implementations. However, on the more theoretical side, the work of Jayanti [9] shows that robust consensus hierarchies must allow an arbitrary number of objects. Here we will assume that processes communicate using an unlimited supply of linearizable objects [8], and as in [3,6,10], we will also assume that there are an unlimited number of processes in the system. Although, in this setting, our impossibility results will still hold in a weaker model where only a bounded number of processes are allowed to run concurrently. (In fact, even if the algorithms can assume that only two processes will ever run at the same time).

Let us say that an implementation is *isolation-bounded* if the following holds: there exists an absolute constant M, such that when the very first method call is executed in complete isolation, it takes at most M steps. Practically all natural algorithms are isolation-bounded, even when an unbounded number of processes are allowed to be concurrent. For example, all algorithms where the step-complexity of a method can be upper-bounded by a function of the maximum contention (number of concurrent processes) encountered are isolation-bounded. We will henceforth consider isolation-bounded implementations.

Consider the test-and-set task [1], a simplification of consensus in which exactly one process knows it is the winner (returns 1) and all other processes know that they are losers (return 0), and assume a corresponding linearizable test-and-set object.

We begin by showing the following results that capture the effects of having registers:

Theorem 2. *It is impossible to implement an isolation-bounded test-and-set object for an unbounded number of processes using any number of (possibly infinitely many) empty queues (or empty stacks).*

The proof of this theorem is interesting as it follows along lines that have, as far as we know, never been used before in deriving shared-memory lower bounds. Essentially, we wish to reduce the general case in which infinitely many processes access infinitely many queues, to the case where infinitely many processes access only finitely many queues in their solo executions. Once reduced, we can use a counting argument to find two processes whose solo executions can be interleaved so that for both processes running in the interleaved execution, their execution is indistinguishable from running alone. To achieve this reduction, we use an argument, akin to diagonalization, to produce an infinite set of processes for which the desired property essentially holds.[1]

On the other hand, if read-write registers are available, one can use the tournament tree construction from [2] to get the following result

Theorem 3. *It is possible to implement an isolation-bounded test-and-set object for an unbounded number of processes using infinitely many consistent set objects (in any initial configuration) and read-write registers.*

These theorems have a few important corollaries. The first of these corollaries demonstrates a fundamental difference between registers and objects like stacks and queues.

Corollary 1. *It is impossible to implement a read-write register in an isolation-bounded way using any number of (possibly infinitely many) empty queues (stacks).*

Interestingly, if the number of processors in the system is bounded, simulations of a read-write register exist [4].

The second corollary is about initialization. Algorithms for consensus usually assume that the objects and registers are initialized in a certain way. In fact, the consensus number of an object can change depending on the initial state. Consider an object with a consensus number at least two that has an additional "invalid" state, unreachable from all other states, such that in the invalid state, all method calls return *null*. Clearly, the object initialized in the invalid state has consensus number one.

But generally, in most initial states the object will have the same consensus number. For instance, as shown in [5], this is always true for states reachable

---

[1] We remark that our proof requires the axiom of countable choice, which we will assume without comment when necessary.

from each other.[2] Our second corollary shows that perhaps surprisingly, for some objects the difference in the synchronization power in these initial states can still be quite significant:

Corollary 2. *It is impossible to implement a queue (a stack) containing one element in its initial state using any number of (possibly infinitely many) empty queues (stacks) in an isolation-bounded way.*

2   Consistent Sets and Two Consensus

Let us define a class of objects, that we will call *consistent sets*. Each consistent set object represents a data-structure of items and implements two linearizable methods: insert($item$) and remove(). We say that a consistent set object *contains* an item, if the item has not been removed since its last insertion in the set. Assume that $s_1, s_2, \ldots, s_m$ are the items contained in some consistent set object, whereby s_1 was inserted before s_2, etc, before s_m. The remove() operation returns one of the items s_i, selected based on a fixed function F, i.e. $s_i = F(s_1, s_2, \ldots, s_m)$. If $m = 0$, then a special value *null* (which can never be an item contained in the set) is returned instead. A consistent set object can be initiliazed to an empty state (containing 0 items), or with any finite number of items pre-inserted in an arbitrary fixed order.

Each consistent set object has its function F, defined for all possible item sequences that satisfies the following two *consistency* properties:
- If there exist (possibly empty) sequences of items L, M, R, such that $F(L, s_i, M, s_j, R) = s_i$, then there do not exist item sequences (represented by dots), so that $F(\ldots, s_i, \ldots, s_j, \ldots) = s_j$.
- If there exist (possibly empty) sequences of items L, M, R, such that $F(L, s_i, M, s_j, R) = s_j$, then there do not exist possible item sequences (represented by dots), so that $F(\ldots, s_i, \ldots, s_j, \ldots) = s_i$.

The exact choice of function F determines precise semantics of the data-structure. For instance, a first-in-first-out queue, a stack and a priority queue are all consistent set objects and correspond to particular choices of F: for a queue $F(s_1, \ldots, s_m) = s_1$, for stack F picks s_m and for a priority queue it picks the item with the maximum (minimum) priority.

Lemma 1. *It is possible to solve wait-free two process consensus using any consistent set object O, initialized with a finite number of arbitrary items in an arbitrary order.*

Proof. Let W be an item that is different from all initial items in O. We claim that the algorithm described in pseudo-code on Figure 1 solves wait-free consensus for two processes. It is straightforward to show wait-freedom, so it suffices to demonstrate that the algorithm solves consensus. It is also straightforward to show that each process returns either its own value or the other process's value.

---

[2] In the above example where the consensus number changed, no state was reachable from the invalid state.

Variables:
$Proposed[2] = \{\bot\};$
$O;$

1 **procedure** decide$(v, id = 0)$
2     $Proposed[0] \leftarrow v$
3   **if** $Proposed[1] = \bot$ **then**
4   **return** v
5   **while**$(true)$
6         $item \leftarrow O.\text{remove}()$
7   **if** $item = W$ **then**
8   **return** v
9   **if** $item = null$ **then**
10   **return** $Proposed[1]$

Algorithm 1: Pseudo-code for process 0

1 **procedure** decide$(v, id = 1)$
2     $O.\text{insert}(W)$
3     $Proposed[1] \leftarrow v$
4   **if** $Proposed[0] \neq \bot$ **then**
5   **return** $Proposed[0]$
6   **while**$(true)$
7         $item \leftarrow O.\text{remove}()$
8   **if** $item = W$ **then**
9   **return** v
10   **if** $item = null$ **then**
11   **return** $Proposed[0]$

Algorithm 2: Pseudo-code for process 1

Fig. 1. Two process consensus using a consistent set object O and registers

For $i \in \{0, 1\}$, let v_i denote the value that process i gets as input. Suppose for the sake of contradiction that the processes return different values. There are two cases.

Process i returns v_i, for $i \in \{0, 1\}$: By inspection, the only way that process 1 can return v_1 is if it returns at line 9, that is, it enters the while loop then removes W. There are two sub-cases. Suppose process 0 returns on line 4, so that it returned since it saw $Proposed[1] = \bot$, and returns v_0. By inspection, this is only possible if this occurs before process 1 executes line 3, which implies that process 0 executes line 2 before process 1 executes line 4, which implies that when process 1 reads $Proposed[0]$ on line 4, it will see v_0, and thus will return it, which is a contradiction. Alternatively, process 0 could return on line 8, but this would imply that on line 7, in some iteration of the loop, removes W. Since W is only inserted once into the consistent set, this is a contradiction, since process 1 must remove it as well.

Process i returns v_{1-i}, for $i \in \{0, 1\}$: By inspection, the only way that process 0 can return v_1 is if it returns on line 10, that is, it sees an empty consistent set. There are again two sub-cases, since process 1 can return v_0 in one of two ways. Suppose process 1 returns on line 5. Then by that point in the execution, process 1 has already executed $O.\text{insert}(W)$. Then, when process 0 enters the while loop, it is guaranteed to eventually remove W since it is the only process removing elements from the consistent set, so it will return v_0 as well, which is a contradiction. Thus, suppose process 1 returns on line 11. But this happens after process 1 performs $O.\text{insert}(W)$, and neither process can see W while removing elements from the consistent set until the set is empty, which is a contradiction.

Let us next consider the synchronization power of consistent sets without registers.

Lemma 2. *It is possible to solve wait-free two process consensus using any two consistent set objects O_0 and O_1, initialized with a finite number of arbitrary items in an arbitrary order.*

Proof. The algorithm is described on Figure 2. Recall F is the function which uniquely defines the consistent set. We have two consistent set objects: O_0, where process O inserts to, and O_1, where process 1 inserts to. Inserted elements are pairs of form $\{P_i, v_i\}$ and $\{Q_i, v_i\}$, where v_i is the input of process i, and P_i or Q_i are two different prefixes, such that the corresponding pairs are not the same as any of the initial items in sets O_i.

We claim that the algorithm solves consensus. As with the proof of Lemma 1, let v_i be the input of the process i, for $i \in \{0, 1\}$. It is again straightforward to see that the algorithm is wait-free. Thus it suffices to prove that the processes will return the same value. Suppose for the sake of contradiction that the processes return different values. Notice by the definition of a consistent set, if a process's call to remLW(O) returns $\{L, v\}$, then there must have been a previous remove operation performed on O which returned the unique other element e inserted into O with e.second $= v$ and e.first $\in \{P_0, P_1, Q_0, Q_1\}$. Moreover, if e was removed due to a remLW operation, that operation would return $\{W, v\}$.

There are two cases.

Process i returns v_i, for $i \in \{0, 1\}$: By inspection, there is one way for process 0 to return v_0, which is to return on line 7, which implies that a_0.first $= W$ and $a_1 = null$. That $a_1 = null$ implies that process 0 executes line 4 before process 1 executes line 13, which implies that $b_0 \neq null$. Since a_0.first $= W$, this implies that b_0.first $= L$. Moreover, since $a_1 = null$, this implies that b_1.first $= W$, which is a contradiction, as then process 1 cannot return v_1.

Process i returns v_{1-i}, for $i \in \{0, 1\}$: By inspection, there is one way for process 1 to return v_0, which is for it to fail the if statement on line 17. To fail this if statement means that b_0.first $= L$ and b_1.first $= W$ (since $b_1 \neq null$). Since b_0.first $= L$, this implies that process 1 finishes line 15 after process 1 finishes line 5, and it also implies that a_0.first $= W$. This implies that process 0 finishes executing line 4 before process 1 starts executing line 16, so the only way that b_1.first $= W$ is if $a_1 = null$, thus process 0 will return v_0 as well.

Any algorithm for two-consensus (including the algorithms above) can be used to solve test-and-set for two processes, simply by having each process return 1 instead of its own value and 0 otherwise.

Let us call a state of an instance of any consistent set object O *lucky*, if it contains only a single copy of some item W.

Lemma 3. *It is possible to implement a test-and-set object for an unbounded number of processes using a single consistent set object O initialized in a lucky state.*

Proof. The algorithm for each process is to simply remove items from O until observing W or *null*. In the first case, the process returns 1 and in the second

Variables:
$O_0, O_1;$

```
1 procedure decide(v, id = 0)
2     O₀.insert({P₀, v})
3     O₀.insert({Q₀, v})
4     a₁ ← remLW(O₁)
5     a₀ ← remLW(O₀)
6     if a₀.first = W and
         a₁ = null then
7         return v
8     else
9         return a₁.second
```

Algorithm 3: Pseudo-code for process 0

```
1 procedure remLW(O)
2     while(true)
3         t ← O.remove()
4         if t = null then
5             return null
6         if t.first ∈ {Pᵢ, Qᵢ} then
7             v = t.second
8             if F({Pᵢ, v}, {Qᵢ, v}) = t
             then
9                 return {W, v}
10            else
11                return {L, v}
12 procedure decide(v, id = 1)
13     O₁.insert({P₁, v})
14     O₁.insert({Q₁, v})
15     b₀ ← remLW(O₀)
16     b₁ ← remLW(O₁)
17     if b₀.first ≠ L or b₁.first = L
         then
18         return v
19     else
20         return b₀.second
```

Algorithm 4: Pseudo-code for process 1

Fig. 2. Two process consensus using two consistent sets objects O_0 and O_1

case, it returns 0. By the semantics of the data-structure, one and only one process will remove W and return 1. Moreover, that process can in fact be linearized as the winner of the test-and-set, i.e. as the first to call the test-and-set() method (since otherwise, another method call must have completed strictly earlier and that it would have removed the unique element W).

Lemma 4. *There exists a consistent set object O, such that it is possible to solve wait-free two process consensus with O initialized in a lucky state.*

Proof. A first-in-first-out queue is such an object. The algorithm for each process is to first enqueue its own item and then keep dequeuing until either observing W or *null*. In the first case, the process returns own value. Otherwise, it returns the value of the other process (we show below how), and the exact argument from Lemma 3 finishes the correctness proof.

To show how the process knows the value to return, consider the process p that observes *null* at time t. Since the other process has dequeued W by time t, it must have already enqueued its value, which comes later than all original items of O (including W) in the first-in-first-out order. The other item with this property is the input value of p itself. Therefore, the last two items dequeued by p must be the input values of the processes, p knows its own value and can simply tell the value of the other process.

Given these insights, the following result may be surprising:

Theorem 1. *It is impossible to solve wait-free two process consensus using a single consistent set object O initialized in an empty state.*

Proof. Assume the contrary. Then the existence of the consensus protocol implies that there also exists a wait-free test-and-set implementation for two processes using just a single consistent set object O initialized in an empty state. For each process $i \in \{0, 1\}$ there exists a solo execution where process i runs in isolation and returns 1 after some finite number t_i of steps. Let E_0 and E_1 be these solo executions. Each step in these executions is either an insert($item$) or remove() call on O.

We obtain a contradiction by constructing a schedule where both processes are executed, but never observe any difference from their solo executions, i.e. the execution of process i is indistinguishable from E_i from its prospective. Formally, given a serial execution E_i which only makes method calls to O, and a linearized execution E containing E_i and other method calls from other processes to O, we say that E_i is *indistinguishable* from E if for every remove operation in E_i, it gets the same response as it does in E. Clearly, if process i has solo execution E_i and E is an execution which is indistinguishable from E_i, it must return 1 in E, so if an execution E is indistinguishable from two solo executions, we derive a contradiction.

To construct this interleaving, we use induction on total number of steps in E_0 and E_1 to prove the existence of the interleaved execution. We say the first ℓ steps of an execution form an ℓ-*prefix*.

The following proposition provides the base case for induction.

Proposition 1. *If for one of the processes, say for process j, $t_j = 0$ holds, then it is possible to interleave the executions E_0 and E_1 such that the interleaved execution is indistinguishable from the solo execution for each process.*

Proof. The number of steps in solo execution E_j is 0, so we start by running process j which immediately returns as in E_j and does not change the state of the object O. Thus we then complete the interleaved execution by running process $1 - j$ until it returns, and because the starting state of O is empty as in E_{1-j}, this execution also precisely matches E_{1-j}.

For inductive step, assume we know that if the total number of steps in two solo executions E_0 and E_1 is less than k, then it is possible to interleave them such that the interleaved execution is indistinguishable from the solo execution for each process.

We now consider several cases, each requiring a different treatment. By adjusting formulations it is possible to merge some cases, but the particular structure is chosen for clarity. Let the total number of steps in E_0 and E_1 be k.

Case 1: A mute prefix: An ℓ-prefix for a solo execution for process i is called *mute* if O remains empty after the prefix is executed by process i in isolation.

Proposition 2. *If one of the executions, say execution E_j contains a non-empty mute prefix, then it is possible to interleave the executions E_0 and E_1 such that* *the interleaved execution is indistinguishable from the solo execution for each process.*

Proof. We start the interleaved execution by letting process j execute the mute prefix of E_j. This is possible because we actually run process j in isolation, so it simply executes the mute prefix exactly as in E_j. Afterwards, by definition of the mute prefix, O is empty. Moreover, the total number of steps in the solo executions that the rest of the interleaved execution should match has strictly decreased. Therefore, we can use the inductive hypothesis for the same E_{1-j} and E_j without the non-empty prefix to construct the rest of the interleaved execution.

Thus we may assume that the solo execution E_i for process $i \in \{0, 1\}$ does not contain a mute prefix and it consists of non-zero number of steps. Define $f_i(\ell)$ to be the item that would be removed by a remove() call right after executing an ℓ-prefix of E_i in isolation.

Case 2: A barrier: For $i \in \{0, 1\}$, let $s_1, s_2, \ldots, s_m$ be the items that are inserted and removed from O during the solo execution E_i by process i, in order of their insertion. Let g_i be the item that would be removed the last if we first inserted all of these items in O in order, and then removed them one-by-one. Note that this does not have to be s_m. We call $f_i(\ell)$ a *barrier* if $F(f_i(\ell), g_{1-i}) = g_{1-i}$.

Example 1. The motivating example of a barrier is when O is a priority queue which returns elements with high priority first. Consider the situation where process 0 (say) inserts a number of elements into the priority queue with priority ≤ 1 then some elements with priority 2 in its solo execution, and process 1 inserts many elements into O with priorities either 2 or 3 in its solo execution. Then, the prefix of process 0 which consists of it inserting elements with priority ≤ 1 forms a barrier, and such a prefix is natural to consider because this essentially acts like a mute prefix to process 1 in that process 1 will never see anything from this prefix, and mute prefixes are easy to induct on.

To reason about this case, we need a technical property about the behavior of consistent sets which is obvious for simple objects such as queues, stacks, and priority queues.

Proposition 3. *Consider a serial execution E consisting of calls to a consistent set object O. Let s be some element inserted and subsequently removed during E, and let E' be the execution constructed by removing insert(s) and the remove() which returned s. Then the output of all other remove() operations in E' is unchanged.*

Proof. We will actually prove a slightly stronger statement: that at any point in the execution E, if O contains s, at that same point in time in E', the state of O is identical except with s removed, and if O does not contain s. then at the same point in time in E', the state of O is exactly the same. This clearly implies our claim.

To prove this stronger statement, we proceed by contradiction. Let R_1 be the first operation after which the states of O in E and E' do not follow this invariant. By inspection this must be a remove operation. Denote the remove() which returned s by R. Clearly the behavior of O at any state before insert(s) occurs is the same in E and E', so R_1 must happen after the insertion of s. Similarly, if R_1 was after R in E, then by the invariant, before R the state of O in E and E' is identical. Thus the last remaining case is if R_1 was scheduled before R in E but after insert(s). Suppose in E it returns some element s' and in E' it returns some element $s'' \neq s'$. Let $A = s_1, \ldots, s_\ell$ be the list of objects in present in O ordered by insertion time if we execute E but pause right before executing R_1. Clearly this is of the form L, s', M, s'', R or L, s'', M, s', R for some L, M, R, where s is in either L, M, or R. W.l.o.g. assume that it is of the former type, and assume $s \in L$ (the other cases are identical). We know that $F(A) = s'$. Form L' by removing s from L, and let $A' = L', s', M, s'', R$. Then by consistency, $F(A') \neq s''$. But by the invariant, before R_1, the state of O in E' was exactly A', which is impossible. This proves the proposition.

Now we have the tools to do the induction in the presence of a barrier:

Proposition 4. *If one of the executions, say execution E_j, contains a barrier $f_j(\ell)$, then it is possible to interleave the executions E_0 and E_1 such that the interleaved execution is indistinguishable from the solo execution for each process.*

Proof. Consider the largest ℓ so that the ℓ-prefix of E_j is a barrier. We start building the desired interleaved execution by executing the ℓ-prefix p_j of E_j. This leaves a number of items in O, so in particular $f_j(\ell)$ is well-defined. Now, let us *trim* the remaining piece of E_j: we get rid of all remove() operations that in the solo execution remove items inserted in p_j. Thus, the *trimmed* schedule $\tilde{E}_j$ does not contain the l-prefix of E_j and any later remove() operations that in the solo execution return items inserted during the l-prefix. By the above proposition, every remove() operation in $\tilde{E}_j$ returns the same thing it did in E_j. In particular, none of them return *null* because none of them could have returned null in E_j as otherwise E_j would have had a mute prefix.

Because the number of operations in $\tilde{E}_j$ is strictly smaller than in E_j, using our inductive hypothesis let us construct an indistinguishable interleaved execution X for executions $\tilde{E}_j$ and E_{1-j} assuming that O started in an empty state. Note that execution X is only indistinguishable if O is initially empty. Moreover, we do not immediately get any guarantees for the original execution E_j.

However, we will show that it is possible to interleave the trimmed operations from E_j back into X to create X' so that $p_j X'$ is a valid interleaving of E_0 and E_1 and is indistinguishable to both processes from their solo executions. Assume the opposite, and consider first time t at which we are unable to indistinguishably schedule the next operation without violating the above invariant. Since the only operations which provide feedback are remove() operations, we can assume without the loss of generality that the next operations to be scheduled for both processes are both remove() operations.

Suppose at time t, the next operation scheduled in X is by process $1 - j$. The operation has to be a remove() that returns some item s instead of another item $r \neq s$ that would be returned at this point in E_{1-j}. By our assumption, all previous operations have been indistinguishable, so O has to contain item r at time t. Also, r is clearly inserted by process $1 - j$, since it is removed by process $1-j$ in the solo execution E_{1-j}. If s was inserted during X (and not in p_j), since we still insert the items according to X in the new interleaved execution, during the corresponding remove() operation in X items s and r would certainly be contained in O in the exact same order as during the above remove() operation in the interleaved execution. But since X is indistinguishable from E_{1-j}, the removal in X returns r and not s, contradicting the consistency of O.

If s was inserted during p_j, let us w.l.o.g. assume that $f_j(\ell)$ was inserted after s and g_{1-j} after r. We will show that $F(s,r) = r$, a contradiction since that means that the remove operation at time t would return r instead of s, as s is inserted before r in the execution of interest since it was inserted during p_j. Consider $u = F(s, f_j(\ell), r, g_{1-j})$.[3] We know $F(r, g_{1-j}) = r$ by the definition of g_{1-j}, so $u \neq g_{1-j}$ by the definition of consistent sets. Similarly, since $F(f_j(\ell), g_{1-j}) = g_{1-j}$ since $f_j(\ell)$ is a barrier, we know $u \neq f_j(\ell)$. Finally, $F(s, f_j(\ell)) = f_j(\ell)$ by definition of $f_j(\ell)$, so we know that $u \neq s$. Thus, $u = r$, and so by the properties of consistent sets we conclude that $F(s,r) = r$.

Now assume that the next operation according to X is by process j. The next operation to be scheduled for E_j must be a remove (which may have been trimmed). Call this operation R. By assumption, it removes some item s instead of an item $r \neq s$ which would be removed in E_j at this step. If s was inserted by process j, then in solo execution E_j process j should have observed items s and r in O in the same order as here, but removed r, contradicting the consistency property.

Thus suppose s was inserted by process $1 - j$. We claim that R must have been trimmed, since otherwise R is the next remove operation in execution X. But then, since all the items present in O at this point in X must also be present in O in this point in the execution we are building, since we have included all the actions of X up to this point in our execution, this implies by the definition of consistent set objects, that in X, R must also remove s, contradicting the indistinguishability of X from solo executions.

But if R was trimmed and would at this point return some s inserted by process $1 - j$, we claim that there exists a $\ell' > \ell$ so that the ℓ'-prefix of E_j would also be a barrier, which contradicts our choice of ℓ. Indeed, let r be the item that R, the last remove() up to this point in the solo execution E_j, removes and let v be the item that would be removed if we executed another remove() right after E_j (v has to exist, otherwise the whole execution E_j is a mute prefix). Since the removal of r is trimmed, insert(r) must be in the p_j. Assume without the loss of generality that $f_j(l)$ is inserted after r and before v in E_j and consider

---

[3] The other cases are symmetric: we would consider $F(f_j(l), s, r, g_{1-j})$, $F(s, f_j(\ell), g_{1-j}, r)$ or $F(f_j(\ell), s, g_{1-j}, r)$.

$F(r, f_j(l), v)$.[4] $F(r, f_j(l)) = f_j(l)$ must hold by the definition of $f_j(l)$, and since the last trimmed removal also observed v but removed r, $F(r, v) = r$ holds. By the definition of a barrier, $F(f_j(l), g_{1-j}) = g_{1-j}$, and so combining these three facts and using consistency like before we get $F(r, f_j(l), v, g_{1-j}) = g_{1-j}$ which again by consistency of F implies that $F(v, g_{1-j}) = g_{1-j}$. Thus if we take the prefix of E_j up to and including R, we get another barrier which has length strictly larger than ℓ, which is a contradiction. This completes the proof of the proposition.

Case 3: No Mute Prefixes or Barriers.: The rest of the proof of the main theorem is relegated to the full version[5] so as not to lose the reader in details and considers the case when none of the executions E_i ($i \in \{0, 1\}$) contains a mute prefix or a barrier. The application of the inductive hypothesis (albeit twice) and the trimming technique is still required, but the partitioning of executions and the proof details differ.

3 Unbounded Number of Objects

Theorem 2. *It is impossible to implement an isolation-bounded test-and-set object for an unbounded number of processes using any number of (possibly infinitely many) empty queues (or empty stacks).*

Proof. Let us assume contrary and consider an isolation-bounded algorithm that implements test-and-set for an unbounded number of processes with initially empty queues. Because of isolation-boundedness, any process that runs in isolation from the initial state can take at most a fixed number of steps, say M, each being an insert($item$) or remove() operation on one of the queues, before returning 1.

Associate to each process p the ordered list s_q of the M steps it would take if it ran in isolation. We call this quantity the *signature* of p. Suppose each queue is touched by finitely many signatures. Let Q_1 be any queue which is touched, say by process p. Then p's signature touches at most M queues, call them $Q_1, \ldots, Q_M$. At most finitely many other processes can touch these same queues, so there must be a process q whose signature does not touch any of the Q_i. Running p then q gives us an immediate contradiction, since their actions on the queues they touch do not interact at all, and thus they cannot distinguish between running together and running in isolation, and must both return 1.

Thus we can assume that there exists a queue Q_1 such that an operation on this queue occurs in infinitely many signatures. Let $\mathcal{P}_1$ denote the set of processes whose signatures contain an operation on Q_1. Next, if there is a queue Q_2 such that an operation on it occurs in infinitely many signatures from $\mathcal{P}_1$, we consider this infinite subset $\mathcal{P}_2 \subseteq \mathcal{P}_1$. Inductively, we build sets $\mathcal{P}_i \subseteq \mathcal{P}_{i-1} \subseteq \ldots \subseteq \mathcal{P}_1$ and choose queues Q_i, until the process terminates. This can only happen at most

---

[4] Otherwise, considering the respective order works analogously.

[5] Available under publications at http://groups.csail.mit.edu/mag/

M times, since the members of $\mathcal{P}_M$ (if they exist) must in isolation perform the maximum number of allowed operations (i.e. M operations), namely on the queues $Q_1, \ldots, Q_M$. Thus, we end up with an infinite set of signatures $\mathcal{P}_m$ ($m \leq M$), such that each of the signatures contains an operation on each Q_j ($1 \leq j \leq m$), and for every other queue, an operation on it is contained only in a finite number of signatures from processes in P_m. We let $\mathcal{Q} = \{Q_1, \ldots, Q_m\}$.

We can now find an infinite subset $\mathcal{P} \subseteq \mathcal{P}_m$, such that if two processes from $\mathcal{P}$ have signatures which involve operations on a shared queue, this queue has to be one of our selected queues $\mathcal{Q}$. We do so inductively: choose $p_1 \in \mathcal{P}_m$ arbitrarily. This process's signature touches at most $M - 1$ queues not in $\mathcal{Q}$. Moreover, finitely many other processes in $\mathcal{P}_m$ have signatures which touch these queues by the construction of $\mathcal{P}_m$. Thus we can choose a $p_2 \in \mathcal{P}_m$ which does not touch any of these queues, and then we recurse to find p_i for all i, and we let $\mathcal{P} = \{p_i\}_{i=1}^\infty$. It is straightforward to verify that this set has the desired property.

Let us now focus on the processes in $\mathcal{P}$ and consider only the operations they perform on queues $\mathcal{Q}$. Clearly, each process performs at most M such operations when run in isolation. Each operation is either insert($item$) or remove() on some Q_j, thus there are $2m$ different types of operations. There are only finitely many different possibilities to order at most M operations of $2m$ different types, and infinitely many processes in $\mathcal{P}$, thus by the pigeon-hole principle, we can find two processes $p, q \in \mathcal{P}$, such that their signatures both involve the same operations on the same queues in $\mathcal{Q}$ in exactly the same order. Moreover, they may perform actions on queues not in $\mathcal{Q}$, but by the construction of $\mathcal{P}$, the sets of queues they touch outside of $\mathcal{Q}$ are disjoint.

Let us execute p and q in the following "lock-step" fashion: we let p take steps until the first operation on some Q_j, then we let q take its steps until it performs the same type of operation on the same Q_j, etc, until they both finish. At any point in the execution when q has just taken a step, we claim that the following invariant holds: none of the processes have observed a difference from their solo executions, and each queue Q_j contains items that p inserted and items that q inserted, interleaved one-by-one. Moreover, if we only consider the items inserted by one of the processes, say p, they are the same items and in the same order as in the solo execution of p.

p and q could only observe a difference after a remove() call on one of the queues Q_j, because other queues are accessed by only one process. Now, the invariant holds initially, and if the next operation on some Q_j is insertion (necessarily the same queue for both processes, but they may insert different items), we let p insert, then q insert, so the invariant holds afterwards. If it is a removal from some Q_j for both processes, then since the items of p and q are interleaved but consistent with respective solo executions, first removal by p will return the item p previously inserted (or $null$) and does not observe a difference, then q does the same with its item.

Thus, we are able to execute p and q, both of which cannot distinguish the execution from a solo execution and return 1 contradicting the correctness of the test-and-set implementation.

A very similar argument works for the stack, except when running processes in lock-step, if the operation is a remove(), we should reverse the order and let q execute first.

On the other hand, if we have registers available implementing test-and-set becomes possible.

Theorem 3. *It is possible to implement an isolation-bounded test-and-set object for an unbounded number of processes using infinitely many consistent set objects (in any initial configuration) and read-write registers.*

Proof. The adaptive tournament tree from [2] is an algorithm that implements isolation-bounded test-and-set for an arbitrary number of concurrent processes.[6] It requires registers and a black-box test-and-set primitive for two processes. Using Lemma 2, we can do test-and-set for two processes with just two consistent set objects initialized with a finite number of arbitrary items in an arbitrary order (or with one object and registers, per Lemma 1). This two process test-and-set object can be directly plugged into the [2] construction as the building block. The other crucial building block is a splitter object [11], which is easily consructed using registers. The algorithm is isolation-bounded, since any process running in isolation from the initial state stops in the first splitter and participates only in a few two-process test-and-sets.

Corollary 1. *It is impossible to implement a read-write register in an isolation-bounded way using any number of (possibly infinitely many) empty queues (stacks).*

Proof. Assume contrary. Then we can use the same algorithm as in Theorem 3 to implement a test-and-set object for an unbounded number of processes, except we replace each register in the construction with an isolation-bounded register implementation out of empty queues. The resulting test-and-set construction would then only use empty queues and would be isolation-bounded, because both the original implementation and the new register implementation are isolation-bounded. In fact, if the constant bounds on the number of steps are c_1 and c_2, the bound for the new construction would be $c_1 c_2$. Such a construction, however, contradicts Theorem 2.

Corollary 2. *It is impossible to implement a queue (a stack) containing one element in its initial state using any number of (possibly infinitely many) empty queues (stacks) in an isolation-bounded way.*

Proof. By Lemma 3, a single consistent set object initialized in a lucky state can implement a wait-free test-and-set object for unbounded number of processes. A queue is a consistent set object and a state with a single item is a lucky state. By inspection, the test-and-set algorithm from Lemma 3 using a queue with a single element is isolation-bounded (an initial isolated run involves just

---

[6] We consider non-randomized version of the construction.

one removal). Therefore, being able to implement a queue with a single item would immediately allow implementing an isolation-bounded test-and-set object for an unbounded number of processes, which by Theorem 2 is impossible using any number of empty queues.

Acknowledgements. Support is gratefully acknowledged from the National Science Foundation under grants CCF-1217921, CCF-1301926, and IIS-1447786, the Department of Energy under grant ER26116/DE-SC0008923, and the Oracle and Intel corporations.

The authors would like to thank Eli Gafni and Yehuda Afek for helpful conversations and feedback.

References

1. Afek, Y., Gafni, E., Tromp, J., Vitányi, P.M.B.: Wait-free test-and-set. In: Segall, A., Zaks, S. (eds.) WDAG 1992. LNCS, vol. 647, pp. 85–94. Springer, Heidelberg (1992)
2. Alistarh, D., Attiya, H., Gilbert, S., Giurgiu, A., Guerraoui, R.: Fast Randomized Test-and-Set and Renaming. In: Lynch, N.A., Shvartsman, A.A. (eds.) DISC 2010. LNCS, vol. 6343, pp. 94–108. Springer, Heidelberg (2010)
3. Attiya, H., Bar-Noy, A., Dolev, D., Peleg, D., Reischuk, R.: Renaming in an asynchronous environment. Journal of the ACM (JACM) 37(3), 524–548 (1990)
4. Bazzi, R.A., Neiger, G., Peterson, G.L.: On the use of registers in achieving wait-free consensus. Distributed Computing 10(3), 117–127 (1997)
5. Borowsky, E., Gafni, E., Afek, Y.: Consensus power makes (some) sense! In: Proceedings of the Thirteenth Annual ACM Symposium on Principles of Distributed Computing, pp. 363–372. ACM (1994)
6. Gafni, E., Merritt, M., Taubenfeld, G.: The concurrency hierarchy, and algorithms for unbounded concurrency. In: Proceedings of the Twentieth Annual ACM Symposium on Principles of Distributed Computing, pp. 161–169. ACM (2001)
7. Herlihy, M.: Wait-free synchronization. ACM Transactions on Programming Languages and Systems (TOPLAS) 13(1), 124–149 (1991)
8. Herlihy, M.P., Wing, J.M.: Linearizability: A correctness condition for concurrent objects. ACM Transactions on Programming Languages and Systems (TOPLAS) 12(3), 463–492 (1990)
9. Jayanti, P.: Robust wait-free hierarchies. Journal of the ACM (JACM) 44(4), 592–614 (1997)
10. Merritt, M., Taubenfeld, G.: Computing with infinitely many processes. In: Herlihy, M.P. (ed.) DISC 2000. LNCS, vol. 1914, pp. 164–178. Springer, Heidelberg (2000)
11. Moir, M., Anderson, J.H.: Fast, long-lived renaming (Extended abstract). In: Tel, G., Vitányi, P. (eds.) WDAG 1994. LNCS, vol. 857, pp. 141–155. Springer, Heidelberg (1994)

Scalable Wake-up of Multi-channel Single-Hop Radio Networks

Bogdan S. Chlebus[1,*], Gianluca De Marco[2], and Dariusz R. Kowalski[3]

[1] University of Colorado Denver, Denver, Colorado, USA
[2] Università di Salerno, Fisciano, Italy
[3] University of Liverpool, Liverpool, United Kingdom

Abstract. We consider waking up a single-hop radio network with multiple channels. There are n stations connected to b channels without collision detection. Some k stations may become active spontaneously at arbitrary times, where k is unknown, and the goal is for all the stations to hear a successful transmission as soon as possible after the first spontaneous activation. We present a deterministic algorithm for the general problem that wakes up the network in $\mathcal{O}(k \log^{1/b} k \log n)$ time. We prove a lower bound that any deterministic algorithm requires $\Omega(\frac{k}{b} \log \frac{n}{k})$ time. We give a deterministic algorithm for the special case when $b > d \log \log n$, for some constant $d > 1$, which wakes up the network in $\mathcal{O}(\frac{k}{b} \log n \log(b \log n))$ time. This algorithm misses time optimality by at most a factor of $\log n \log b$. We give a randomized algorithm that wakes up the network within $\mathcal{O}(k^{1/b} \ln \frac{1}{\epsilon})$ rounds with the probability of at least $1 - \epsilon$, for any unknown $0 < \epsilon < 1$. We also consider a model of jamming, in which each channel in any round may be jammed to prevent a successful transmission, which happens with some known parameter probability p, independently across all channels and rounds. For this model, we give a deterministic algorithm that wakes up the network in $\mathcal{O}(\log^{-1}(1/p)k \log n \log^{1/b} k)$ time with the probability of at least $1 - 1/\text{poly}(n)$.

Keywords: multiple access channel, radio network, multi-channel, wake-up, randomized algorithms, distributed algorithms.

1 Introduction

We consider wireless networks organized as a group of stations connected to a number of channels. Each channel provides the functionality of a single-hop radio network. A station can use any of these channels to communicate directly with any other station.

This topology is called multi-channel in the literature. The assumption usually made is that a station can connect to at most one channel at a time for either transmitting or listening. We depart from this restriction and consider the

* This work was supported by the NSF Grant 1016847.

M.K. Aguilera et al. (Eds.): OPODIS 2014, LNCS 8878, pp. 186–201, 2014.
© Springer International Publishing Switzerland 2014

apparently stronger model in which a station can use all the available channels simultaneously and independently from each other, some for transmitting and others for listening. On the other hand, channels do not provide collision detection, which makes the model weaker than a multi-channel with carrier sensing capabilities.

The algorithmic problem that we consider is to wake up the network. Initially, all the stations are dormant but connected and listening to all the channels. Some stations become active spontaneously and want the whole network to be activated and synchronized. The first successful transmission on any channel suffices to accomplish this goal.

We use the following parameters to characterize a multi-channel network. We denote by n the number of stations and b is the number of shared channels. At most k stations become active spontaneously at arbitrary times and join an execution with the goal to wake up the network. When the first message is heard on some channel then the network is considered woken up and synchronized. Stations know n and b, but the number k is an unknown parameter used only to characterize the scalability of a given wake up algorithm.

Our results. We present a deterministic algorithm which wakes up the network in $\mathcal{O}(k \log^{1/b} k \log n)$ rounds. We give a deterministic wake-up algorithm for the special case of sufficiently many channels, which operates in $\mathcal{O}(\frac{k}{b} \log n \log(b \log n))$ time when $b > \lg(128 b \lg n)$. We prove a lower bound of $\Omega(\frac{k}{b} \log \frac{n}{k})$ rounds, which are required by any deterministic algorithm. In view of this lower bound, the algorithm of time performance $\mathcal{O}(\frac{k}{b} \log n \log(b \log n))$ misses time optimality by at most a factor of $\log n \log b$. We give a randomized algorithm that wakes up the network within $\mathcal{O}(k^{1/b} \ln \frac{1}{\epsilon})$ rounds with the probability of at least $1 - \epsilon$, for any unknown $0 < \epsilon < 1$. We also consider a model of jamming, in which each channel in any round may be jammed to prevent a successful transmission, which happens with some known parameter probability p, independently across all channels and rounds. For this model, we give a deterministic algorithm that wakes up the network in $\mathcal{O}(\log^{-1}(1/p) k \log n \log^{1/b} k)$ time with the probability of at least $1 - 1/\mathrm{poly}(n)$.

For a multiple access channel, Jurdziński and Stachowiak [36] gave two randomized algorithms, one working in $\mathcal{O}(\log^2 n)$ time steps with high probability with respect to n, and another working in $\mathcal{O}(k)$ time steps with high probability with respect to k. Our randomized algorithm for multi-channel networks has performance sublinear in k for even just two channels.

Our deterministic general algorithm to wake up the network runs in time that is $\mathcal{O}(k \log^{1/b} k \log n)$. When $b = \Omega(\log \log n)$ then wake-up is performed in $\mathcal{O}(k \log n)$ time. This is similar to the time bound $\mathcal{O}(k + k \log(n/k))$ given by Komlós and Greenberg [37] to resolve conflict for access to the channel among any k stations that start an execution in the same round.

Previous and related work. Shi et al. [41] considered the model of a multi-channel in which a node can simultaneously obtain different messages on different channels, while each channel is a single-hop radio network. They studied the information-exchange problem, in which some ℓ nodes start with a rumor each and the goal is to disseminate all rumors across all stations. They gave and algorithm of time performance $\mathcal{O}(\log \ell \log \log \ell)$ with n channels available. Most of the previous work on algorithms for multi-channel single-hop radio networks used the model defined as a collection of multiple-access channels such that a node has to choose a channel per round to participate in communication in this particular channel either as a listener or transmitter. Variants to this model with adversarial disruptions of channels were also considered. To the best out our knowledge, [41] was the only previous paper that used the strong model in which nodes can use all the available channels simultaneously and independently.

Next we review work done for the multi-channel model in which a station can use at most one channel for communication at a time. Dolev et al. [27] studied a parametrized variant of gossip for multi-channel radio networks. They gave oblivious deterministic algorithms for an adversarial setting in which a malicious adversary can disrupt one channel per round. Daum et al. [19] considered leader election and Dolev et al. [26] gave algorithms to synchronize a network, both papers about an adversarial setting in which the adversary can disrupt a number of channels in each round, this number treated as a parameter for performance bounds.

Information exchange has been investigated extensively for multi-channel wireless networks. The problem is about some ℓ nodes initialized with a rumor each and the goal is either to disseminate the rumors across the whole network or, when the communication environment is prone to failures, to have each node learn as many rumors as possible. Gilbert et al. [32] gave a randomized algorithm for the scenario when an adversary can disrupt a number of channels per round, this number being an additional parameter in performance bounds. Holzer et al. [35] and [34] gave deterministic and randomized algorithms to accomplish the information-exchange task in time $\mathcal{O}(\ell)$, for ℓ rumors and for suitable numbers of channels that make this achievable. This time bound $\mathcal{O}(\ell)$ is optimal when multiple rumors cannot be combined into compound messages. Wang et al. [43] considered information-exchange in a model when collision detection is available and rumors can be combined into compound messages. They gave an algorithm of time performance $\mathcal{O}(\ell/b + n \log^2 n)$, for ℓ rumors and b channels.

A multi-channel single-hop network is a generalization of a multiple-access channel, which consists of just one channel. For recent work on algorithms for multiple-access channels, see [3,4,5,6,7,11,12,17,38].

The problem of waking up a radio network was first investigated by Gąsieniec et al. [31] in the case of multiple access channels, see [23,24,25,36] for more on a related work. A broadcast from a synchronized start in a radio network was considered in [8,14,15,16,21,22,39]. The general problem of waking up a multi-hop radio network was studied in [9,10,13].

A lower bound for a multiple access channel was given by Greenberg and Winograd [33]. Lower bounds for multi-hop radio networks we proved by Alon et al. [1], Clementi et al. [15], Farach-Colton et al. [29] and Kushilevitz and Mansour [40].

Ad-hoc multi-hop multi-channel networks were studied by Alonso et al. [2], Daum et al. [18] and [20], Dolev et al. [28], and So and Vaidya [42].

2 Technical Preliminaries

The model of a multi-channel single-hop radio network is defined as follows. There are n nodes attached to a spectrum of b frequencies. We use the term "station" and "node" interchangeably. The set of all stations is denoted by V. Each frequency determines a multiple access channel. All these b channels operate concurrently and independently from each other. All stations listen to all channels all the time and obtain the same feedback from each channel. A station can transmit on any set of channels at any time. A station obtains the respective feedback from each channel separately and concurrently.

When a station successfully receives a message transmitted on some channel then we say that the station *hears* the message. When no station transmits on a channel then the channel is *silent*. When more than one stations transmit on one channel such that their transmissions overlap then we say that a *collision* occurs on this channel during the time of overlap.

The semantics of channels. When a station transmits on a channel and no collision occurs during the transmission on this channel then each station hears the transmitted message. When a station transmits a message and a collision occurs during the transmission on the channel of transmission then no station hears this transmitted message. Channels operate independently, in particular, there could be a collision on one channel and at the same time a message may be heard on some other channel. There is no collision detection, which means that when a station listens to a channel then it receives the same feedback when the channel is silent and when a collision occurs on this channel.

Transmissions on all channels are synchronized. This means that an execution of an algorithm is partitioned into rounds of equal length so that each transmission occurs in some round. Each station has its private clock which is ticking at the rate of rounds. Rounds begin and end at the same time on all channels. When we refer to a round number then this refers to the indiction of some station's private clock and this station is understood. Messages are scaled to duration of rounds so that transmitting a message takes a whole round. Two transmissions overlap in time precisely when they are performed in the same round.

Spontaneous activations and waking up the network. Initially, all stations are *passive*, in that they do not execute any communication algorithm, and in particular do not transmit any messages on any channel. Passive stations listen to all channels all the time, in that when a message is heard on a channel then all passive stations hear it. At a point in time, some stations become *activated* spontaneously and afterwards they are *active*. Passive stations may keep getting activated spontaneously after the round of the first activations. A specific scenario of timings of certain stations being activated is called an *activation pattern*.

An activated station resets its private clock to zero at the round of activation. When a station becomes active, it starts from the first round of its private clock to execute an algorithm with the goal to wake up the whole network. This goal of *waking up* the network is accomplished in the first round when some active station transmits on some channel as the only station transmitting in this round on this particular channel. This moment is understood as all passive stations receiving a signal to wake up and proceed with executing a predetermined communication algorithm. The moment of wake-up can be used to synchronize local clocks so that they begin to reflect the coordinated time.

Performance of a wake-up algorithm is measured as the number of rounds measured from the first spontaneous activation to the round of the first message heard on the network. We use an additional parameter k, which is a natural number such that $1 \leq k \leq n$ and denotes an upper bound on the number of stations that may get activated spontaneously in an execution. Performance bounds of wake-up algorithms employ the following three variables: n, b, and k. A parameter of a system or executions is *known* when it can be used in codes of algorithms. The numbers n and b are assumed to be known while the parameter k is unknown.

Definitions. Next we summarize definitions used throughout the paper.

Definition 1 (Global time). *The term* time step *refers to the time as measured by an external observer. We call this time* global. *The first round of spontaneous activation of some station becomes the first time step of this global time. The time step in which a station u becomes activated spontaneously is denoted by σ_u. The set of stations that are active by time step t is denoted by $W(t)$.*

We consider oblivious algorithms that have schedules of transmission precomputed for each station. Each such a schedule is represented as a sequence of 0s and 1s. The schedules are organized as rows of a binary matrix for the sake of visualization and discussion.

Definition 2 (Transmission arrays). *Let ℓ be positive integer treated as a parameter. An array $\mathcal{T}$ of entries of the form $T(u, \beta, j)$, where $u \in V$ is a station, β such that $1 \leq \beta \leq b$ is a channel, and integer j is such that $0 \leq j \leq \ell$, is a transmission array when each entry is either a 0 or a 1. The parameter $\ell = \ell(\mathcal{T})$ is called the* length *of array $\mathcal{T}$. Entries of a transmission array $\mathcal{T}$ are called* transmission bits *of $\mathcal{T}$. The number j is the* position *of a transmission bit $T(u, \beta, j)$.*

Every station $u \in V$ is provided with a copy of all entries $T(u, *, *)$ of some transmission array $\mathcal{T}$ as a way to instantiate the code of a wake-up algorithm.

Definition 3 (Schedules). *For a transmission array $\mathcal{T}$, a station u and channel β, the sequence of entries $T(u, \beta, j)$, for $j = 1, \ldots, \ell$, is called a (u, β)-schedule and is denoted $\mathcal{T}(u, \beta, *)$. A (u, β)-schedule $\mathcal{T}(u, \beta, *)$ defines the following schedule of transmissions for station u: it transmits on channel β in the jth round exactly when $T(u, \beta, j) = 1$.*

When a station u became active then it executes the following algorithm:

Algorithm WAKE-UP$(u, \mathcal{T})$:
Execute the schedule of transmissions determined by $\mathcal{T}(u, \beta, *)$ on each channel β.

If u is active in time step t then u perceives this time step t as round $t - \sigma_u$. The concept of a transmission that wakes up the network is defined as follows.

Definition 4 (Isolation). *A station v is β-isolated at time step t when $v \in W(t)$ and when both $T(v, \beta, t - \sigma_v) = 1$ and $T(u, \beta, t - \sigma_u) = 0$, for every $u \in W(t) \setminus \{v\}$. A station v is isolated at time step t when v is β-isolated at time step t for some channel β where $1 \leq \beta \leq b$.*

For a given transmission array, an *isolated position* is a pair (t, β) of time step t and channel β such that there is a β-isolated station at time step t. Note that an isolated position (t, β) means that successful wake-up has occurred by time t.

We impose structure on a transmission array by partitioning it into sections of increasing length.

Definition 5 (Stages). *Let c be a positive integer and let $\varphi(0) = 0$ and $\varphi(i) = c2^i \cdot i^{1/b} \lg n$, for positive integers i. The ith section of a (u, β)-schedule $T(u, \beta, *)$, for $1 \leq i \leq \lg n$, consists of all subsequences*

$$T(u, \beta, \varphi(i)), T(u, \beta, \varphi(i) + 1), \ldots, T(u, \beta, \varphi(i + 1) - 1)$$

of consecutive transmission bits. A station executing the ith section of its schedules is said to be in stage i. The stations that are in stage i at a time step j are denoted by $W_i(j)$.

The constant c in Definition 3 is be determined later as needed. The identity $\bigcup_{i=1}^{\lg n} W_i(j) = W(j)$ holds for every time step j, because an active station is in some stage. The length of the ith section for any (u, β)-schedule is $\varphi(i+1) - \varphi(i)$, which is at least as large as $\varphi(i)$.

Definition 6 (Balanced time steps). *For a stage ω, where $1 \leq \omega \leq \lg k$, a time step j is ω-balanced when the following hold: (a) $2^\omega \leq |W_\omega(j)| \leq 2^{\omega+2}$ and (b) $|W_i(j)| = 0$, for all stages i such that $i > \omega$.*

Definition 7 (Balanced time intervals). *Let ω be a stage, where $1 \leq \omega \leq \lg k$. A time interval $[t_1, t_2]$ of size $\varphi(\omega - 1)$, is said to be ω-balanced, if every time step $j \in [t_1, t_2]$ is ω-balanced. An interval is called balanced when there exists a stage ω, for $1 \leq \omega \leq \lg k$, such that it is ω-balanced.*

For a time step j, we define $\Psi(j)$ as follows:

$$\Psi(j) = \sum_{\omega=1}^{\lg k} \frac{|W_\omega(j)|}{2^i}.$$

Definition 8 (Light time intervals). *Let ω be a stage, where $1 \le \omega \le \lg k$. An ω-balanced time interval $[t_1, t_2]$ is called ω-light when (1) the inequality $\left| \bigcup_{i=1}^{\omega} W_i(j) \right| \le 2^{\omega+4}$ holds for every time step $j \in [t_1, t_2]$, and (2) interval $[t_1, t_2]$ contains at least $\varphi(\omega - 2)$ time steps j such that*

$$1 \le \Psi(j) \le 128 \cdot \omega . \tag{1}$$

An interval is called light *when there exists a stage ω, for $1 \le \omega \le \lg n$, such that it is ω-light.*

We will use transmission arrays in which entries are independent random variables.

Definition 9 (Regular randomized transmission arrays). *A randomized transmission array $\mathcal{T}$ has the structure of a transmission array. Transmission bits $T(u, \beta, j)$ are not fixed but instead are independent Bernoulli random variables. Let u be a station and β denote a channel. For $1 \le i \le \lg n$, the entries of the ith section of the (u, β)-schedule are stipulated to have the following probability distribution, for $j = \varphi(i), \ldots, \varphi(i+1) - 1$:*

$$\Pr(T(u, \beta, j) = 1) = 2^{-i} \cdot i^{-\beta/b} .$$

We say that the number of channels b is n-*large*, or simply *large*, or they there are n-*many channels*, when the inequality $b > \lg(128b \lg n)$ holds. We set $\varphi(i) = c \cdot (2^i/b) \lg n \lg(128b \lg n)$ for such b, where c is a sufficiently large constant to be specified later. Recall the notation $\Psi(j) = \sum_{i=1}^{\lg k} \frac{|W_i(j)|}{2^i}$ that, for a time step j. For n-many channels, we use a modified version of a light time interval (see Definition 8), where condition (2) is replaced by the following one:

$$1 \le \Psi(j) \le 128 \cdot \lg n. \tag{2}$$

For a channel β, we use the notation $\beta^* = \beta \bmod \lg(128b \lg n)$.

Definition 10 (Modified randomized transmission arrays). *A modified randomized transmission array $\mathcal{T}$ has the structure of a transmission array. Transmission bits $T(u, \beta, j)$ are not fixed but instead are independent Bernoulli random variables. Let u be a station and β denote a channel. For $1 \le i \le \lg n$, the entries of the ith section of the (u, β)-schedule are stipulated to have the following probability distribution, for $j = \varphi(i), \ldots, \varphi(i+1) - 1$:*

$$\Pr(T(u, \beta, j) = 1) = b \cdot 2^{-i-\beta^*} .$$

A randomized transmission array, whether regular or modified, is used to represent a randomized wake-up algorithm. To decide if a station u transmits on channel β in the jth round, this station first carries out a Bernoulli trial with the probability of success as stipulated in the definition of the respective randomized array, and transmits when the experiment results in success. Regular arrays are used in the general case and modified arrays when there are n-many channels.

Definition 11 (Waking arrays). *A transmission array $\mathcal{T}$ is said to be waking when for every k such that $1 \le k \le n$ and a light interval $[t_1, t_2]$ such that $|W(t)| \le k$, whenever $t_1 \le t \le t_2$, there exist both a time step $j \in [t_1, t_2]$ and a station $w \in W(j)$ such that w is isolated at time step j.*

The length of a waking array is the worst-case time bound on performance of the wake-up algorithm determined by this transmission array.

We denote by $\mathcal{F}^n_k$ the family of sets with exactly k elements out of n possible elements, interpreted as k-sets of stations taken from all n stations. For $\lambda \le \kappa \le n$, a family $\mathcal{F} \subseteq \mathcal{F}^n_\kappa$ is said to be (n, κ, λ)-*intersection free* if $|F_1 \cap F_2| \ne \lambda$ for every F_1 and F_2 in $\mathcal{F}^n_\kappa$. The following fact is an upper bound on the size of intersection-free families.

Fact 1 ([30]) *For any (n, κ, λ)-intersection free family $\mathcal{F}$ the following inequality holds true:*

$$|\mathcal{F}| \le \binom{n}{\lambda} \cdot \frac{\binom{2\kappa - \lambda - 1}{\kappa}}{\binom{2\kappa - \lambda - 1}{\lambda}} \,,$$

assuming that $2\lambda + 1 \ge \kappa$ and $\kappa - \lambda$ is a prime power. □

3 A Lower Bound for Deterministic Algorithms

We prove a lower bound on time performance of any deterministic wake-up algorithm. We assume that all stations start simultaneously and have access to a global clock. This means that the lower bound is valid in a much stronger setting than the one for which we design efficient algorithms.

We define a *query* to be a set of ordered pairs (x, β) for $x \in V$ and $1 \le \beta \le b$. An interpretation of a pair $(x, \beta) \in Q$, for a query Q, is that station x is to transmit on channel β at the time step assigned for the query. In this section, an algorithm $\mathcal{A}$ is represented as a sequence of queries $\mathcal{A} = \{Q_1, \dots, Q_t\}$. The index i of a query Q_i in such a sequence $\mathcal{A}$ is interpreted as the time step assigned for the query. We use the notation $Q_{i,\beta} = \{x \in V : (x, \beta) \in Q_i\}$, for a query Q_i. This represents the subset of all stations that at time step i transmit on channel β.

We use the Iverson's bracket $[\mathcal{P}]$, where $\mathcal{P}$ is a statement that is either true or false, defined as follows: $[\mathcal{P}] = 1$ if $\mathcal{P}$ is true and $[\mathcal{P}] = 0$ if $\mathcal{P}$ is false. We use the notation $\lg x$ for $\log_2 x$.

Lemma 1. *Let $\mathcal{A} = \{Q_1, Q_2, \dots, Q_t\}$ be a sequence of queries representing an algorithm. There exists a sub-family $\mathcal{S} \subseteq \mathcal{F}^n_k$ with at least $|\mathcal{F}^n_k|/2^{bt}$ elements such that any two sets $A, B \in \mathcal{S}$ satisfy $[A \cap Q_{i,\beta} \ne \emptyset] = [B \cap Q_{i,\beta} \ne \emptyset]$ for all i and β such that $1 \le \beta \le b$ and $1 \le i \le t$.*

Proof. The proof is by induction on t. The base of induction relies on the identity $\mathcal{S}(0) = \mathcal{F}^n_k$. For the inductive step, assume that the claim holds for i such that $0 \le i < t$. Let $\mathcal{S}(i+1)$ be a largest sub-family of $\mathcal{S}(i)$ with the property that for all sets A and B in $\mathcal{S}(i+1)$, the following equality holds for every $1 \le \beta \le b$:

$$[A \cap Q_{i+1,\beta} \ne \emptyset] = [B \cap Q_{i+1,\beta} \ne \emptyset] \,.$$

The inequity $|\mathcal{S}(i+1)| \geq |\mathcal{S}(i)|/2^b$ holds by the pigeonhole principle.   □

Lemma 2. *Let* $\mathcal{A} = \{Q_1, Q_2, \ldots, Q_t\}$ *be an algorithm, where* $t \leq \frac{k}{2b} \lg \frac{n}{k} - \frac{k+1}{b}$. *There exist two sets* $A, B \subseteq \mathcal{F}_k^n$ *such that the following are satisfied:*

(a) $|A \cap B| = k/2$,
(b) $[A \cap Q_{i,\beta} \neq \emptyset] = [B \cap Q_{i,\beta} \neq \emptyset]$, *for every* $1 \leq \beta \leq b$ *and* $1 \leq i \leq t$.

Proof. By Lemma 1, there exists a sub-family $\mathcal{S} \subseteq \mathcal{F}_k^n$ of at least

$$|\mathcal{S}| \geq |\mathcal{F}_k^n|/2^{bt} = \binom{n}{k}/2^{bt} \tag{3}$$

elements in $\mathcal{F}_k^n$ such that $[A \cap Q_{i,\beta} \neq \emptyset] = [B \cap Q_{i,\beta} \neq \emptyset]$, for every $A, B \in \mathcal{S}$, $1 \leq \beta \leq b$ and $1 \leq i \leq t$. Therefore, any two sets A and B in $\mathcal{S} \subseteq \mathcal{F}_k^n$, satisfy condition (b).

It remains to show that there are at least two sets in $\mathcal{S}$ satisfying also condition (a), that is, intersecting in a set of $k/2$ elements. We use Fact 1 for $\kappa = k$ and $\lambda = k/2$ to obtain that any sub-family of $\mathcal{F}_k^n$ containing sets that have pairwise intersections of size different from $k/2$ has at most these many elements:

$$\binom{n}{k/2} \cdot \binom{(3/2)k-1}{k} / \binom{(3/2)k-1}{k/2} = \binom{n}{k/2} \cdot \frac{1}{2}.$$

It follows that it is sufficient to show that the following inequality holds:

$$|\mathcal{S}| > \binom{n}{k/2} \cdot \frac{1}{2}.$$

We show it, starting from (3), in the following manner:

$$|\mathcal{S}| \geq \binom{n}{k}/2^{bt} \geq 2^{k \lg(n/k)-bt} \geq 2^{(k/2)\lg(2ne/k)} - 1 = \left(\frac{2ne}{k}\right)^{k/2} \cdot \frac{1}{2} > \binom{n}{k/2} \cdot \frac{1}{2},$$

where in the last step in the derivation we used the inequality $\binom{n}{k} < \left(\frac{ne}{k}\right)^k$. Therefore, there exist two sets in $\mathcal{S}$ with an intersection with $k/2$ elements, which completes the proof of (a).   □

Theorem 1. *Any deterministic algorithm that solves the wake-up problem on a multi-channel network with b channels requires* $\Omega(\frac{k}{b} \log \frac{n}{k})$ *time steps.*

Proof. We show that for any family $\mathcal{A} = \{Q_1, Q_2, \ldots, Q_t\}$ of queries, where t satisfies the following inequality:

$$t \leq \frac{k}{2b} \lg \frac{n}{k} - \frac{k+1}{b}, \tag{4}$$

there exists a k-set X such that $X \cap Q_i = \emptyset$, for all $i = 1, 2, \ldots, t$. To this end, let $\mathcal{A} = \{Q_1, Q_2, \ldots, Q_t\}$ be an algorithm such that (4) holds. Let A and B be two sets $A, B \subseteq \mathcal{F}_k^n$, with the properties as stated in Lemma 2. Set $A' = A \setminus B$ and $B' = B \setminus A$. Observe that if A and B have properties (a) and (b) of Lemma 2 then the following holds for A' and B':

(a*) $|A'| = |B'| = k/2$,

(b*) $A' \cap B' = \emptyset$,

(c*) $[A' \cap Q_{i,\beta} \neq \emptyset] = [B' \cap Q_{i,\beta} \neq \emptyset]$, for every $1 \leq \beta \leq b$ and $1 \leq i \leq t$.

We set $X = A' \cup B'$ to obtain that (a*) and (b*) imply $|X| = k$. Moreover, from (c*) it follows that $X \cap Q_{i,\beta} = \emptyset$, for all $1 \leq \beta \leq b$ and $1 \leq i \leq t$. This implies that $X \cap Q_i = \emptyset$, for all $i = 1, 2, \ldots, t$. Consider an execution in which the stations in X are simultaneously activated spontaneously as the only stations activated spontaneously. Then during the first t time steps after activations, no station in X transmits on any channel. We conclude that if an algorithm $\mathcal{A}$ always wakes up the network then (4) cannot be the case. □

4   A General Deterministic Algorithm

The purpose of this section is to show the following fact:

Theorem 2. *There exists a deterministic waking array of $\mathcal{O}(n \log n \log^{1/b} k)$ length guaranteeing wake-up in $\mathcal{O}(k \log n \log^{1/b} k)$ time for any number $k \leq n$ of activated stations.*

This fact is proved by the probabilistic method. We want to show that there is a transmission array of length $\mathcal{O}(k \log^{1/b} k \log n)$ such that if all awoken stations execute Protocol WAKE-UP, then there exists a time slot $t = \mathcal{O}(k \log^{1/b} k \log n)$ such that exactly one station transmits at time t. The transmission array with such desired property is defined by way of Definition 11. For a given wake-up pattern, by an isolated position we understand a pair (t, β) of time step t and channel β such that there is a β-isolated station at time slot t. Note that an isolated position (t, β) means that successful wake-up has occurred by time t.

Lemma 3. *Let c in the definition of φ be bigger than some sufficiently large constant. There exists an waking array of length $2cn \lg n \lg^{1/b} k$ such that, for any transmission array, there is an integer $0 \leq \omega \leq \lg k$ with the following properties:*

(1) *There are at least $c \cdot 2^{\omega-259} \lg n$ isolated positions by time $c \cdot 2^{\omega+1} \lg n \lg^{1/b} k$.*

(2) *At least $c \cdot 2^{\omega-259} \lg n$ isolated positions occur at time steps with at least 2^{ω} but no more than $2^{\omega+4}$ activated stations.*

Proof of Theorem 2: There is an isolated position for every activation pattern by time $\mathcal{O}(k \log n \log^{1/b} k)$. This follows from point (1) of Lemma 3. To see this, notice that otherwise the ω-light interval, which is also ω-balanced, would have at least $2^{\omega} > k$ stations activated, by Definitions 6 and 7, contradicting the assumption of the theorem.

Channels with random jamming. Assume that at each time step and on every channel there is a jamming error with probability $0 \leq p < 1$, independently over time steps and channels. The case $p = 0$ is covered by Theorem 2.

Theorem 3. *For a given error probability $0 < p < 1$, there exists a waking array of $\mathcal{O}(\log^{-1}(1/p)n \log n \log^{1/b} k)$ length guaranteeing wake-up in time that is $\mathcal{O}(\log^{-1}(1/p)k \log n \log^{1/b} k)$, for any number $k \leq n$ of spontaneously activated stations.*

Proof. Let us set $c = c' \cdot \lg^{-1}(1/p)$ for sufficiently large constant c', and consider any activation pattern. By Lemma 3, at least $c \cdot 2^{\omega-259} \lg n$ isolated positions occur by time $c \cdot 2^{\omega+1} \lg^{1+1/b} n$ and by that time no more than $2^{\omega+4}$ stations are activated Each such isolated position can be jammed independently with probability p. Therefore, the probability that all these positions are jammed, and thus no successful transmission occurs by time

$$c \cdot 2^{\omega+1} \lg n \lg^{1/b} k = \mathcal{O}(\log^{-1}(1/p)k \log n \log^{1/b} k) \ ,$$

is at least

$$p^{c \cdot 2^{\omega-259} \lg n} = \exp\left(c' \cdot \lg^{-1}(1/p) \cdot 2^{\omega-259} \lg n \cdot \ln p\right) \ ,$$

This is smaller than $1/\text{poly}(n)$ for sufficiently large constant c'. Here we use the fact that $\frac{\ln p}{\lg(1/p)}$ is a negative constant for $p \in (0, 1)$. When estimating the time of a successful wake-up we relied on the fact that 2^ω, which is the lower bound on the number of activated stations by Lemma 3(2), must be smaller than k, by the assumption. □

5   A Deterministic Algorithm for Sufficiently Many Channels

The main result of this section is as follows:

Theorem 4. *There exists a waking array of $\mathcal{O}((n/b) \log n \log(b \log n))$ length, for $b > \lg(128b \lg n)$, which completes wake-up in $\mathcal{O}((k/b) \log n \log(b \log n))$ time for any number $k \leq n$ of spontaneously activated stations.*

The proof of this fact is by way of showing the existence of a waking array, as defined in Definition 11, for a section length defined as $\varphi(i) = c \cdot (2^i/b) \lg n \lg(128b \lg n)$.

Lemma 4. *Let c in the definition of φ be bigger than some sufficiently large constant. There exists a waking array of length $2c(n/b) \lg n \lg(128b \lg n)$ such that for any activation pattern, there is an integer $0 \leq \omega \leq \lg n$ with the following properties:*

(1) *There are at least $c \cdot 2^{\omega-6} \lg n$ isolated positions by the time step of number $c \cdot (2^{\omega+1}/b) \lg n \lg(128b \lg n)$.*
(2) *These positions occur at time step with at least 2^ω but no more than $2^{\omega+4}$ activated stations.*

Proof of Theorem 4: There is an isolated position by $\mathcal{O}((k/b) \log n \lg(128b \lg n))$ time for every activation pattern. This follows from point (1) of Lemma 4. Indeed, otherwise the ω-light interval, which is also ω-balanced, would have at least $2^\omega > k$ stations activated, by Definitions 6 and 7, contrary to the assumptions.

Channels with random jamming. Assume that at each time step and on every channel there is a jamming error with probability $0 \le p < 1$, independently over time steps and channels. The case $p = 0$ is subsumed by Theorem 2.

Theorem 5. *There exists a waking array of $\mathcal{O}(\log^{-1}(1/p)(n/b) \log n \log(b \log n))$ length, for a probability p such that $0 < p < 1$, which guarantees wake-up in $\mathcal{O}(\log^{-1}(1/p)(k/b) \log n \log(b \log n))$ time with the probability of at least $1 - 1/\text{poly}(n)$.*

Proof. Let us set $c = c' \cdot \lg^{-1}(1/p)$, for sufficiently large constant c', and consider any activation pattern. By Lemma 4, $c \cdot 2^{\omega-6} \lg n$ isolated positions occur by time $c \cdot (2^{\omega+1}/b) \lg n \lg(128b \lg n)$ and by that time no more than $2^{\omega+4}$ stations are activated. Each such isolated position can be jammed independently with probability p. Therefore, the probability that all these positions are jammed, and thus no successful transmission occurs by time $c \cdot (2^{\omega+1}/b) \lg n \lg(128b \lg n) = \mathcal{O}(\log^{-1}(1/p)(k/b) \log n \log(b \log n))$, is at least

$$p^{c \cdot 2^{\omega-6} \lg n} = \exp \left(c' \cdot \lg^{-1}(1/p) \cdot (2^{\omega-6}/b) \lg n \cdot \ln p \right) ,$$

which is smaller than $1/\text{poly}(n)$ for sufficiently large constant c'. Here we use the fact that $\frac{\ln p}{\lg(1/p)}$ is a negative constant for $p \in (0,1)$. When bounding time of a successful wake-up to occur, we rely on the fact that 2^ω, which is the lower bound on the number of activated stations by Lemma 4(2), must be smaller than k by the assumption. □

6   A Randomized Algorithm

In this Section, we present a randomized wake-up algorithm, which is complementary to deterministic algorithms considered so far. The code for a station u is as follows:

Algorithm CHANNEL-SCREENING(u):
For $\beta = 1, 2, \ldots, b$ transmit a message on channel β with probability $k^{-\beta/b}$.

Lemma 5. *Let t be a time step and let $1 \le \beta \le b$ be such that bounds*

$$k^{(\beta-1)/b} \le |W(t)| \le k^{\beta/b}$$

hold. Algorithm CHANNEL-SCREENING guarantees that the probability of hearing a message at time step t on channel β is at least $1/2ek^{1/b}$.

Proof. Let $E(\beta, t)$ be the event of a successful transmission on channel β at time t. The probability that a station $w \in W(t)$ transmits at time t on channel β while all the others remain silent is

$$\Pr(E(\beta, t)) \geq \frac{|W(t)|}{k^{\beta/b}} \left(1 - \frac{1}{k^{\beta/b}}\right)^{|W(t)|-1}$$

$$\geq \frac{k^{(\beta-1)/b}}{k^{\beta/b}} \left(1 - \frac{1}{k^{\beta/b}}\right)^{k^{\beta/b}},$$

where the last inequality follows from the assumption that

$$k^{(\beta-1)/b} \leq |W(t)| \leq k^{\beta/b} .$$

Hence

$$\Pr(E(\beta, t)) \geq \frac{1}{2ek^{1/b}} ,$$

which completes the proof. □

An estimate the number of rounds needed to make the probability of failure smaller than a threshold ϵ is as follows:

Theorem 6. *Algorithm* CHANNEL-SCREENING *on b channels succeeds in waking up the network, with at most k active stations out of n in* $\mathcal{O}(k^{1/b} \ln(1/\epsilon))$ *time with the probability of at least* $1 - \epsilon$.

Proof. Let us consider a set of contiguous time steps T. For $1 \leq \beta \leq b$, let

$$T_\beta = \{t \in T \mid k^{(\beta-1)/b} \leq |W(t)| \leq k^{\beta/b}\} .$$

Let $\bar{E}(t)$ be the event of an unsuccessful time step t, in which no station transmits as the only transmitted on any channel, and let $\bar{E}(\beta, t)$ be the event of an unsuccessful time step t on channel β, with $1 \leq \beta \leq b$. By Lemma 5, the probability of having a sequence of $\lambda = |T|$ unsuccessful time steps can be estimated as follows:

$$\Pr\left(\bigcap_{t \in T} \bar{E}(t)\right) \leq \Pr\left(\bigcap_{t \in T_1} \bar{E}(1, t)\right) \cdot \Pr\left(\bigcap_{t \in T_2} \bar{E}(2, t)\right) \cdots \Pr\left(\bigcap_{t \in T_b} \bar{E}(b, t)\right)$$

$$\leq \left(1 - \frac{1}{2ek^{1/b}}\right)^{|T_1|} \cdot \left(1 - \frac{1}{2ek^{1/b}}\right)^{|T_2|} \cdots \left(1 - \frac{1}{2ek^{1/b}}\right)^{|T_b|}$$

$$\leq \left(1 - \frac{1}{2ek^{1/b}}\right)^\lambda$$

$$\leq \epsilon,$$

for $\lambda \geq 2ek^{1/b} \ln(1/\epsilon)$. □

References

1. Alon, N., Bar-Noy, A., Linial, N., Peleg, D.: A lower bound for radio broadcast. Journal of Computer and System Sciences 43(2), 290–298 (1991)
2. Alonso, G., Kranakis, E., Sawchuk, C., Wattenhofer, R., Widmayer, P.: Probabilistic protocols for node discovery in ad hoc multi-channel broadcast networks. In: Pierre, S., Barbeau, M., An, H.-C. (eds.) ADHOC-NOW 2003. LNCS, vol. 2865, pp. 104–115. Springer, Heidelberg (2003)
3. Anantharamu, L., Chlebus, B.S.: Broadcasting in ad hoc multiple access channels. In: Moscibroda, T., Rescigno, A.A. (eds.) SIROCCO 2013. LNCS, vol. 8179, pp. 237–248. Springer, Heidelberg (2013)
4. Anantharamu, L., Chlebus, B.S., Kowalski, D.R., Rokicki, M.A.: Deterministic broadcast on multiple access channels. In: Proceedings of the 29th IEEE International Conference on Computer Communications (INFOCOM), pp. 1–5 (2010)
5. Anantharamu, L., Chlebus, B.S., Kowalski, D.R., Rokicki, M.A.: Medium access control for adversarial channels with jamming. In: Kosowski, A., Yamashita, M. (eds.) SIROCCO 2011. LNCS, vol. 6796, pp. 89–100. Springer, Heidelberg (2011)
6. Anantharamu, L., Chlebus, B.S., Rokicki, M.A.: Adversarial multiple access channel with individual injection rates. In: Abdelzaher, T., Raynal, M., Santoro, N. (eds.) OPODIS 2009. LNCS, vol. 5923, pp. 174–188. Springer, Heidelberg (2009)
7. Bieńkowski, M., Klonowski, M., Korzeniowski, M., Kowalski, D.R.: Dynamic sharing of a multiple access channel. In: Proceedings of the 27th International Symposium on Theoretical Aspects of Computer Science (STACS). Leibniz International Proceedings in Informatics, vol. 5, pp. 83–94. Schloss Dagstuhl–Leibniz-Zentrum fuer Informatik (2010)
8. Chlebus, B.S., Gąsieniec, L., Gibbons, A., Pelc, A., Rytter, W.: Deterministic broadcasting in ad hoc radio networks. Distributed Computing 15(1), 27–38 (2002)
9. Chlebus, B.S., Gąsieniec, L., Kowalski, D.R., Radzik, T.: On the wake-up problem in radio networks. In: Caires, L., Italiano, G.F., Monteiro, L., Palamidessi, C., Yung, M. (eds.) ICALP 2005. LNCS, vol. 3580, pp. 347–359. Springer, Heidelberg (2005)
10. Chlebus, B.S., Kowalski, D.R.: A better wake-up in radio networks. In: Proceedings of the 23rd ACM Symposium on Principles of Distributed Computing (PODC), pp. 266–274 (2004)
11. Chlebus, B.S., Kowalski, D.R., Rokicki, M.A.: Maximum throughput of multiple access channels in adversarial environments. Distributed Computing 22(2), 93–116 (2009)
12. Chlebus, B.S., Kowalski, D.R., Rokicki, M.A.: Adversarial queuing on the multiple access channel. ACM Transactions on Algorithms 8(1), 5:1–5:31 (2012)
13. Chrobak, M., Gąsieniec, L., Kowalski, D.R.: The wake-up problem in multihop radio networks. SIAM Journal on Computing 36(5), 1453–1471 (2007)
14. Chrobak, M., Gąsieniec, L., Rytter, W.: Fast broadcasting and gossiping in radio networks. Journal of Algorithms 43(2), 177–189 (2002)
15. Clementi, A.E.F., Monti, A., Silvestri, R.: Distributed broadcast in radio networks of unknown topology. Theoretical Computer Science 302(1-3), 337–364 (2003)
16. Czumaj, A., Rytter, W.: Broadcasting algorithms in radio networks with unknown topology. Journal of Algorithms 60(2), 115–143 (2006)
17. Czyżowicz, J., Gąsieniec, L., Kowalski, D.R., Pelc, A.: Consensus and mutual exclusion in a multiple access channel. IEEE Transaction on Parallel and Distributed Systems 22(7), 1092–1104 (2011)

18. Daum, S., Ghaffari, M., Gilbert, S., Kuhn, F., Newport, C.C.: Maximal independent sets in multichannel radio networks. In: Proceedings of the 32nd ACM Symposium on Principles of Distributed Computing (PODC), pp. 335–344 (2013)

19. Daum, S., Gilbert, S., Kuhn, F., Newport, C.C.: Leader election in shared spectrum radio networks. In: Proceedings of the 31st ACM Symposium on Principles of Distributed Computing (PODC), pp. 215–224 (2012)

20. Daum, S., Kuhn, F., Newport, C.: Efficient symmetry breaking in multi-channel radio networks. In: Aguilera, M.K. (ed.) DISC 2012. LNCS, vol. 7611, pp. 238–252. Springer, Heidelberg (2012)

21. De Marco, G.: Distributed broadcast in unknown radio networks. In: Proceedings of the 9th ACM-SIAM Symposium on Discrete Algorithms (SODA), pp. 208–217 (2008)

22. De Marco, G.: Distributed broadcast in unknown radio networks. SIAM Journal on Computing 39(6), 2162–2175 (2010)

23. De Marco, G., Kowalski, D.R.: Contention resolution in a non-synchronized multiple access channel. In: Proceedings of the 27th IEEE International Parallel and Distributed Processing Symposium (IPDPS), pp. 525–533 (2013)

24. De Marco, G., Kowalski, D.R.: Searching for a subset of counterfeit coins: Randomization vs determinism and adaptiveness vs non-adaptiveness. Random Structures and Algorithms 42(1), 97–109 (2013)

25. De Marco, G., Pellegrini, M., Sburlati, G.: Faster deterministic wakeup in multiple access channels. Discrete Applied Mathematics 155(8), 898–903 (2007)

26. Dolev, S., Gilbert, S., Guerraoui, R., Kuhn, F., Newport, C.C.: The wireless synchronization problem. In: Proceedings of the 28th ACM Symposium on Principles of Distributed Computing (PODC), pp. 190–199 (2009)

27. Dolev, S., Gilbert, S., Guerraoui, R., Newport, C.: Gossiping in a multi-channel radio network. In: Pelc, A. (ed.) DISC 2007. LNCS, vol. 4731, pp. 208–222. Springer, Heidelberg (2007)

28. Dolev, S., Gilbert, S., Khabbazian, M., Newport, C.: Leveraging channel diversity to gain efficiency and robustness for wireless broadcast. In: Peleg, D. (ed.) DISC 2011. LNCS, vol. 6950, pp. 252–267. Springer, Heidelberg (2011)

29. Farach-Colton, M., Fernandes, R.J., Mosteiro, M.A.: Lower bounds for clear transmissions in radio networks. In: Correa, J.R., Hevia, A., Kiwi, M. (eds.) LATIN 2006. LNCS, vol. 3887, pp. 447–454. Springer, Heidelberg (2006)

30. Frankl, P., Füredi, Z.: Forbidding just one intersection. Journal of Combinatorial Theory, Series A 39(2), 160–176 (1985)

31. Gąsieniec, L., Pelc, A., Peleg, D.: The wakeup problem in synchronous broadcast systems. SIAM Journal on Discrete Mathematics 14(2), 207–222 (2001)

32. Gilbert, S., Guerraoui, R., Kowalski, D.R., Newport, C.: Interference-resilient information exchange. In: Proceedings of the 28th IEEE International Conference on Computer Communications (INFOCOM), pp. 2249–2257 (2009)

33. Greenberg, A.G., Winograd, S.: A lower bound on the time needed in the worst case to resolve conflicts deterministically in multiple access channels. Journal of the ACM 32(3), 589–596 (1985)

34. Holzer, S., Locher, T., Pignolet, Y.A., Wattenhofer, R.: Deterministic multi-channel information exchange. In: Proceedings of the 24th ACM Symposium on Parallelism in Algorithms and Architectures (SPAA), pp. 109–120 (2012)

35. Holzer, S., Pignolet, Y.A., Smula, J., Wattenhofer, R.: Time-optimal information exchange on multiple channels. In: Proceedings of the 7th ACM International Workshop on Foundations of Mobile Computing (FOMC), pp. 69–76 (2011)

36. Jurdziński, T., Stachowiak, G.: Probabilistic algorithms for the wake-up problem in single-hop radio networks. Theory of Computing Systems 38(3), 347–367 (2005)
37. Komlós, J., Greenberg, A.G.: An asymptotically fast nonadaptive algorithm for conflict resolution in multiple-access channels. IEEE Transactions on Information Theory 31(2), 302–306 (1985)
38. Kowalski, D.R.: On selection problem in radio networks. In: Proceedings of the 24th ACM Symposium on Principles of Distributed Computing (PODC), pp. 158–166 (2005)
39. Kowalski, D.R., Pelc, A.: Broadcasting in undirected ad hoc radio networks. Distributed Computing 18(1), 43–57 (2005)
40. Kushilevitz, E., Mansour, Y.: An $\Omega(D \log(N/D))$ lower bound for broadcast in radio networks. SIAM Journal on Computing 27(3), 702–712 (1998)
41. Shi, W., Hua, Q.-S., Yu, D., Wang, Y., Lau, F.C.M.: Efficient information exchange in single-hop multi-channel radio networks. In: Wang, X., Zheng, R., Jing, T., Xing, K. (eds.) WASA 2012. LNCS, vol. 7405, pp. 438–449. Springer, Heidelberg (2012)
42. So, J., Vaidya, N.H.: Multi-channel MAC for ad hoc networks: handling multi-channel hidden terminals using a single transceiver. In: Proceedings of the 5th ACM International Symposium on Mobile Ad Hoc Networking and Computing (MobiHoc), pp. 222–233 (2004)
43. Wang, Y., Wang, Y., Yu, D., Yu, J., Lau, F.: Information exchange with collision detection on multiple channels. Journal of Combinatorial Optimization (2014)

A Disruption-Resistant MAC Layer
for Multichannel Wireless Networks*

Henry Tan, Chris Wacek, Calvin Newport, and Micah Sherr

Georgetown University
Washington, DC, USA
{ztan,cwacek,cnewport,msherr}@cs.georgetown.edu

Abstract. Wireless networking occurs on a shared medium which renders communication vulnerable to disruption from other networks, environmental interference, and even malicious jammers. The standard solution to this problem is to deploy coordinated spread spectrum technologies that require pre-shared secrets between communicating devices. These secrets can be used to coordinate hopping patterns or spreading sequences. In this paper, by contrast, we study the local broadcast and unicast problems in a disrupted multichannel network with *no* pre-shared secrets between devices. Previous work in this setting focused on the special case of a single pre-designated sender in a single hop network topology. We consider in this paper, for the first time, upper and lower bounds to these problems in multihop topologies with multiple senders. To validate the potential real world application of our strategies, we conclude by describing a general purpose MAC protocol that uses the algorithms as key primitives, and validates its usefulness with a proof-of-concept implementation that runs the protocol on commodity hardware.

Keywords: wireless, broadcast, jamming.

1 Introduction

Wireless networks operate over a shared medium. This leaves them vulnerable to message loss due to (often unpredictable) disruption, such as contention from other networks, unrelated electromagnetic noise, and in some cases even malicious jamming. The standard solution to these issues in real deployments is to use *coordinated spread spectrum strategies*, such as frequency hopping spread spectrum (FHSS) [18, 27, 28], in which devices evade disruption by hopping between channels, and direct-sequence spread spectrum (DSSS) [5, 10], in which devices modulate their signal over additional frequencies to gain robustness.

A key property of these existing spread spectrum strategies is that they require the communicating devices to use pre-shared secrets (i.e., to synchronize hopping or signal spreading). In recent years, however, researchers from both the theory and systems communities have noted the need for reliable spread spectrum strategies that work in

---

* This work is supported in part by NSF grants CNS-1149832, CNS-1064986, CNS-1223825, CNS-1445967 and CCF 1320279, and the Ford Motor Company University Research Program. The findings and opinions described in this paper are those of the authors, and do not necessarily reflect the views of the funding parties.

M.K. Aguilera et al. (Eds.): OPODIS 2014, LNCS 8878, pp. 202–216, 2014.
© Springer International Publishing Switzerland 2014

the *absence* of pre-shared secrets (the so-called, *uncoordinated shared spectrum* setting) [6, 9, 14–16, 19, 24–26]. Such uncoordinated algorithms are useful to the increasingly common case where an ad hoc collection of devices need to reliably coordinate in a crowded environment. The existing work cited above studies how to generalize both FHSS and DSSS strategies to work without pre-shared secrets. In most cases, however, it focuses on the scenario of a single sender in a single hop network. Generalizing these strategies to work in general network topologies with arbitrary message arrivals—e.g., what would be required to implement a general-purpose MAC layer—was identified as an open question. This paper answer it.

Results. We prove new upper and lower bounds for two key communication primitives: *broadcast* (a device must deliver a message to all its neighbors in the network topology) and *unicast* (a device must deliver a message to a single known neighbor). As in existing work [6, 19, 25, 26], we model a crowded band of the wireless spectrum with n nodes having access to $C \geq 1$ channels. In each round, each node can participate on a single channel. An adversary can choose up t channels (for some fixed $t < C$) to *disrupt* locally at each receiver, preventing communication. This adversary incarnates the diversity of different sources of unpredictable interference that plague real deployments. We model the network topology with a graph, where the nodes correspond to the devices and edges to links. We assume that for both broadcast and unicast, messages arrive at arbitrary nodes at arbitrary times, and we require that randomized algorithms solve the relevant problem with high probability, abbreviated to w.h.p., in n.

We begin by describing a randomized broadcast algorithm that delivers a message, with high probability, from a sender to its (unknown) neighbors in the network in $O\left(\frac{C}{C-t}\Delta(\log(\Delta/C)+1)\log n\right)$ rounds, where Δ is the maximum degree in the topology graph (and therefore a measure of the *worst case* amount of nearby contention). The core strategy in this algorithm and its analysis is to use uncoordinated frequency hopping for a sufficient amount of time to ensure that nearby nodes have an opportunity to receive the message—regardless of the behavior of the disruption adversary. Note that we add 1 to the $\log(\Delta/C)$ factor in the asymptotic complexity to avoid a factor of 0 when $\Delta = C$ (in our algorithm, the relevant term replaced with 1 when $\Delta = C$).

Notice, in many cases, the actual amount of nearby contention might be much smaller than the worst-case. Motivated by this reality, we proceed with our primary technical result: a randomized *adaptive* broadcast algorithm that assumes a natural geographic constraint on the topology (see Section 2), and in exchange guarantees to solve broadcast for each node u in $O\left(\frac{Ct}{C-t}\delta_u \log^3 n(\log(\Delta/C)+1)\right)$ rounds, where δ_u describes the actual amount of contention local to u (as noted: in practice δ_u might be much smaller than Δ). The core strategy in this algorithm is to have broadcasting devices participate in repeated iterations of local leader election competitions. If a device succeeds in becoming a leader, its local contention is small, and it can terminate confident that it successfully delivered its message. To obtain the δ_u factor in the time complexity, we prove that the leader election competition is fair—a given sender's probability of winning is inversely proportional to the number of nearby competing senders.

We then describe a randomized unicast algorithm that guarantees delivery of a message from u to a known neighbor v in $O\left(\frac{Ct'}{C-t'}\log\Delta\log C\log n\right)$ rounds, where t' bounds the *actual* amount of disruption at v. This algorithm uses a similar uncoordinated frequency hopping strategy as our non-adaptive broadcast solution. The algorithm adapts to the actual amount of disruption (and not the worst-case) by testing different estimates. This strategy works because the algorithm is sending to a known receiver, and it can therefore use acknowledgments to know when it succeeded. In the interest of space, we defer many of our proofs and pseudocode to the full version [8].

We conclude our theoretical analysis by proving that our broadcast bounds are optimal within polylogarithmic factors for large and small δ (contention) values. We then turn our attention to our claimed practical motivation: that these primitives can aid the design of uncoordinated but reliable MAC layers. To validate this claim, we describe a general purpose MAC protocol that uses our algorithms as key primitives. The MAC protocol implements a name service that reliably discovers nearby devices, and then provides broadcast and unicast communication. It also guarantees a form of link layer authentication that does not rely on a public key infrastructure: once a pair of neighboring honest devices begin to communicate, a malicious node cannot spoof messages on this link. We then describe a proof-of-concept implementation of this link layer protocol using commodity 802.11 hardware, and the Click Modular Router [11] and FreeMAC [23] (a modified Atheros 802.11 driver) software. Our testbed evaluation validates that our algorithmic strategies can be implemented in practice and yield reliability (at the cost of performance).

Related Work. The most relevant related work on uncoordinated shared spectrum protocols focuses on delivering messages from a designated source to receivers in a single-hop version of our disrupted multi-channel network model. This research direction began with Strasser et al.'s UFH algorithm [25], which delivers k small message fragments from a single sender to one or more local receivers in $O\left(\frac{C^2}{C-t}k\log(n)\right)$ rounds with high probability. Erasure coding and clever use of the channels when t is small improved this cost to $O\left(\frac{Ct}{C-t}(k+\log(n))\right)$ [24, 26], while our recent work improved this result further to $O\left(\frac{C}{C-t}(k+\log(n))\right)$ (under certain assumptions) by recruiting successful receivers to help propagate the message faster [6].

A related problem in the theory literature studies reliable local communication in a network with a *single* channel and a *resource-bounded* adversary causing disruption. Constant-competitive throughput for local communication is possible in this setting for both single hop [1, 7, 20] and multihop [21] networks. These results leverage different techniques than those used in this paper (which focuses on uncoordinated frequency hopping), but are motivated by the same real world issues surrounding shared spectrum.

2 Model and Problems

We model contended shared spectrum using the *t-disrupted* model, which is parameterized by $0 \leq t < C$, and describes a wireless network with C communication channels, up to t of which might be *disrupted* locally at each receiver in each round, preventing communication at that receiver on those channels. As detailed below, the disruption decisions are made by a bounded adversary the incarnates the diversity of unpredictable interference possible in shared spectrum settings.

Network Topology. To describe the network topology, we fix an undirected graph $G = (V, E)$ with diameter D and a maximum degree upper-bounded by a known parameter

Δ, where the vertices in V correspond to the $n = |V|$ wireless devices, which we call *nodes*. Let $N(u)$, for $u \in V$, be the neighbors of u in G. We assume nodes know only a polynomial upper bound on n and do not know G. The t-disrupted model is parameterized by $\mathcal{C}$ (the number of available channels) and t (the number of channels that can be concurrently disrupted), where $0 \le t < \mathcal{C}$, and both parameters are known to the nodes. To simplify our asymptotic notation in some of the results that follow, we assume without loss of generality that $\Delta \ge \mathcal{C}$.[1]

Executions. We assume executions in our model proceed in synchronous rounds labeled $1, 2, 3...$. To capture unpredictable disruption we assume an adversary can disrupt up to t channels per node, per round. Formally, for each round r and node u, let $adv(u, r)$ be an array of size $\mathcal{C}$ describing how the adversary behaves on each channel with respect to u in r. Let $[\mathcal{C}] = 1, 2, ..., \mathcal{C}$ be the set of available channels. For each $c \in [\mathcal{C}]$: $adv(u, r)[c] = \perp$ indicates the adversary does not affect c with respect to u in this round; $adv(u, r)[c] = \pm$, on the other hand, indicates that the adversary disrupts c. We similarly define $disp(u, r) = \{c : adv(i, r)[c] \ne \perp\}$. We require that $|disp(u, r)| \le t$. If $c \in disp(u, r)$, we say c is *disrupted* w.r.t. u and r.

To define communication behavior, fix a node u and round r. At the beginning of r, u chooses a channel from $[\mathcal{C}]$ to participate on, deciding either to *transmit* or *listen*. If u decides to transmit a message, then it cannot also receive a message in r (i.e., the channels are half-duplex). If u listens, the outcome depends on the adversary and its neighbors. In more detail, if $adv(u, r) = \pm$, u receives nothing. If $adv(u, r) = \perp$ and exactly one neighbor of u transmits on c during r, u receives this message. Otherwise, if multiple neighbors transmit, u receives one of the messages, chosen arbitrarily, or nothing (concurrent broadcasts may be lost to undetectable collision). It follows, therefore, that nodes in this model must handle *both* contention from their own network and disruption from outside. When modeling unicast communication from a node u to a single known neighbor v in the graph, we assume the presence of link layer acknowledgements that allow u to discover when v has successfully received its message.[2]

We define $t' \le t$, with respect to an execution, to be the maximum value of $|disp(u, r)|$ over all u and r. That is, t' is the actual amount of concurrent disruption experienced in an execution, whereas t is the worst-case possibility. We typically assume nodes do not know t' in advance. Similarly, when studying executions of *communication algorithms* (i.e., algorithms in which nodes are passed messages to communicate to nearby nodes), we define δ_u for each u, to be the total number of nodes in $N(u)$ that are passed a message to communicate in the execution. Clearly, $\delta_u \le |N(u)| \le \Delta$.

Finally, we bound our adversary's power by assuming it is an arbitrary randomized algorithm that generates $adv(u, r)$ for all u at the beginning of each r. When defining this array, the adversary can leverage knowledge of G, the algorithm being run by the

---

[1] We say this holds w.l.o.g. because Δ is an upper bound. To explicitly handle the case of smaller Δ in our time complexity results, it is necessary only to replace the linear Δ factor in the broadcast bound with the slightly more messy notation: $\max\{\Delta, \mathcal{C}\}$.

[2] We omit in this model the case where v receives a message but u does not receive the corresponding acknowledgment due to disruption—we assume a full channel is disrupted or it is not disrupted at all. That is, we always assume the worst-case, that if there is any disruption for a given transmission, everything is lost. This simplifies the analysis of our algorithms.

nodes, and the history of the execution through $r - 1$. It does not, however, know in advance u's random choices for r.

Graph Restrictions. When studying multihop radio networks, it is common to assume some geographic constraint on the communication topology. For the adaptive broadcast algorithm in this paper, we assume the constraint introduced by Daum et al. [3], that generalizes many of the constraints typically assumed in the wireless algorithms literature. In more detail, let $\mathcal{R} = \{R_1, R_2, ..., R_k\}$ be a partition of the nodes in G into regions such that the sub-graph of G induced by each region R_i is a clique. The corresponding *clique graph* (or *region graph*) is a graph $G_\mathcal{R}$ with one node u_i for each $R_i \in \mathcal{R}$, and an edge between u_i and u_j iff $\exists v \in R_i, w \in R_j$ such that v and w are connected in G; we write $R(v)$ for the region that contains v. In this paper, when we say a graph G satisfies the *regional clique decomposition* property, we mean that it can be partitioned into cliques $\mathcal{R}$ such that the maximum degree of $G_\mathcal{R}$ is upper bounded by some constant parameter. Notice, this model generalizes many common geometric network models, including unit ball graphs with constant doubling dimension [12], which was shown [22] to generalize (quasi) UDGs [2, 13].

The Broadcast and Unicast Problems. In this paper, we study upper and lower bounds for the *broadcast* and *unicast* problems in the t-disrupted model. Both problems assume the presence of a *message process* that passes broadcast/unicast messages to the network nodes. We place no restrictions on this message process besides the requirement that it waits for a node to indicate it is done processing its current message before passing the next message (i.e., we do not address queuing issues in this paper).

The *broadcast* problem requires a node u, when passed a message m from the message process, to attempt to deliver m to the nodes in $N(u)$. When it is done, it indicates this to the message process. We say a given algorithm implements a broadcast service with latency T rounds if the following two properties hold *with high probability* (i.e., with probability at least $1 - n^{-c}$, for a provided constant $c \geq 1$) for each time a node u is passed a message m: (1) u finishes transmitting m within T rounds of being passed m; (2) every node in $N(u)$ receives m during this interval.

The *unicast* problem requires a node u, when passed a message m and node $v \in N(u)$, to deliver m to v. The definition of u implementing a unicast service with latency T rounds is the same as for broadcast except it only need deliver the message to v.

3   Upper and Lower Bounds

In this section we present our upper and lower bounds for the broadcast (both non-adaptive and adaptive variants) and unicast problems. Due to space constraints, we defer some of the proofs and pseudocode to the full version.

3.1   Non-adaptive Broadcast Algorithm

We call our first broadcast algorithm *non-adaptive* as its time complexity is defined with respect to the worst-case contention. It works as follows: Each node u groups rounds into *phases*, each of which contains $\max(\lceil \log(\Delta/C) \rceil, 1)$ rounds. When u is passed a message m to send, it waits until the beginning of the next phase, then considers itself *active* for the next $T_p = \Theta(\frac{C}{C-t}\Delta \log n)$ phases (we define the constants for T_p later). After these phases are done, u considers its broadcast complete and returns to *inactive* status. During each round r, node u, regardless of whether it is active or inactive, chooses a channel on which to participate with uniform independent randomness. If u is active, it decides to transmit m with probability $\frac{1}{2^k}$, where $k = (r \mod \max(\lceil \log(\Delta/C)\rceil, 1)) + 1$; otherwise it listens. If u is inactive, it always listens. When u receives a broadcast message m' from another node for the first time, it passes it up to the higher layer message process.

Analysis. In the following analysis, and those that follow, we make use of these basic probability facts: (1) for $p \leq \frac{1}{2}: (1-p) \geq (\frac{1}{4})^p$; and (2) for $p > 0: (1-p) < e^{-p}$. Fix some node u and some phase q. Let A_u be the set of active nodes that neighbor u in q (notice, A_u is fixed over q). Let p_u be the probability that u receives a message during q. We start by bounding p_u, treating separately the case where A_u is large (and therefore, many channels are likely to be occupied by active neighbors), and A_u is small (and therefore, few channels are occupied—requiring more time to find active neighbors).

Lemma 1. *Assume* $|A_u| > C$. *It follows:* $p_u \geq \frac{C-t}{8C}$.

Lemma 2. *Assume* $|A_u| = C^{1-\epsilon}$, *for* $0 \leq \epsilon \leq 1$. *It follows:* $p_u \geq \frac{C-t}{8C^{1+\epsilon}}$.

We now combine our lemmas to prove our main theorem.

Theorem 1. *Our algorithm implements a broadcast service with a latency of* $O\left(\frac{C}{C-t}\Delta(\log(\Delta/C)+1)\log n\right)$ *rounds.*

Proof. The latency follows from the definition of our algorithm, which attempts to deliver a message for $T_p = O(\frac{C}{C-t}\Delta \log n)$ phases, each containing $\max(\lceil \log(\Delta/C)\rceil, 1)$ rounds. We are left to show that during this active period the message is delivered to all neighbors with probability at least $1 - 1/n$.

Fix some node v that is passed message m to broadcast. We will show that during the T_p phases in which v is active with m, every neighbor of v receives m, with high probability. To do so, we start by analyzing a particular neighbor u. For each of the T_p phases during which v is active with m, we call the phase *crowded* if u has more than C active neighbors. Similarly, we call the phase *sparse* if u has $C^{1-\epsilon}$ active neighbors, for some $0 \leq \epsilon \leq 1$. Notice, it must be the case that either: (1) at least half of v's T_p active phases are crowded; or (2) at least half of v's T_p active phases are sparse.

Case 1: At least half the active phases are crowded. It follows that there are $T'_p \geq T_p/2 = \Theta(\frac{C}{C-t}\Delta \log n)$ phases where $|A_u| > C$. In each such phase, Lemma 1 tells us that u receives *some* message with probability at least $\frac{C-t}{8C}$. Because active nodes behave uniformly, u receives v's message in particular with probability at least $\frac{C-t}{8C|A_u|} \geq \frac{C-t}{8C\Delta}$. We conclude that node u fails to receive v's message in all T'_p crowded phases with probability p_f, bounded as: $p_f \leq \left(1 - \frac{C-t}{8C\Delta}\right)^{T'_p} < (1/e)^{(c_1/8)\log n}$.

For any constant $c \geq 1$, we can choose a sufficiently large constant c_1 for the definition of T_p (and thus T'_p) such that this failure probability is no more than n^{-c}. (We will fix the particular c we need later in the proof.)

Case 2: At least half the active phases are sparse. In each of these T'_p phases, $|A_u| = C^{1-\epsilon}$ for some $0 \leq \epsilon \leq 1$. The problem here is that this ϵ value can change from phase to phase as neighbors of u become active and inactive. To simplify notation, for the rth sparse phase, let ϵ_r be the definition of ϵ and let A^r_u be the set of u's active neighbors. Following the same general approach as for case 1, we now apply Lemma 2 to determine that in phase r, u will receive *some* message with probability at least $\frac{C-t}{8C^{1+\epsilon_r}}$. In particular, due to uniformity, u will receive v's message with probability at least: $\frac{C-t}{8C^{1+\epsilon_r}|A^r_u|} = \frac{C-t}{8C^{1+\epsilon_r}C^{1-\epsilon_r}} = \frac{C-t}{8C^2}$. Therefore, u fails to receive v's message in all T'_p phases with probability p_f, bounded as: $p_f \leq \left(1 - \frac{C-t}{8C^2}\right)^{T'_p} < (1/e)^{\frac{c_1\Delta}{8C}\log n}$.

Since we assumed in Section 2 that $\Delta \geq C$, we can replace Δ/C with some $c' \geq 1$, and conclude by choosing c_1 to be sufficiently large such that this failure probability is no more than n^{-c}, for any constant c. We can now tweak our constants in T_p to ensure that in both cases above, our failure probability is no more than n^{-2}. It follows that u receives m with probability at least $1 - n^{-2}$. A union bound over v's neighbors tells us that every neighbor of v receives m with probability at least $1 - n^{-1}$, as needed.

3.2   Adaptive Broadcast Algorithm

The broadcast algorithm presented in the previous section has a time complexity that depends on the worst-case amount of local contention (as captured by Δ). In practice, Δ might be large compared to the actual amount of contention at a node u (i.e., δ_u). Here we present a broadcast algorithm that requires the network topology satisfy the regional clique decomposition property defined in our model section, and in return is able to replace a Δ with a δ_u factor in its complexity. It does so, however, at the cost of an additional factor of $t \log^2 n$. The adaptive solution, therefore, is applicable when the actual amount of contention is at least a factor of $t \log^2 n$ improved over the worst-case.

Algorithm Description.   Our adaptive broadcast algorithm has each node u divide time into *iterations*. An iteration I is composed of 2 phases, a *knockout* phase followed by a *cleanup* phase. Nodes begin *inactive*. When a node obtains a message m for broadcast, it waits until the beginning of the next iteration before becoming *active*.

In the knockout phase of I, active nodes compete to become local leaders. To simplify our analysis, we present the subroutine run in this phase as a combination of a subroutine designed for a disruption-free single channel (DFSC) model (i.e., our model with $C = 1$ and $t = 0$), and a simulator capable of simulating any DFSC subroutine in our more general t-disrupted setting. The DFSC knockout subroutine works as follow. Every node u divides time into $\max(\lceil \log(\Delta/C) \rceil, 1)$ epochs, each consisting of $\alpha \log n$ rounds, where α is a constant fixed in the proof below. For each epoch $e \in \{1, ..., \max(\lceil \log(\Delta/C) \rceil, 1)\}$, if u is active, it transmits with probability $p_e = 2^{-(\log \Delta - e + 1)}$ for each round in e. (i.e., there is one epoch for each probability in the sequence $1/\Delta, 2/\Delta, ..., 1/C$.) In every round, if u chooses not to transmit, including the case where it is inactive, it listens. If u receives a message from another node, it becomes *inactive* for the rest of the iteration and restarts active in the next iteration.

The simulator strategy we use to simulate the DFSC knockout subroutine in our t-disrupted model works as follows. The simulator uses only the first $\hat{C} = \min\{2t, C\}$ channels. For each node u and simulated round, the simulator uses $T_s = \frac{Ct}{C-t}\beta \log n$

real rounds, for a constant β we fix in the proof below. It begins by determining u's behavior for the simulated round according to the routine being simulated; i.e., it decides whether u transmits, and if so, what message m to transmit. It then spends T_s rounds executing u's decision. In each of these real rounds, u chooses a channel with uniform randomness. If it decided to transmit, it transmits m with probability 1, otherwise, it listens. At the end of T_s, the simulator simulates u's reception behavior. There are three cases: if u transmitted in this simulated round, then u simulates receiving nothing. If u decided to listen in this simulated round, and does not receive any messages during the T_s real rounds, it simulates receiving nothing. If u decided to listen in this simulated round, and received at least 1 message during the T_s real rounds, then it simulates receiving one of these messages (chosen arbitrarily if there are multiple).

During the subsequent cleanup phase in a given iteration, each node u runs the non-adaptive broadcast algorithm described in Section 3.1 with $\Delta = \mathcal{C}$, participating as an active node only if it began this phase active. If u begins the clean-up phase active, then after this phase concludes, it completes its broadcast and becomes inactive.

Analysis of the Adaptive Broadcast Algorithm. Our analysis below requires that we show certain properties hold throughout a full execution. For technical reasons, therefore, we must assume that executions are bounded by a number of rounds polynomial in n: say, $O(n^k)$ for some constant $k \geq 1$. In the following, let $\mathcal{R} = \{R_1, R_2, ..., R_\ell\}$ be the set of $\ell \leq n$ non-empty regions provided by our assumed regional clique decomposition property on G. When analyzing the knockout subroutine in the DFSC model, we make use of the following helpful notation. For round r, let e_r be the epoch number from $\{1, ..., \max(\lceil \log(\Delta/\mathcal{C}) \rceil, 1)\}$ associated with that round, and let $A(r)$ be the nodes active in the network at the beginning of round r (i.e., nodes broadcasting a message). For region $R_i \in \mathcal{R}$, let $W_i(r) = \sum_{v \in A(r) \cap R_i} p_{e_r}$; i.e., the sum of transmission probabilities of active nodes in R_i in round r (recall $p_j = 2^{-(\log \Delta - j + 1)}$ is the transmission probability used in the jth epoch). Finally, let $N^+(u) = N(u) \cup \{u\}$.

We begin by studying the behavior of our DFSC knockout subroutine when executed in the DFSC model. Lemmas 3, 4 and 5 all apply to this scenario. Recall in the following statements that α is a constant used in defining the epoch length for this subroutine. We begin below by adapting a strategy introduced in [17] to prove that for each region R_i, W_i self-regulates to never grow too large.

Lemma 3. *For sufficiently large α, in an execution of the DFSC knockout subroutine, the following holds w.h.p.: for every round r and region $R_i \in \mathcal{R}$: $W_i(r) < 2$.*

We now leverage this result to show that every active node survives an execution of the knockout subroutine with probability proportional to its local contention.

Lemma 4. *For all $\alpha \geq 1$, the following holds for each node u starting a given execution of the DFSC knockout subroutine active: the probability $\hat{p}_u$ that u transmits alone before any node in $N(u)$, or that no node in $N(u)$ transmits, is bounded as $\hat{p}_u \in \Omega(1/\delta_u)$.*

Proof. Let u be some node that starts an execution of this subroutine active, C be the subset of u's neighbors that are also active, $C^+ = C \cup \{u\}$, and $x = |C^+|$. We cannot argue directly about the behavior of nodes in C during this execution because their transmission behavior might also depend on their own neighbors (e.g., if a node in C is knocked out by one of u's two-hop neighbors, then this obviously affects its transmission probability). We will instead argue about their behavior in the absence of other nodes. In particular, for each $v \in C^+$ let b_v be the binary string where bit ℓ, indicated $b_v[\ell]$, equals 1 if and only if v would broadcast in round ℓ of iteration I, given its random bits,[3] under the assumption that v has not received any messages through the preceding $\ell - 1$ rounds of the iteration. Let $B^+ = \{b_v \mid v \in C^+\}$.

We now bound the probability that certain properties of this set of strings, each generated with independent randomness, hold. In particular, let $r_p = \min\{r' \mid \exists v \in C^+ : b_v[r'] = 1\}$; the first round with a transmission. If all strings in B^+ contain only 0's, we say that r_p is *undefined*. If r_p *is* defined, we say that its value is *good* if it occurs in a round corresponding to a broadcast probability less than $1/x$ (recall $x = |C^+|$), and is *bad* if it occurs in a higher probability round. We can now partition the space of possible outcomes for the generation of B^+ into three mutually exclusive portions: (1) r_p is undefined; (2) r_p is defined and good; (3) r_p is defined and bad.

We first prove that the size of the outcome space that satisfies condition 3 is small. In particular, for r_p to be defined for a bad epoch means that there was some previous epoch, $\hat{e}$, for which $p_{\hat{e}} = 1/(kx)$, for some $1 < k \leq 2$, and yet no node generated a 1 for the corresponding rounds in its bit string. Given that there are x nodes, the probability that no node generates a 1 for all $\alpha \log n$ rounds corresponding to this epoch $\hat{e}$ is $(1 - \frac{1}{kx})^{x\alpha \log n} \leq \frac{1}{n^{\alpha/2}}$. In other words: it is a small probability.

Now consider the case where r_p is defined and good. We want to calculate the size of the portion in the outcome space where $b_u[r_p] = 1$, and $b_v[r_p] = 0$ for all $v \in C$. To do so, we note that by the definition of *good*, the probability associated with r_p can be expressed as $1/(kx)$ for some $k > 1$. Let p_1 be the probability that exactly *one node* in C^+ selects 1 in a round associated with this probability $p_1 = \binom{x}{1}\frac{1}{kx}(1 - \frac{1}{kx})^{x-1} > \frac{x}{kx}(\frac{1}{4})^{\frac{x}{kx}} = \frac{1}{k}(\frac{1}{4})^{\frac{1}{k}}$. Similarly, let p_i be the probability that exactly i nodes in C^+ transmit in a round associated with this good probability. We upper bound this value as follows: $p_i = \binom{x}{i}\left(\frac{1}{kx}\right)^i \left(1 - \frac{1}{kx}\right)^{x-i} < x^i \left(\frac{1}{kx}\right)^i = \frac{1}{k^i}$ Let p_{2+} be the probability of more than 1 node in C_+ transmits in this round: $p_{2+} \leq \sum_{i=2}^{x} \frac{1}{k^i} \leq \frac{1}{k(k-1)}$.

Finally, we consider both the case where k is small and large.

For $1 \leq k \leq 2$: Define p_0 as the probability that no nodes in C^+ transmit in r_p. We note that $p_0 = (1 - \frac{1}{kx})^x \geq (\frac{1}{4})^{\frac{1}{k}} \geq \frac{1}{4}$, $p_1 \geq \frac{1}{4}$ and $p_{2+} = 1 - p_1 - p_0 \leq \frac{1}{2}$.

For $k > 2$: $p_1 \geq \frac{1}{2k}$ and $p_{2+} \leq \frac{1}{k}$ (by our above equation).

It follows that for both possibilities for the value of k, the probability of two or more transmitters in r_p is no more than twice as likely as a single transmitter. Given that there is at least a single transmitter, $p_1 \geq 2p_{2+}$ implies that $p_1 \geq \frac{1}{3}$. Finally, if there is a single transmitter, by the uniform probabilities used in each round, the probability that this transmitter is u is $1/x$. We have shown, therefore, that in the portion where r_p is defined and good, $\frac{1}{3x}$ of this portion has u transmitting alone before any node in C.

We can now pull together the pieces. The event whose probability we are bounding with $\hat{p}_u$ contains the fraction of the outcome space in which condition (1) from above

---

[3] To define b_v we treat randomness here such that v has a pre-determined collection of random bits from which it extracts the needed randomness for its probabilistic choices.

holds, as well as the fraction where condition (2) holds and u broadcasts alone before any of its neighbors. We proved that the fraction where condition (3) holds is small, i.e., $\leq n^{-\alpha/2}$, and therefore most of the outcome space (e.g., at least a constant fraction, $\geq 1 - n^{-1}$ for $\alpha > 2$) is dominated by conditions (1) and (2). Furthermore, we proved that at least a $\frac{1}{3x}$ fraction of the condition (2) portion has u broadcast alone before its neighbors. Combined, $p_{(1)} + p_{(2)} * \frac{1}{3x} \geq (p_{(1)} + p_{(2)}) * \frac{1}{3x}$ and $p_{(1)} + p_{(2)} \geq 1 - n^{-1}$ provide our lemma statement.

We now show that if u transmits alone in its local neighborhood, it has a constant probability of finishing the subroutine execution active (as there is a constant probability that this message knocks out your neighbors for the remainder of the execution).

Lemma 5. *For sufficiently large α, if node u transmits in an execution of the DFSC knockout subroutine before any other node in $N(u)$ transmits, then with constant probability u ends this execution active.*

We now shift our focus from the behavior of the knockout subroutine in the DFSC model to our broadcast algorithm as a whole. If a node ends the knockout phase of a given iteration active, it has a constant probability of successfully completing broadcast in the subsequent clean-up phase. The key insight is that it is unlikely that more than $\approx \mathcal{C}$ nearby nodes finish this phase active, allowing the non-adaptive algorithm, run with maximum contention $\mathcal{C}$, to succeed.

Lemma 6. *If u is active at the end of the knockout phase of some iteration I of the non-adaptive broadcast algorithm, then with high probability every node that neighbors u receives u's message by the end of I.*

Our final step before our final theorem, is to prove that our simulator routine can successfully simulate a subroutine designed for the DFSC setting. The core insight in the below proof is that we use a significantly large number of real rounds per simulated round to ensure that if a single neighbor v of some u decides to broadcast in a simulated round (the key case), u will hear from v with high probability during the corresponding T_s real rounds.

Theorem 2. *With high probability, the DFSC simulation subroutine correctly simulates the DFSC model using $O(\frac{\mathcal{C}t}{\mathcal{C}-t} \log n)$ rounds for each simulated round.*

We can now pull together the pieces for our final theorem statement. The key insight is that there are two good cases with respect to an active node u coming out of the knockout phase: (1) no node in u's neighborhood transmitted (implying that there are only a small number of such nodes), or (2) u transmitted alone before any of its neighbors (implying, by our above lemma, that u is the only active node in its neighborhood with constant probability). In either of these cases, the clean-up phase has a good chance of helping u succeed. Because we previously proved that one of these two conditions occurs with probability in $\Omega(1/\delta_u)$, we can show that $O(\delta_u \log n)$ iterations is enough to ensure success for u with high probability.

Theorem 3. *The adaptive broadcast algorithm implements a broadcast service in the general model with latency of $O(\frac{\mathcal{C}t}{\mathcal{C}-t} \delta_u \log^3 n(\log(\Delta/\mathcal{C}) + 1))$ rounds.*

3.3 Broadcast Lower Bounds

We now prove that our broadcast algorithms are close to optimal (which we define to mean *within polylogarithmic factors in the network size*) for most parameter values. We begin with a bound that shows our non-adaptive broadcast algorithm is close to optimal for a node u when δ_u is large (i.e., close to Δ). We then adapt a result from [3] (see Theorem 4.1) to prove that our adaptive algorithm is close to optimal when δ_u is small (i.e., constant). A requirement of the algorithm is that it is regular, a definition which we take from [3]. An algorithm is regular is there exists a sequence of pairs (F_i, b_i) for $i \in \{1, 2, ...\}$ where each F_i is a probability distribution over $[\mathcal{C}]$ and b_i is a probability s.t. for each node u and local round r, u will select a frequency from F_r to transmit on with probability b_i, until it receives a message. Once it receives message, its behavior is no longer constrained.

Theorem 4. *Fix some algorithm $\mathcal{A}$ that implements a broadcast service with latency $f(n, \delta_u, \mathcal{C}, t)$, for each node u. It follows that $f(n, \delta_u, \mathcal{C}, t) = \Omega(\frac{\mathcal{C}}{\mathcal{C}-t}\delta_u)$.*

Theorem 5 (Adapted from [3]). *Fix some regular algorithm $\mathcal{A}$ that implements a broadcast service with latency $f(n, \delta_u, \mathcal{C}, t)$, for each node u. It follows that $f(n, \delta_u, \mathcal{C}, t) = \Omega(\frac{\mathcal{C}t}{\mathcal{C}-t}\log n)$.*

3.4 Unicast Algorithm

In the unicast problem, a node u attempts to deliver a message to a single *known* neighbor v. Because the destination is now known, we assume u can leverage *link layer* acknowledgments to determine when it is successful in its delivery. This added power allows us to adapt the time complexity to the actual amount of local disruption (captured by t') as opposed to the worst case (captured by t).

Algorithm Description. As in non-adaptive broadcast, we assume that each node u groups rounds into *phases* of size $\lceil \log \Delta \rceil$ rounds. It then divides these phases into *groups*, each containing $\lceil \log \mathcal{C} \rceil$ phases. When node u is passed a pair (m, v), indicating that it should send message m to neighbor v, it becomes *active* at the beginning of the next phase. It will remain active until it receives an acknowledgment from v (in practice we might also add a timeout equal to the worst case broadcast latency from Section 3.1). During round r of phase k of some group, u chooses a channel with uniform probability from the first $\min\{2^k, \mathcal{C}\}$ channels. If u is active, it transmits (m, v) with probability $1/2^r$, otherwise it listens. If i is *inactive*, it always listens. Larger k corresponds to larger guesses for t' (as it requires u to choose from among more channels). If at any point, u receives (m', u) from some neighbor v, u replies immediately with an acknowledgment.

Theorem 6. *Our algorithm implements a unicast service with a latency of $O(\frac{\mathcal{C}t'}{\mathcal{C}-t'} \log \Delta \log \mathcal{C} \log n)$ rounds.*

4 Evaluation: A Disruption-Resistant Link Layer Protocol

We demonstrate the utility of our algorithms with a link layer protocol that is robust against unpredictable disruption and operates on commodity 802.11 hardware. Our protocol consists of (1) a *name service* that maintains the list of local neighbors, performs

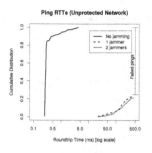
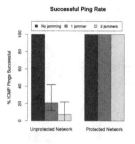
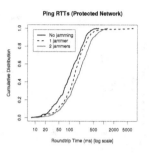

Fig. 1. Cumulative distribution of ICMP ping RTTs in an *unprotected* network

Fig. 2. Percentage of successful ICMP ping requests (with min/max ranges)

Fig. 3. Cumulative distribution of ICMP ping RTTs in a *protected* network

key exchange and key derivation, and serves as an address resolution protocol, based on the broadcast algorithm, and (2) an *authenticated communication service* that encrypts, signs, authenticates and sends messages, based on the unicast algorithm.

The disruption-resistant link layer leverages the broadcast protocol to conduct Diffie-Hellman (DH) [4] exchanges between nodes. Since the broadcast protocol guarantees, w.h.p., the reception of messages in the presence of a disrupting adversary, the adversary cannot prevent honest nodes' announced DH public keys from being received by their neighbors. When transmitting messages, a node includes a nonce that is encrypted with the receiver's public DH key. Only the intended receiver can decrypt the nonce and reply with an acknowledgment that contains the nonce, confirming that the message was received as no other node could have generated this response.

We developed a proof-of-concept implementation *using commodity 802.11 hardware* to demonstrate our protocol's efficacy as a robust communication primitive. Our implementation is built on top of the Click Modular Router [11] and FreeMAC [23], a modified Atheros 802.11 driver. We present the full details of our link layer protocol and its implementation in the full version. In the next section, we evaluate its ability to provide reliable communication even in the presence of malicious jammers.

4.1 Experimental Setup

Our testbed consists of 4 nodes $n1 \dots n4$ and two *jammers* $j1, j2$. These reside in the same 4-by-6 meter room, run Linux 2.6, are equipped with PCMCIA Atheros 802.11b/g wireless controllers with AR5212 chipsets, and use the 802.11b bit rate of 2 Mbit/s.

Network Configurations. Since our protocol is a general link layer protocol, we compare it against standard 802.11 ad hoc mode (with acknowledgments and incremental backoff enabled), which we refer to as the *unprotected network*. We run our reliable MAC-layer protocols in the *protected network* setting. In both settings, we restrict the choice of channels to the non-overlapping frequencies: 2412Hz, 2437Hz, and 2462Hz.

Network Jamming. To maximize disruption, jammers continuously send packets. Jammers do not obey 802.11 backoff requirements in either network. We assume the adversary has knowledge of the communication protocols in use and adapts its jamming

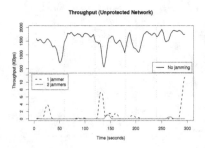

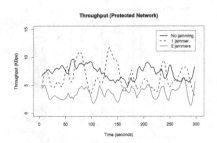

Fig. 4. Average throughput over time when operating in *unprotected* mode, with and without jammers

Fig. 5. Average throughput over time when operating in *protected* mode, with and without jammers

strategy accordingly. For the unprotected network, jammers operate on the channel used by the ad hoc network. For the protected network, jammers hop between the channels used by the protocol in a round robin fashion. When there are two jammers, they jam two of the three channels at any given time.

Metrics. The nodes periodically transfer large files over HTTP (i.e., via TCP) and send ICMP ping requests to measure roundtrip times (RTTs). Nodes $n1$ and $n3$ respectively receive large files over HTTP from nodes $n2$ and $n4$ in the presence of jammers $j1$ and $j2$. Ping measurements are conducted between node pairs $(n1, n2)$ and $(n3, n4)$.

We explore the performance of our disruption-resistant protocols by measuring the *roundtrip time* of ICMP-based ping measurements, the percentage of lost ping messages, and the effective *throughput* of the file transfers. To increase contention, the two pairs of nodes communicate simultaneously. All experiments are repeated five times.

4.2 Performance Results

Latency and Packet Loss. As shown in Figure 1, 802.11 ad hoc mode maintains ping RTTs of less than 0.250ms when there are no jammers. However, the performance of the network significantly degrades when even a single jammer is active. For instance, when only $j1$ is active, the minimum RTT of *successful* pings rises to 21ms, and the median increases to 153ms. The jammers are also able to cause significant packet loss in the unprotected network. As illustrated in Figure 2, with one (resp. two) active jammers, only 20% (resp. 5%) of pings are successful.

For comparison, the performance of ICMP pings in protected networks is shown in Figure 3. The protected network produces larger ping times – the median RTT without jamming is 109.5ms. However, performance degrades only slightly with jamming: in the worst case in which both $j1$ and $j2$ attempt to disrupt communication, the median RTT increases to 191.5ms. Importantly, as shown in Figure 2, *all pings are successful*.

Throughput. Figures 4 and 5 respectively show the throughput over time for the file transfer on the unprotected and protected networks, with and without jamming. Results are shown for 5 consecutive experiments (separated with dashed lines).

Without jamming, ad hoc mode outperforms our protocol. However, *with even a single jammer, the median throughput of the unprotected network drops to zero.* In contrast, while the protected network experiences a modest decrease when jammers become active, it maintains the ability to transmit data effectively. Without jammers, median throughput for the protected network is 7.4KBps; with one and two jammers, the respective throughputs drops to 6.1 and 3.2KBps. Although our protocols do not perform as efficiently as standard 802.11 *in the absence of contention,* our MAC layer achieves significantly lower loss rates and reasonable throughput when jammers are present. In environments where reliability is of utmost importance (e.g., in first responder networks), our protocols provide strong delivery guarantees.

5   Conclusion

This paper describes and proves correct uncoordinated communication algorithms for general noisy shared spectrum networks. It also describes a reliable link layer protocol based on these primitives, and performs a preliminary testbed evaluation. This work indicates that algorithmic techniques can be used to enable reliable communication even in settings,e.g., low power, where such reliability is otherwise hard to achieve.

References

[1] Awerbuch, B., Richa, A., Scheideler, C.: A Jamming-Resistant MAC Protocol for Single-Hop Wireless Networks. In: PODC (2008)

[2] Barriére, L., Fraigniaud, P., Narayanan, L.: Robust position-based routing in wireless ad hoc networks with unstable transmission ranges. In: DIAL M, pp. 19–27 (2001)

[3] Daum, S., Gilbert, S., Kuhn, F., Newport, C.: Leader Election in Shared Spectrum Radio Networks. In: PODC (2012)

[4] Diffie, W., Hellman, M.E.: New Directions in Cryptography. IEEE Transactions on Information Theory 22(6), 644–654 (1976)

[5] Foerster, J.: The Performance of a Direct-Sequence Spread Ultra-Wideband System in the Presence of Multipath, Narrowband Interference, and Multiuser Interference. In: IEEE Conference on Ultra Wideband Systems and Technologies (2002)

[6] Ghaffari, M., Gilbert, S., Newport, C., Tan, H.: Optimal Broadcast in Shared Spectrum Radio Networks. In: Baldoni, R., Flocchini, P., Binoy, R. (eds.) OPODIS 2012. LNCS, vol. 7702, pp. 181–195. Springer, Heidelberg (2012)

[7] Gilbert, S., Guerraoui, R., Newport, C.: Of Malicious Motes and Suspicious Sensors: On the Efficiency of Malicious Interference in Wireless Networks. Theoretical Computer Science 410(6-7), 546–569 (2009)

[8] Tan, H., Wacek, C., Newport, C., Sherr, M.: A Disruption-Resistant MAC Layer for Multichannel Wireless Networks,
http://people.cs.georgetown.edu/~cnewport/publications.html

[9] Jin, T., Noubir, G., Thapa, B.: Zero Pre-Shared Secret Key Establishment in the Presence of Jammers. In: MOBIHOC (2009)

[10] Kavehrad, M., Ramamurthi, B.: Direct-Sequence Spread Spectrum with DPSK Modulation and Diversity for Indoor Wireless Communications. IEEE Transactions on Communications 35(2), 224–236 (1987)

[11] Kohler, E., Morris, R., Chen, B., Jannotti, J., Kaashoek, M.: The Click Modular Router. ACM Transactions on Computer Systems (TOCS) 18(3), 263–297 (2000)

[12] Kuhn, F., Moscibroda, T., Wattenhofer, R.: On the locality of bounded growth. In: PODC, pp. 60–68 (2005)

[13] Kuhn, F., Wattenhofer, R., Zollinger, A.: Ad hoc networks beyond unit disk graphs. Wireless Networks 14(5), 715–729 (2008)

[14] Liu, A., Ning, P., Dai, H., Liu, Y.: USD-FH: Jamming-Resistant Wireless Communication using Frequency Hopping with Uncoordinated Seed Disclosure. In: MASS (2010a)

[15] Liu, A., Ning, P., Dai, H., Liu, Y., Wang, C.: Defending DSSS-Based Broadcast Communication Against Insider Jammers via Delayed Seed-Disclosure. In: ACSAC (2010b)

[16] Liu, Y., Ning, P., Dai, H., Liu, A.: Randomized Differential DSSS: Jamming-Resistant Wireless Broadcast Communication. In: INFOCOM (2010c)

[17] Moscibroda, T., Wattenhofer, R.: Maximal independent sets in radio networks. In: PODC, pp. 148–157. ACM (2005)

[18] Navda, V., Bohra, A., Ganguly, S., Rubenstein, D.: Using channel hopping to increase 802.11 resilience to jamming attacks. In: INFOCOM (2007)

[19] Pöpper, C., Strasser, M., Čapkun, S.: Jamming-Resistant Broadcast Communication without Shared Keys. In: USENIX Security Symposium (2009)

[20] Richa, A., Scheideler, C., Schmid, S., Zhang, J.: Competitive and Fair Throughput for Co-Existing Networks Under Adversarial Interference. In: PODC (2012)

[21] Richa, A., Scheideler, C., Schmid, S., Zhang, J.: Competitive Throughput in Multi-Hop Wireless Networks Despite Adaptive Jamming. Distributed Computing 26(3), 159–171 (2013)

[22] Schmid, S., Wattenhofer, R.: Algorithmic models for sensor networks. In: Proc. 14th Int. Workshop on Parallel and Distributed Real-Time Systems, pp. 1–11 (2006)

[23] Sharma, A., Belding, E.M.: FreeMAC: Framework for Multi-channel MAC Development on 802.11 Hardware. In: ACM Workshop on Programmable Routers for Extensible Services of Tomorrow (PRESTO) (2008)

[24] Slater, D., Tague, P., Poovendran, R., Matt, B.: A Coding-Theoretic Approach for Efficient Message Verification over Insecure Channels. In: WiSec (2009)

[25] Strasser, M., Capkun, S., Popper, C., Cagalj, M.: Jamming-Resistant Key Establishment using Uncoordinated Frequency Hopping. In: IEEE Symposium on Security and Privacy (2008)

[26] Strasser, M., Pöpper, C., Čapkun, S.: Efficient Uncoordinated FHSS Anti-Jamming Communication. In: MOBIHOC (2009)

[27] Xu, W., Wood, T., Trappe, W., Zhang, Y.: Channel surfing and spatial retreats: Defenses against wireless denial of service. In: ACM Workshop on Wireless Security (2004)

[28] Xu, W., Trappe, W., Zhang, Y.: Channel Surfing: Defending Wireless Sensor Networks from Interference. In: ACM IPSN (2007)

Distributed Computing by Mobile Robots: Solving the Uniform Circle Formation Problem

Paola Flocchini[1], Giuseppe Prencipe[2], Nicola Santoro[3], and Giovanni Viglietta[3]

[1] University of Ottawa
flocchin@site.uottawa.ca
[2] University of Pisa
giuseppe.prencipe@unipi.it
[3] Carleton University
santoro@scs.carleton.ca, viglietta@gmail.com

Abstract. Consider a set of $n \neq 4$ simple autonomous mobile robots (decentralized, asynchronous, no common coordinate system, no identities, no central coordination, no direct communication, no memory of the past, deterministic) initially in distinct locations, moving freely in the plane and able to sense the positions of the other robots. We study the primitive task of the robots arranging themselves equally spaced along a circle not fixed in advance (UNIFORM CIRCLE FORMATION). In the literature, the existing algorithmic contributions are limited to restricted sets of initial configurations of the robots and to more powerful robots. The question of whether such simple robots could deterministically form a uniform circle has remained open. In this paper, we constructively prove that indeed the UNIFORM CIRCLE FORMATION problem is solvable for any initial configuration of the robots without any additional assumption. In addition to closing a long-standing problem, the result of this paper also implies that, for pattern formation, asynchrony is not a computational handicap, and that additional powers such as chirality and rigidity are computationally irrelevant.

1 Introduction

Consider a set of punctiform computational entities, called *robots*, located in $\mathbb{R}^2$ where they can freely move. Each entity is provided with a local coordinate system and operates in *Look-Compute-Move* cycles. During a cycle, a robot obtains a snapshot of the positions of the other robots, expressed in its own coordinate system (*Look*); using the snapshot as an input, it executes an algorithm (the same for all robots) to determine a destination (*Compute*); and it moves towards the computed destination (*Move*). After a cycle, a robot may be inactive for some time.

To understand the nature of the distributed universe of these mobile robots and to discover its computational boundaries, the research efforts have focused on the minimal capabilities the robots need to have to be able to solve a problem. Thus, the extensive literature on distributed computing by mobile robots

M.K. Aguilera et al. (Eds.): OPODIS 2014, LNCS 8878, pp. 217–232, 2014.

has almost exclusively focused on very simple entities operating in strong adversarial conditions. The robots we consider are *anonymous* (without ids or distinguishable features), *autonomous* (without central or external control), *oblivious* (no recollection of computations and observations done in previous cycles), *disoriented* (no agreement among the individual coordinate systems, nor on unit distance and chirality). In particular, the choice of individual coordinate systems, the activation schedule, the duration of each operation during a cycle, and the length traveled by a robot during its movement are determined by an adversary; the only constraints on the adversary are fairness (i.e., for every time t and each robot r there exists $t' > t$ when r is active), finiteness (i.e., the duration of each activity and inactivity is arbitrary but finite), and minimality (i.e., there exists $\delta > 0$, unknown to the robots, such that if the destination is at distance at most δ the robot will reach it, else it will move at least δ towards the destination, and then it may be unpredictably stopped by the adversary). For this type of robots, depending on the activation schedule and timing assumptions, three main models have been studied in the literature: the *asynchronous* one ($\mathcal{A}$SYNC), where no assumptions are made on synchronization among the robots' cycles nor their duration, and the *semi-synchronous fully synchronous* models, denoted by $\mathcal{S}$SYNC and $\mathcal{F}$SYNC, respectively, where the robots, oblivious and disoriented, however operate in synchronous rounds, and each round is "atomic": all robots active in that round terminate their cycle by the next round; the only difference is whether all robots are activated in every round ($\mathcal{F}$SYNC), or, subject to some fairness condition, a possibly different subset is activated in each round ($\mathcal{S}$SYNC). All three models have been intensively studied (e.g., see [1,2,3,5,6,7,8,9,10,15,16,17,23,24]; for a detailed overview refer to the recent monograph [13]).

The research on the *computability* aspects has focused almost exclusively on the fundamental class of GEOMETRIC PATTERN FORMATION problems. A *geometric pattern* (or simply *pattern*) P is a set of points in the plane; the robots *form* the pattern P at time t if the configuration of the robots (i.e., the set of their positions) at time t is similar to P (i.e., coincident with P up to scaling, rotation, translation, and reflection). A pattern P is *formable* if there exists an algorithm that allows the robots to form P within finite time and no longer move, regardless of the activation scheduling and delays (which, recall, are decided by the adversary) and of the initial placement of the robots in distinct points. Given a model, the research questions are: to determine if a given pattern P is formable in that model; if so, to design an algorithm that will allow its formation; and, more in general, to fully characterize the set of patterns formable in that model. The research effort has focused on answering these questions for $\mathcal{A}$SYNC and the less restrictive models both in general (e.g., [5,15,16,22,23,24]) and for specific classes of patterns (e.g., [1,7,8,10,11,12,19,20]).

Among specific patterns, a special research place is occupied by two classes: `Point` and `Uniform Circle`. The class `Point` is the set consisting of a single point; point formation corresponds to the important GATHERING problem requiring all robots to gather at a same location, not determined in advance (e.g., see [2,3,4,18,21]). The other important class of patterns is `Uniform Circle`: the points of the pattern form the vertices of a regular n-gon, where n is the number of robots (e.g., [1,6,7,8,10,11,12,20]).

In addition to their relevance as individual problems, the classes `Point` and `Uniform Circle` play another important role. A crucial observation, by Suzuki and Yamashita [23], is that formability of a pattern P from an initial configuration Γ in model $\mathcal{M}$ depends on the relationship between $\rho_{\mathcal{M}}(P)$ and $\rho_{\mathcal{M}}(\Gamma)$, where $\rho_{\mathcal{M}}(V)$ is a special parameter, called *symmetricity*, of a multiset of points V, interpreted as robots modeled by $\mathcal{M}$. Based on this observation, it follows that the only patterns that *might* be formable from any initial configuration in $\mathcal{F}$SYNC (and thus also in $\mathcal{S}$SYNC and $\mathcal{A}$SYNC) are single points and uniform circles. It is rather easy to see that both points and uniform circles can be formed in $\mathcal{F}$SYNC, i.e. if the robots are fully synchronous. After a long quest by several researchers, it has been shown that GATHERING is solvable (and thus `Point` is formable) in $\mathcal{A}$SYNC (and thus also in $\mathcal{S}$SYNC) [2], leaving open only the question of whether `Uniform Circle` is formable in these models. In $\mathcal{S}$SYNC, it was known that the robots can *converge* towards a uniform circle without ever forming it [7]. Some recent results indicate that the robots can actually form a uniform circle in $\mathcal{S}$SYNC. In fact, by concatenating the algorithm of [19], for forming a biangular configuration, with the one of [11], for circle formation from an equiangular starting configuration, it is possible to form a uniform circle starting from any initial configuration in $\mathcal{S}$SYNC; notice that the two algorithms can be concatenated only if the robots are semi-synchronous. Hence, the only outstanding question is whether it is possible to form a uniform circle in $\mathcal{A}$SYNC.

In spite of the simplicity of its formulation and the repeated efforts by several researchers, the existing algorithmic contributions are limited to restricted sets of initial configurations of the robots and to more powerful robots. In particular, it has been proven that, with the additional property of *chirality* (i.e., a common notion of "clockwise"), the robots can form a uniform circle [12], and with a very simple algorithm; the fact that `Uniform Circle` is formable in $\mathcal{A}$SYNC +*chirality* follows also from the recent general result of [16]. The difficulty of the problem stems from the fact that the inherent difficulties of asynchrony, obliviousness, and disorientation are amplified by their simultaneous presence.

In this paper we show that indeed the UNIFORM CIRCLE FORMATION problem is solvable for any initial configuration of $n \neq 4$ robots without any additional assumption, closing a problem open for over a decade. This result also implies that, for GEOMETRIC PATTERN FORMATION problems, *asynchrony* is not a computational handicap, and that additional powers such as *chirality* and *rigidity*[1] are computationally irrelevant.

2 Definitions

For a finite set $S \subset \mathbb{R}^2$ of $n > 2$ points, we define the *Smallest Enclosing Circle*, or *SEC*, to be the circle of smallest radius such that every point of S lies on the

---

[1] A move is *rigid* if it is not interrupted before reaching the destination point.

circle or in its interior. For any S, SEC is easily proven to exist and to be unique. Three other circles will play a special role: these are concentric with SEC, and have radiuses that are $1/2$, $1/3$, and $1/4$ the radius of SEC. They are denoted by $SEC/2$, $SEC/3$, and $SEC/4$, respectively.

The *angular distance*, with respect to point x, between two points p and q (distinct from x) is the measure of the smallest angle between $\angle pxq$ and $\angle qxp$, and is denoted by $\theta_x(p,q)$. The *sector* defined by two points a and b is the locus of points c such that $\theta_x(a,c) + \theta_x(c,b) = \theta_x(a,b)$. Whenever x is not specified, it is assumed to be the center of the SEC of a well-understood set of points.

Given a finite set S, the positions of its points around some point $x \notin S$, taken clockwise, naturally induce a cyclic order on S. If several points of S lie on the same ray emanating from x, their relative order is induced by their distance from x, starting from the nearest point.

Let $p_0 \in S$ be any point, and let $p_i \in S$ be the $(i+1)$-th point in the cyclic order around $x \notin S$, starting from p_0. Let $\alpha_x^{(i)} = \theta_x(p_i, p_{i+1})$, where the indices are taken modulo n. Then, $(\alpha_x^{(i)})_{0 \leqslant i < n}$ is called the *angle sequence* induced by p_0. (Of course, depending on the choice of $p_0 \in S$, there may be at most n different angle sequences with respect to x.) Letting $\beta_x^{(i)} = \alpha_x^{(n-i)}$, for $0 \leqslant i < n$, we call $(\beta_x^{(i)})_{0 \leqslant i < n}$ the *reverse angle sequence* induced by p_0. We let $\widetilde{\alpha}_x$ and $\widetilde{\beta}_x$ be, respectively, the lexicographically smallest angle sequence and the lexicographically smallest reverse angle sequence of S. Also, we denote by μ_x the lexicographically smallest between $\widetilde{\alpha}_x$ and $\widetilde{\beta}_x$, and by $\mu_x^{(i)}$ the i-th element of μ_x. If $p \in S$ is any point inducing μ_x as a clockwise or counterclockwise angle sequence, we say that p is a *lex-first* point of S (with respect to x), and we denote by $\mathcal{L}_1$ the set of all lex-first points. Let p be a lex-first point of S and suppose that μ_x is the clockwise (resp. counterclockwise) angle sequence induced by p. Let p' be the first point after p in the clockwise (resp. counterclockwise) order around x that is not collinear with x and p. Then, p' is said to be a *lex-second* point of S (with respect to x), and we denote by $\mathcal{L}_2$ the set of all lex-second points. If x is not specified, it is assumed to be the center of the SEC of S.

The following definitions apply whenever the symbols used are well defined, i.e., if and only if no point of S lies in the center of SEC. S is *co-radial* if $\mu^{(0)} = 0$. In a co-radial set, every two points at angular distance 0 are said to be *co-radial* with each other. The number of distinct clockwise angle sequences of S (with respect to the center of its SEC) is called the *period* of S. It is easy to verify that the period is always a divisor of n.

We will be distinguishing among different types of configurations, defined below (see also Figure 3). S is said to be *Equiangular* if its period is 1, *Biangular* if its period is 2, and *Aperiodic* if its period is n. In a *Biangular* set, any two points at angular distance $\mu^{(0)}$ are called *neighbors*, and any two points at angular distance $\mu^{(1)}$ are called *quasi-neighbors*. If a *Biangular* configuration is not co-radial, it is called *Simple biangular*. An *Aperiodic* configuration can be *Uni-aperiodic* if $\widetilde{\alpha} \neq \widetilde{\beta}$, and *Bi-periodic* if $\widetilde{\alpha} = \widetilde{\beta}$. A set S that is not *Aperiodic* is said to be *Uni-periodic* if $\widetilde{\alpha} \neq \widetilde{\beta}$, and *Bi-periodic* if $\widetilde{\alpha} = \widetilde{\beta}$. S is *Regular* if its points are the vertices of a regular n-gon.

Algorithm UNIFORM CIRCLE FORMATION

Find first match of observed configuration:
1. **Regular:** Do nothing;
2. **Central:** Execute CENTRAL;
3. **Equiangular:** Execute EQUIANGULAR;
4. **Pre-regular:** Execute PRE-REGULAR;
5. **Pre-equiangular:** Execute PRE-EQUIANGULAR;
6. **Landmark-co-radial:** Execute LANDMARK-CO-RADIAL;
7. **Post-periodic:** Execute POST-PERIODIC;
8. **Antipodal-referees:** Execute ANTIPODAL-REFEREES;
9. **Simple Biangular:** Execute SIMPLE biangular;
10. **Periodic:** Execute PERIODIC;
11. **Post-aperiodic:** Execute POST-APERIODIC;
12. **Aperiodic:** Execute APERIODIC;

Fig. 1. The UNIFORM CIRCLE FORMATION algorithm

We say that point $p \in S$ is *homologous* to point $q \in S$ if the angle sequence induced by p is equal to the angle sequence induced by q, or to its reverse. In particular, if it is equal to the angle sequence induced by q (and not necessarily to its reverse), p and q are said to be *analogous*. Homology and analogy are equivalence relations on S, and the equivalence classes that they induce on S are called *homology classes* and *analogy classes*, respectively. In a *Uni-periodic* set of period k, all homology classes are *Equiangular* sets of size n/k. In a *Bi-periodic* set of period k, each homology class is either a *Biangular* set of size $2n/k$, or an *Equiangular* set of size n/k or $2n/k$. In a *Uni-aperiodic* set, the homology classes consist of one point; in a *Bi-aperiodic*, they consist of either one or two points.

S is said to be *Double-biangular* if it is *Bi-periodic* with period 4 and has exactly two homology classes.

S is *Pre-regular* if there exists a regular n-gon (called the *supporting polygon*) such that, for each pair of adjacent edges, one edge contains exactly two points of S (possibly on its endpoints), and the other edge's relative interior contains no point of S [8]. There is a natural correspondence between points of S and vertices of the supporting polygon: the *matching vertex* v of point $p \in S$ is such that v belongs to the edge containing p, and the segment vp contains no other point of S. If two points of S lie on the same edge of the supporting polygon, then they are said to be *companions*.

Finally, S is *Central* if one of its points lies at the center of SEC.

3 The Algorithm

3.1 High-Level Description

The general idea of the algorithm is that first some robots identify themselves as *referees* (in spite of anonymity) and maintain their role until they are the only ones not in their final position. The referees univocally determine special points, the *landmarks*, which, in turn, define a set of half-lines from the centre of SEC, the *targets*, partitioning the plane in n equal sectors. Each robot is assigned a different target. By positioning themselves on the targets, the robots reach an *Equiangular* configuration, and they ultimately form a uniform circle.

Algorithm UNIFORM CIRCLE FORMATION (see Figure 1) consists of an ordered set of tests to determine the class of the current configuration; this determines which action is going to be taken by a robot in order to implement the general strategy described above. The universe of possible configurations is decomposed by the algorithm into several classes. Some of the classes (i.e., *Regular*, *Central*, *Equiangular*, *Pre-regular*, *Simple biangular*, *Periodic*, *Aperiodic*) have been defined in Section 2; the others will be defined in the following, along with the description of the corresponding actions. It is easy to see that all possible configurations are covered by these classes, simply because any configuration is either *Periodic* or *Aperiodic*. Hence, if all other tests fail, one of these two necessarily succeeds.

We stress that some configurations belong to more than one class, and so the order in which such classes are tested by the algorithm matters. For instance, a *Pre-regular* configuration may easily be also *Aperiodic*. The reason why *Pre-regular* is tested before *Aperiodic* is that, when the robots execute procedure PRE-REGULAR and the configuration remains *Pre-regular* but it also accidentally becomes *Aperiodic*, we want all robots to keep executing the same procedure, without letting some of them "erroneously" start executing procedure APERIODIC. Of course, now the opposite problem may arise: when the robots are executing procedure APERIODIC, they may accidentally form a *Pre-regular* configuration. However, as it will be apparent in later sections, this event is much less likely, and it is easier to predict and handle by the algorithm in such a way that, if a *Pre-regular* configuration may be formed accidentally during the execution, then all robots agree to stop in that configuration and consequently start executing procedure PRE-REGULAR in a synchronized fashion.

3.2   Basic Tools

The above high-level description gives an idea of the general *intended behavior* of the robots. Asynchrony and special configurations can easily make the algorithm deviate from this behavior. The rules and movements of the robots are carefully designed so to handle any deviation, and they are quite complex. In particular, two tools are employed: cautious moves and special circles.

Cautious Move. If a robot's movement can potentially create some configuration that would be treated by other observing robots in an inconsistent way (i.e., a configuration of a class tested *before* the current one by the algorithm), the rule will prescribe the robot to stop in the first point that might create it. We call these points *critical points*. Thus in some procedures of the algorithm, robots are specifically required to perform an operation called *cautious move*; this method is invoked when there is a set of robots that need to move on disjoint paths, each of which contains finitely many critical points. It is assumed that, as the robots move along their paths, the set of critical points does not change.

In a cautious move, first the set of critical points is expanded with a set of "auxiliary" critical points: if a robot has a critical point on its path, located at distance d from the endpoint of the path (where the distance is measured along the path itself), then each other robot whose path is not shorter than d acquires a new critical point at distance d from the end of its path. The last point along each robot's path is also taken as a critical point.

Then, each robot r whose remaining path is longest moves forward along its path by the greatest possible amount, with the following constraints:

- r's destination point must not be past the next critical point (auxiliary or not);
- if r is currently lying on a critical point (auxiliary or not), its destination point must be at most halfway toward the next critical point (auxiliary or not) along its path;
- if the remaining path of r has length d, and there is another robot whose remaining path has length $d' < d$, then r's destination point must be at most d' away from the endpoint of r's path (in other words, robots do not "pass each other" in one turn).

On the other hand, the robots whose remaining path is not longest wait.

Special Circles. In the algorithm we use specific concentric circles: SEC, SEC/2, SEC/3, and SEC/4. This is done first of all to facilitate the recognition of the current configuration and coordinate the operations of the robots. For example, SEC/4 is used in *Periodic* while SEC/3 is used in *Aperiodic*. More importantly, these circles are used to *avoid* the accidental formation of certain configurations. In particular, as long as some robots are on or inside SEC/3, a *Pre-regular* configuration may never be formed: this is crucial in the proof of correctness of the algorithm. Note that we assume the robots can perform "circular movements" when the destination point is along one of these circles, but, at the cost of slightly modifying the algorithm, it is possible to let the robots move only along straight lines.

3.3   The Initial Tests

The first four tests performed by the algorithm are the simplest ones. The algorithm first checks if a uniform circle has been formed; if so, no further action is taken. Otherwise, it checks if there is a robot at the centre of SEC. In this case, that robot moves, avoiding collisions, to become co-radial with the robots on one of the most populated radiuses, and stopping before SEC/4. This action (procedure CENTRAL) transforms the configuration in one of class *Aperiodic*. In the third test, the algorithm checks if the configuration is *Equiangular*; if so, all robots move radially towards SEC eventually evolving into a *Regular* configuration. In the fourth test, if the configuration is *Pre-regular*, each robot moves towards its matching vertex in the supporting polygon. This action, called procedure PRE-REGULAR is precisely the technique described in [8] to move from a

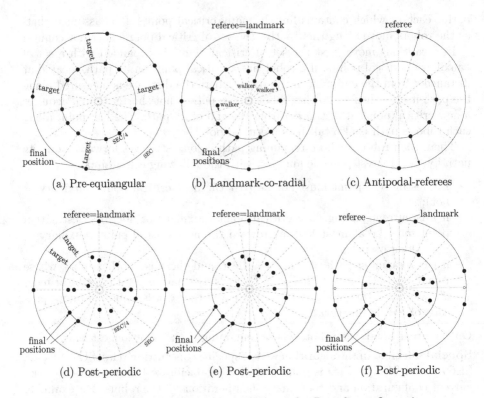

(a) Pre-equiangular   (b) Landmark-co-radial   (c) Antipodal-referees

(d) Post-periodic   (e) Post-periodic   (f) Post-periodic

Fig. 2. Examples of the possible evolutions of a *Periodic* configuration

Biangular configuration into a *Regular* one; during the action the configuration remains *Pre-regular* and it eventually evolves into *Regular*.

3.4   The Intermediate Tests

Having failed the initial tests, the next sequence of tests is for the classes of configurations defined below, which can occur as the initial configuration, or as an evolution from a *Periodic* configuration. Along with the definitions, the actions to perform in each configuration are given.

Pre-equiangular. There are robots both on SEC and on SEC/4, and nowhere else. The robots on SEC are at least three, and those on SEC/4 are forming an "almost" *Regular* configuration; that is, a *Regular* with one missing point for each robot on SEC. The missing points may be arranged in two different ways. They may form a "regular pairs" arrangement, in which there are pairs of missing points in adjacent positions, in such a way that the pairs are equally spaced around SEC/4; otherwise, they form a "regular pairs" arrangement in which exactly one element of each pair has been removed. There is a bijection between robots on SEC and missing points, determined by the minimum total distance the robots on SEC must travel to occupy them (Figure 2(a) shows an arrangement on SEC/4 of the second type).

In this case, the robots on SEC rotate towards their targets, which are uniquely determined by the positions of the robots on SEC/4. With this action, called procedure PRE-EQUIANGULAR, the robots eventually reach an *Equiangular* configuration.

Landmark-co-radial. The robots on SEC form an *Equiangular* set, and these are the referees, which also coincide with the landmarks. The landmarks define the n target half-lines, in such a way that either all landmarks lie on some targets (as in Figure 2(b)), or they lie on bisectors of adjacent targets. All the non-referee robots are on or inside SEC/4: each robot on SEC/4 is on a target; the only ones *strictly* inside are those co-radial with the referees, and at most one robot (called *walker*) for each referee. The central targets of each sector defined by two adjacent referees are all occupied by robots on SEC/4 in such a way that, for each landmark, the open sector Γ defined by the nearest target in the clockwise direction that is occupied by a robot on SEC/4 and the nearest one in the counterclockwise direction contains as many robots as targets. Moreover, Γ contains at most one walker, and the targets in Γ that lie to the left of the landmark differ by at most one unit from those to the right. A *co-radial Biangular* configuration falls in this class, too.

In this configuration, the intended behavior is to "resolve" all the robots that are co-radial to the referees, and have them move to their targets, reaching an *Equiangular* or a *Pre-equiangular* configuration (depending whether the referees are already on their targets or not).

Note that in a *Landmark-co-radial* the only unoccupied targets correspond to the groups of co-radial robots of the landmarks and to at most one robot per landmark, the *walker*, which is moving towards a target. The co-radial robots move in turns. If there is no walker in the sector Γ (as defined above) around a landmark, the most internal non-referee that is co-radial with that landmark rotates toward the farthest away target among those in Γ, becoming a *walker*. When a walker reaches its target, it moves radially to reach SEC/4. If all the non-referees are on their targets, they all lie on SEC/4, and the configuration happens to be *Antipodal-referees* (see below), then the two non-referees closest to the landmarks move toward SEC (thus "forcing" the configuration to transition into an *Antipodal-referees* configuration that is not a *Landmark-co-radial* anymore, which is tested after *Landmark-co-radial* by the algorithm). Otherwise, the configuration becomes either *Equiangular* or *Pre-equiangular*, as intended.

Post-periodic. The robots on SEC form an *Equiangular* or a *Simple biangular* set, and they are the referees. All other robots lie on SEC/4 or inside of it. If the referees are *Equiangular*, the landmarks coincide with the referees and they all have the same number of co-radial robots, which lie *strictly* inside SEC/4. If the referees are *Biangular*, the landmarks are the midpoints of neighboring referees, and no robot is co-radial with any landmark. The robots that are not co-radial with the landmarks are equidistributed among the sectors defined by the landmarks.

In this configuration, the targets are calculated with respect to the landmarks, depending on the parity of the robots that are co-radial with each landmark (including the referees): if they are odd, then the landmarks lie on some targets (Figure 2(d)); if they are even (which includes zero), the landmarks lie on bisectors of adjacent targets (Figures 2(e) and 2(f)). Note that, if such co-radials are odd, the referees must be *Equiangular*. Each robot may be associated with a unique target, or to two possible targets (in case of left-right symmetry of its view).

The intended behavior in a *Post-periodic* configuration is to have all robots move onto SEC/4 on their respective targets, except for the robots that are co-radial with some landmark, thus reaching a *Landmark-co-radial* configuration. To do so, the non-referees that are not co-radial with the landmarks and that can reach SEC/4 without colliding with other robots, move radially toward it. If none can do it and there are co-radial robots that are not co-radial with any landmark, the most internal of these co-radials rotates in an arbitrary direction of 1/4 of the minimum non-zero element in μ. If all the non-referees that are not co-radial with the landmarks are already on SEC/4, they orderly rotate on SEC/4 until they reach their targets (which are now uniquely determined). This is done in such a way that only the robots that can reach their target without colliding with other robots move. Each move is cautious, with critical points corresponding to *Landmark-co-radial* and *Pre-equiangular* configurations. At this point, the configuration becomes: *Landmark-co-radial* if the referees are *Equiangular* and have co-radial robots; *Pre-equiangular* if the referees are *Biangular* and they are not on their targets; or *Equiangular* if the referees are *Equiangular* and there are not co-radial robots, or if they are *Biangular* and already on their targets.

Antipodal-referees. There are two antipodal robots on SEC, which are the referees. On SEC/4 there are (possibly among others) $n - 4$ robots that are forming a *Regular* configuration with some missing points. More precisely there are two antipodal pairs of adjacent missing points, such that each referee is equidistant to two adjacent missing points. Furthermore, there are two other robots co-radial with two non-adjacent missing points, which lie between SEC/4 and SEC (possibly on SEC/4 or on SEC). Note that this configuration is uniquely identifiable and has period either n or $n/2$ (see Figure 2(c)). In this configuration, the robots closest to the referees (one for each referee) move towards SEC, eventually reaching a *Pre-equiangular* configuration.

Simple biangular. In this case, the intended behavior of the robots is to reach a *Pre-regular* configuration by moving toward SEC according to the cautious move protocol, with critical points on SEC/4 (where a *Landmark-co-radial* or a *Pre-equiangular* may be formed), and additional critical points where *Pre-regular* configurations may be formed (see Theorem 2). If the robots already on SEC belong to the same analogy class, the other robots in the same class move first.

3.5   The Periodic Test

Periodic. If the procedure PERIODIC is executed, it means that the configuration is *Periodic*, and additionally it does not belong to any of the classes described above. In this case, the intended behavior is to elect the referees, define the landmarks, have the referees move onto SEC and the non-referees move into SEC/4, reaching a *Post-periodic* configuration. In trying to do this, the robots can find themselves in a variety of different configurations, and the algorithm might switch to several different cases.

Let k be the period. If there exist robots with exactly n/k homologous robots, then the lex-first among these robots are chosen to be the referees, as well as the landmarks. If this is not the case, all homology classes must have size exactly $2n/k$. If the robots in $\mathcal{L}_1$ are not *Equiangular* (and therefore they are strictly *Biangular*), they are chosen to be the referees; otherwise the referees are the robots in $\mathcal{L}_2$. Note that in both cases the referees form a *Simple biangular* set; the landmarks are selected to be the midpoints of neighboring referees. Hence, by construction, the landmarks are always n/k points forming an *Equiangular* set (with respect to the center of the SEC of all robots), and they define n/k sectors, each containing the same number of robots in its interior.

If the configuration is *Double-biangular*, no referee is on SEC, and some non-referees are not on SEC, then all the non-referees move radially to reach SEC. Otherwise, if there are referees not on SEC, they move radially to reach SEC. If all the referees are on SEC, the other robots move radially inward until they reach SEC/4 or its interior. All non-referee robots that are co-radial with some landmark move *strictly* inside SEC/4. The non-referee robots move in turns, in such a way that only homologous robots can move together. Specifically, the non-referees that belong to homology classes of size n/k move first.

In all cases, all movements are cautious, with critical points on SEC/4 (which may yield a transition into *Landmark-co-radial*, *Antipodal-referees*, or *Pre-equiangular*), and those determined by *Pre-regular* configurations.

When this is done, the configuration becomes *Post-periodic*, with some exceptions: if the robots not co-radial with the referees are already on their targets on SEC/4, except at most one per landmark, the configuration becomes *Landmark-co-radial*; if the only robots not on their targets are the referees, and the referees are more than two, the configuration becomes *Pre-equiangular*; if the only robots not on their targets are the referees, and the referees are only two, the configuration becomes *Antipodal-referees*.

3.6   The Aperiodic Tests

In this last set of tests, *Post-aperiodic* and *Aperiodic* configurations are addressed. Similarly to the previous cases, the intended behavior of the actions is to elect the referees and to identify landmarks and targets. From the *Aperiodic* configuration, the intended behavior is to reach a *Post-aperiodic* configuration and, from there, an *Equiangular* configuration.

Post-aperiodic. There are either one or two robots on SEC/3, which are the referees. All other robots are found between SEC/2 and SEC. If there are two referees, they are not antipodal (i.e., their midpoint is not the center of SEC).

In this configuration, the actions taken by the robots (procedure POST-APERIODIC) are as follows. If there are two referees, and all the non-referees are on SEC forming a *Regular* set with two adjacent missing points, the two referees rotate on SEC/3 until they become co-radial with the missing points, and the configuration becomes *Equiangular*. Otherwise, the targets are identified by the referees on SEC/3, and a unique target is assigned to each robot. The non-referees that can move radially to SEC without colliding, do so. If there are non-referees that cannot radially move to SEC (because other robots are in their way), then the most internal non-referees rotate of 1/4 of the minimum non-zero element of μ to remove the co-radiality. If all the non-referees are on SEC and there is only one referee, the non-referees cautiously rotate to their respective targets, in such a way that SEC never changes and no two robots collide, and using *Simple biangular* and *Periodic* configurations as critical points. If the targets are reached, the configuration becomes *Equiangular*. Finally, if there are two referees, and all non-referees are on SEC, not forming a *Regular* set with two adjacent missing points, the non-referees rotate on SEC with a cautious move as in the previous case, with additional critical points given by the configurations in which the robots on SEC form a *Regular* set with two adjacent missing points. In this last case, the configuration may become *Simple biangular*, *Periodic*, *Equiangular*, or remain *Post-aperiodic*.

Aperiodic. The procedure APERIODIC is executed if the current configuration fails all previous tests. If the configuration is *co-radial uni-aperiodic*, then the lex-first is unique, and must have co-radial robots. In this case the referee is the most internal among the robots that are co-radial with the lex-first. If the configuration is *non-co-radial uni-aperiodic*, the lex-first is still unique, but it may be necessary to keep SEC intact. If this is not the case, the lex-first is the referee, otherwise the referee is the lex-second (it is easy to see that, if $n \geqslant 5$, one of these two robots can be removed without altering SEC).

If the configuration is *co-radial bi-aperiodic*, let r and r' be, among the robots that are co-radial with the lex-first robots, the most internal ones, respectively. If r and r' are not aligned with the center of SEC, then they are chosen to be the referees. Otherwise, the referees are the first two robots in the lexicographically minimum order (which are homologous) that can be *safely* removed without altering SEC (assuming that all robots that can reach SEC radially are already on SEC), and such that they are the most internal robots among their co-radials. (Note that, in some configurations, these referees happen to be the same robot. In these cases, the referee is unique.)

Finally, if the configuration is *non-co-radial bi-aperiodic*, the referees are the first two (just one, in some special cases) homologous robots that are not aligned with the center of SEC, and such that, when all robots are on SEC, they can be removed without changing SEC.

The non-referees that are inside or on SEC/2 move out of SEC/2. Those that can reach SEC without colliding, do so. They take turns in such a way that only homologous robots can move together (hence at most two), and they move radially outward, performing a cautious move with critical points on SEC/4, SEC/3, SEC/2, SEC, and those determined by the *Pre-regular* configurations (see Theorem 3). During these movements, the configuration may become

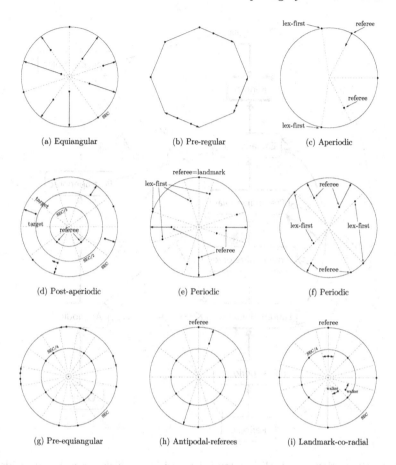

(a) Equiangular

(b) Pre-regular

(c) Aperiodic

(d) Post-aperiodic

(e) Periodic

(f) Periodic

(g) Pre-equiangular

(h) Antipodal-referees

(i) Landmark-co-radial

Fig. 3. Some examples of configurations

either *Post-periodic*, *Landmark-co-radial*, *Antipodal-referees*, or *Pre-equiangular* (when they pass through SEC/4 or when they reach SEC), or *Post-aperiodic* (when they pass through SEC/3), or *Pre-regular*. Otherwise, the configuration stays *Aperiodic*. If all the non-referees are outside SEC/2 and none of them can move to SEC without colliding, the referees move and reach SEC/3. They use a cautious move with SEC/4 as a critical point, and those determined by the *Pre-regular* configurations (see Theorem 3). The configuration may become *Post-periodic*, *Landmark-co-radial*, *Antipodal-referees*, or *Pre-equiangular* (when the robots reach SEC/4), or *Pre-regular*. Otherwise it becomes *Post-aperiodic*, as intended.

4   Correctness

The correctness proof is quite lengthy and can be found in [14]. We give here only an intuition of the main ingredients.

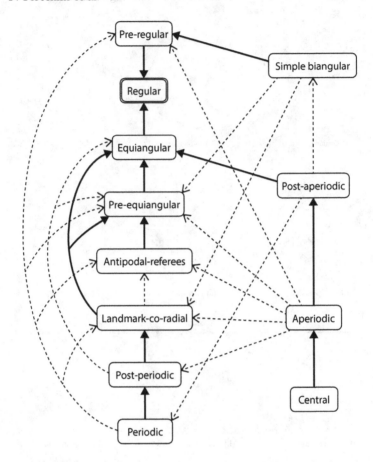

Fig. 4. Configurations, with their intended transitions (thick arrows) and incidental transitions (dashed arrows)

To prove correctness, we need to analyze all possible transitions between configurations. Some transitions come as a result of the "intended" behavior of the robots executing the algorithm; other transitions come as "accidental" byproducts of the execution. The proof is then a detailed examination of all the possible executions of the algorithm in the space of robots' configurations, paying special attention to the transitions that may arise as critical points of cautious moves. In the following, let $\mathcal{R} = \{r_1, \cdots, r_n\}$ denote a swarm of $n > 4$ robots. let $r_i(t)$ denote the location of robot r_i at time $t \geqslant 0$, and $\mathcal{R}(t) = \{r_1(t), \cdots, r_n(t)\}$.

We first prove that robots executing the cautious move protocol indeed behaves as intended.

Theorem 1. *Let a swarm of n robots execute a cautious move with critical point set $\bigcup_{i=1}^{k} C_i$, with $|C_i| = n$ for $1 \leqslant i \leqslant k$, from an initial configuration in which no robot is moving. Then, during the cautious move, whenever the robots are found in a configuration C_i, they all stop in that configuration.*

We then analyze the behaviour of the algorithms with respect to the critical points; in particular, to the *Pre-regular* case, which turns out to be the hardest to treat.

Lemma 1. *If S is* Pre-regular, *then it is not* Central, Post-periodic, Landmark-co-radial, Antipodal-referees, Pre-equiangular, *nor* Post-aperiodic.

Theorem 2. *Let* $\mathcal{R}(0)$ *be a* Simple biangular *configuration,* $n > 4$, *and let the robots execute procedure* SIMPLE BIANGULAR *with suitable critical points. Then, the robots eventually reach a* Pre-regular *configuration, and they stop as soon as they reach it.*

Theorem 3. *Let* $\mathcal{R}(0)$ *be an* Aperiodic *configuration,* $n > 4$, *and let the robots execute procedure* APERIODIC. *Then, as soon as they reach a* Pre-regular *or a* Simple biangular *or a* Aperiodic *configuration, they all stop in that configuration.*

The previous set of Theorems guarantees the correct execution of cautious moves. We conclude showing that the directed graph of configurations and their transitions (depicted in Figure 4) contains no cycles, and the only sink is the *Regular* configuration.

Lemma 2. *If* $n > 4$, *no transition is possible other than those illustrated in Figure 4.*

Theorem 4. *The* UNIFORM CIRCLE FORMATION *problem is solvable by* $n \neq 4$ *robots in* $\mathcal{ASYNC}$.

The case $n = 4$ is still open.

Acknowledgments. The authors would like to thank Marc-André Paris-Cloutier for many helpful discussions and insights, and Peter Widmayer and Vincenzo Gervasi for having shared some of the fun and frustrations emerging from investigating this problem. Partially supported by project ARS TechnoMedia (MIUR of Italy).

References

1. Chatzigiannakis, I., Markou, M., Nikoletseas, S.: Distributed circle formation for anonymous oblivious robots. In: 3rd Workshop on Efficient and Experimental Algorithms, pp. 159–174 (2004)
2. Cieliebak, M., Flocchini, P., Prencipe, G., Santoro, N.: Distributed computing by mobile robots: Gathering. SIAM Journal on Computing 41(4), 829–879 (2012)
3. Cohen, R., Peleg, D.: Convergence properties of the gravitational algorithms in asynchronous robots systems. SIAM Journal on Computing 34, 1516–1528 (2005)
4. Cohen, R., Peleg, D.: Convergence of autonomous mobile robots with inaccurate sensors and movements. SIAM Journal on Computing 38, 276–302 (2008)
5. S. Das, P. Flocchini, N. Santoro, and M. Yamashita Forming sequences of geometric patterns with oblivious mobile robots. Distributed Computing (to appear, 2014)
6. Défago, X., Konagaya, A.: Circle formation for oblivious anonymous mobile robots with no common sense of orientation. In: 2nd ACM Int. Workshop on Principles of Mobile Computing (POMC), pp. 97–104 (2002)

7. Défago, X., Souissi, S.: Non-uniform circle formation algorithm for oblivious mobile robots with convergence toward uniformity. Theoretical Computer Science 396(1-3), 97–112 (2008)
8. Dieudonné, Y., Labbani-Igbida, O., Petit, F.: Circle formation of weak mobile robots. ACM Trans. on Autonomous and Adaptive Systems 3(4), 16:1–16:20 (2008)
9. Dieudonné, Y., Levé, F., Petit, F., Villain, V.: Deterministic geoleader election in disoriented anonymous systems. Theoretical Computer Science 506, 43–54 (2013)
10. Dieudonné, Y., Petit, F.: Swing words to make circle formation quiescent. In: Prencipe, G., Zaks, S. (eds.) SIROCCO 2007. LNCS, vol. 4474, pp. 166–179. Springer, Heidelberg (2007)
11. Dieudonné, Y., Petit, F.: Squaring the circle with weak mobile robots. In: Hong, S.-H., Nagamochi, H., Fukunaga, T. (eds.) ISAAC 2008. LNCS, vol. 5369, pp. 354–365. Springer, Heidelberg (2008)
12. Flocchini, P., Prencipe, G., Santoro, N.: Self-deployment algorithms for mobile sensors on a ring. Theoretical Computer Science 402(1), 67–80 (2008)
13. Flocchini, P., Prencipe, G., Santoro, N.: Distributed Computing by Oblivious Mobile Robots. Synthesis Lectures on Distributed Computing Theory. Morgan & Claypool (2012)
14. Flocchini, P., Prencipe, G., Santoro, N., Viglietta, G.: Distributed Computing by Mobile Robots: Solving the Uniform Circle Formation Problem. arXiv:1407.5917 [cs.DC] (2014)
15. Flocchini, P., Prencipe, G., Santoro, N., Widmayer, P.: Arbitrary pattern formation by asynchronous oblivious robots. Theoretical Computer Science 407(1-3), 412–447 (2008)
16. Fujinaga, N., Yamauchi, Y., Kijima, S., Yamashita, M.: Asynchronous pattern formation by anonymous oblivious mobile robots. In: Aguilera, M.K. (ed.) DISC 2012. LNCS, vol. 7611, pp. 312–325. Springer, Heidelberg (2012)
17. Kamei, S., Lamani, A., Ooshita, F., Tixeuil, S.: Asynchronous mobile robot gathering from symmetric configurations without global multiplicity detection. In: Kosowski, A., Yamashita, M. (eds.) SIROCCO 2011. LNCS, vol. 6796, pp. 150–161. Springer, Heidelberg (2011)
18. Izumi, T., Souissi, S., Katayama, Y., Inuzuka, N., Défago, X., Wada, K., Yamashita, M.: The gathering problem for two oblivious robots with unreliable compasses. SIAM Journal on Computing 41(1), 26–46 (2012)
19. Katreniak, B.: Biangular circle formation by asynchronous mobile robots. In: Pelc, A., Raynal, M. (eds.) SIROCCO 2005. LNCS, vol. 3499, pp. 185–199. Springer, Heidelberg (2005)
20. Miyamae, T., Ichikawa, S., Hara, F.: Emergent approach to circle formation by multiple autonomous modular robots. J. Robotics and Mechatr. 21(1), 3–11 (2009)
21. Oasa, Y., Suzuki, I., Yamashita, M.: A robust distributed convergence algorithm for autonomous mobile robots. In: IEEE Int. Conference on Systems, Man and Cybernetics, pp. 287–292 (1997)
22. Sugihara, K., Suzuki, I.: Distributed algorithms for formation of geometric patterns with many mobile robots. J. Robot. Syst. 3(13), 127–139 (1996)
23. Suzuki, I., Yamashita, M.: Distributed anonymous mobile robots: Formation of geometric patterns. SIAM Journal on Computing 28(4), 1347–1363 (1999)
24. Yamashita, M., Suzuki, I.: Characterizing geometric patterns formable by oblivious anonymous mobile robots. Theoretical Computer Science 411(26-28), 2433–2453 (2010)

Approximation Algorithms for the Set Cover Formation by Oblivious Mobile Robots[*]

Tomoko Izumi[1], Sayaka Kamei[2], and Yukiko Yamauchi[3]

[1] College of Information Science and Engineering, Ritsumeikan University, Shiga, Japan
izumi-t@fc.ritsumei.ac.jp
[2] Graduate School of Engineering, Hiroshima University, Hiroshima, Japan
s-kamei@se.hiroshima-u.ac.jp
[3] Graduate School of Information Science and Electrical Engineering, Kyushu University,
Fukuoka, Japan
yamauchi@inf.kyushu-u.ac.jp

Abstract. Given n robots and n target points on the plane, *the minimum set cover formation* (SCF) problem requires the robots to form a set cover by the minimum number of robots. In previous formation problems by mobile robots, such as gathering and pattern formation, the problems consist only of the mobile robots, and there are no points fixed in the environment. In addition, the problems do not require a control of the number of robots constructing the formation. In this paper, we first introduce the formation problem in which robots move so that they achieve a desired deployment with the minimum number of robots for a given set of positions of fixed points.

Since the minimum set cover problem with disks in the centralized settings is NP-hard, our goal is to propose approximation algorithms for the minimum SCF problem. First, we show a minimal SCF algorithm from any initial configuration in the asynchronous system. Moreover, we propose an 8-approximation SCF algorithm in the semi-synchronous system for an initial configuration with a low symmetricity. This approximation algorithm achieves $2(1 + 1/l)^2$ approximation ratio for an initial configuration with the lowest symmetricity ($l \geq 1$).

Keywords: Oblivious mobile robots, set cover formation, approximation algorithms, distributed algorithms.

1 Introduction

Background. Studies about cooperations among autonomous mobile robots with weak capability attract much attention in the distributed computing community. In most of those studies, it is assumed that robots are *oblivious* (i.e., no memory to keep the history of the execution), *anonymous* (i.e., no ID to distinguish them), and *uniform* (i.e, all robots execute the same algorithm). In addition, it is also assumed that each robot has no direct means of communication. To interact with other robots, each robot observes the environment in its local coordinate system, which includes the positions of

[*] This work was supported in part by KAKENHI No. 26330015 and a Grant-in-Aid for Scientific Research on Innovative Areas gMolecular Roboticsh (No. 24104519) of The Ministry of Education, Culture, Sports, Science, and Technology, Japan.

M.K. Aguilera et al. (Eds.): OPODIS 2014, LNCS 8878, pp. 233–247, 2014.

other robots. Since observation is done in its local coordinate system, the perception of the environment is different for each of the robots. A robot determines its next destination based only on the observed positions in its local coordinate system. A robot executes the algorithm by repeating a Look-Compute-Move cycle: The robot observes the environment (Look), computes its next destination according to the developed algorithm (Compute), and moves to the destination (Move).

From the theoretical aspect, it is an interesting problem to clarify the class of cooperation tasks which autonomous robots can achieve, and to reveal the weakest capabilities of the robots to accomplish the task. Therefore, the problems, such as gathering and pattern formation, have been actively studied because they have the fundamental difficulties of coordination such as agreement and symmetry breaking. Common factors to these problems are that they consider only mobile robots and not any point fixed in the environment, and that they require the formation with all of the robots. That is, given an initial set of positions of robots, the formation problem requires the robots to form a specific pattern by using all of the robots, such as a single point (i.e., gathering) or a shape (i.e., pattern formation), and any size and location of the pattern is acceptable.

Our contribution.   In this paper, we first consider a kind of the formation problems in which there are fixed points in addition to the robots on the plane. Specifically, the *set cover formation* (SCF) problem is introduced in this paper as one of cooperation tasks among mobile robots. In the SCF problem, robots and target points are initially located on the two-dimensional Euclidean plane. The target points are visible and static. The goal of the SCF problem is to allocate the robots so that for every target point s, there is at least one robot within a circle centered at s with radius r (in this paper, we consider the case that $r = 1$). The minimum SCF problem requires to cover a set of target points by the minimum number of robots.

Since the centralized minimum set cover problem with disks is NP-hard [4], the robots cannot get the optimal positions to cover given target points. So, in this paper, we consider approximation algorithms for the minimum SCF problem. First, we propose a minimal SCF algorithm in the asynchronous system, which constructs the set cover by the minimal number of robots from any initial configuration. Next, we consider the minimum SCF problem in a semi-synchronous system for an initial configuration with a low symmetricity m defined by the initial positions of the target points and robots. To guarantee the approximation ratio, we use the *shifting strategy* introduced in [10]. We propose an approximation minimum SCF algorithm which achieves $2(1 + 1/l)^2$ approximation ratio in the semi-synchronous system for arbitrary $l \geq 1$ when $m = 1$ holds, that is, the robots and target sensors are all fully distinguished. This approximation ratio is twice the ratio in the centralized setting [10]. The challenging work is to solve the SCF problem for an initial configuration with $m \geq 2$. In this paper, we show that this approximation minimum SCF algorithm achieves 8 approximation ratio for an initial configuration with $m = 1, 2$, or 4 in the semi-synchronous system.

To deal with such formation problem in which robots move so that they construct a desired structure with the minimum number of them for a given set of positions of fixed points is a new challenging task in the field of the distributed computing for the mobile robots. The existence of fixed points in the environment may help the anonymous robots accomplish the task because they may be able to distinguish themselves from each other based on the positions of fixed points. However, the SCF problem has the difficulties caused by the existence of fixed points: While the previous pattern formation problem allows the scaling, translation and rotation of a desired shape, the SCF problem does not allow it, and a solution for the SCF problem depends on positions of fixed points. The oblivious robots must agree on a common solution based on the observation results in their local coordinate systems. In addition, the difficulties introduced by obliviousness and asynchrony of robots also remain in the SCF problem.

Related work. The SCF problem is considered as a variant of the pattern formation problem, which has been studied extensively. In the pattern formation problem, robots are required to form F' which is a result of transforming a given pattern F by the translations, rotations, and uniform scaling. The main interest in the pattern formation problem is to clarify the effect of the asynchrony on the solvability. There are three systems as for the asynchrony of robots, fully-synchronous (FSYNC), semi-synchronous (SSYNC), and asynchronous (ASYNC) systems. The class of formable patterns F from an initial configuration I is characterized in [11,12]. To define the formable class, they introduced the *symmetricity* $\rho(P)$ of set P of points: $\rho(P)$ is 1 if there is a point at center $c(P)$ of the smallest enclosing circle $SC(P)$ of P, and otherwise, $\rho(P)$ is the number of different angles α ($\in [0, 2\pi)$) such that rotating P by α around $c(P)$ results in P. In [11,12], it is shown that a pattern F is formable from an initial configuration I if and only if $\rho(I)$ divides $\rho(F)$ except F is a single point (i.e., gathering). For an instance satisfying the above condition, the pattern formation algorithms in the FSYNC and SSYNC system was proposed in [12]. The pattern formation in the ASYNC system was introduced in [3], and in [6], they proposed an asynchronous pattern formation algorithm that forms F from any initial configuration I such that $\rho(I)$ divides $\rho(F)$.

In [6], they used an *embedded pattern formation* algorithm in [5]. The embedded pattern formation problem requires for every visible point in a pattern, exactly one robot moves to the point. They showed that there is an embedded pattern formation algorithm which forms any given pattern from any initial configuration in the ASYNC system.

The differences of the (embedded) pattern formation and SCF problem are summarized in terms of the existence and type of fixed points, and the request of consisting the pattern of the minimum number of robots. In the pattern formation problem, there are no fixed points and the pattern is formed by all of the robots. The embedded pattern formation problem considers the fixed points to which the robots move, but it requires all of the robots to the fixed points. Unlike the embedded pattern formation problem, in the SCF problem, the robots are given a set of fixed points which should be covered, not points to be moved. Thus, to guarantee the minimality or approximation ratio for the minimum SCF problem, the robots must calculate and agree on a solution for the set cover, and some of the robots move to the positions in the solution.

In the centralized settings, the minimum set cover problem with disks and many of its variants are known to be NP-hard [4]. This problem is to find the minimum number of disks to cover a given set of points on the plane. The first polynomial time approximation algorithm was proposed by Hochbaum and Mass in [10]. They proposed a crucial strategy to guarantee the approximation ratio in the covering problem, called the *shifting strategy*. By using the shifting strategy recursively, they proposed a $(1 + 1/l)^2$-approximation algorithm with $O(n^{4l^2})$ time, where n is the number of target points and

$l \geq 1$ is the integer parameter of the algorithm. After that, the studies about the minimum set cover problem focus on the improvement of the time complexity. Feder and Green [2] and independently Gonzalez [9] improved the time complexity to $O(n^{4l})$. Some researches tried to get better time complexity at the expense of the approximation ratio [1,7,8]. Notice that since a set of target points is given on a global coordinate system in the centralized setting problem, these algorithms cannot be applied directly to the SCF problem.

Road map. This paper is organized as follows. In Section 2, we present the system model and define the SCF problem. Section 3 and Section 4 present a minimal and approximation SCF algorithm respectively. We conclude this paper in Section 5.

2   Preliminaries

2.1   System Model

The system consists of a set of n autonomous mobile robots $R = \{r_0, r_1, \cdots, r_{n-1}\}$ and a set of n visible target points $S = \{s_0, s_1, \cdots, s_{n-1}\}$, where $|R| = |S| = n$ [1]. The robots and target points are modeled as points located on the two-dimensional Euclidean plane. For two points x and y, $\overline{xy}$ denotes the line segment whose endpoints are x and y, and $|xy|$ denotes the Euclidean distance of line segment $\overline{xy}$. r Robots are *anonymous* and *oblivious*: Anonymous robots have no identifier to distinguish them, and execute the same algorithm. An oblivious robot cannot explicitly remember the history of its execution. In addition, no device for direct communication is equipped. The cooperation of robots is done in an implicit manner: Each robot observes the environment and obtains the positions of other robots and target points. The robots can distinguish the positions of robots and target points. The robot has no capability of multiplicity detection. Note that, however, the robot can detect a robot r_i on a target point s_j. The positions of the target points are fixed and unchanged during an execution of the algorithm.

Each robot executes the deployed algorithm in *computational cycles* (or briefly *cycles*). At the beginning of a cycle, the robot observes the current environment, i.e., the positions of other robots and target points (Look), and determines the destination point based on the deployed algorithm (Compute). Then, the robot moves to the computed destination (Move). In this paper, we assume the *rigid movement model*, in which each

---

[1] Our algorithms achieve the minimal and approximated SCF even if there are more than n robots (i.e., $|R| \geq |S| = n$): Since n robots are enough to cover n target points, the redundant robots go far enough away from the center $c(S)$ of the smallest enclosing circle $SC(S)$ so that they do not interfere with the other robots' movements. The redundant robots can be detected because the number of the redundant robots is $|R| - n = c \cdot m$ for the symmetricity m (defined in Section 4.2) and an integer c. That is, the farthest m robots go far away from $c(S)$, and then the next farthest m robots go far away from $c(S)$. These movements are repeated until the number of robots within a certain distance from $c(S)$ becomes n. In the algorithm, the robots consider the positions of n robots within a certain distance from $c(S)$ to calculate their next destination. Since any two robots do not go to the same position in our algorithms, the redundant robots keep their positions far enough away from $c(S)$ and do not join in the processing of the SCF algorithm. In the paper, we assume that $|R| = |S| = n$ holds for simplicity.

robot reaches a destination point at the end of each Move phase. There are three systems for the synchronization of the execution of cycles. In the asynchronous (ASYNC) system, each robot executes each phase of Look, Compute and Move asynchronously. On the other hand, in the fully-synchronous (FSYNC) system, all robots execute the cycle simultaneously. The semi-synchronous (SSYNC) system is the one between the ASYNC and FSYNC systems. More precisely, in the SSYNC system, an execution is divided into consecutive time steps. At each time instant, the *scheduler* selects the set of robots to perform a cycle. The selected robots by the scheduler perform one cycle simultaneously. Note that in the SSYNC system, a robot never observe the environment while another robot is moving.

Each robot has no access to the global coordinate system, and the observation of the environment is represented as the set of points in its *local coordinate system*. The local coordinate system of a robot is the Cartesian coordinate system whose origin is the current position of the robot. There is no agreement on the direction of local coordinate systems among robots. That is, the robots do not have common knowledge about the direction of $x - y$ axis of their local coordinate systems. However, they agree on the orientation of their coordinate systems. As for the scale, we assume that all robots agree on the unit length [2]. For the convenience of explanations, we introduce the global coordinate system. Notice that the global coordinate system is introduced only for the convenience of explanations, and thus each robot cannot be aware of it. The positions of robot r_i and target point s_j in the global coordinate system are denoted by $p(r_i)$ and $p(s_j)$ respectively, or simply r_i and s_j when no confusion occurs. For a set P of positions, let $SC(P)$ be the smallest enclosing circle of P and $c(P)$ be the center of $SC(P)$. The radius of $SC(P)$ is denoted by $radius(P)$. The circle $RC(r_i)$ is the circle centered at r_i with radius 1.

2.2 The Set Cover Formation (SCF) Problem

The set cover formation (SCF) problem requires some of the robots to construct a set cover. At first, we define a set cover for set S of target points.

Definition 1 (Set cover). *A set P of points is called a set cover for S if for each $s \in S$, there is a $p \in P$ such that $|sp| \leq 1$. A set cover P for S is minimum if $|P| = \min_{P' \in \mathcal{P}} |P'|$ holds, where $\mathcal{P}$ is the set of set covers for S, and P is minimal if no proper subset of P is a set cover for S.*

To distinguish the robots in a set cover, we introduce the two states of robots, *active* and *asleep*. However, it is assumed that the robot has no memory to keep its state of the execution. So, we do not allow the robots to execute any action after they change their states to asleep: Initially, the states of all robots are set to active. Then, if some robots change their states to asleep during the execution, they never move from the current position, and never return to the active state. Note that since the state of robot is internal information of the robot, the other robots cannot know the state of the robot by the observation.

---

[2] This assumption is required to detect the coverage of the target points by the robots.

A *configuration* is denoted by the set of $p(r_i)$ for each $r_i \in R$, the set of $p(s_j)$ for each $s_j \in S$ and the states of all robots. We define $C(t)$ as the configuration at time t. Let $P(C(t))$ be a set of positions of all robots in a configuration $C(t)$, and $\tilde{P}(C(t))$ be a set of positions of active robots in $C(t)$. For short, $P(C(t))$ and $\tilde{P}(C(t))$ are denoted by $P(t)$ and $\tilde{P}(t)$. For a target point $s \in S$ and robot r, we say that s is covered by r if $|rs| \leq 1$ holds. In an initial configuration $C(0)$, we assume that the positions of robots and target points are scattered. That is, initially, the robot observes n robots and n target points, and no robot stays on a target point in S. The SCF problem is defined as follow.

Definition 2 (Set cover formation (SCF) problem). *Given an initial configuration $C(0)$ consisting of the sets $P(0)$ and S, an algorithm $\mathcal{A}$ solves the set cover formation (SCF) problem for S, if for all possible executions of $\mathcal{A}$ from $C(0)$, there exists time t' such that for all $t > t'$, $\tilde{P}(t)$ is a set cover for S. The minimum (resp. minimal) set cover formation problem requires that $\tilde{P}(t)$ is a minimum (resp. minimal) set cover for S.*

By the definition of system model, since the number of robots and the number of target points are the same, the SCF problem is easily solved by moving each robot to each position of target points. In [5], Fujinaga et al. proposed the embedded pattern formation algorithm by the ASYNC robots, called CWM algorithm, which moves the robots to any given visible points from any initial configuration in the *non-rigid movement model*. In the non-rigid movement model, a Move phase may finish when a robot is still on the way to its next destination[3]. In the CWM algorithm, each robot r_i finds an optimum matching M_i between the robots and the given points, and moves in a straight line to its matched point. They showed that all robots compute the same matching (i.e., $M_i = M_j$ for any robot r_i, r_j), and that the matching M_i never change during the straight movement of the robots. Thus, by using the CWM algorithm, for any set S of target points, exactly one robot can reach each $s \in S$ from any initial configuration, and the robot keeps its state active. In a configuration $C(t)$ after all robots move to their matched points, the set $\tilde{P}(t)$ ($=P(t)$) constructs a set cover for S. However, the strategy using the CWM algorithm does not guarantee the quality of the set cover, even the minimality.

3 Minimal Set Cover Formation Algorithm

In this section, we propose a minimal SCF algorithm from any initial configuration by the ASYNC oblivious robots.

The algorithm is simple: For each target point s_j, the robot calculates the position q_j such that s_j is on the segment $\overline{q_j c(S)}$ and $|q_j s_j| = 1$ holds (Fig. 1). Note that for any s_i, s_j ($i \neq j$), $q_i \neq q_j$ holds, and that since the position q_j is determined only by the positions of the target points in S and s_j, q_j does not change even if the robots move. In addition, all the robots agree on q_j even if they calculate q_j in its local coordinate system. Let F be the set of the calculated positions ($|F| = n$). For F and the set of robots, we apply the CWM algorithm. That is, each robot r_j goes to the matched position

---

[3] Precisely, a robot stops on y or on a point z on $\overline{xy}$ such that $|xz| \geq \epsilon$ (≥ 0) holds at the end of a move phase, where x is the previous position, y is the destination position and ϵ is (unknown) minimum moving distance.

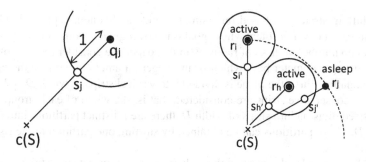

Fig. 1. A calculated position q_j for s_j **Fig. 2.** Determination of active or asleep

$q_j \in F$ in a straight direction. Since all robots agree on the position of every q_j, the robots detect the termination of the movements of all robots. After all robots stop on the matched positions, they determine their states, active or asleep, in the following manner: Each robot calculates the states of all robots in the order of increasing distance $|r_i c(S)|$. Let $s_{i'}$ be the target point on the intersection with $\overline{r_i c(S)}$ and the circumference of circle $RC(r_i)$, which is the circle centered at r_i with radius 1. If the target point $s_{i'}$ is covered by an active robot r_h ($h \neq i$) which satisfies $|r_h c(S)| < |r_i c(S)|$, then robot r_i becomes asleep, otherwise r_i keeps its state active. The states of the robots r_i and r_j which $|r_i c(S)| = |r_j c(S)|$ holds can be determined independently because the target point $s_{i'}$ of r_i is not within a circle $RC(r_j)$ (See Fig. 2). Since the state of every robot is determined uniquely based only on the distances from $c(S)$ to the robots, all robots agree on the same assignment of the states.

We can show that the above algorithm solves the minimal SCF problem because every target point s_j is covered by at least one robot and an active robot r_i has at least one target point $s_{i'}$ which is not covered by any active robots. However, we omit the detailed proof because of the page limitation.

Theorem 1. *In the ASYNC system and the rigid movement model, there is an algorithm for the minimal SCF problem from any initial configuration* [4].

4 An Approximation Algorithm for the Minimum SCF Problem

4.1 The Shifting Strategy

In [10], an algorithm, called the *shifting strategy*, is proposed as an approximation scheme to solve the minimum set cover problem with disks in a centralized setting, where the positions of the target points are given to an algorithm in the global coordinate system. Most of the previous works about the minimum set cover problem with disks use this strategy to guarantee the approximation ratio of their algorithms. In this paper, we also use the shifting strategy, so we explain it in this subsection.

---

[4] Notice that Theorem 1 also holds in the non-rigid movement model because the CWM algorithm works in the model.

The shifting strategy is based on a simple divide-and-conquer approach. Consider the minimum set cover problem whose goal is to cover given n points on the plane with the minimum number of disks of radius r (in this paper, we consider the case of $r = 1$). Let the shifting parameter be l, which is an integer parameter to control the accuracy of approximation. First, the plane is divided into vertical strips of width $D = 2r$. Then, groups of l consecutive strips are considered, that is, the width of each group is $l \cdot D$. For any fixed division into strips of width D, there are l distinct partitions into strips of width $l \cdot D$. These partitions can be obtained by shifting one partition to the right over distance D.

Let $\mathcal{A}$ be a given local algorithm that solves the minimum set cover for any group of width $l \cdot D$. Thus, the union of the disks output by $\mathcal{A}$ as a solution for each of the groups is a feasible solution for the problem with n points. We repeat the same strategy for l distinct partitions, and choose the feasible solution of the minimum cardinality among the l solutions. For the algorithm using the shifting strategy, the following crucial lemma is proven in [10].

Lemma 1 (The Shifting Lemma [10]). *Let $AR_{\mathcal{A}}$ be the approximation ratio of a local algorithm $\mathcal{A}$. Then, given n target points, the approximation ratio for the minimum set cover problem with disks is less than or equal to $AR_{\mathcal{A}} \left(1 + \frac{1}{l}\right)$.*

On the two-dimensional plane, the shifting strategy is applied twice; first, the plane is divided into vertical strips, and then each strips of width $l \cdot D$ is divided into horizontal strips. As a result, the plane is divided into the squares of side length $l \cdot D$. The local algorithm $\mathcal{B}$ provides the minimum set cover for each of the squares. Then, we have $AR_{\mathcal{B}}(1 + 1/l)^2$ approximation ratio for the minimum set cover problem with n points.

In [10], an optimal set cover algorithm for the square is also proposed. The algorithm searches the optimal solution exhaustively. The square of side length $l \cdot D$ can be covered by $\lceil l\sqrt{2} \rceil^2$ disks of radius $D/2$. Moreover, any disk that covers at least two points can be assumed to have two of these points on its circumference. Thus, the number of possible positions of disks is finite. By checking all possible arrangements, we find the optimal covering for the square, that is, we achieve $AR_{\mathcal{B}} = 1$. We call the algorithm which uses the local optimal algorithm $\mathcal{B}$ and the shifting strategy SS-OPT algorithm. That is, the approximation ratio of SS-OPT algorithm for the minimum set cover problem with disks is $(1 + 1/l)^2$.

4.2 An Approximation Algorithm

In this section, we consider the approximation algorithm for the minimum SCF problem in the SSYNC system and rigid movement.

To classify given initial configurations, we define the geometric symmetricity $\rho(q, Q)$ for a point q and a set Q of points. The geometric symmetricity is introduced in [12], but in this paper, we modify the definition because we deal with the two sets of points, $P(t)$ and S.

Definition 3 (The geometric symmetricity $\rho(q, Q)$). *Let q be a point and Q be a set of distinct points on a plane. Then, if $q \in Q$, the symmetricity $\rho(q, Q)$ is 1, otherwise, $\rho(q, Q)$ is the number of different angles α ($0 \le \alpha < 2\pi$) such that rotating Q by α around q results in Q.*

From the definition of $\rho(q, Q)$, we can also say that Q can be divided into regular k-gons with co-center q, and $\rho(q, Q)$ is the maximum of such k [11] (when $\rho(q, Q) = 1$, we say that each point consists 1-gon). We denote $\rho(c(S), S)$, $\rho(c(S), P(t))$ and $\rho(c(S), S \cup P(t))$ by ρ_S, $\rho_{P(t)}$, and $\rho_{S \cup P(t)}$ simply. Figure 3-(a) is an example of configurations where $\rho_S = \rho_{P(t)} = \rho_{S \cup P(t)} = 4$. Let $m = \min\{\rho_S, \rho_{P(0)}, \rho_{S \cup P(0)}\}$.

The essential difficulties of the minimum SCF problem are that each robot must agree on a set cover with other robots based on observation obtained in its local coordinate system, and that each oblivious robot must detect the termination of construction of a desired set cover by the robots. For the problem of agreement, our idea is to construct a temporal $x - y$ axis for each of divided area by some robots. Then, every robot in each of the divided area agrees on a common solution for the set cover based on the temporal axis by using SS-OPT algorithm. To guarantee the construction of the temporal $x - y$ axis, in this section, we assume that a given initial configuration $C(0)$ satisfies the condition that m is 1, 2, or 4.

The detection of the construction of a desired set cover is also crucial problem for the SCF problem. The robots may construct a feasible set cover that does not guarantee the quality of solution initially. Since the robot is oblivious and it cannot know the state of other robots (i.e, active or asleep), it must decide whether the current configuration constructs a desired set cover, which guarantees an approximation ratio, or not based only on the positions of robots.

To distinguish the desired configurations from the other configurations, we define a *safe cover*. By definition, S can be partitioned into n/ρ_S regular ρ_S-gons $S_0, S_1, \ldots,$ $S_{n/\rho_S - 1}$. The definition $\rho(c(S), S)$ is the same as the symmetricity $\rho(S)$ defined in [5]. So, as in [5], all robots agree on an order of these regular ρ_S-gons, such that the distance of the points in S_i from $c(S)$ is no greater than that of S_{i+1}. We call the target points in $S_{n/\rho_S - 1}$ *outer points*. Let $T(t) \subseteq P(t)$ be the sets of positions of robots on the target points and $Out(t) \subseteq T(t)$ be the sets of positions of robots on the outer points in $C(t)$. We say that a configuration $C(t)$ constructs a *safe cover* if the following conditions are satisfied in $C(t)$:

1. All robots are in the smallest enclosing circle $SC(S)$ of S (including the circumference of the circle),
2. $|Out(t)| > 0$, and
3. $P(t) \setminus (T(t) \setminus Out(t))$ constructs a feasible set cover for S.

Note that initially $T(0) = \emptyset$, and $Out(0) = \emptyset$ hold, so even if the robots construct a feasible set cover in initial configuration, they detect the configuration does not construct a safe cover. In Fig. 3, the configuration in (f) constructs the safe cover while the configurations in (a)-(e) do not.

The definition of safe cover indicates the idea of our algorithm: We construct a set cover only by the robots not on the target points (except the outer points). That is, when a configuration constructs a safe cover, the robots on target points become asleep, and all robots stop executing the algorithm. However, there is a case such that all of the robots should be active to cover all points in S. That is, each robot covers at least one target point and keeps its state active. In this case, every robot must exist not on the target point. Even for this case, we must detect the termination of the construction of the desired set cover. To solve this problem, in our algorithm, at least one robot moves to an outer point (i.e., $|Out(t)| > 0$), and if necessary, the robot on the outer point keeps its state active.

The outline of our algorithm in the SSYNC system is as follows: First, some robots move to specific positions that are outside of the circle $SC(S)$ to define an unique lattice, which is used to partition of the plane into squares for the shifting strategy. The robots which define the unique lattice are called *axial robots*. Then, each of the other robots moves to the matched target points by the CWM algorithm. After that, every robot on the target point uses SS-OPT algorithm to get a solution A for the square in which the robot stays. To construct a safe cover, we modify A and get a feasible solution A', whose size is twice of A in the worst case (i.e., $|A'| \leq 2|A|$). Each robot computes a matching between the robots and the positions in A', and the matched robot for each position q in A' moves to q. At the last, the axial robots move to the corresponding outer points. To distinguish the above phases of the algorithm, the axial robot changes its position.

In what follows, we explain details of our approximation SCF algorithm. The algorithm consists of five phases. We assume that there are no target points and no robots on $c(S)$ initially. In the case where $c(S)$ is occupied by a target point or robot initially, we need additional preprocessing phase, which is described later. Let $radius(S)$ be the radius of $SC(S)$. An axial robot is a robot r_i which is the farthest one from $c(S)$ and satisfies $|r_i c(S)| > radius(S)$. If the farthest robots stay in the circle $SC(S)$ then we say that there are no axial robots. Let $\#r_a(t)$ be the number of axial robots, and $d_{r_a}(t)$ be the distance between the axial robots and $c(S)$ in configuration $C(t)$.

If the configuration constructs a safe cover, then the active robot on the target point, which is covered by at least one robot not on the target point, becomes asleep. The following phases are executed only when the configuration does not construct a safe cover (Notice that a safe cover is not constructed until the end of Phase 5). Remember that the system is SSYNC and movements of robots are rigid.

Phase 1: Phase 1 is executed when the configuration $C(t)$ satisfies the two conditions that $T(t) = \emptyset$, and $\#r_a(t) = 0$, or $4 < \#r_a(t)$, or $d_{r_a}(t) < radius(S)+3$. When $1 \leq \#r_a(t) \leq 4$ and $d_{r_a}(t) = radius(S) + 2$ hold, each robot checks the conditions for Phase 5 and if the conditions are satisfied then it executes Phase 5, not Phase 1.

In this phase, at most m robots become the axial robots by moving to the positions q which $|qc(S)| \geq radius(S) + 3$ holds. Initially, the number of the farthest robots may be more than 4. By definition, however, at least one of sets S, $P(0)$, and $S \cup P(0)$ can be partitioned into the regular m-gons centered at $c(S)$. Thus, every robot can agree on at most m farthest robots from $c(S)$ by checking the positions of S, $P(t)$ and $S \cup P(t)$ in this order, and select them as candidates for axial robots. The candidate robot r_i moves to the position q such that the current position $p(r_i)$ of r_i is on the segment $\overline{qc(S)}$ and $|qc(S)| = \max\{|p(r_i)c(S)| + 1, radius(S) + 3\}$ (Fig. 3-(b)). Since the system is SSYNC, it is possible that some of the candidates does not move while the others move to the positions q which $|qc(S)| \geq radius(S) + 3$ holds. Even in this case, since we assume that m is 1, 2, or 4, the angle between segments $\overline{r_i c(S)}$ and $\overline{r_j c(S)}$ is $\pi/2$ or π for any two axial robots r_i and r_j after Phase 1.

Phase 2: Phase 2 is executed when $1 \leq \#r_a(t) \leq 4$ and $d_{r_a}(t) \geq radius(S) + 3$, and there is a robot not on the target points except the axial ones. In this phase, the robots except

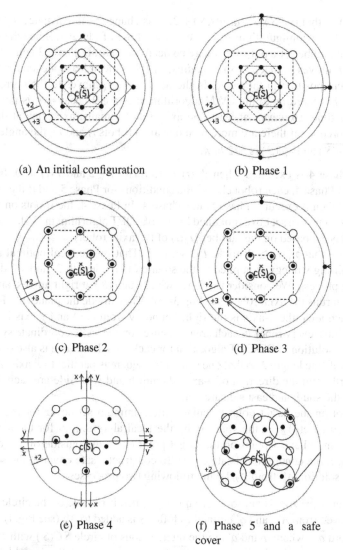

(a) An initial configuration

(b) Phase 1

(c) Phase 2

(d) Phase 3

(e) Phase 4

(f) Phase 5 and a safe cover

Fig. 3. An example of executions of our approximation algorithm, where the black and white circles represent the robots and target points respectively

the axial robots move to the target points by using the CWM algorithm. For each axial robot r_i, let $s_{i'}$ be the nearest outer point from r_i (if there are more than one points with the same distance, the tie is broken in the clockwise manner around r_i). Let B be the set of the nearest outer point for every axial robot. For the set $S \setminus B$, the robots except the axial ones execute the CWM algorithm. At the end of Phase 2, all of the robots except the axial robots stay on the target points (Fig. 3-(c)).

Phase 3: Phase 3 is executed when $1 \leq \sharp r_a(t) \leq 4$, $d_{r_a}(t) \geq radius(S) + 3$, and the robots except the axial ones stay on the target points. The axial robot r_i goes straight toward $c(S)$ so that $|r_i c(S)| = radius(S) + 2$. This change of the distance $d_{r_a}(t)$ is used to distinguish the configurations of Phase 2 from those of Phase 4. Since the system is SSYNC, the number of axial robots may be decreased during Phase 3 (Fig. 3-(d)); the axial robot r_i moves so that $|r_i c(S)| = radius(S) + 2$ while at least one axial robot waits on the current position in the cycle. In the next cycle, r_i moves to the corresponding outer point $s_{i'}$ according to the CWM algorithm in Phase 2 because r_i is not the axial robot at the time. However, at least one axial robot remains axial one at the end of Phase 3. Moreover, if there are more than one axial robots r_i and r_j, the angle between segments $\overline{r_i c(S)}$ and $\overline{r_j c(S)}$ is $\pi/2$ or π.

Phase 4: Phase 4 is executed when $d_{r_a}(t) = radius(S) + 2$ and $T(t) \neq \emptyset$. Before the execution of Phase 4, each robot checks the conditions for Phase 5 and if the conditions are satisfied then it executes Phase 5, not Phase 4. In Phase 4, the robots on the target points move to form a set cover obtained by the SS-OPT algorithm in [10]. The shifting parameter l is set based on the number $\sharp r_a(t)$ of the axial robots.

In the case that $\sharp r_a(t) \geq 2$ holds, l is set to 1. That is, the local optimal algorithm $\mathcal{B}$ in the shifting strategy is applied to the square of side length 2, and we do not use the shifting process. We consider two lines which run $c(S)$ at right angles, and at least one of which runs the axial robot (see Fig. 3-(e)). They divide the plane into four areas. Then, in each area, the x-axis is the right boundary from $c(S)$ and y-axis is the other boundary. Thus, every robot in each area can agree on the same coordinate system, and get the same solution of SS-OPT algorithm for each area. Each area is also divided into the squares of side length 2 so that one side of the square is parallel to x-axis. Assuming that the north is in the direction of y-axis, the north and west side are included in the square, but the south and east side are not.

The robot on the target point calculates the minimum set cover for the square in which it stays. Let $A_{\mathcal{B}}$ be the solution by the local algorithm $\mathcal{B}$ for the square. The solution $A_{\mathcal{B}}$ may have a point on the target point or on the outside of $SC(S)$, which does not meet the conditions of safe cover. To construct a safe cover in the square, we construct a solution $A'_{\mathcal{B}}$ from $A_{\mathcal{B}}$ by the following two processes:

1. First, set $A'_{\mathcal{B}} = A_{\mathcal{B}}$. Then, for every $q \in A'_{\mathcal{B}}$ which is outside of the circle $SC(S)$, q is removed from $A'_{\mathcal{B}}$, and q' defined as follows is added to $A'_{\mathcal{B}}$ (see Fig.4): Consider the chord $\overline{aa'}$ where a and a' are the intersections of circle $SC(S)$ with $RC(q)$. Let $\overline{cq}$ be the perpendicular line toward $\overline{aa'}$ from q where c is on $\overline{aa'}$, and let c' be the intersection of $SC(S)$ with $\overline{cq}$. The position q' is a position on segment $\overline{cc'}$ which holds $|cq'| = |cc'|/2^k$, where $k = 0, 1, \ldots$ is the minimum integer such that q' is not on the target point. The target points covered by q are also covered by q' because there is no target point outside of $SC(S)$.

2. Next, for every $q \in A'_{\mathcal{B}}$ on the target point, q is removed from $A'_{\mathcal{B}}$, and two positions q_1 and q_2 defined as follows are added to $A'_{\mathcal{B}}$ (see Fig.5)[5]. We set q_1 and q_2 so that the target points covered by q are covered by q_1 and q_2. The circle $RC(q)$ is divided

---

[5] Here, we assume that there are more than two target points in the circle $RC(q)$. If there are two target points in the circle, the middle point between the two points is added to $A'_{\mathcal{B}}$ instead of q. If there is one target point, that is, it is on q, the point q' which is on $\overline{qc(S)}$ and satisfies $|qq'| = 1/2$ is added to added to $A'_{\mathcal{B}}$ instead of q.

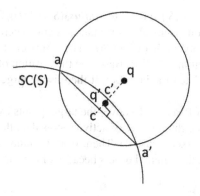

Fig. 4. A position q'

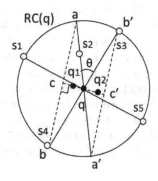

Fig. 5. Positions q_1 and q_2

by chords each of which runs the target point s_i in $RC(q)$ and q. Since the number of target points is finite, there is the minimum angle $\theta > 0$ in the angles between any two chords (tie is broken in the angle to $c(S)$). Let $\overline{aa'}$ and $\overline{bb'}$ be the chords between which the angle is θ. The intersection of segment $\overline{ab}$ (resp. $\overline{a'b'}$) with the perpendicular line from q toward $\overline{ab}$ is denoted by c (resp. c'). The position q_1 (resp. q_2) is a position on segment $\overline{qc}$ (resp. $\overline{qc'}$) which holds $|qq_1| = |qc|/2^k$ (resp. $|qq_2| = |qc'|/2^k$), where $k = 0, 1, \ldots$ is the minimum integer such that q_1 (resp. q_2) is not on the target point. Notice that the target points covered by q in $A_\mathcal{B}$ are also covered by q_1 and q_2 in $A'_\mathcal{B}$ because there are no target points in the sectors with central angle θ surrounded by chords $\overline{aa'}$ and $\overline{bb'}$ and circle $RC(q)$.

The total number of disks in $A'_\mathcal{B}$ is at most twice of $A_\mathcal{B}$. By the above described way, each robot gets the unique solution $A'_\mathcal{B}$ in the square, and the nearest robot to $q \in A'_\mathcal{B}$ in the square moves to q. If the robot on the target point detects that the robots not on the target points construct the solution $A'_\mathcal{B}$, it becomes asleep.

In the case that $\sharp r_a(t)$ is 1, all robots agree on a single coordinate system. Thus, the robots get the same solution of **SS-OPT** algorithm globally. By the same way as the case that $\sharp r_a(t) \geq 2$, each robot calculates the solution, modifies it, and it moves to the position in the modified solution or becomes asleep.

Phase 5: Phase 5 is executed when $1 \leq \sharp r_a(t) \leq 4$ and $d_{r_a}(t) = radius(S) + 2$ hold, and the axial robots detect that the other robots not on the target points construct the solution of **SS-OPT** algorithm. This detection is possible by the axial robots because the axes defined by the axial robots do not change from Phase 2. In Phase 5, the axial robot r_i moves to the corresponding outer point $s_{i'}$, and if $s_{i'}$ is covered by another robot not on the target point then r_i becomes asleep, otherwise it keeps its state active (Fig. 3-(f)).

In the case that $c(S)$ is occupied by a target point or robot initially, we need additional preprocessing phase to determine the candidate of axial robots: If there are no robots on $c(S)$ (i.e., a target point exists on $c(S)$), one robot r which is matched with $c(S)$ by the

CWM algorithm moves to $c(S)$. The robot r on $c(S)$ moves to $radius_0(S \cup P(t))/2$ distance from $c(S)$ in the northern direction on its local coordinate system, where $radius_0(S \cup P(t))$ is the radius of the largest empty circle of $S \cup P(t)$. The empty circle means that its interior does not include any point in $S \cup P(t)$. Based on the position of r, every robot can select one robot as the candidate of axial robot. If there is the target point s_i on $c(S)$, s_i is covered by r in Phase 4.

The robots construct a set cover at the end of Phase 5. To cover the target points by the robots not on the target points, we need twice the number of active robots than the size of solution obtained by SS-OPT algorithm (see Phase 4). Then, from Lemma 1, we get the following theorem, however, we omit the detailed proofs because of the page limitation.

Theorem 2. *Let l be the integer parameter of the algorithm. In the SSYNC system and the rigid movement model, there is an approximation SCF algorithm which guarantees 8 approximation ratio for any initial configuration where $m = 1, 2,$ or 4, and $2 \cdot (1 + 1/l)^2$ approximation ratio for any initial configuration where $m = 1$.*

5 Conclusion

In this paper, we have introduced the set cover formation (SCF) problem by the oblivious mobile robots, which is the first formation problem in which fixed points exist on the plane and which requires minimization of the number of robots constructing the pattern. We have proposed the minimal SCF algorithm from any initial configuration in the ASYNC system, and an 8-approximation SCF algorithm from a configuration with a low symmetricity in the SSYNC system.

For the minimum SCF problem, one of interesting open problems is to reveal the relationship between the approximation ratio and the symmetricity of an initial configuration. We guess that the upper/lower bounds of the approximation ratio for the minimum SCF problem depend on the symmetricity. That is, our future work is to propose an approximation algorithm for the SCF problem for any symmetricity m, and to reveal the lower bounds for the SCF problem. In addition, the SCF problem in a weak assumption such as initial non-scattered configuration (i.e., some robots and target points may be on the same position initially) and non-rigid movement is also an interesting problem. A challenging task in the research area about distributed mobile robots is to consider another formation problem with fixed points in which the robots are required to construct a desired structure for the fixed points, such as a (connected) dominating set and a spanning tree.

References

1. Bronnimann, H., Goodrich, M.T.: Almost Optimal Set Covers in Finite VC-Dimension. Discrete & Computational Geometry 14(1), 463–479 (1995)
2. Deder, T., Greene, D.: Optimal Algorithms for Approximate Clustering. In: Proc. of STOC 1988, pp. 434–444 (1988)

3. Flocchini, P., Prencipe, G., Santoro, N., Widmayer, P.: Arbitrary Pattern Formation by Asynchronous, Anonymous, Oblivious Robots. Theoretical Computer Science 407, 412–447 (2008)
4. Fowler, R.J., Paterson, M.S., Tanimoto, S.L.: Optimal packing and covering in the plane are NP-complete. Information Processing Letter 3(12), 133–137 (1981)
5. Fujinaga, N., Ono, H., Kijima, S., Yamashita, M.: Pattern Formation through Optimum Matching by Oblivious CORDA Robots. In: Lu, C., Masuzawa, T., Mosbah, M. (eds.) OPODIS 2010. LNCS, vol. 6490, pp. 1–15. Springer, Heidelberg (2010)
6. Fujinaga, N., Yamauchi, Y., Kijima, S., Yamashita, M.: Asynchronous Pattern Formation by Anonymous Oblivious Mobile Robots. In: Aguilera, M.K. (ed.) DISC 2012. LNCS, vol. 7611, pp. 312–325. Springer, Heidelberg (2012)
7. Franceschetti, M., Cook, M., Bruck, J.: A Geometric Theorem for Approsimate Disk Covering Algorithms, Report ETR035, Caltech (2001)
8. Fu, B., Chen, Z., Abdelguerfi, M.: An Almost Linear Time 2.8334-Approximation Algorithm for the Disc Covering Problem. In: Kao, M.-Y., Li, X.-Y. (eds.) AAIM 2007. LNCS, vol. 4508, pp. 317–326. Springer, Heidelberg (2007)
9. Gonzalez, T.F.: Covering a set of points in multidimensional space. Information Processing Letters 40(4), 181–188 (1991)
10. Hochbaum, D.S., Mass, W.: Approximation Schemes for Covering and Packing Problems in Image Processing and VLSI. Journal of the ACM 32(1), 130–136 (1985)
11. Suzuki, I., Yamashita, M.: Distributed Anonymous Mobile Robots: Formation of geometric patterns. SIAM J. of Comput. 28(4), 1347–1363 (1999)
12. Yamashita, M., Suzuki, I.: Characterizing Geometric Patterns Formable by Oblivious Anonymous Mobile Robots. Theoretical Computer Science 411, 2433–2453 (2010)

Fast Collisionless Pattern Formation by Anonymous, Position-Aware Robots[*]

Tamás Lukovszki[1] and Friedhelm Meyer auf der Heide[2]

[1]Faculty of Informatics, Eötvös Loránd University, Budapest, Hungary
lukovszki@inf.elte.hu
[2]Heinz Nixdorf Institute and Department of Computer Science,
University of Paderborn, Germany
fmadh@uni-paderborn.de

Abstract. We consider a scenario of n identical autonomous robots on a 2D grid. They are memoryless and do not communicate. Their initial configuration does not have to be connected. Each robot r knows its position $p_r \in \mathbf{Z}^2$. In addition, each robot knows the connected pattern F to be formed. F may be given by a set of n points in $\mathbf{Z}^2$, or may be only partially described, e.g., by "form a connected pattern", or "build a connected formation with minimum diameter" (Collisonless Gathering). We employ the Look-Compute-Move (LCM) model, and assume that in a time step each robot is able to move to an unoccupied neighboring grid vertex, thus guaranteeing that two robots will never collide, i.e., occupy the same position. The decision where to move solely depends on the configuration of its 2-hop neighborhood in the grid $\mathbf{Z}^2$.

First we consider a helpful intermediate problem - we call it the *Lemmings problem* - where collision at one single point g, known to all robots, is allowed and the goal is that all robots gather at g. We present an algorithm solving this problem in $2n + D - 1$ time steps, where D denotes the maximum initial distance from any robot to g. This time bound is easily shown to be optimal up to a constant factor.

Based on this strategy, forming a connected pattern can be done within the same time bound. Forming a connected pattern F needs additional considerations. We show how to do so in time $O(n + D^*)$, where D^* denotes the diameter of the point set consisting of the initial configuration and F. For Collisionless gathering we obtain the same time bound, up to constant factors. This significantly improves upon the previous upper bound of $O(nD)$ for this problem presented in [5].

Keywords: Autonomous mobile robots, pattern formation, gathering.

1 Introduction

We consider various pattern formation problems by n identical autonomous robots on a 2D grid. They are memoryless (or use only $O(1)$ bits of persistent

---

[*] Supported by the German Research Foundation (DFG) within the Collaborative Research Center "On-The-Fly Computing" (SFB 901).

M.K. Aguilera et al. (Eds.): OPODIS 2014, LNCS 8878, pp. 248–262, 2014.

memory) and operate without explicit communication. They have computation and locomotion capabilities and limited visibility range. They are represented by discs of unit diameter. Each robot r knows its position $p_r \in \mathbf{Z}^2$ but not the position of the other robots. In addition, each robot knows the connected pattern F to be formed. F may be given by a set of n points in $\mathbf{Z}^2$, or as a predicate, e.g., "form a horizontal line segment", or may be only partially described, e.g., by "form a connected pattern", or "build a connected formation with minimum diameter" (Collisonless Gathering). All robots have a common coordinate system. Each robot has a visibility range of 2 units, i.e. it can see the robots within its local range of 2 units. With other words, the robots only have information about their 1- and 2-hop grid neighbors. The robots are able to move only on the edges of the grid. They all move synchronously with unit speed, s.t. they travel an edge of the grid in one time unit.

The robots operate corresponding to the *Look-Compute-Move* (LCM) model. In one cycle, a robot takes a snapshot of its current visibility range (Look), makes a decision to stay idle or to move to one of the neighboring vertices (Compute), and in the latter case makes an instantaneous move to this neighbor (Move). We assume that the LCM cycles are synchronous at each robot. Collision not allowed during the algoritms, i.e. in each time step each vertex of $\mathbf{Z}^2$ can be occupied by at most one robot. The motion ends when the robots form the connected pattern F. From now on we will use the terms node and robot interchangeably.

1.1 Our Contribution

First we consider a helpful intermediate *gathering* (or *point formation*) problem - we call it the *Lemmings problem* - where collision at one single point g, known to all robots, is allowed. The goal is that all robots gather at g. We consider *oblivious* robots, i.e. the robots do not remember results from any of the previous computations. We present an algorithm solving this problem, called *x-y-routing*, where the robots only need local knowledge about their 2-hop neighborhood in the grid $\mathbf{Z}^2$. We show that the *x-y*-routing method can be used to guarantee the gathering of all robots at g in $2n + D - 1$ time steps, where D is the maximum initial hop distance of a robot from g. We prove that this time bound is optimal up to a constant factor.

After this we investigate the gathering problem of n oblivious robots, where no collision is allowed at g and the robots have to form a connected configuration containing g. We show that the *x-y*-routing solves this problem in $n + D - 1$ time steps. This significantly improves the previous upper bound of $O(nD)$ on this problem presented in [5].

After this we consider *finite state* robots, i.e., the robots can use $O(1)$ bits of persistent memory for the computation. We show, how the set of n robots can be arranged to form a connected axis parallel line segment containing a given point g, known to all robots, in $3n + D + 3$ steps.

Finally, for finite state robots, we show how an arbitrary connected pattern F, known to all robots, can be formed in time $O(n + D^*)$, where D^* denotes the diameter of the point set consisting of the initial configuration and F. In case when all robots know n, this solution can also be applied for solving the focused coverage problem on the 2D grid. This results in $O(n + D)$ covering time. If the number of robots n is not known for the robots, then best known upper bound on this problem is $O(S)$, presented in [2], where S is the sum of initial distances of the mobile sensors from g.

This paper is organized as follows. Section 2 gives an overview of related work. In Section 3 we describe the x-y-routing algorithm, which plays a key role in our solutions. In Section 4 we define the Lemmings problem, where all robots must be gathered at the given point g. We prove the lower bound of $\Omega(D + n)$ time steps on the running time of each discrete, synchronous algorithm solving this problem. After this we prove that the x-y-routing algorithm solves this problem in at most $2n + D - 1$ time steps. After this, in Section 5 we study the gathering problem, where no collision is allowed at g and the robots have to form a connected configuration containing g. We show that the x-y-routing solves this problem in $n + D - 1$ time steps. In Section 6 we show how the set of n robots can be arranged to form a connected axis parallel line segment containing a given point g in $3n + D + 3$ time steps. In Section 7 we show, how a connected pattern F, known to all robots, can be formed in time $O(n + D^*)$, where D^* denotes the diameter of the point set consisting of the initial configuration and F. Section 8 summarizes the work.

2   Related Work

Cohen and Peleg [4] presented an asynchronous algorithm to gather oblivios robots at the center of gravity. Their algorithm uses the LCM (Look-Compute-Move) discrete cycle based model to move their robots. They mathematically proved upper and lower bounds on the convergence speed of their solution.

Cord-Landwehr et al. [6] described an easy-to-check property of target functions that guarantee convergence and gives upper time bounds. This property holds for the target function in [4] and improves the upper bound on the speed of the convergence.

Czyczowicz et al. [7] considered the gathering problem for few fat robots, where the robots are modeled by unit disks. The goal was to gather the robots, such that the union of the unit disks is connected at the end. Collisions of the robots are not allowed during the gathering. A main problem which had to be solved here is that the line sight of a robot may be blocked by the extent of other robots.

Cord-Landwehr et al. [5] studied the problem of gathering mobile robots with an extent at a given position as dense as possible to form a disk of minimum radius around the gathering point. The authors present an algorithm for the continuous case and the discrete case, where the robots are moving on a grid. They prove an $O(nD)$ upper bound for the gathering time, where n is the number of robots and D is the distance of the farthest robot from the gathering point. They empirically studied the continuous case, where in they report a few deadlock situations in the simulations.

For the gathering problem of mobile robots many different variants exist differing in levels of synchronization, computational power of the robots, memory, range of visibility, agreement on coordinate system. For a survey we refer to [3].

Another related problem in distributed robotics is the Pattern Formation problem, where a group of mobile robots have to form a desired geometric pattern. The pattern can be given as set of points in the plane (by their coordinates) or as a predicate (e.g. "form a circle"). A common requirement is that the robots have distinct initial positions and that the number of points in the pattern and the number of robots are the same. Suzuki and Yamashita [13] [14] investigated the question what kinds of patterns can be formed by a group of autonomous, anonymous and homogenous mobile robots that do not communicate, but they are able to observe each others movements. In [13] [14] the authors have shown that without agreeing a common coordinate system, a pattern can be formed if and only if it is purely symmetrical, i.e., a regular polygon (or a point), or a set of regular concentric polygons. They also have shown that by agreeing on a coordinate system, the robots can form any geometric pattern. Flocchini et al. [9] have shown that if each robot has a compass needle that indicates North (the compass needles are parallel), then any odd number of robots can form an arbitrary pattern, but an even number, in the worst case, cannot. If each robot has two independent compass needles, say North and East, then any set of robots can form any pattern. Pattern Formation by robots with limited visibility has been studies in [15].

A further related problem is the Focused Coverage self-deployment problem in mobile sensor networks, where an area with maximum radius around a Point of Interest (POI) must be covered without sensing holes. This problem was introduced by Li at al. [12], [10], [11]. They solved the problem by driving the mobile sensors along an equilateral triangle tessellation graph centered at the POI. They showed that their algorithms terminate in finite time. The convergence time has also been evaluated by simulations. Blázovics and Lukovszki [2] presented a collision free algorithm solving the focused coverage problem in $O(S)$ time, where S is the sum of initial distances of the mobile sensors from the POI. The theoretical results has been also validated by simulations.

Another related fundamental problem is the Filling problem (see [1]), in which a given region must be covered by robots. In this problem the robots are initially not in the region, they enter the the space one by one, from a point called "door". When a robot enters the door, it must disperse itself in the region. The goal is to cover the entire region. Barrameda et al. [1] have proven that the Filling problem can be solved with limited visibility, for any simple orthogonal space, i.e., a polygonal region without holes with sides either parallel or orthogonal, with a single door, by finite-state robots with a common coordinate system and common unit of distance in finite time.

For an excellent overview on distributed computing by mobile robots we refer to the the book by Flocchini et al. [8].

3   Collisionless Routing Towards a Point g

Let V be a set of n robots placed at the vertices of the rectangular grid $\mathbb{Z}^2$ and $g \in \mathbb{Z}^2$ a gatheting vertex. Each robot knows its own position and the position of g. We assume that the gathering vertex g has coordinates $(0,0)$. We use the synchronous Look-Compute-Move model, i.e. all robots robots perform the Look, Compute, and Move steps synchronously. In each time step, each robot is able to move to a neighboring vertex of the rectangular grid or it stays in place. We assume that each robot can see its local environment within two hops. Thus, when a robot r chooses a neigboring vertex p to which it wants to move, r sees all robots that are potentially able to move to p in the same time step. The knowedge about the 2-hop neighborhood makes possible to decide locally, which robot can move to a certain vertex, such that no collision occurs.

3.1   The x-y Routing Algorithm

Now we present the routing algorithm. In each time step each robot wants to decrease its hop distance to the gathering vertex g, such that it moves in x-direction until it has the same x-coordinate than g. After this, it moves towards g in y-direction until it reaches g. A path of a robot emerging in this way is called an x-y-path, which terminates in g. For each robot r, at time t let $nexthop(r,t)$ be the neighboring vertex of r on the x-y-path towards g. For a robot r at g, we define $nexthop(r,t) = g$. If in a time step t the vertex $p = nexthop(r,t)$ is occupied by another robot, then the robot r must stay in place. If in a time step t there are two (or more) robots r and r' with $nexthop(r,t) = nexthop(r',t) = p$, then the robot with smaller y-distance from the gathering vertex g has higher priority, if r and r' has the same y-distance from g, then the robot with greater x-coordinate has higher priority. The robot with highest priority, say r, is allowed to move to vertex p and the other robot(s) must stay in place. Fomally, let r and r' be two robots with coordinates (r_x, r_y) and (r'_x, r'_y), respectively, s.t. in time t $nexthop(r,t) = nexthop(r',t) = p$. Then $priority(r,p) > priority(r',p)$ $\iff |r_y| < |r'_y|$ or ($p = g$ and $r_y > 0$ and $r'_y < 0$) or ($|r_y| = |r'_y|$ and $r_x > 0$).

Algorithm 1. x-y-routing(r)

---

while r has not yet reached g **do**
    $p \leftarrow nexthop(r,t)$
    if p is unoccupied **and** $\nexists$ another robot r' with $nexthop(r',t) = p$, s.t. r' has higher priority than r **then**
        r moves to p
    else
        r stays in place
    end if
    $t \leftarrow t + 1$
end while

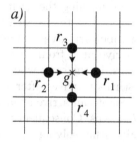 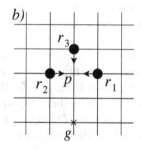

Fig. 1. Priorties of the robots having the same point p as $nexthop(.)$. *a)* If $p = g$, there are at most four robots r_1, r_2, r_3, r_4 having g as $nexthop(.)$. In this case $priority(r_1, g) > priority(r_2, g) > priority(r_3, g) > priority(r_4, g)$. *b)* If $p \neq g$, there are at most three robots r_1, r_2, r_3 having p as $nexthop(.)$. In this case $priority(r_1, g) > priority(r_2, g) > priority(r_3, g)$.

The above rule guarantees that, for each unoccupied vertex p, the robot which can occupy it in the next step – if any – is unique, and the robots are able to make this decision based on its local knowledge about their 2-hop neighborhood.

The x-y-routing algorithm is oblivious and the decision of each robot where to move solely depends on the configuration of the 2-hop neighborhood in the 2D grid.

4   The Lemmings Problem

We assume that at the beginning the robots are placed on different vertices of the grid $\mathbb{Z}^2$. They try to move on the edges towards the gathering vertex g. The goal is to gather all the robots at g. Collision is only allowed at the gathering vertex g, i.e. we allow that g can be occupied by more than one robots at the same time. The only modification in Algoritm 1 is that Algoritm 1 treats the gathering vertex g as it would be always an unoccupied vertex. After performing the algorithm all robots will reside on vertex g.

4.1   Lower Bound

First we show a lower bound of $\Omega(n + D)$ time steps on the Lemmings problem, where n robots must be gathered at g, where D is the maximum initial hop distance of a robot from g. We prove the lower bound for robots with infinite visibility range. Clearly, this bound also holds for robots with limited visibility.

Theorem 1. *Let $V = \{v_1, ..., v_n\}$ be the set of n robots with infinite visibility placed on different vertices of the grid $\mathbb{Z}^2$. Each algorithm, solving the synchronous discrete Lemmings problem needs $\Omega(n + D)$ time steps, where D is the maximum initial hop distance of a robot from g.*

Proof. Since each robot must arrive at g after performing the gathering algorithm and each robot can move to a neighboring grid vertex in each time step, at least D steps are necessary until the furthest robot arrives at g. On the other hand, in each time step only one robot can arrive at g from each direction. Therefore, each algorithm needs at least $n/4$ steps. In the case, when the initial distance of all robots from g is D, then the first robot arrives after D steps. In this step and in each further step at most four robot can arrive at g. Thus, the last robot needs at least $D + n/4 - 1 = \Omega(n + D)$ steps. Consequently, $\Omega(n + D)$ is a lower bound on the running time of each algorithm solving the problem. $\square$

4.2 Upper Bound

Now we turn to the analysis of Algorithm 1. The only modification in Algoritm 1 to solve the Lemmings problem is that Algoritm 1 treats the gathering vertex g as it would be always an unoccupied vertex.

First we consider the special but important case, where all robots are initially placed on one x-y-path P terminating in g. We show that after $n + D - 1$ steps all robots reach g. We use this result for proving the time bound on the Lemmings problem in the general case.

Lemma 1. *Let $V = \{v_1, ..., v_n\}$ be the set of n oblivious robots placed on the same x-y-path $P \subset \mathbb{Z}^2$ terminating in g, such that all robots are placed on different vertices. Let D be the maximum initial hop distance of a robot from g, i.e. $D = \max_{r \in V}(d(r, g))$. Then by performing Algorithm 1 all robots reach g in $n + D - 1$ steps.*

Proof. We prove the claim by induction on the number of robots n.

For $n = 1$, the claim holds obviously, the robot v_1 gets strictly closer to g in each step, until it reaches g. Therfore, for $n = 1$, the algoritm terminates in $1 + D - 1 = D$ steps.

Now, assume that the claim holds for $n-1$ robots. Let $\{v_1, ..., v_n\}$ be the set of robots ordered by their initial hop distances from g, s.t. $d(v_1, g) < d(v_2, g) < ... < d(v_n, g)$. By the induction hypothesis v_{n-1} reaches g within $d(v_{n-1}) + (n-1) - 1 = d(v_{n-1}) + n - 2$ steps. Let $t_1, ..., t_k$ be the time steps in them v_{n-1} moved towards g. In time step t_k it reaches g. Then in time steps $t_1 + 1, ..., t_k + 1$ the robot v_n can also move towards g. In these time steps the distance between v_n and g decreases by $d(v_{n-1}, g)$ units. Therefore, after $t_k + 1$ steps the distance between v_n and g becomes at most $d(v_n, g) - d(v_{n-1}, g) = \delta$. In time steps $t > t_k$ the robot v_n can move towards g in each step. Therefore, v_n reaches g in $t_k + 1 + \delta$ steps. By the induction hypothesis, $t_k \leq d(v_{n-1}, g) + n - 2$. Therefore, v_n reaches g within $d(v_{n-1}, g) + n - 2 + 1 + \delta = d(v_n, g) + n - 1$ steps, which proves the claim. $\square$

Now we turn to the general case, where the n robots are arbitrarily placed on different vertices of $\mathbb{Z}^2$. We show that after $2n + D - 1$ steps all robots reach g.

Theorem 2. *Let $V = \{v_1, ..., v_n\}$ be the set of n oblivious robots placed on different vertices of $\mathbb{Z}^2$. Let g be the gathering vertex and D be the maximum initial*

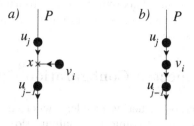

Fig. 2. Joining of robot v_i to the path P

hop distance of a robot from g, i.e. $D = \max_{r \in V}(d(r, g))$. *Then by performing Algorithm 1 all robots reach g in $2n + D - 1$ steps.*

Proof. Let $v_1, v_2, ..., v_n$ be the robots ordered regarding the time they arrive at g. The x-y-path of each robot is unique, each such path terminates in g, and the union of the x-y-paths define a tree.

Consider the robot v_n arriving at g as the last one. Let P be the x-y-path from the initial position of v_n to g. Let $U = \{u_1, ..., u_k\}$ with $u_k = v_n$ be the set of robots that are initially on the tree path P. If we remove all robots in $V \setminus U$ from the scene, then by Lemma 1, all robots in U would reach g within $n + D' - 1$ steps, where D' is the initial distance of u_k from g.

Now consider the steps of v_n in the presence of the robots in $V \setminus U$. The key observation is that a robot $v_i \in V \setminus U$ can increase the arrival time of u_k at g by at most two time steps, that are (i) the step t_i, in which v_i reaches the path P, i.e., $nexthop(v_i, t_i) = x \in P$ and v_i moves to x, and (ii) the step immediately after v_i has joined the path P (see Figure 2):

Case (i): Consider the time step t_i, when the hop distance of v_i and P is one and after peforming this step v_i belongs to P (Figure 2.a). Let x be the vertex of P with $x = nexthop(v_i, t_i)$. If there is a robot $u_j \in P$ with $nexthop(u_j, t_i) = x$ with lower priority than v_i then u_j must wait in this step. This will increase the arriving time of v_n to g by one time unit.

Case (ii): The time step $t_i + 1$, imediately after v_i has moved to $x \in P$ (Figure 2.b). In this time step u_j must wait because v_i is immediatly in front of u_j in P, i.e. v_i occupies the vertex $nexthop(u_j, t_i + 1)$. Starting with this time step, the robot v_i behaves exactly the same way as u_j would behave without the existence of v_i. Thus, v_i reaches g at the same time u_j would reach g without the existence of v_i. If no other robot joins the path P on the front of u_j, then u_j reaches g two time units after v_i reached g, i.e. at most two time units later than it would reach without the existence of v_i. The robots $v_{j'}$, $j' = j + 1, ..., k$, arrive at g by at most two time steps later than they would arrive without the existence of v_i.

Conequently, joining of all robots of $V \setminus U$ increases the arrival time of u_k by at most $2|V \setminus U| = 2(n - k)$ time units. Therefore, u_k arrives at g and the algorithm terminates in at most $D' + k - 1 + 2(n - k) \leq D' + 2n - 1 \leq D + 2n - 1$ steps, where D is the maximum initial distance of a robot from g. $\square$

Remark 1. By Theorem 1 and Theorem 2 the x-y-routing algorithm solves the discrete Lemmings problem, where all robots must be gathered at g in optimal time, up to a constant factor.

5   Forming a Connected Configuration Containing g

Now we turn to the discrete gathering problem, where no collision is allowed at g and the robots have to form a connected configuration containing g. We show that the x-y-routing algorithm (Algoritm 1) solves this problem in $n + D - 1$ steps.

Theorem 3. *Let $V = \{v_1, ..., v_n\}$ be the set of n oblivious robots placed on different vertices of $\mathbb{Z}^2$. Let g be the gathering vertex and D be the maximum initial hop distance of a robot from g, i.e. $D = \max_{r \in V}(d(r, g))$. Then by performing Algorithm 1 the robots form a connected configuration containing g in $n + D - 1$ steps.*

Proof. For each robot $v \in V$, let p_v be the initial position of v and P_v the x-y-path from p_v to g. Let $T = \bigcup_{v \in V} P_v$ be the tree defined as the union of the x-y-paths. For each robot $v \in V$, let p_v^* be the closest position of v to g during the algorithm. Since the hop distance of a robot never increases during the algorithm, p_v^* is the final position of v. Let P_v^* the x-y-path from p_v^* to g and $T^* = \bigcup_{v \in V} P_v^*$.

We show that (i) T^* contains g, (ii) T^* has no unoccupied vertex, i.e. the robots form a connected configuration, and (iii) each vertex of T^* becomes occupied in $n + D - 1$ steps.

(i) Obviously, T^* contains g. A robot v with smallest initial distance form g will occupy g, which will be the final position v.

(ii) Assume for contradiction, that T^* has an unoccupied vertex. Let x be an unoccupied vertex of T^* with maximum hop distance from g. Let $U = \{v \in V : x \in P_v\}$ be the set of robots whose x-y-paths contain x. Then the final position of at least one of the robots in U is further from g than x, otherwise x would not be contained in T^*. Since x is an unoccupied vertex of T^* with maximum hop distance from g, there is at least one robot $u \in U$ with distance one from x, s.t. $d(p_u^*, g) = d(x, g) + 1$. Then one of these robots occupies x in the next step.

(iii) Now we show that each vertex of T^* becomes occupied in $n + D - 1$ steps. By induction, we prove that after $i + D$ steps, $0 \le i \le n - 1$, all vertices of T^* with hop distance at most i from g are occupied. Thus, after $i + D$ steps, the robots on those vertices have reached their final positions.

For $i = 0$, the claim holds, since the robot which occupies g will never be stopped by another robot and its initial distance from g is at most D. Therefore, after D steps it occupies g.

Assume that the induction hypothesis holds for i, $0 \le i < n - 1$. Let V_i be the set of robots with final position of hop distance at most i from g. If $V_i = V$ we are done. Otherwise, consider a robot v whose final position will be at a vertex x with distance $i + 1$ from g.

We say that a robot u stops v in a certain step t if either $nexthop(v,t)$ is occupied by u or $nexthop(v,t)$ is unoccupied and u occupies it in step t. Observe, that during the algorithm none of the robots in $V \setminus V_i$ can stop v before v reaches its final position. To see this, assume that a robot $w \in V \setminus V_i$ stops v in a certain step t. Then in step $t+1$ the robot w becomes on the front of v in the x-y-path from v to g, i.e. w gets strictly closer to g than v and the distance of w to g remains strictly lower than the distance of v. Therfore, the final distance of w from g will also be strictly lower than the final distance of v, which is by assumption $i+1$.

Therefore, v can be only stopped by robots in V_i before v reaches its final position. Let u be the robot whose final position q is on the x-y-path of v to g and the hop distance $d(q,g) = i$. Let t be the time step in which u reaches q. By the induction hypothesis $t \leq i + D$. If v was stopped by u in some time step $t' < t$, then after this time step v "follows" u, and thus, in time step $t+1 \leq D+i+1$ the robot v also reaches its final position. If v was never stopped by u then v was able to get closer to g in each of the t steps. Since the distance between the initial position of v and its final position x is at most D, the robot v reaches x in at most $D \leq D+i+1$ steps. This completes the proof of the the the induction hypothesis for $i+1$.

Since the hop distance between g and any vertex of T^* is at most $n-1$, each robot reaches its final position within $n + D - 1$ steps and the claim of the theorem follows. $\square$

6   Forming an Axis Parallel Line Segment Containing g as an End Point

Given a point g, known for each robot. The goal is to arrange the n robots in a connected horizontal line segment containing g as an end point. Now we consider so called finite state robots, i.e. with $O(1)$ persistent bits of memory (see e.g. in [1]). The visibility range of the robots is limited to the 2-hop neighborhood in the 2D grid. We show how the robots can form a connected horizontal line L containing g as its left end point in $3n + D + 3$ steps. A vertical line segment can be formed in a similar way.

6.1   Forming the Horizontal Line Segment L

Forming the connected horizontal line segment L containing g as its left end point consists of 3 phases for each robot. Each of the 3 phases are oblivious, in each step the decision of each robot where to move solely depends on the configuration of the 2-hop neighborhood in the 2D grid. The only persistent memory used by a robot is to store, which phase of the algorithm it currently executes. The robots execute the following phases:

1. Let (g_x, g_y) be the coordinates of g and let g' be the point with coordinates $(g_x - 1, g_y + 1)$. Each robot with initial y-coordinate greater than g_y moves

one step upwards and each robot with initial y-coordinate less or equal than g_y moves one step downwards. During this step no collision can arrise. At the end of this step we obtain a horizontal stripe H of height 2 containing the horizontal line coincident with g and the horizontal line coincident with g'.

2. Execute the Lemmings algorithm with sink g'. More precisely, execute Algorithm 1 with sink g', such that g' can be occupied in a step t, if g' is unoccupied at the beginning of step t. When a robot occupies g' in a certain step, it moves one hop to the right from g' to the point $g'' = (g_x, g_y + 1)$ in the next step and starts phase 3.

3. Build L from the source g'' as follows. Let (x, y) be the current coordinates of a robot. Until the the vertex at $(x, y - 1)$ is occupied, move to the right. Otherwise, occupy the vertex at $(x, y - 1)$ and terminate the algoritm of that robot.

Theorem 4. *Let $V = \{v_1, ..., v_n\}$ be the set of n finite state robots placed on different vertices of $\mathbb{Z}^2$. Let g be a point, known for all robots. Then by the above algorithm the robots form a connected horizontal line segment L with left end point g in $D + 3n + 3$ steps, where D is the maximum initial hop distance of a robot from g.*

Proof. After phase 1, the horizontal stripe H of height 2 containing the horizontal lines coincident with g and g' is free of robots. Let ℓ be the vertical line coincident to g'. In phase 2, non of the robots visits any vertex in $H \setminus \ell$. In phase 3 each robot only visits vertices in $H \setminus \ell$. Therefore, no collision can occure.

Phase 1 of the algorithm takes 1 time unit. The only difference between phase 2 of the formation of L and the Lemmings algorithm is that g' can be occupied in a step t, if g' is unoccupied at the beginning of step t. The robot, which has occupied g' will move away from g to the left in the next step. and starts phase 3. It is easy to check that all arguments of the proof of Theorem 2 also apply for this case and the time bound $2n + D' - 1$ stated in Theorem holds, where $D' = D + 2$ is the maximum hop distance of a robot from g' at the beginning of phase 2. Therefore, each robot finishes phase 1 and 2 after $2n + D + 3$ steps. It is easy to check that each robot spends at most n time steps in phase 3. Consequently, the robots build the connected line segment L by the algorithm in $3n + D + 3$ steps.

7   Forming an Arbitrary Connected Pattern F

Now we show how the Lemmings algorithm can be used for forming an arbitrary connected pattern of n vertices in the 2D grid. We consider finite state robots. The visibility range of the robots is limited to the 2-hop neighborhood in the 2D grid.

Given a connected pattern F of size n. Let B be the axis-parallel bounding box of F and (x, y) the upper left corner of B. Let g be the point with coordinates $(x - 1, y + 1)$ and g' be the point with coordinates $(x, y + 1)$ (see Figure 3). Let T be a spanning tree of F. Consider the rooted tree T^* with root g that starts

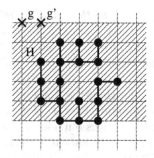

Fig. 3. After phase 1, the horizontal stripe H (shaded region) of height $h+1$ containing the axis parallel bounding box B' of $F \cup \{g'\}$ is free of robots. In phase 2, each robot only visits vertices outside B'. In phase 3 each robot only visits vertices in B'.

with a shortest path from g through g' to a closest node g^* of F (This path does not contain further nodes of F) and contains a rooted version of T rooted at g^*.

Assume that n robots appear in g, one after another. Empty steps where no new robot appears are allowed, all robots have appeared after some number R of steps. The point g plays a similar role as the "door" in the Filling problem [1], but g is not contained in F. As further difference, F can be any connected pattern in our case. Let $d(T^*)$ be the depth of T^*. It is easy to check that a depth-first filling of T^* (more precisely, the depth-first filling of T) can be executed by the appearing robots, so that, after $O(R + d(T^*))$ many steps, each node of F is occupied by one robot, and no collisions appeared during these steps.

7.1   Forming F

Forming the connected pattern F consists of 3 phases for each robot. The only persistent memory used by a robot is to store, which phase of the algorithm it currently executes. The robots execute the following phases:

1. Let x_{min} and x_{max} be the minimum and maximum x-coordinate of F, respectively. Let $h = x_{max} - x_{min} + 1$ be the height of the connected pattern F. Each robot with initial position above or on the lower horizontal boarder of F moves $h + 1$ many steps upwards. During this $h + 1$ steps there are no collisions. At the end we obtain a horizontal stripe H of height $h + 1$ containing the axis parallel bounding box B' of $F \cup \{g'\}$ free of robots (see Figure 3).

2. Execute the Lemmings algorithm with sink g. More precisely, execute Algorithm 1, such that g can be occupied in a step t, if it is unoccupied at the beginning of step t. When a robot occupies g in a certain step, it moves one hop to the right from g to the point g' in the next step and starts phase 3.

3. Build F from the source g' by depth-first filling of T, using the arrivals of the robots in g during the Lemmings algorithm as input stream.

Theorem 5. *Let $V = \{v_1, ..., v_n\}$ be the set of n finite state robots placed on different vertices of $\mathbb{Z}^2$. Let F be a connected formation, known for all robots. Then the robots form F in time $O(n + D^*)$, where D^* denotes the diameter of the point set consisting of the initial configuration and F.*

Proof. After phase 1 of the algorithm, the horizontal stripe H of height $h + 1$ containing the axis parallel bounding box B' of $F \cup \{g'\}$ is free of robots. In phase 2, each robot only visits vertices outside B'. In phase 3 each robot only visits vertices in B'. Therfore, no collision can occure.

Phase 1 takes $h + 1 = O(D^*)$ steps. The only difference between phase 2 of the pattern formation and the Lemmings algorithm is that g can be occupied in a step t, if it is unoccupied at the beginning of step t. When a robot occupies g in a certain step, it moves one hop to the right from g to the point g' in the next step and starts phase 3. It is easy to check that all arguments of the proof of Theorem 2 also apply for this case and the time bound $2n + D' - 1$ stated in Theorem holds, where D' is the maximum hop distance of a robot from g at the beginning of phase 2. Therefore, each robot finishes phase 2 after $O(n + D^*)$ steps. It is easy to check that each robot spends $O(n)$ time in phase 3. Consequently, the formation F becomes built by the algorithm in $O(n + D^*)$ steps.

7.2   Focused Coverage Problem, When n Is Known for All Nodes

A closely related problem to the pattern formation problem is the focused coverage self-deployment problem in mobile sensor networks, where an area with maximum radius around a Point of Interest (POI) must be covered without sensing holes. This problem was introduced in [10], [11],[12]. The authors solved the problem by driving the mobile sensors along an equilateral triangle tessellation graph centered at the POI. They showed that their algorithms terminate in finite time. Subsequently, in [2] a collision free algorithm has been presented solving the focused coverage problem in $O(S)$ time, where S is the sum of initial distances of the mobile sensors from the POI.

Now we consider the focused coverage problem on the 2D grid instead of the an equilateral triangle tessellation graph. Additionally, we assume that the number of sensors n is known for each sensor node. Then each node can compute the connected hole free formation F with maximum radius centered at the POI. Then our algorithm for connected pattern formation can be applied, which solves the problem in $O(n + D^*)$ time, where D^* denotes the diameter of the point set consisting of the initial configuration and F.

Corollary 1. *Let $V = \{v_1, ..., v_n\}$ be the set of n finite state mobile sensor nodes placed on different vertices of $\mathbb{Z}^2$. Assume that all nodes know the POI, n, and the configutation of its 2-hop neighborhood in $\mathbb{Z}^2$. Then by applying the connected pattern formation algorithm, the focused coverage problem can be solved in time $O(n + D^*)$, where D^* denotes the diameter of the point set consisting of the initial configuration of the nodes and the POI.*

8 Summary

We have investigated the Pattern Formation problem on a 2D grid in the synchronous Look-Compute-Move model. First we have considered the helpful intermediate problem, called the Lemmings problem, where all robots have to be gathered at a given point g, and thus at the single point g collision is allowed, we proved a lower bound of $\Omega(n + D)$ time steps on the running time of each discrete synchronous algorithm solving this problem.

We have introduced the x-y-routing algorithm for solving this problem, where the nodes only need local knowledge about their 2-hop neighborhood in the grid $\mathbf{Z}^2$. Based on this knowledge the nodes are able to move towards the gathering point g without collision, such that after at most $2n + D - 1$ time steps all robots reach g, where D is the maximum hop distance of a robot from g. Thus, the running time of the algorithm is optimal up to a constant factor.

We have shown that the x-y-routing method also solves the gathering problem in $n + D - 1$ time steps, where no collision is allowed at g and the robots have to form a connected configuration containing g. This significantly improves the previous upper bound of $O(nD)$ for this problem presented in [5]

Furthermore, we have shown how a the robots can form a connected axis parallel line segment L containing g as an end point in $3n + D + 3$ steps.

Finally, we have investigated the problem of forming a connected pattern F, which is known to all robots. We have shown how to build F in time $O(n + D^*)$, where D^* denotes the diameter of the point set consisting of the initial configuration and F.

In case when all robots know n, this solution can also be applied for solving the focused coverage problem on the 2D grid. This results in $O(n + D)$ covering time. If the number of robots n is not known for the robots, then best known upper bound on this problem is $O(S)$, presented in [2], where S is the sum of initial distances of the mobile sensors from g.

References

1. Barrameda, E.M., Das, S., Santoro, N.: Deployment of asynchronous robotic sensors in unknown orthogonal environments. In: Fekete, S.P. (ed.) ALGOSENSORS 2008. LNCS, vol. 5389, pp. 125–140. Springer, Heidelberg (2008)
2. Blázovics, L., Lukovszki, T.: Fast localized sensor self-deployment for focused coverage. In: Flocchini, P., Gao, J., Kranakis, E., Meyer auf der Heide, F. (eds.) ALGOSENSORS 2013. LNCS, vol. 8243, pp. 83–94. Springer, Heidelberg (2014)
3. Cieliebak, M., Flocchini, P., Prencipe, G., Santoro, N.: Distributed computing by mobile robots: Gathering. SIAM J. Comput. 41(4), 829–879 (2012)
4. Cohen, R., Peleg, D.: Convergence properties of the gravitational algorithm in asynchronous robot systems. SIAM J. Comput. 34(6), 1516–1528 (2005)
5. Cord-Landwehr, A., et al.: Collisionless gathering of robots with an extent. In: Černá, I., Gyimóthy, T., Hromkovič, J., Jefferey, K., Královič, R., Vukolić, M., Wolf, S. (eds.) SOFSEM 2011. LNCS, vol. 6543, pp. 178–189. Springer, Heidelberg (2011)

6. Cord-Landwehr, A., et al.: A new approach for analyzing convergence algorithms for mobile robots. In: Aceto, L., Henzinger, M., Sgall, J. (eds.) ICALP 2011, Part II. LNCS, vol. 6756, pp. 650–661. Springer, Heidelberg (2011)
7. Czyzowicz, J., Gasieniec, L., Pelc, A.: Gathering few fat mobile robots in the plane. Theor. Comput. Sci. 410(6-7), 481–499 (2009)
8. Flocchini, P., Prencipe, G., Santoro, N.: Distributed Computing by Oblivious Mobile Robots. Synthesis Lectures on Distributed Computing Theory. Morgan & Claypool Publishers (2012)
9. Flocchini, P., Prencipe, G., Santoro, N., Widmayer, P.: Arbitrary pattern formation by asynchronous, anonymous, oblivious robots. Theor. Comput. Sci. 407(1-3), 412–447 (2008)
10. Li, X., Frey, H., Santoro, N., Stojmenovic, I.: Localized sensor self-deployment for guaranteed coverage radius maximization. In: IEEE International Conference on Communications, ICC 2009, pp. 1–5 (2009)
11. Li, X., Frey, H., Santoro, N., Stojmenovic, I.: Focused-coverage by mobile sensor networks. In: 6th IEEE International Conference on Mobile Adhoc and Sensor Systems (MASS), pp. 466–475 (2009)
12. Li, X., Frey, H., Santoro, N., Stojmenovic, I.: Strictly localized sensor self-deployment for optimal focused coverage. IEEE Trans. Mob. Comput. 10(11), 1520–1533 (2011)
13. Suzuki, I., Yamashita, M.: Distributed anonymous mobile robots. In: Proc. 3rd International Colloquium on Structural Information and Communication Complexity (SIROCCO), pp. 313–330 (1996)
14. Suzuki, I., Yamashita, M.: Distributed anonymous mobile robots: Formation of geometric patterns. SIAM J. Comput. 28(4), 1347–1363 (1999)
15. Yamauchi, Y., Yamashita, M.: Pattern formation by mobile robots with limited visibility. In: Moscibroda, T., Rescigno, A.A. (eds.) SIROCCO 2013. LNCS, vol. 8179, pp. 201–212. Springer, Heidelberg (2013)

Tradeoffs between Cost and Information
for Rendezvous and Treasure Hunt

Avery Miller and Andrzej Pelc*

Université du Québec en Outaouais, Gatineau, Canada
avery@averymiller.ca, pelc@uqo.ca

Abstract. Rendezvous and treasure hunt are two basic tasks performed by mobile agents in networks. In rendezvous, two agents, initially located at distinct nodes of the network, traverse edges in synchronous rounds and have to meet at some node. In treasure hunt, a single agent has to find a stationary target (treasure) situated at an unknown node. The network is modeled as an undirected connected graph whose nodes have distinct identities. The cost of a rendezvous algorithm is the worst-case total number of edge traversals performed by both agents until meeting. The cost of a treasure hunt algorithm is the worst-case number of edge traversals performed by the agent until the treasure is found. If the agents have no information about the network, the cost of both rendezvous and treasure hunt can be as large as $\Theta(e)$ for networks with e edges.

We study tradeoffs between the amount of information available *a priori* to the agents and the cost of rendezvous and treasure hunt. Following the paradigm of algorithms with advice, this information is provided to the agents at the start of their navigation by an oracle knowing the network, the starting positions of the agents, and, in the case of treasure hunt, the node where the treasure is hidden. The oracle assists the agents by providing them with a binary string called *advice*, which can be used by each agent during the algorithm execution. In the case of rendezvous, the advice given to each agent can be different. The length of the string given to the agent in treasure hunt and the sum of the lengths of strings given to the agents in rendezvous is called the *size of advice*.

Our goal is to find the smallest size of advice which enables the agents to solve rendezvous and treasure hunt at a given cost C in a network with e edges. This size turns out to depend on the initial distance D and on the ratio $g = e/C$, which is the *relative cost gain* due to advice. For arbitrary graphs, we give upper and lower bounds of $O(D \log(Dg))$ and $\Omega(D \log g)$, respectively, on the optimal size of advice. Hence, our bounds leave only a logarithmic gap in the general case. For the class of trees we give tight upper and lower bounds of $\Theta(D \log g)$.

Keywords: rendezvous, treasure hunt, advice, deterministic algorithm, mobile agent, cost.

* Partially supported by NSERC discovery grant and by the Research Chair in Distributed Computing at the Université du Québec en Outaouais.

M.K. Aguilera et al. (Eds.): OPODIS 2014, LNCS 8878, pp. 263–276, 2014.

1   Introduction

Model and Problems. Rendezvous and treasure hunt are two basic tasks performed by mobile agents in networks. In rendezvous, two agents, initially located at distinct nodes of the network, traverse network edges in synchronous rounds and have to meet at some node. In treasure hunt, a single agent has to find a stationary target (called treasure) situated at an unknown node of the network. The network might model a labyrinth or a system of corridors in a cave, in which case the agents might be mobile robots. The meeting of such robots might be motivated by the need to exchange previously collected samples, or to agree how to share a future cleaning or decontamination task. Treasure hunt might mean searching a cave for a resource or for a missing person after an accident. In other applications we can consider a computer network, in which the mobile entities are software agents. The meeting of such agents might be necessary to exchange data or share a future task of checking the functionality of network components. Treasure hunt in this case might mean looking for valuable data residing at some node of the network, or for a virus implanted at some site.

The network is modeled as a simple undirected connected graph whose nodes have distinct identities. Ports at a node of degree d are numbered $0, \ldots, d-1$. Agents execute a deterministic algorithm, such that, at each step, they choose a port at the current node. When an agent enters a node, it learns the entry port number, the label of the node and its degree. The agents may have distinct labels or be anonymous. The cost of a rendezvous algorithm is the total worst-case number of edge traversals performed by both agents until meeting. The cost of a treasure hunt algorithm is the worst-case number of edge traversals performed by the agent until the treasure is found. If the agents have no information about the network, the cost of both rendezvous and treasure hunt can be as large as $\Theta(e)$ for networks with e edges. This is clear for treasure hunt, as all edges (except one) need to be traversed by the agent to find the treasure in the worst case. The same lower bound for rendezvous follows from Proposition 1 in the present paper. On the other hand, if D is the distance between the initial positions of the agents, or from the initial position of the agent to the treasure, a lower bound on the cost of rendezvous and of treasure hunt is D.

In this paper, we study tradeoffs between the amount of information available *a priori* to the agents and the cost of rendezvous and treasure hunt. Following the paradigm of algorithms with advice [1, 11, 13, 19–25, 27, 28, 34, 39], this information is provided to the agents at the start of their navigation by an oracle that knows the network, the starting positions of the agents and, in the case of treasure hunt, the node where the treasure is hidden. The oracle assists the agents by providing them with a binary string called *advice*, which can be used by the agent during the algorithm execution. In the case of rendezvous, the advice given to each agent can be different. The length of the string given to the agent in treasure hunt and the sum of the lengths of strings given to both agents in rendezvous is called the *size of advice*.

Our Results. Using the framework of advice permits us to quantify the amount of information needed for an efficient solution of a given network problem (in our case, rendezvous and treasure hunt) regardless of the type of information that is provided. Our goal is to find the smallest size of advice which enables the agents to solve rendezvous and treasure hunt at a given cost C in a network with e edges. This size turns out to depend on the initial distance D (between the agents in rendezvous, and between the agent and the treasure in treasure hunt) and on the ratio $g = e/C$, which is the *relative cost gain* due to advice. For arbitrary graphs, we give upper and lower bounds of $O(D \log(Dg))$ and $\Omega(D \log g)$, respectively, on the optimal size of advice. Hence our bounds leave only a logarithmic gap in the general case. For the class of trees, we give matching upper and lower bounds of $\Theta(D \log g)$. Our upper bounds are obtained by constructing an algorithm for all graphs (respectively, for all trees) that works at the given cost and with advice of the given size, while the lower bounds are proved by exhibiting networks for which it is impossible to achieve the given cost with smaller advice.

Missing proofs will appear in the full version of the paper.

Related Work. Treasure hunt, network exploration and rendezvous in networks are interrelated problems that have received much attention in recent literature. Treasure hunt has been investigated in the line [10, 26], in the plane [7] and in other terrains [31]. Treasure hunt in anonymous networks (without any knowledge of the network) was studied in [38, 40] with the goal of minimizing cost.

The related problem of graph exploration by mobile agents (often called robots) has been intensely studied as well. The goal of this task is to visit all of the nodes and/or traverse all of the edges of a graph. Many papers, e.g., [16, 35] studied the scenario where the graph to be explored is labeled and undirected, and the agent can traverse edges in both directions. In [35], it was shown that a graph with n nodes and e edges can be explored in time $e + O(n)$. In [37], a log-space construction of a deterministic exploration for all graphs with a given bound on size was shown.

The problem of rendezvous has been studied both under randomized and deterministic scenarios. In the framework of networks, it is usually assumed that the nodes do not have distinct identities. An extensive survey of randomized rendezvous in various models can be found in [4], cf. also [2, 3, 5]. Deterministic rendezvous in networks has been surveyed in [36]. Several authors considered geometric scenarios (in an interval of the line, e.g., [9], or in the plane, e.g., [6]).

For the deterministic setting, many authors studied the feasibility and time complexity of rendezvous of agents that move in rounds [16, 29, 38]. In [33] the authors studied tradeoffs between the time of rendezvous and the number of edge traversals by both agents. The amount of memory required by the agents to achieve deterministic rendezvous was studied in [14] for general graphs. The amount of memory needed for randomized rendezvous in the ring was discussed, e.g., in [30]. Several authors investigated asynchronous rendezvous in the plane [12] and in networks [8, 15, 17].

Providing nodes or agents with information of arbitrary type that can be used to perform network tasks more efficiently has been proposed in [1, 11, 13, 19–25,

27, 28, 32, 34, 39]. This approach was referred to as algorithms with *advice*. The advice is given either to nodes of the network or to mobile agents performing some network task. Most of the authors studied the minimum size of advice required to solve the respective network problem in an efficient way.

In [28], given a distributed representation of a solution for a problem, the authors investigated the number of bits of communication needed to verify the legality of the represented solution. In [20], the authors compared the minimum size of advice required to solve two information dissemination problems using a linear number of messages. In [22], it was shown that a constant amount of advice enables the nodes to carry out the distributed construction of a minimum spanning tree in logarithmic time. In [19], the advice paradigm was used for online problems. In the case of [34], the issue was not efficiency but feasibility: it was shown that $\Theta(n \log n)$ is the minimum size of advice required to perform monotone connected graph clearing. In [24], the authors studied the problem of topology recognition with advice given to nodes.

Among the papers using the paradigm of advice, [13, 21, 32] are closest to the present work. In [13], the authors investigated the minimum size of advice that has to be given to unlabeled nodes (and not to the agent) to permit graph exploration by an agent modeled as a k-state automaton. In [21], the authors established the size of advice that has to be given to an agent completing exploration of trees, in order to break competitive ratio 2. In [32], the authors studied the minimum size of advice that must be provided to labeled agents, in order to achieve rendezvous at minimum possible cost, i.e., at cost $\Theta(D)$, where D is the initial distance between the agents. They showed that this optimal size of advice for rendezvous in n-node networks is $\Theta(D \log(n/D) + \log \log L)$, where the labels of agents are drawn from the set $\{1, \ldots, L\}$. This paper differs from the present one in two important aspects. First, as opposed to the present paper, in [32], agents get identical advice, and nodes of the network are unlabeled. Second, instead of looking at tradeoffs between cost and the size of advice, as we do in the present paper, the focus of [32] was on the size of advice sufficient to achieve the lowest possible cost.

2   Preliminaries

In this section we show that, in the context of advice, treasure hunt and rendezvous are essentially equivalent. More precisely, the following proposition shows that the minimum advice sufficient to solve both problems at a given cost in the class of graphs with $\Theta(e)$ edges and with the initial distance $\Theta(D)$ is the same, up to constant factors. Throughout the paper a *graph* means a simple connected undirected graph. The number of nodes in the graph is denoted by n, and the number of edges is denoted by e. All logarithms are to base 2.

Proposition 1. *Let $D \leq e$ be positive integers.*

1. *If there exists an algorithm TH that solves treasure hunt at cost C with advice of size A in all graphs with e edges and with initial distance D between the agent and the treasure, then there exists an algorithm RV that solves*

rendezvous at cost C with advice of size $O(A)$ in all graphs with e edges and with initial distance D between the agents.

2. If there exists an algorithm RV solving rendezvous at cost C with advice of size less than A in all graphs with $2e + 1$ edges and with initial distance $2D + 1$ between the agents, then there exists an algorithm TH that solves treasure hunt at cost at most C with advice of size less than A in all graphs with e edges and with initial distance D between the agent and the treasure.

In view of Proposition 1, in the rest of the paper we can restrict attention to the problem of treasure hunt. All of our results, both the upper and the lower bounds, also apply to the rendezvous problem.

3 Treasure Hunt in Arbitrary Graphs

In this section, we proceed to prove upper and lower bounds on the advice needed to solve treasure hunt in arbitrary graphs. These bounds are expressed in terms of D, which is the distance between the treasure and the initial position of the agent, and in terms of the ratio $g = e/C$, where e is the number of edges in the graph and C is an upper bound on the cost of the algorithm. This ratio is the relative cost gain due to advice. We first provide an algorithm that solves treasure hunt using $O(D \log(Dg))$ bits of advice, and then prove that any deterministic algorithm for this task uses at least $\Omega(D \log g)$ bits of advice.

ALGORITHM. Consider a graph G and a node s of G, which is the initial position of the agent. Fix an integer $z \geq 1$. We describe a binary advice string of length Dz and an algorithm that uses this advice when searching for the treasure. To construct the advice, the idea is to find a shortest path P from s to the treasure, and then to produce D advice substrings to guide the agent along this path. However, since Dz bits may not be enough to exactly describe the D edges of the path, the advice will specify a subset of ports that the agent should try. In particular, the advice consists of D binary substrings $A_0, \ldots, A_{D-1}$, each of length z. For each $i \in \{0, \ldots, D-1\}$, the substring A_i is created by considering the node v_i on path P that is at distance i from s in G. The set of ports at this node is partitioned into numbered sectors (i.e., subintervals) of size $\lfloor deg(v_i)/2^{z-1} \rfloor$, and A_i is taken to be the z-bit binary representation of the number of the sector containing the port that leads to the next node v_{i+1} on path P towards the treasure.

Below, we provide pseudocode that describes how the advice is created. First, Algorithm 1 finds a shortest path P from s to the treasure. The path consists of node/port pairs (v_i, p_i) for each $i \in \{0, \ldots, D-1\}$, where $v_0 = s$ and, for each $i \in \{0, \ldots, D-2\}$, port p_i leads from node v_i to node v_{i+1}. Each such pair (v_i, p_i) is passed to Algorithm 2, which divides the set of ports at v_i into numbered sectors of equal size, determines to which sector port p_i belongs, and outputs the binary representation of this sector number as a string A_i.

The resulting sequence of substrings $(A_0, \ldots, A_{D-1})$ is encoded into a single advice string to pass to the algorithm. More specifically, the sequence is encoded by doubling each digit in each substring and putting 01 between substrings. This permits the agent to unambiguously decode the original sequence, and to calculate the value of D by looking at the number of separators 01. Denote by $Concat(A_0, \ldots, A_{D-1})$ this encoding and let $Decode$ be the inverse (decoding) function, i.e. $Decode(Concat(A_0, \ldots, A_{D-1})) = (A_0, \ldots, A_{D-1})$. As an example, $Concat((01), (00)) = (0011010000)$. Note that the encoding increases the total number of advice bits by a constant factor. The advice string, calculated by Algorithm 1 using the strings A_i supplied by Algorithm 2, is $\mathcal{A} = Concat(A_0, \ldots, A_{D-1})$. The advice string $\mathcal{A}$ is given to the agent.

Algorithm 1. CreateAdvice(G,s)

1: Find a shortest path P in G from node s to the node containing the treasure.
2: **for** $i = 0, \ldots, D-1$ **do**
3:      $v_i \leftarrow$ node on path P at distance i from s
4:      $p_i \leftarrow$ port number leading from v_i to node on path P at distance $i+1$ from s
5:      $A_i \leftarrow$ EncodeSectorNumber(v_i, p_i)
6: Output $Concat(A_0, \ldots, A_{D-1})$

Algorithm 2. EncodeSectorNumber(v, $port$)

1: $SectorSize \leftarrow \lfloor deg(v)/2^{z-1} \rfloor$
2: $SectorNumber \leftarrow \lfloor port/SectorSize \rfloor$
3: // $port$ is contained in the range $\{SectorNumber \cdot SectorSize, \ldots, (SectorNumber + 1) \cdot SectorSize - 1\}$
4: return the z-bit binary representation of $SectorNumber$

Next, we describe FindTreasure, which is the agent's algorithm given an advice string $\mathcal{A} = Concat(A_0, \ldots, A_{D-1})$. For the purpose of description only, we define the *trail* of the agent, which is a stack of edges that it has previously traversed. The stack gets popped when the agent backtracks. The agent performs a walk in G starting at node s. In each step of the algorithm, the agent chooses an edge to add to the trail, or it backtracks along the trail edge that it added most recently. The number of edges in the agent's trail will be used to measure the agent's progress. In particular, when the agent is located at a node v and there are i edges in the agent's trail, we will say that the agent is at *progress level i*. The agent keeps track of its current progress level by maintaining a counter that is incremented when it adds a trail edge and decremented when it backtracks.

The agent maintains a table containing the labels of the nodes that it has visited, and, for each node label, the smallest progress level at which the agent visited the node so far. When the agent arrives at a node v from a lower progress level and does not find the treasure, it checks if its current progress level i is lower than the progress level stored in the table for node v. If this is not the case, or if $i = D$, then the agent backtracks by going back along the edge it just arrived on. Otherwise, the agent uses the advice substring A_i in the following way: it divides the set of port numbers into sectors (i.e., intervals of port numbers) of size $\lfloor deg(v)/2^{z-1} \rfloor$, numbers the sectors, and then interprets A_i as the binary representation of an integer that specifies one of these sectors. For each port number in the specified sector, the agent takes the port and arrives at some neighbour w of v. The agent terminates if it finds the treasure at node w, or, otherwise, repeats the above at node w. If, after trying all ports at node v in the specified sector, the treasure has not been found, the agent backtracks.

Note that the advice was created with the goal of 'steering' the agent in the right direction, i.e., along path P, but we can only guarantee that this will happen when the agent is located at nodes on path P. In fact, an even stronger condition must hold: for any node v on path P at distance i from s, we can only guarantee that the advice will be helpful if the agent is located at node v at progress level i, since this is when the agent reads the advice substring A_i. In other words, it is possible that the agent visits a node v on P at the 'wrong' progress level, in the sense that it won't use the advice that was created specifically for v. This is why it is not sufficient to simply have the agent backtrack whenever it arrives at a previously-visited node, since during its previous visit, it may have used the wrong advice. Moreover, we must ensure that the algorithm gracefully deals with the situation where the agent is at a node w at progress level j, but the advice substring A_j specifies ports that do not exist at w. In our algorithm, the agent ignores any port numbers that are greater than or equal to the current node's degree.

To summarize, in our algorithm, the agent searches for the treasure in a depth-first manner, but it cannot perform DFS (even only to distance D) because the cost would be too large. Instead, the agent takes only a fraction of ports at each node, but may possibly have to pay for it by traversing the same edge several times (while in DFS every edge is traversed at most twice). As our analysis will show, this gives a decrease of the overall cost, especially when the advice is large.

The pseudocode of the search conducted by algorithm FindTreasure is described by Algorithm 3. It shows how the agent takes a step in the graph, i.e., for each $i \in \{0, \ldots, D-1\}$, how it uses A_i to move from a node at progress level i to a node at progress level $i+1$.

Algorithm 3. TakeStep$(\mathcal{A}, v, i, prev)$

$\mathcal{A}$ is the advice string, v is the node where the agent is currently located, i is the current progress level, $prev$ is the node from which the agent arrived

1: **If** treasure is located at v **then** Stop
2: **If** $(i < D)$ AND $(i < $ CurrentMin$(v))$ **then**
3:     UpdateTable(v, i)
4:     $(A_0, \ldots A_{D-1}) \leftarrow Decode(\mathcal{A})$
5:     $sector \leftarrow$ GetSector(v, A_i)
6:     **for** each port p in $sector$ **do**
7:         if $p < deg(v)$ **then**
8:             take port p
9:             $w \leftarrow$ the node reached after taking port p
10:            call TakeStep$(\mathcal{A}, w, i+1, v)$
11: Return to node $prev$

In order to initiate the search, Algorithm 3 is called at node s with progress level 0 (and $prev = s$). Algorithm 4, used as a subroutine in Algorithm 3, shows how the agent decodes substring A_i to obtain a range of port numbers. We assume that we have two functions related to the agent-maintained table of visited nodes: UpdateTable(v, i) that writes i into the entry for node v as the smallest progress level at which the agent has ever visited node v, and CurrentMin(v) that reads the entry of the table for node v. Each table entry is initialized to ∞.

Algorithm 4. GetSector$(v, SectorNumberEncoding)$

1: $z \leftarrow$ number of bits in $SectorNumberEncoding$
2: $SectorSize \leftarrow \left\lfloor \frac{deg(v)}{2^z - 1} \right\rfloor$
3: $SectorNumber \leftarrow$ integer value of $SectorNumberEncoding$
4: return $\{SectorNumber \cdot SectorSize, \ldots, (SectorNumber + 1) \cdot SectorSize - 1\}$

ANALYSIS. In what follows, let P be the path from s to the treasure that is used to create the advice $\mathcal{A} = Concat(A_0, \ldots, A_{D-1})$. Suppose that P consists of the nodes $v_0, \ldots, v_D$, where, for each $i \in \{0, \ldots, D\}$, v_i is at distance i from s, and the treasure is located at node v_D. Also, for each $i \in \{0, \ldots, D-1\}$, let p_i be the port at node v_i that leads to node v_{i+1}.

To prove the correctness of the algorithm, we first consider an arbitrary node v_i on path P and suppose that the agent is at progress level i. Clearly, this occurs at least once during the execution of FindTreasure since the agent is initially located at v_0 at progress level 0. One of the ports at v_i that are specified by the advice substring A_i leads to node v_{i+1}, but the agent may try some other of these ports first. We can show that either the agent finds the treasure by recursively calling TakeStep after taking one of these other ports, or, the agent eventually takes the port that leads to node v_{i+1}. Extending this by induction, we obtain the following result.

Lemma 1. *For any $i \in \{0, \ldots, D-1\}$, consider the first time that the agent is located at node v_i at progress level i. During the execution of TakeStep$(\mathcal{A}, v_i, i, w)$, for some node w, the agent finds the treasure.*

By Lemma 1 with $i = 0$, the agent finds the treasure during the first execution of TakeStep, hence FindTreasure is correct. Next, we proceed to find an upper bound on the cost of algorithm FindTreasure in terms of a fixed upper bound on the amount of advice provided.

Lemma 2. *When provided with Dz bits of advice, FindTreasure has cost at most $O(De/2^z)$.*

Proof. It suffices to count the total number of times that line 8 of TakeStep is called. This is because the cost incurred by backtracking (i.e., line 11 of TakeStep) is at most 1 for each execution of TakeStep, which amounts to an overall multiplicative factor of at most 2. So, we consider the number of times that line 8 of TakeStep is called at an arbitrary node v. The number of times that the **for** loop at line 6 is iterated is at most $deg(v)/2^{z-1}$, since the size of the range returned by GetSector is $\lfloor deg(v)/2^{z-1} \rfloor$. The number of times that the **if** statement on line 2 evaluates to true is bounded above by D: parameter i never has value larger than D, and the call to UpdateTable on line 3 ensures that the sequence of values of i such that $i <$ CurrentMin(v) is strictly decreasing. Therefore, the total number of times that line 8 is executed is bounded above by $D \cdot deg(v)/2^{z-1}$. Hence, the total number of calls to TakeStep is bounded above by $\sum_v D \cdot deg(v)/2^{z-1} = D \cdot (2e/2^{z-1}) \in O(De/2^z)$. $\qquad\square$

Finally, we fix an upper bound C on the cost of FindTreasure and re-state Lemma 2 as an upper bound on the amount of advice needed to solve treasure hunt at cost C.

Theorem 1. *Let G be any graph with e edges, and let $D \le e$ be the distance from the initial position of the agent to the treasure. Let C be any integer such that $D \le C \le e$. The amount of advice needed to solve treasure hunt at cost at most C is at most $O(D \log(Dg))$ bits.*

LOWER BOUND. The following lower bound follows immediately from Theorem 4, which is proven by constructing a tree for which treasure hunt requires $\Omega(D \log g)$ bits of advice. This theorem will be proven in Section 4.

Theorem 2. *Let $D \le C \le e$. There exists a graph G with $\Theta(e)$ edges, and a position of the treasure at distance D from the initial position of the agent, such that treasure hunt at cost C requires $\Omega(D \log g)$ bits of advice.*

The gap between the upper bound given by Theorem 1 and the lower bound given by Theorem 2 is at most a factor logarithmic in D. Moreover, it should be noted that our bounds become asymptotically tight whenever D is polynomial in the gain $g = e/C$.

4   Treasure Hunt in Trees

We now proceed to prove upper and lower bounds on the advice needed to solve treasure hunt in trees. Unlike in the case of arbitrary graphs, where our upper and lower bounds may differ by a logarithmic factor, for trees we obtain matching upper and lower bounds. Again, our bounds will be expressed in terms of D, which is the distance between the treasure and the initial position of the agent, and in terms of the ratio $g = e/C = (n-1)/C$, where e is the number of edges in the tree, n is the number of nodes, and C is an upper bound on the cost of the algorithm. Also, for any two nodes a, b, we will denote by $d(a, b)$ the distance between a and b in the tree, i.e., the number of edges in the simple path between them.

UPPER BOUND. To obtain our upper bound, we will use FindTreasure that was defined and proven correct in Section 3 for arbitrary graphs. In this section, we provide an analysis of the algorithm specifically for the case of trees, which gives a strictly better upper bound. We start with the following technical lemma, which shows that, if we take the agent's initial position as the root of the tree, the agent's progress level and the agent's current depth in the tree (i.e., its current distance from the root) do not differ. Essentially, this is because there is only one simple path from the agent's initial position to each node, and the algorithm ensures that the agent's trail does not contain the same edge multiple times.

Lemma 3. *Consider algorithm* `FindTreasure` *executed in any tree. Suppose that, for some neighbouring nodes v and prev,* `TakeStep`$(\mathcal{A}, v, i, prev)$ *is executed at node v. If line 2 evaluates to true, then progress level* $i = d(s, v)$.

Next, we proceed to find an upper bound on the cost of `FindTreasure` in trees in terms of a fixed upper bound on the amount of advice provided. The proof is analogous to the proof of Lemma 2, the main difference being that we do not need to multiply by a factor of D in order to account for the different paths that the agent could use to reach a given node.

Lemma 4. *Let* $D < n$ *be the distance between the treasure and the initial position of the agent in an n-node tree. Algorithm* `FindTreasure` *executed with Dz bits of advice has cost at most* $O(n/2^z)$.

Finally, we fix an upper bound C on the cost of `FindTreasure` and re-state Lemma 4 as an upper bound on the amount of advice needed to solve treasure hunt in trees at cost C.

Theorem 3. *Let* $D \leq C \leq e = n - 1$. *The amount of advice needed to solve treasure hunt on trees of size n with cost at most C is at most* $O(D \log g)$ *bits.*

LOWER BOUND. In order to prove a lower bound on the amount of advice needed to solve treasure hunt at cost at most C, we first construct a certain class of trees such that, for an arbitrary treasure hunt algorithm, there exists a tree on which the algorithm incurs a high cost.

Lemma 5. *Consider any treasure hunt algorithm A that takes Dz bits of advice. For any positive integer k, there exists a tree of size* $Dk + 1$ *such that A has cost* $\Omega(D + \frac{Dk}{2^z})$ *on this tree.*

Proof. We consider a collection $\mathcal{T}(D, k)$ of *caterpillar trees*, each constructed as follows. Take a path graph P consisting of $D + 1$ nodes $v_0, \ldots, v_D$, where v_i and v_{i+1} are adjacent, for every $i \in \{0, \ldots D - 1\}$. For each $i \in \{0, \ldots, D - 1\}$, add $k - 1$ nodes to the graph such that each of them has degree 1 and is adjacent only to node v_i. The resulting graph is a tree on $Dk + 1$ nodes. For each node v in this tree, the ports at v are labeled with the integers $\{0, \ldots, deg(v) - 1\}$ so that, for each $i \in \{0, \ldots, D - 2\}$, the port numbers at both ends of the edge $\{v_i, v_{i+1}\}$ are equal. Finally we fix node labels as follows. Each node v_i has label $i(k + 2)$, and each leaf adjacent to v_i has label $i(k + 2) + j + 1$, where the port number at v_i leading to it is j. Notice that all labels are distinct.

For each $i \in \{0, \ldots, D-1\}$, let p_i be the port number at v_i corresponding to the edge $\{v_i, v_{i+1}\}$. Each tree in $\mathcal{T}(D, k)$ is uniquely identified by the sequence $(p_0, \ldots, p_{D-1})$ because the label of each leaf is determined by the port number (at the adjacent node v_i) leading to it. It follows that the number of distinct caterpillar trees in $\mathcal{T}(D, k)$ is k^D. Figure 1 gives a diagram of a caterpillar tree in $\mathcal{T}(D, k)$ and shows how nodes are labeled.

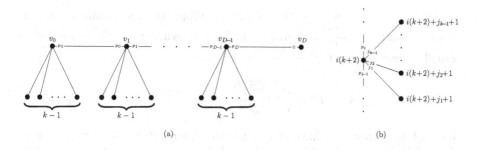

(a)   (b)

Fig. 1. (a) A caterpillar tree in $\mathcal{T}(D, k)$ with ports on path P labeled. (b) The labels of the $k - 1$ added leaves adjacent to v_i are shown. Node v_i is labeled $i(k + 2)$.

Consider any fixed caterpillar tree $G \in \mathcal{T}(D, k)$. We set the starting node of the agent to be v_0 and place the treasure at node v_D. To find the treasure, the agent must traverse the D edges of path P. Suppose that, for some $i \in \{0, \ldots, D-1\}$, the agent is located at node v_i. If the agent takes port p_i, it will arrive at node v_{i+1}, and we say that this edge traversal is *successful*. We may assume that the agent does not return to node v_i, i.e., away from the treasure, because such a move would only increase the cost of the algorithm. Further, the agent can detect when it has found the treasure and terminate immediately.

By the Pigeonhole Principle, there is a subset S of $\mathcal{T}(D, k)$ consisting of at least $\frac{|\mathcal{T}(D,k)|}{2^{Dz}} = \frac{k^D}{2^{Dz}}$ caterpillar trees for which the agent is given the same advice string. We proceed to find an upper bound on the size of such a set S.

When an agent's step is not successful (that is, when located at node v_i, it chooses a port other than p_i) it arrives at a leaf adjacent to v_i. In this case, we say that the agent *misses*. After a miss, the agent's next step is to return to node v_i. Let $miss_{i,G}$ be the number of times that the agent takes a port other than p_i when located at node v_i in G. The *cost at node v_i*, denoted by $cost_{i,G}$, is $2miss_{i,G}+1$, since there are two edge traversals for each miss and one successful edge traversal. Hence, the total cost of any treasure hunt algorithm in G is $\sum_{i=0}^{D-1} cost_{i,G} = D + 2\sum_{i=0}^{D-1} miss_{i,G}$.

Now, suppose that every execution of algorithm A has cost at most C. For every $G \in \mathcal{T}(D, k)$, we know that $C \geq D + 2\sum_{i=0}^{D-1} miss_{i,G}$, so $\sum_{i=0}^{D-1} miss_{i,G} \leq (C - D)/2$. Consider any two different trees $G, G' \in \mathcal{T}(D, k)$ such that the agent is given the same advice string for both of them. Let i be the smallest index such that the port at v_i leading to v_{i+1} is different in G and G'. Then the behaviour of the agent prior to visiting v_i for the first time is the same in G and in G'. Hence $miss_{i,G'} \neq miss_{i,G}$. Therefore, the number of trees in S is bounded above by the number of distinct integer-valued D-tuples of positive terms whose sum is at most $(C - D)/2$. (These tuples correspond to sequences $(miss_{0,G}, \ldots, miss_{D-1,G})$). Clearly, this is bounded above by the number of distinct real-valued D-tuples with non-negative terms whose sum is at most $(C - D)/2$, i.e., by the volume of the simplex $\{(a_0, \ldots, a_{D-1}) \in \mathbb{R}^D \mid \sum_{i=0}^{D-1} a_i = (C - D)/2$ and $0 \leq a_i \leq (C - D)/2$ for all $i\}$. From [18], the volume of such a simplex is equal to $\frac{(C-D)^D}{2^D D!}$. Therefore, we have shown that $|S| \leq \frac{(C-D)^D}{2^D D!}$.

Finally, since $\frac{k^D}{2^{Dz}} \leq |S| \leq \frac{(C-D)^D}{2^D D!}$, we get that

$$C \geq D + \sqrt[D]{D! \frac{k^D 2^D}{2^{Dz}}} = D + \sqrt[D]{D!} \frac{2k}{2^z}$$

By Stirling's formula we have $D! \geq \sqrt{D}(D/e)^D$, for sufficiently large D. Hence $\sqrt[D]{D!} \geq D^{1/(2D)} \cdot (D/e)$, where e is the Euler's constant. Since the first factor converges to 1 as D grows, we have $\sqrt[D]{D!} \in \Omega(D)$. Hence the above bound on C implies $C \in \Omega(D + \frac{Dk}{2^z})$. $\qquad\square$

For any treasure hunt algorithm, Lemma 5 with $k = \lceil n/D \rceil$ proves the existence of a tree of size $\Theta(n)$ in which the algorithm has relatively high cost in terms of the fixed amount of advice provided. By instead fixing an upper bound C on the algorithm's cost, we can re-arrange this lower bound to obtain a lower bound on the amount of advice needed. This lower bound matches the upper bound from Theorem 3.

Theorem 4. *Let $D \leq C \leq e = n - 1$. There exists a tree of size $\Theta(n)$, and a position of the treasure at distance D from the initial position of the agent, such that treasure hunt at cost C requires $\Omega(D \log g)$ bits of advice.*

5   Conclusion

We established upper and lower bounds on the minimum size of advice sufficient to solve the problems of rendezvous and of treasure hunt at a given cost. For the class of trees our bounds are tight, up to constant factors. For the class of arbitrary graphs, our bounds (the upper bound $O(D \log(Dg))$ and the lower bound $\Omega(D \log g)$) leave a gap of a logarithmic factor. Closing this gap is a natural open problem. It should be noted, however, that, even for arbitrary graphs, our bounds are asymptotically tight, whenever D is polynomial in the gain $g = e/C$. This is the case, for example, when we want to accomplish treasure hunt or rendezvous at cost $\Theta(\sqrt{n})$ in an n-node graph. Only in special situations, if D is very large with respect to the gain, e.g., for an n-node graph with $\Theta(n^{3/2})$ edges in which the treasure is located at distance $\Theta(\sqrt{n})$, and we seek cost $\Theta(n^{3/2}/\log n)$, the gap is non-constant (it is $\Theta(\log n/\log \log n)$ in this case).

It should also be noted that, in the context of advice, treasure hunt is not only equivalent to rendezvous of two agents, as shown in Proposition 1, but also to rendezvous of many agents, which is often called *gathering*. This task consists in gathering several agents at the same node in the same round. In this case the cost should be defined as the number of edge traversals per agent, and the reduction given by Proposition 1 remains valid.

References

1. Abiteboul, S., Kaplan, H., Milo, T.: Compact labeling schemes for ancestor queries. In: Proc. 12th ACM-SIAM Symp. on Discrete Algorithms (SODA), pp. 547–556 (2001)
2. Alpern, S.: The rendezvous search problem. SIAM J. on Control and Optimization 33, 673–683 (1995)
3. Alpern, S.: Rendezvous search on labelled networks. Naval Reaserch Logistics 49, 256–274 (2002)
4. Alpern, S., Gal, S.: The theory of search games and rendezvous. Int. Series in Operations research and Management Science. Kluwer Academic Publisher (2002)
5. Anderson, E., Weber, R.: The rendezvous problem on discrete locations. Journal of Applied Probability 28, 839–851 (1990)
6. Anderson, E., Fekete, S.: Two-dimensional rendezvous search. Operations Research 49, 107–118 (2001)
7. Baeza-Yates, R.A., Culberson, J.C., Rawlins, G.J.E.: Searching in the plane. Information and Computation 106, 234–252 (1993)
8. Bampas, E., Czyzowicz, J., Gąsieniec, L., Ilcinkas, D., Labourel, A.: Almost optimal asynchronous rendezvous in infinite multidimensional grids. In: Lynch, N.A., Shvartsman, A.A. (eds.) DISC 2010. LNCS, vol. 6343, pp. 297–311. Springer, Heidelberg (2010)
9. Baston, V., Gal, S.: Rendezvous search when marks are left at the starting points. Naval Reaserch Logistics 48, 722–731 (2001)
10. Bose, P., De Carufel, J.-L., Durocher, S.: Revisiting the Problem of Searching on a Line. In: Bodlaender, H.L., Italiano, G.F. (eds.) ESA 2013. LNCS, vol. 8125, pp. 205–216. Springer, Heidelberg (2013)
11. Caminiti, S., Finocchi, I., Petreschi, R.: Engineering tree labeling schemes: A case study on least common ancestor. In: Halperin, D., Mehlhorn, K. (eds.) ESA 2008. LNCS, vol. 5193, pp. 234–245. Springer, Heidelberg (2008)
12. Cieliebak, M., Flocchini, P., Prencipe, G., Santoro, N.: Distributed computing by mobile robots: Gathering. SIAM J. Comput. 41, 829–879 (2012)
13. Cohen, R., Fraigniaud, P., Ilcinkas, D., Korman, A., Peleg, D.: Label-guided graph exploration by a finite automaton. ACM Transactions on Algorithms 4 (2008)
14. Czyzowicz, J., Kosowski, A., Pelc, A.: How to meet when you forget: Log-space rendezvous in arbitrary graphs. Distributed Computing 25, 165–178 (2012)
15. Czyzowicz, J., Labourel, A., Pelc, A.: How to meet asynchronously (almost) everywhere. ACM Transactions on Algorithms 8, article 37 (2012)
16. Dessmark, A., Fraigniaud, P., Kowalski, D., Pelc, A.: Deterministic rendezvous in graphs. Algorithmica 46, 69–96 (2006)
17. Dieudonné, Y., Pelc, A., Villain, V.: How to meet asynchronously at polynomial cost. In: Proc. 32nd ACM Symp. on Principles of Distributed Comp. (PODC), pp. 92–99 (2013)

18. Ellis, R.: Volume of an N-Simplex by Multiple Integration. Elemente der Mathematik 31, 57–59 (1976)
19. Emek, Y., Fraigniaud, P., Korman, A., Rosen, A.: Online computation with advice. Theoretical Computer Science 412, 2642–2656 (2011)
20. Fraigniaud, P., Ilcinkas, D., Pelc, A.: Communication algorithms with advice. Journal of Computer and System Sciences 76, 222–232 (2010)
21. Fraigniaud, P., Ilcinkas, D., Pelc, A.: Tree exploration with advice. Information and Computation 206, 1276–1287 (2008)
22. Fraigniaud, P., Korman, A., Lebhar, E.: Local MST computation with short advice. Theory of Computing Systems 47, 920–933 (2010)
23. Fusco, E., Pelc, A.: Trade-offs between the size of advice and broadcasting time in trees. Algorithmica 60, 719–734 (2011)
24. Fusco, E.G., Pelc, A., Petreschi, R.: Use knowledge to learn faster: Topology recognition with advice. In: Afek, Y. (ed.) DISC 2013. LNCS, vol. 8205, pp. 31–45. Springer, Heidelberg (2013)
25. Gavoille, C., Peleg, D., Pérennes, S., Raz, R.: Distance labeling in graphs. Journal of Algorithms 53, 85–112 (2004)
26. Hipke, C.A., Icking, C., Klein, R., Langetepe, E.: How to find a point on a line within a fixed distance. Disc. App. Math. 93, 67–73 (1999)
27. Katz, M., Katz, N., Korman, A., Peleg, D.: Labeling schemes for flow and connectivity. SIAM Journal of Computing 34, 23–40 (2004)
28. Korman, A., Kutten, S., Peleg, D.: Proof labeling schemes. Distributed Computing 22, 215–233 (2010)
29. Kowalski, D.R., Malinowski, A.: How to meet in anonymous network. In: Flocchini, P., Gąsieniec, L. (eds.) SIROCCO 2006. LNCS, vol. 4056, pp. 44–58. Springer, Heidelberg (2006)
30. Kranakis, E., Krizanc, D., Morin, P.: Randomized Rendez-Vous with Limited Memory. In: Laber, E.S., Bornstein, C., Nogueira, L.T., Faria, L. (eds.) LATIN 2008. LNCS, vol. 4957, pp. 605–616. Springer, Heidelberg (2008)
31. Lopez-Ortiz, A., Schuierer, S.: The ultimate strategy to search on m rays? Theoretical Computer Science 261, 267–295 (2001)
32. Miller, A., Pelc, A.: Fast rendezvous with advice. In: Proc. 10th Int. Symp. on Algorithms and Experiments for Sensor Systems, Wireless Networks and Distributed Robotics (ALGOSENSORS 2014) (2014), Full version at arxiv:1407.1428v1 [cs.DS]
33. Miller, A., Pelc, A.: Time versus cost tradeoffs for deterministic rendezvous in networks. In: Proc. 33rd Annual ACM Symposium on Principles of Distributed Computing (PODC 2014), pp. 282–290 (2014)
34. Nisse, N., Soguet, D.: Graph searching with advice. Theoretical Computer Science 410, 1307–1318 (2009)
35. Panaite, P., Pelc, A.: Exploring unknown undirected graphs. Journal of Algorithms 33, 281–295 (1999)
36. Pelc, A.: Deterministic rendezvous in networks: A comprehensive survey. Networks 59, 331–347 (2012)
37. Reingold, O.: Undirected connectivity in log-space. Journal of the ACM 55 (2008)
38. Ta-Shma, A., Zwick, U.: Deterministic rendezvous, treasure hunts and strongly universal exploration sequences. In: Proc. 18th ACM-SIAM Symposium on Discrete Algorithms (SODA 2007), pp. 599–608 (2007)
39. Thorup, M., Zwick, U.: Approximate distance oracles. J. ACM 52, 1–24 (2005)
40. Xin, Q.: Faster treasure hunt and better strongly universal exploration sequences. In: Tokuyama, T. (ed.) ISAAC 2007. LNCS, vol. 4835, pp. 549–560. Springer, Heidelberg (2007)

Maintaining a Spanning Forest in Highly Dynamic Networks: The Synchronous Case*

Matthieu Barjon, Arnaud Casteigts, Serge Chaumette,
Colette Johnen, and Yessin M. Neggaz

LaBRI, University of Bordeaux

Abstract. Highly dynamic networks are characterized by frequent changes in the availability of communication links. Many of these networks are in general partitioned into several components that keep splitting and merging continuously and unpredictably. We present an algorithm that strives to maintain a forest of spanning trees in such networks, without any kind of assumption on the rate of changes. Our algorithm is the adaptation of a coarse-grain interaction algorithm (Casteigts et al., 2013) to the synchronous message passing model (for dynamic networks). While the high-level principles of the coarse-grain variant are preserved, the new algorithm turns out to be significantly more complex. In particular, it involves a new technique that consists of maintaining a distributed permutation of the set of all nodes IDs throughout the execution. The algorithm also inherits the properties of its original variant: It relies on purely localized decisions, for which no global information is ever collected at the nodes, and yet it maintains a number of critical properties whatever the frequency and scale of the changes. In particular, the network remains always covered by a spanning forest in which 1) no cycle can ever appear, 2) every node belongs to a tree, and 3) after an arbitrary number of edge disappearance, all maximal subtrees immediately restore exactly one token (at their root). These properties are ensured whatever the dynamics, even if it keeps going for an arbitrary long period of time. Optimality is not the focus here, however the number of tree per components – the metric of interest here – eventually converges to one if the network stops changing (which is never expected to happen, though). The algorithm correctness is proven and its behavior is tested through experimentation.

1 Introduction

The current development of mobile and wireless technologies enables direct *ad hoc* communication between various kinds of mobile entities, such as vehicles, smartphones, terrestrian robots, flying robots, or satellites. In all these contexts, the set of communication links depends on distances between entities, thus the network topology changes continuously as the entities move. Not only changes are frequent,

---

* An extended version of this paper is available online [7].

M.K. Aguilera et al. (Eds.): OPODIS 2014, LNCS 8878, pp. 277–292, 2014.
© Springer International Publishing Switzerland 2014

but in general they even make the network partitionned. Clearly, the usual assumption of connectivity does not hold here, although another form of connectivity is often available over time and space (temporal connectivity). Also, the classical view of a network whose dynamics corresponds to *failures* is no longer suitable in these scenarios, where dynamics is the norm rather than the exception.

This induces a shift in paradigm that strongly impacts algorithms. In fact, it even impacts the problems themselves. What does it mean, for instance, to elect a leader in such a network? Is the objective to distinguish a unique global leader, whose leadership then takes place over time and space, or is it to *maintain* a leader in each connected component, so that the decisions concerning each component are taken quickly and locally. The same remark holds for spanning trees. Should an algorithm construct a unique, global tree whose logical edges survive intermittence, or should it build and maintain a *forest* of trees that strive to cover collectively all components in each instant? Both viewpoints make sense, and so far, were little studied in distributed computing (see e.g. [4,12] for temporal trees, [3,11] for maintained trees).

We focus on the second interpretation, which reflects a variety of scenarios where the expected output of the algorithm should relate to the *immediate* configuration (*e.g.* direct social networking, swarming of flying robots, vehicles platooning on the road). A particular feature of this type of algorithms is that they never terminate. More significantly, in highly dynamic networks, they are not even expected to stabilize to an optimal state (here, a single tree per component), unless the changes stop, which never happens. This precludes, in particular, all approaches whereby the computation of a new solution requires the previous computation to have completed.

This paper is an attempt to understand what can still be computed (and guaranteed) when no assumptions are made on the network dynamics: neither on the rate of change, nor on their simultaneity, nor on global connectivity. In other words, the topology is controlled by an almighty adversary. In this seemingly chaotic context, we present an algorithm that strives to maintain as few trees per components as possible, while always guaranteeing some properties.

1.1   Related Work

Several works have addressed the spanning tree problem in dynamic networks, with different goals and assumptions. Burman and Kutten [9] and Kravchik and Kutten [15] consider a self-stabilizing approach where the legal state corresponds to having a (single) minimum spanning tree and the faults are topological changes. The strategy consists in recomputing the entire tree whenever changes occur. This general approach, sometimes called the "blast away" approach, is meaningful if stable periods of time exist, which is not assumed here.

Many spanning tree algorithms rely on random walks for their elegance and simplicity, as well as for the inherent localized paradigm they offer. In particular, approaches that involve multiple coalescing random walks allow for uniform initialization (each node starts with the same state) and topology independence (same strategy whatever the graph). Pionneering studies involving such processes include Bar-Ilan and Zernik [6] (for the problem of election and spanning tree), Israeli and Jalfon [14] (mutual exclusion), and Chapter 14 of Aldous and Fill [2] (for general analysis).

The principle of using coalescing random walks to build spanning trees in mildly dynamic networks was used by Baala et al. [1] and Abbas et al. [5], where tokens are annexing territories gradually by capturing each other. Regarding dynamicity, both algorithms require the nodes to know an upper bound on the cover time of the random walk, in order to regenerate a token if they are not visited during a long-enough period of time. Besides the strength of this assumption (akin to knowing the number of nodes n, or the size of components in our case), the efficiency of the timeout approach decreases dramatically with the rate of topological changes. In particular, if they are more frequent than the cover time (itself in $O(n^3)$), then the tree is constantly fragmented into "dead" pieces that lack a root, and thus a leader.

Another algorithm based on random walks is proposed by Bernard et al. [8]. Here, the tree is constantly redefined as the token moves (in a way that reminds the snake game). Since the token moves only over present edges, those edges that have disappeared are naturally cleaned out of the tree as the walk proceeds. Hence, the algorithm can tolerate failure of the tree edges. However it still suffers from detecting the disappearance of tokens using timeouts based on the cover time, which as we have seen, suits only slow dynamics.

A recent work by Awerbuch et al. [3] addresses the maintenance of *minimum* spanning trees in dynamic networks. The paper shows that a solution to the problem can be updated after a topological change using $O(n)$ messages (and same time), while the $O(m)$ messages of the "blast away" approach was thought to be optimal. (This demonstrates, incidentally, the revelance of *updating* a solution rather than recomputing it from scratch in the case of minimum spanning trees.) The algorithm has good properties for highly dynamic networks. For instance, it considers as natural the fact that components may split or merge perpetually. Furthermore, it tolerates new topological events while an ongoing update operation is executing. In this case, update operations are enqueued and consistently executed one after the other. While this mechanism allows for an arbitrary number of topological events *at times*, it still requires that such burst of changes are only episodical and that the network remains eventually stable for (at least) a linear amount of time in the number of nodes, in order for the update operations to complete and thus the logical tree to be consistent with physical reality.

All the aforementioned algorithms either assume that *global update* operations (e.g. wave mechanisms) can be performed contemporaneously, or at least eventually, or that some node can collect *global information* about the tree structure. As far as dynamics is concerned, this forbids arbitrary and ever going changes to occur in the network.

1.2   The Spanning Forest Principle

A purely localized scheme was proposed by Casteigts et al. [11] for the maintenance of a (non-minimum) spanning forest in unrestricted dynamic networks, using a coarse grain interaction model inspired from graph relabeling systems [17]. It can be described informally as follows. Initially every node hosts a token and is the *root* of its own individual tree. Whenever two roots arrive at the endpoints of a same edge (see merging rule on Figure 1), one of them destroys its tokens and selects the other as parent (*i.e.* the trees are merged). The rest of the time, each token executes a random walk within its own tree in the search for other merging opportunities (circulation rule). Tree relations are flipped accordingly. The fact that the random walk is *confined* to the underlying tree is crucial and different from all algorithms discussed above, in which they were free to roam everywhere without restriction. This simple feature induces very attractive properties for highly dynamic networks. In particular, whenever an edge of the tree disappears, the child side of that edge knows instantly that no token remains on its whole subtree. It can thus regenerate a token (*i.e.* become root) *instantly*, without global concertation nor further information collection. As a result, both merging and splitting of trees are managed in a purely localized fashion.

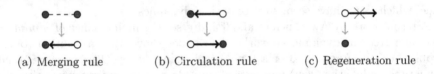

(a) Merging rule   (b) Circulation rule   (c) Regeneration rule

Fig. 1. Spanning forest principle (high-level representation). *Black nodes are those having a token. Black directed edges denote child-to-parent relationships. Gray vertical arrows represent transitions.*

At an abstract graph level, this very simple scheme guarantees that the network remains covered by a spanning forest at any time, in which 1) no cycle can ever appear, 2) maximal subtrees are always directed rooted trees (with a token at the root), and 3) every node always belongs to such a tree, whatever the chaos of topological changes. On the other hand, it is not expected to reach an optimal state where a single tree covers each connected component. Even if the network were to stabilize, convergence to the optimum (though easy to be made certain) would not be expected to occur fast. Whether this general principle could be implemented in a message passing model remained an open question.

1.3   Our Contribution

This paper provides an implementation of the spanning forest principle in the synchronous message-passing model. Due to the loss of atomicity and exclusivity in the interaction, the algorithm turns out to be much more sophisticated than its original counterpart. While still reflecting the very same high-level principle, it faces new problems that require conceptual differences. In particular, the original model prevented a node from both selecting a parent and being selected as parent simultaneously, making it easier to avoid cycle creations. One of the ingredients in the new algorithm to circumvent this problem is an original technique (which we refer to as the *unique score* technique) that consists of maintaining, network-wide, a set of score variables that always remain a permutation of the set of nodes IDs. This mechanism allows us to break symmetry and avoid the formation of cycles in a context where IDs alone could not. The paper is organized as follows. In Section 2, we present the model and notations that we use throughout the paper. Then Section 3 presents the algorithm, whose correctness analysis is outlined in Section 4. (The detailed proofs are in the online version of this paper [7].) Section 5 finally presents some experimental results that validate our algorithm in real context.

2   Model and Notations

The network is represented by an untimed evolving graph [13] $\mathcal{G} = (G_1, G_2, \dots)$, such that $G_i = (V, E_i)$, where V is a static set of vertices and E_i is a dynamically changing set of undirected edges. Following Kuhn et al. [16], we consider a synchronous (thus rounded) computational model, where in each round i, the adversary chooses the set of edges E_i that are present. In our case, this set is arbitrary (*i.e.* the adversary is unrestricted). At the beginning of each round, each node sends a message that it has prepared at the end of the previous round. This message is sent to all its neighbors in E_i, although the list of these neighbors is not know by the node. Then it receives all messages sent by its neighbors (in the same round), and finally computes its new state and the next message. Hence, each round corresponds to three phases (send, receive, compute), which corresponds to a rotation of the original model of [16] where the phases are (compute, send, receive). This adaptation is not necessary, but it allows us to formulate correctness of our algorithm in terms of the state of the nodes *after* each round rather than in the middle of rounds.

We assume that the nodes have a unique identifier taken from a totally ordered set, that is, for any two nodes u and v, it either holds that $ID(u) > ID(v)$ or $ID(u) < ID(v)$. A node can specify what neighbor its message is intended to (although all neighbors will receive it) by setting the target field of that message. Symmetrically, the ID of the emitter of a message can be read in the sender field of that message. Since the edges are undirected, if u receives a message from v at round i, then v also receives a message from u at that round. We call this property the *reciprocity principle* and it is an important ingredient for the correctness of our algorithm.

Using synchronous rounds allows us to represent the progress of the execution as a sequence of *configurations* $(C_0, C_1, C_2, ..., C_i)$, where each C_i corresponds to the state of the system *after* round i (except for C_0, the initial state). Each *configuration* consists of the union of all nodes variables, defined next.

3   The Spanning Forest Algorithm

3.1   State Variables

Besides the ID variable, which we assume is externally initialized, each node has a set of variable that reflects its situation in the tree: status accounts for the possession of a token (T if it has a token, N if it does not); parent contains the ID of this node's parent (⊥ if it has none); children contains the set of this node's children (∅ if it has none). Observe that both variables status and parent are somewhat redundant, since in the spanning forest principle (see Section 1.2) the possession of a token is equivalent to being a root. Our algorithm enforces this equivalence, yet, keeping both variables separated simplifies the description of the algorithm and our ability to think of it intuitively. Variable neighbors contains the set of nodes from which a message was received in the last reception. These neighbors may or may not belong to the same tree as the current node. Variable contender contains the ID of a neighbor that the current node considers selecting as parent in the next round (or ⊥ if there is no such node). Finally, the variable score is the main ingredient of our cycle-avoidance mechanism, whose role is described below.

Initial values: All the nodes are uniformly initialized. They are initially the root of their own individual tree (*i.e.* status $= T$, parent $= \perp$, and children $= \emptyset$). They know none of their neighbors (neighbors $= \emptyset$), have no contenders (contender $= \perp$), and their score is set to their own ID.

3.2   Structure of a Message (and Associated Variables)

Messages are composed of a number of fields: sender is the ID of the sending node; senderStatus its status (either T or N); and score its score when the message was prepared. The field action is one of $\{FLIP, SELECT, HELLO\}$. Informally, $SELECT$ messages are sent by a root node to another root node to signify that it "adopts" it as a parent (merging operation); $FLIP$ messages are sent by a root node to circulate the token to one of its children (circulation operation); $HELLO$ messages are sent by a node by default, when none of the other messages are sent, to make its presence and status known by its neighbors. Finally, target is the ID of the neighbor to which a FLIP or a SELECT message are intended (⊥ for HELLO messages).

Received messages are stored in a variable mailbox, which is a map collection whose *keys* are the senders ID (*i.e.*, a message whose sender ID is u can be accessed as mailbox[u]). In each round, the algorithm makes use of a RECEIVE() function that clears the mailbox and fill it with all the messages received in that round (one for each physical neighbor). A node can thus update the set of its neighbors by fetching the *keys* of its mailbox. Similarly, it can eliminate from its list of children those nodes which are no more neighbor.

As mentioned above, every node prepares at the end of a round the message to be sent at the beginning of the next round. This message is stored in a variable outMessage. We allow the short hand $m \leftarrow (a, b, c, d, e)$ to define a new message m whose emitter is node a (with status b and score e); target is node d; and action is c.

Initial values: The mailbox is initially empty (mailbox $= \emptyset$) and outMessage is initialized to $(ID, T, HELLO, \bot, ID)$.

3.3 Informal Description of the Algorithm

The algorithm implements the general scheme presented in Section 1.2. In this Section we explain how each of the three core operations (*merging*, *circulation*, *regeneration*) is implemented. Then we discuss the specificities of the merging operation in more detail and the problems that arise due to its entanglement with the circulation operation, a fact due to the loss of atomicity in the message passing model. The resulting solution is substantially more sophisticated than its original scheme, and yet it faithfully reflects the same high-level principle. Let us start with some generalities. In each round, each node broadcasts to its neighbors a message containing, among others, its status (T or N) and an action (SELECT, FLIP, or HELLO). Whether or not the message is intended to a specific *target* (which is the case for SELECT and FLIP messages), all the nodes who receive it can possibly use this information for their own decisions. More generally, based on the received information and the local state, each node computes at the end of the round its new status and the local structure of its tree (variables children and parent), then it prepares the next message to be sent. We now describe the three operations. Throughout the explanations, the reader is invited to refer to Figure 2, where an example of execution involving all of them is shown. All details are also given in the listings of Algorithm 1 and 2.

Merging: If a root (*i.e.* a node having a token), say v, detects the existence of a neighbor root with higher score than its own, then it considers that node as a possible contender, *i.e.* as a node that it might select as a parent in the next round. If several such roots exist, then the one with highest score, say u, is chosen. At the beginning of the next round, v sends a *SELECT* message to u to inform it that it is its new parent. Two cases are possible: either the considered edge is still present in that round, or it disappeared in-between both rounds. If it is still present, then u receives the message and adds v to its children list, among others (Line 16). As for v, it sets its parent variable to u and its status to N (Lines 8 and 9). If the edge disappeared, then u does not receive the message, which is lost. However, due to the reciprocity of message exchange, v does not receive a message from u either and thus simply does not executes the corresponding changes. By the end of the round, either the trees are properly merged, or they are properly separated.

Circulation: If a root v does not detect another root with higher score, then it selects one of its children at random, if it has any (see Line 27), otherwise it

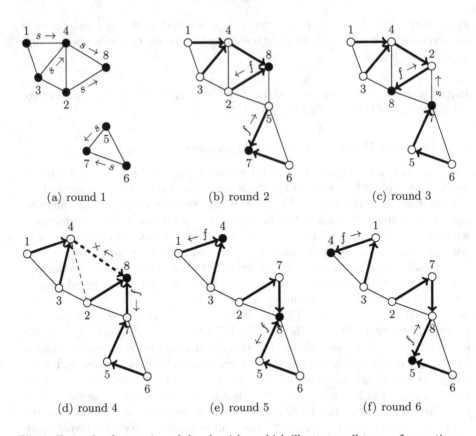

Fig. 2. Example of execution of the algorithm which illustrates all types of operations: parent selection ($s \rightarrow$), token circulation ($f \rightarrow$), and tree disconnection ($\times \leftarrow$). *The first two symbols represent FLIP or SELECT messages to be sent in the next round. Black (resp. white) nodes are those (not) having a token at the beginning of the round. Tree edges are represented by bold directed edges. Dash edges have just disappeared.*

simply remains root. Randomness is not a strict requirement of our algorithm and replacing it with any deterministic strategy would not affect correctness of the algorithm. Once the child is chosen, say u, the root prepares a FLIP message intended to u, and sends it at the beginning of the next round. Two cases are again possible, whether or not the edge $\{u, v\}$ is still present in that round. If it is still present, then u receives the message, it updates its status and adds v to its children list, among others (Lines 15 and Line 16). As for v, it sets its **parent** variable to u and its **status** to N (Lines 8 and 9). If the edge disappeared, then v can detect it as before simply does not executes the corresponding changes. Node u, on the other hand, detects that the edge leading to its current parent disappeared, thus it regenerates a token (discussed next). Notice that in the absence of a merging opportunity, a node receiving the token in round i will immediately prepare a FLIP message to circulate the token in the next round.

Unless the tree is composed of a single node, the tokens are thus moved in each round. In order for them to remain detectable in this case, the status announced in *FLIP* messages is T (whereas it is N for *SELECT* messages).

Regeneration: The first thing a non-root node does after receiving the messages of the current round is to check whether the edge leading to its current parent is still present. If the edge disappeared, then the node regenerates a root directly (Line 7). A nice property of the spanning forest principle is that this cannot happen twice in the same tree. And if a tree is broken into several pieces simultaneously, then each of the resulting subtree will have exactly one node performing this operation.

The Unique Score Technique: Unlike the high-level graph model from [11], in which the merging operation involved two nodes in an *exclusive* way, the non-atomic nature of message passing allows for a *chain* of selection that may involve an arbitrary long sequence of nodes (e.g. a selects b, b selects c, and so on). This has both advantages and drawbacks. On the good side, it makes the initial merging process very fast (see rounds 1 and 2 in Figure 2 to get an example). On the bad side, it is the reason why scores need to be introduced to avoid cycles. Indeed, relying only on a mere comparison of ID to avoid cycles is not sufficient. Consider a chain of selection in round i that ends up at some root node u. Nothing prevents u to have passed the token to a lower-ID child, say v, in the previous round $i - 1$ (that same round when u's status T was overheard by the next-to-last root in the chain). Now, nothing again prevents v to have selected one of the nodes in the selection chain in round i, thereby creating a cycle. The score mechanism prevents such a situation by enforcing that after each FLIP, the new root has a larger score than its predecessor (see Lines 9 and 13 in Algorithm 2). The score mechanism also guarantees that the current set of scores (network-wide) is always a permutation of the initial set of scores. Hence, scores are always unique. All of these elements are crucial ingredients in the proofs of correctness of Section 4.

A Note about Convergence: Each token performs a random walk in its underlying tree. Hence, unless some of the trees are bipartite, the configuration will eventually (and with high probability) stabilize into a single tree per connected component if the network stops changing. Although convergence is not the main focus here, we believe that pathetic scenarios where some trees are bipartite can easily be avoided, by making the tokens stop for a random additional round at the nodes (*lazy* walk). This way, the symmetry of bipartiteness is eventually broken *w.h.p.*

```
1  repeat
2  |   SEND(outMessage);
3  |   mailbox ← RECEIVE();   // Received messages, indexed by sender ID
4  |   neighbors ← mailbox.keys();              // All the senders IDs
5  |   children ← children ∩ neighbors

   |   // Regenerates a token if parent link is lost
6  |   if status=N ∧ parent ∉ neighbors then
7  |   |   BECOME_ROOT();
   |   // Checks if the outgoing FLIP or SELECT (if any) was successful
8  |   if outMessage.action ∈ {FLIP,SELECT} ∧ outMessage.target ∈
   |   neighbors then
9  |   |   ADOPT_PARENT(outMessage)
   |   // Processes the received messages
10 |   contender ← ⊥;
11 |   contenderScore ← 0;
12 |   forall message ∈ mailbox do
13 |   |   if message.target = ID then
14 |   |   |   if message.action = FLIP then
15 |   |   |   |   BECOME_ROOT();
16 |   |   |   ADOPT_CHILD(message);     // called for both FLIP or SELECT
17 |   |   else
18 |   |   |   if message.status = T ∧ message.score > contenderScore then
19 |   |   |   |   contender ← message.ID;
20 |   |   |   |   contenderScore ← message.score;
   |   // Prepares the message to be sent
21 |   outMessage ← ⊥
22 |   if status = T then
23 |   |   if contenderScore > score then
24 |   |   |   PREPARE_MESSAGE(SELECT, contender);
25 |   |   else
26 |   |   |   if children ≠ ∅ then
27 |   |   |   |   PREPARE_MESSAGE(FLIP, random(children));
28 |   if outMessage = ⊥ then
29 |   |   PREPARE_MESSAGE(HELLO, ⊥);
30 ;
```

Algorithm 1. Main Algorithm

```
 1 procedure BECOME_ROOT
 2 │   status ← T;
 3 │   parent ← ⊥;

 4 procedure ADOPT_PARENT(outMessage)
 5 │   status ← N;
 6 │   parent ← outMessage.target;
 7 │   if outMessage.action = FLIP then
 8 │   │   children ← children∖parent;
 9 │   │   score ← min(score, mailbox[parent].score);

10 procedure ADOPT_CHILD(message)
11 │   children.add(message.ID);
12 │   if message.action = FLIP then
13 │   │   score ← max(score, message.score);

14 procedure PREPARE_MESSAGE(action, target)
15 │   switch action do
16 │   │   case SELECT
17 │   │   │   outMessage ← (ID, N, SELECT, target, score);
18 │   │   case FLIP
19 │   │   │   outMessage ← (ID, T, FLIP, target, score);
20 │   │   case HELLO
21 │   │   │   outMessage ← (ID, status, ⊥, ⊥, score);
```

Algorithm 2. Functions called in Algorithm 1.

4 Outline of the Correctness Analysis

This section summarizes the correctness analysis of our algorithm, whose detail can be found in the long version of the paper [7]. We first define a handful of instrumental concepts that help minimize the number of properties to be proven. Then, as we start formulating the key properties to be proved, we adopt concise notations regarding the state of the system. Precisely, we denote by $(i^-)u.varname$ (resp. $(i^+)u.varname$) the value of variable $varname$ at node u before (resp. after) round i. Notice that for any node u, round i, and variable $varname$, we have $(i^+)u.varname = ((i+1)^-)u.varname$. We use whichever notation is the most convenient in the given context.

4.1 Helping Definitions

These definitions are not specific to our algorithm, they are general graph concepts that simplify the subsequent proofs.

Definition 1 (Pseudotree and pseudoforest). *A directed graph whose vertices have outdegree at most 1 is a* pseudoforest. *A vertex whose outdegree is 0 is called a* root. *The weakly connected components of a pseudoforest are called* pseudotrees.

Lemma 1. *A pseudotree has at most one root.*

Proof. By definition, a pseudotree $\mathcal{T} = (V_{\mathcal{T}}, E_{\mathcal{T}})$ is connected, thus $|E_{\mathcal{T}}| \geq |V_{\mathcal{T}}| - 1$. If $\mathcal{T}$ has several roots, then at least two nodes in $V_{\mathcal{T}}$ have no outgoing edge. Since the others have at most one, we must have $|E_{\mathcal{T}}| \leq |V_{\mathcal{T}}| - 2$, which is a contradiction.   □

Lemma 2. *If a pseudotree $\mathcal{T}$ contains a root r, then it has no cycle.*

Proof. Let $V_1 \subset \mathcal{T}$ be the set of nodes at distance 1 from $V_0 = \{r\}$. Since r has outdegree 0, there is an edge from each node in V_1 to r. Since $\mathcal{T}$ is a pseudotree, these nodes have no other outgoing edge than those ending up in V_0. The same argument can be applied inductively, all nodes at distance i having no other outgoing edges than those ending up in V_{i-1}.   □

Definition 2 (Correct tree and correct forest). *At the light of Lemma 1 and 2, we define a* correct tree *(or simply a* tree*) as a pseudotree in which a root can be found. We naturally define a* correct forest *(or simply a* forest*) as a pseudoforest whose pseudotrees are trees.*

Finally, because forests are considered in a spanning context, we say that a pseudoforest $\mathcal{F}$ is a correct forest *on graph G* iff $\mathcal{F}$ is a correct forest *and $\mathcal{F}$ is a subgraph of G.* Defining correct trees as pseudotrees in which a root can be found is the key. When the moment arrives, this will allow us to reduce the correctness of our algorithm to the presence of a root in each pseudotree.

4.2   Consistency

Forest Consistency: At the end of a round, the state of an edge (whether it belongs to a tree, and if so, in what direction) must be consistently decided at both endpoints:

Definition 3 (forest consistency). *The configuration C_i is forest consistent if and only if for all nodes u, $(i^+)u.parent = v \Leftrightarrow u \in (i^+)v.children$.*

The proof of forest consistency is inductively established by Theorem 1, based on consistency of the initial configuration (Lemma 3) and the maintenance the consistency over the rounds (Lemma 18). Forest consistency allows us to reduce the output of interest of the algorithm after each round i to the mere **parent** variable.

Graph Consistency: At the end of round i, the values of all **parent** variables should be consistent with the underlying graph G_i.

Definition 4 (graph consistency). *The configuration C_i is graph consistent if and only if for all nodes u, $(i^+)u.parent = v \Rightarrow \{u, v\} \in E_i$.*

This property is established by Corollary 1. Graph consistency allows us to say that the output of the algorithm forms a pseudoforest on G_i.

Definition 5 (Resulting forest). *Given a round $i \geq 1$, occurring on graph G_i, the graph $\mathcal{F}_i = (V, E_{\mathcal{F}_i})$ such that $E_{\mathcal{F}_i} = \{(u, v) : \{u, v\} \in E_i, (i^+)u.parent = v\}$ is called the* pseudoforest *resulting from round i.*

State Consistency: As explained in Section 3.1, the variables `parent` and `status` are somewhat redundant, since the possession of a token is synonymous with being a root. The equivalence between both variables after each round is established in Lemma 4. The main advantage of this equivalence is that it allows us to formulate and prove a large number of lemmas based on whichever of the two variables is the most convenient (and intuitive) for the considered property.

4.3   Correctness of the Forest

In this section, we prove that the resulting forest is always correct (Definition 2). To achieve that goal, we first define a validity criterion at the node level, which recursively ensures the correctness of the pseudotree this node belongs to thanks to Definition 2 (*i.e.* the existence of a root implies correctness).

Definition 6. *A node u is said to be valid at the beginning of round i if either $(i^-)u.status = T$ or $(i^-)u.parent$ is valid.*

The correctness of the whole forest can thus be established through showing that, first, it is initially correct (Lemma 3) and, second, if it is correct after round i, then it is correct after round $i + 1$ (Theorem 2). The latter is difficult to prove, and it involves a number of intermediate steps that correspond to a case analysis based on every action a node can perform (sending FLIP messages, SELECT messages, etc.).

We first prove that a node u that sends a successful FLIP to v in a round, is valid at the end of that round (lemma 23) because at the end of that round v is a root. The proof relies on the fact that during a given round, a node cannot receive a FLIP and send a SELECT or a FLIP (lemma 20).

We then prove some necessary properties on the `score` variable at each node. For instance, a node changes its score at most once during a round (Lemma 25 and 26). Also, the set of all scores are a permutation of the node identifiers after each round (Lemma 27).

Then we prove that a node that sends a successful SELECT in a round i, is valid at the end of that round (Lemma 36). This part is the most technical and is the one that proves that chains of selection can not create cycles thanks to the property that score variables remain a permutation of all nodes IDs.

Finally, we prove that all roots at the beginning of a round are still valid at the end of the round (lemma 37). Therefore, if all nodes are valid at the beginning of round, then they are also valid at the end of the round (theorem 2). Since they are initially valid (Lemma 3), we conclude by induction on the number of rounds.

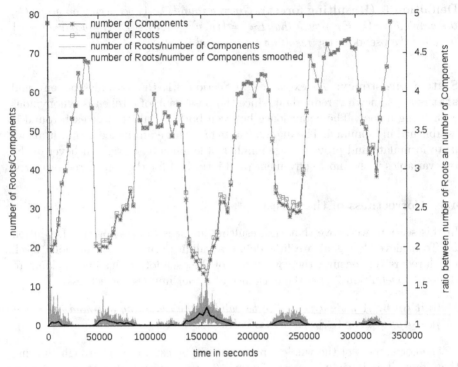

Fig. 3. Number of roots per connected components, assuming 10 rounds per second

5   Simulation on Real World Traces (Infocomm 2006)

We verified the applicability of our algorithm to real world situations. The algorithm was implemented in the JBotSim simulator [10] and tested upon the Infocomm06 dataset [18]. This dataset is a record of the possible interactions between people during the Infocomm'06 conference. The resulting graph has the following characteristics: the number of nodes is 78 and the average node degree is 1.3. It should also be noted that an edge can appear at any time but the fact that it is still present is thereafter only tested every 120 seconds; this means that the presence time of an edge is a multiple of 120 seconds. We assumed that 10 rounds can be performed per seconds (Figure 3), which seems a reasonable assumption. The results show the average number of trees per connected component, averaged over 100 runs.

These results show that the number of trees per connected component, averaged over time, is very close to 1 (about 1.027). Furthermore, the algorithm achieves an optimal configuration of a single spanning tree per connected component about 47% of the time, which is encouraging. These results also illustrate, incidentally, that the algorithm works correctly in such a scenario.

Acknowledgment. This work was partially supported by ANR projects DISPLEX-ITY and ASTRID-Maturation, as well as DGA scholarship No 2013 60 0074.

References

1. Abbas, S., Mosbah, M., Zemmari, A.: Distributed computation of a spanning tree in a dynamic graph by mobile agents. In: Proc. of IEEE Int. Conference on Engineering of Intelligent Systems (ICEIS), pp. 1–6 (2006)
2. Aldous, D., Fill, J.: Reversible markov chains and random walks on graphs (2002)
3. Awerbuch, B., Cidon, I., Kutten, S.: Optimal maintenance of a spanning tree. J. ACM 55(4), 18:1–18:45 (2008)
4. Awerbuch, B., Even, S.: Efficient and reliable broadcast is achievable in an eventually connected network. In: Proceedings of the Third Annual ACM Symposium on Principles of Distributed Computing, pp. 278–281. ACM (1984)
5. Baala, H., Flauzac, O., Gaber, J., Bui, M., El-Ghazawi, T.: A self-stabilizing distributed algorithm for spanning tree construction in wireless ad hoc networks. Journal of Parallel and Distributed Computing 63, 97–104 (2003)
6. Bar-Ilan, J., Zernik, D.: Random leaders and random spanning trees. In: Bermond, J.-C., Raynal, M. (eds.) WDAG 1989. LNCS, vol. 392, pp. 1–12. Springer, Heidelberg (1989)
7. Barjon, M., Casteigts, A., Chaumette, S., Johnen, C., Neggaz, Y.M.: Maintaining a spanning forest in highly dynamic networks: The synchronous case. CoRR, abs/1410.4373 (2014)
8. Bernard, T., Bui, A., Sohier, D.: Universal adaptive self-stabilizing traversal scheme: Random walk and reloading wave. J. Parallel Distrib. Comput. 73(2), 137–149 (2013)
9. Burman, J., Kutten, S.: Time optimal asynchronous self-stabilizing spanning tree. In: Pelc, A. (ed.) DISC 2007. LNCS, vol. 4731, pp. 92–107. Springer, Heidelberg (2007)
10. Casteigts, A.: The JBotSim library. CoRR, abs/1001.1435 (2013), See also the project website at http://jbotsim.sourceforge.net
11. Casteigts, A., Chaumette, S., Guinand, F., Pigné, Y.: Distributed maintenance of anytime available spanning trees in dynamic networks. In: Cichoń, J., Gębala, M., Klonowski, M. (eds.) ADHOC-NOW 2013. LNCS, vol. 7960, pp. 99–110. Springer, Heidelberg (2013)
12. Casteigts, A., Flocchini, P., Mans, B., Santoro, N.: Shortest, fastest, and foremost broadcast in dynamic networks. CoRR, abs/1210.3277 (2014)
13. Ferreira, A.: On models and algorithms for dynamic communication networks: The case for evolving graphs. In: Proc. ALGOTEL (2002)
14. Israeli, A., Jalfon, M.: Token management schemes and random walks yield self-stabilizing mutual exclusion. In: Proceedings of the Ninth Annual ACM symposium on Principles of Distributed Computing, pp. 119–131. ACM (1990)
15. Kravchik, A., Kutten, S.: Time optimal synchronous self stabilizing spanning tree. In: Afek, Y. (ed.) DISC 2013. LNCS, vol. 8205, pp. 91–105. Springer, Heidelberg (2013)

16. Kuhn, F., Lynch, N., Oshman, R.: Distributed computation in dynamic networks. In: Proceedings of the 42nd ACM symposium on Theory of computing (STOC), pp. 513–522. ACM (2010)

17. Litovsky, I., Metivier, Y., Sopena, E.: Graph relabelling systems and distributed algorithms. In: Handbook of Graph Grammars and Computing by Graph Transformation. Citeseer (2001)

18. Scott, J., Gass, R., Crowcroft, J., Hui, P., Diot, C., Chaintreau, A.: Crawdad trace cambridge/haggle/imote/infocom (January 31, 2006), http://crawdad.cs.dartmouth.edu/cambridge/haggle/imote/infocom

A Communication-Efficient Self-stabilizing Algorithm for Breadth-First Search Trees

Ajoy K. Datta[1], Lawrence L. Larmore[1], and Toshimitsu Masuzawa[2]

[1] Department of Computer Science, University of Nevada, Las Vegas, USA
{ajoy.datta,lawrence.larmore}@unlv.edu
[2] Graduate School of Information Science and Technology, Osaka University, Japan
masuzawa@ist.osaka-u.ac.jp

Abstract. A self-stabilizing algorithm converges to its designated behavior from an arbitrary initial configuration. It is standard to assume that each process maintains communication with all its neighbors. We consider the problem of self-stabilizing construction of a breadth first search (BFS) tree in a connected network of processes, and consider algorithms which are not given the size of the network, nor even an upper bound on that size. It is known that an algorithm that constructs a BFS tree must allow communication across every edge, but not necessarily in both directions. If m is the number of undirected edges, and hence the number of directed edges is $2m$, then every self-stabilizing BFS tree algorithm must allow perpetual communication across at least m directed edges. We present an algorithm with reduced communication for the BFS tree problem in a network with unique identifiers and a designated root. In this algorithm, communication across all channels is permitted during a finite prefix of a computation, but there is a reduced set of directed edges across which communication is allowed forever. After a finite prefix, the algorithm uses only $m + n - 1$ directed edges for communication, where n is the number of processes in the network and m is the number of edges.

1 Introduction

A *self-stabilizing distributed system* [8] can eventually recover its intended behavior without external intervention even when started from an arbitrary configuration (or global state). Thus, a self-stabilizing distributed system attains high tolerance to transient faults and high adaptability to dynamic topology changes of networks. Self-stabilization is usually implemented by combining a *transient-fault detection* mechanism and a *fault correction* mechanism. The former guarantees that an alarm is raised at a process if the distributed system is at an illegitimate configuration. The latter is initiated by the alarmed process to bring the system to a legitimate configuration.

The fault correction mechanism can cost dearly in time and/or communication, since it must make the distributed system recover from *any possible* configuration.

M.K. Aguilera et al. (Eds.): OPODIS 2014, LNCS 8878, pp. 293–306, 2014.

Thus, the prime concern in efficiency of self-stabilization considered so far is efficiency of the fault correction mechanism, for example, the maximum time (*stabilization time*) required to recover legitimacy from any configuration, and the maximum number of bits exchanged until the system reaches a legitimate configuration from any configuration.

The cost of the transient-fault detection mechanism is another crucial difference in cost between self-stabilizing protocols and non-self-stabilizing classical protocols, since the latter (starting from predetermined initial configurations) need no transient-fault detection. In a self-stabilizing protocol, the transient-fault detection mechanism requires each process to keep communicating forever with some of its neighboring processes to check consistency; otherwise, a process may initially start and remain forever at a state inconsistent with those of its neighboring processes, and the protocol cannot recover legitimacy.

Cost of the transient-fault detection mechanism has not caused much concern so far. Most self-stabilizing protocols proposed up to now require every process to communicate with all of its neighboring processes repeatedly and forever to check consistency among them. This leads to a high communication load in networks and makes self-stabilizing protocols unacceptable in some real situations.

It is worth noting that efficiency of transient-fault detection mechanism is one of the most important concerns of self-stabilization. In self-stabilizing distributed systems, the transient-fault detection mechanism operates almost all the time, but the fault correction mechanism operates only when necessary. Thus, efficiency of the transient-fault detection mechanism normally dominates the efficiency of self-stabilizing distributed systems.

Our Contribution. In this paper, we present a silent and self-stabilizing communication-efficient algorithm, ROOT-UID, for constructing a *breadth-first search* (BFS) tree in a connected network G with the UID property, *i.e.*, where processes have unique IDs, and where there is a designated root process, *Root*. Throughout this work, we let n be the number of processes in our network, and m the number of edges. ROOT-UID is $\Diamond\text{-}(m + n - 1)$-communication-efficient, meaning that, eventually, communication is needed across only $m + n - 1$ of the $2m$ directed edges of the network. More specifically, there is 2-way communication forever across each of the BFS tree edges, but communication in only one direction across each cross edge. ROOT-UID uses the identifiers to decide which direction is used: if $\{x, y\}$ is a cross edge (*i.e.*, an edge which does not connect a process with its parent) and $x.id > y.id$ then, eventually, x can read the variables of y, but y cannot read the variables of x.

We introduce two techniques designed to cope with the peculiar problems of communication efficient computation. We introduce the *three color control scheme*, which is used to "wake up" processes when necessary, and "put them back to sleep" when their job is done. This scheme is important because during some finite prefix of a computation, processes may need to read all their neighbors, but after that prefix, only a subset of neighbors.

We also introduce the concept of "net polarity" by which ROOT-UID detects whether the putative BFS tree contains all processes of the network. Suppose, for example, that, at the initial configuration (which is arbitrary) there is a *false BFS tree*, a subgraph T' rooted at $Root$, which locally looks exactly like a BFS tree, but does not contain all processes. Yet because it "looks good" locally, all processes of T' "think" that a final configuration has been reached, and are resting. In order to restart the BFS tree construction phase of the algorithm, some process of T' must become aware that T' does not contain all processes. We further suppose that the ID of every process in T' is smaller than the ID of any of its neighbors not in T', which implies that, since all processes of T' are resting, none of them can read any neighbor not in T'. The problem is, how can the resulting deadlock be broken? That is, how can the processes of T' detect that T' does not contain all processes in the network?

Related Work. Anguilera *et al.* [1] introduce the concept of *communication efficiency* in implementation of a failure detector Ω^1 in partially synchronous systems. Following their work, there was further investigation of the possibility of communication efficient implementations of failure detector Ω (*e.g.*, [2,3,4,11]). The aim of communication-efficiency is to reduce the number of indefinitely communicating process pairs [1,2,3,11], and to reduce the number of processes that broadcast indefinitely [4].

Dolev and Schiller [9] introduce the *communication adaptive* property of self-stabilizing protocols. A self-stabilizing protocol is communication adaptive if the communication load of the transient-fault detection mechanism is low, while that of the fault correction mechanism is high. They present a communication adaptive self-stabilizing protocol for group membership service. Its communication complexity per asynchronous cycle is $O(nm \log N)$ bits before convergence to a legitimate configuration, and reduces to $O(n^2 \log N)$ bits after convergence, where n and m are the numbers of processes and links respectively, and N is an upper bound on the number of processes.

Delporte-Gallet *et al.* [6] consider self-stabilizing leader election that can tolerate process crashes as well as transient faults. They present an algorithm in the fully-synchronous system that uses only $n - 1$ unidirectional links to carry messages repeatedly and forever.

Devismes *et al.*[7] introduce communication efficiency with a *local* criterion. They consider, as communication-efficiency, the number of neighbors that each process communicates with forever as well as the total number of communicating process pairs. They investigate communication-efficiency for the maximal independent set problem and the maximal matching problem.

The most closely related previous work is [13], which considers communication-efficiency of self-stabilizing spanning-tree construction and gives possibility and impossibility results. They show that there exists a self-stabilizing communication-efficient algorithm for spanning tree construction, where only $n - 1$ process pairs maintain communication indefinitely, provided a unique root is designated. Their algorithm constructs an arbitrary spanning tree, not a BFS tree.

---

[1] Roughly speaking, the failure detector Ω eventually provides all processes with the identifier of a unique correct process (*i.e.*, a leader).

In the same paper, Masuzawa *et al.*give an important lower bound. If there is no designated root process, even if the UID property holds, any communication efficient self-stabilizing algorithm for spanning tree construction must use every edge in one or the other direction, yielding a lower bound of m on the communication efficiency of spanning tree construction when there is no designated root, where m is the number of edges in the network.

Takimoto *et al.* [14] consider communication-efficiency in wireless networks where a communication primitive is a local broadcast that allows a process to send a message to all of its neighboring processes. They aim to reduce the number of processes that keep broadcasting forever. The results of [13] and [14] are also summarized in [12].

Kutten and Zinenko [10] investigate the possibility of self-stabilizing protocols that are communication-efficient, both during and after convergence to legitimacy. They use *randomness* to achieve communication efficiency, and present communication efficient self-stabilizing protocols for spanning tree construction, distributed reset, and unison.

Outline. In Section 2, we define our model, and describe some of the common features of our algorithms. In Section 3, we present our algorithm ROOT-UID, a communication-efficient algorithm for constructing a BFS tree in a connected network with the UID property, given that the network has a distinguished root process. In 3.4 we explain the *three color control scheme*, which is used during error recovery. In Section 3.6, we explain the need for *net polarity* variables, which we use to ensure that each edge is used for communication in at least one direction. Section 4 concludes the paper.

2   Preliminaries

2.1   Self-stabilization

We use the composite atomicity shared memory model of computation [8]. The *program* of each process consists of a finite set of *actions* of the following form: $<$ *label* $>$ $<$ *guard* $>$ $\longrightarrow$ $<$ *statement* $>$. The *guard* of an action in the program of a process x is a Boolean expression involving the variables of x and its neighbors. The *statement* of an action of x updates one or more variables of x. An action can be executed only if it is *enabled*, *i.e.,* its guard evaluates to true. A process is said to be *enabled* if at least one of its actions is enabled. We use the *distributed daemon*. If one or more processes are enabled, the daemon *selects* at least one of these enabled processes to execute an action. We also assume that daemon is *unfair*, *i.e.,* it selects an arbitrary non-empty set of enabled processes at a step, if there is at least one. Thus, the daemon need never select a given enabled process unless it becomes the only enabled process. We measure time complexity in *rounds* elapsed before the first legitimate configuration. A *round* is defined to be a minimal computation sequence during which every process initially enabled is selected or becomes disabled by the end of the round [8].

Self-Stabilization. We say that an algorithm $\mathcal{A}$ for a problem $\mathbb{P}$ is *self-stabilizing* if there is a given class of configurations of $\mathcal{A}$, which we call the *legitimate* configurations of $\mathcal{A}$, such that the following conditions hold: (i) **Closure:** If a computation of $\mathcal{A}$ starts in a legitimate configuration, all subsequent configurations of that computation are legitimate. (ii) **Convergence:** Starting from an arbitrary configuration, the configuration of any computation of $\mathcal{A}$ contains a legitimate configuration. (iii) **Correctness:** If configuration is legitimate, the output conditions of $\mathbb{P}$ are satisfied. Note that Closure and Convergence together imply that every computation is eventually legitimate. We say that a configuration of a distributed algorithm $\mathcal{A}$ is *final* if, at that configuration, no process is enabled to execute any action of $\mathcal{A}$. We say that a self-stabilizing distributed algorithm $\mathcal{A}$ is *silent* if every computation of $\mathcal{A}$ contains a legitimate final configuration.

2.2 Communication Efficiency

Informally, an algorithm is *communication efficient* [7,13] if, eventually, not all links are used for communication between processes. We say that a computation of $\mathcal{A}$ is r-communication-efficient if the number of *directed* edges of the network used by the computation does not exceed r. That is, for every process x there is a set $R(x) \subseteq N(x)$, where $N(x)$ is the set of neighbors of x, such that x reads only its own variables and those of members of $R(x)$ during the computation, where $\sum |R(x)| \leq r$. We say that an algorithm $\mathcal{A}$ is $\Diamond$-r-communication-efficient, or *eventually* r-communication-efficient, if every computation of $\mathcal{A}$ has a suffix which is r-communication-efficient.

In an implementation of a distributed algorithm, not only must a process which executes an action have evaluated the guard of that action to TRUE, but processes which do *not* execute must evaluate guards of all actions to FALSE. More specifically, at any particular step, if correctness of $\mathcal{A}$ depends on a process x not executing a particular action, x must evaluate the guard of that action to be FALSE. If the value of that guard cannot be computed without using the values of a neighboring process y, then x must read y at that step. This fact imposes an interesting condition on a communication efficient algorithm. In any computation, except for a finite prefix, every process must evaluate every guard using only its own variables and those of $R(x)$, rather than the variables of all neighbors.

3 BFS Tree Computation with Distinguished Root and UID

In this section, we give a distributed algorithm, ROOT-UID, which constructs a BFS tree in a connected network G with the UID property and a distinguished process. ROOT-UID is silent and self-stabilizing under the unfair daemon, converges in $O(n)$ rounds, and is $\Diamond$-$(m+n-1)$-communication efficient. Eventually, each tree edge is used in both directions, but each cross edge is used in just one direction. More specifically, if $\{x, y\}$ is a cross edge, and $x.id < y.id$, then $x \in R(y)$ and $y \notin R(x)$, *i.e.*, eventually, y reads x, but x does not read y.

3.1 Simple BFS Construction

ROOT-UID is a communication efficient implementations of the algorithm SIMPLE-TREE given below, a silent self-stabilizing BFS tree construction algorithm for any connected network with a distinguished root, which has only two variables, *level* and *parent*. We assume that $Root.levl = 0$ and $Root.parent = \perp$ are fixed. SIMPLE-TREE is not communication efficient; each process reads all of its neighbors at every step.

Figure 3.1: Code of SIMPLE-TREE **for process** $x \neq Root$.

Function: $Level(x) = 1 + \min\{y.level : y \in N(x)\}$

Actions:

3.1.1. $x.level \neq Level(x)$ $\longrightarrow$ $x.level \leftarrow Level(x)$
3.1.2. $(x.level \neq 1 + x.parent.level) \wedge (y \in N(x)) \wedge (1 + y.level = x.level) \longrightarrow x.parent \leftarrow y$

Remark 1. On a connected network G with a distinguished process *Root*, SIMPLE-TREE is silent and self-stabilizing, and converges in $O(d)$ rounds, where d is the diameter of G.

Proof Sketch. After t rounds, $x.level = ||x, Root||$ if $||x, Root|| \leq t$, and within one more round, $x.parent$ will be correct. $\qquad\qquad\square$

3.2 Variables and Functions of ROOT-UID

In ROOT-UID, each process x has the following variables.

1. $x.parent \in N(x) \cup \{\perp\}$.
2. $x.level$, non-negative integer, the *level* of x, which is eventually equal to the distance from x to the root of the BFS tree.
3. $x.color \in \{0, 1, 2\}$, the *color* of x, which is used in the three-color control scheme, which we describe in detail in Section 3.4. If $x.color = 0$, then x is *resting*, while x is *alert* if $x.color \in \{1, 2\}$.
4. $x.polarity[y] \in \{-1, 1\}$, the *polarity* of the directed edge (x, y), for each $y \in N(x)$. We say that y is a *restricted* neighbor of x is $x.polarity[y] = -1$, otherwise y is an *unrestricted* neighbor.
5. $x.is\_child[y]$ for $y \in N(x)$, a Boolean array, where $x.is\_child[y]$ means that y is a child of x.
   Both $x.polarity[\,]$ and $x.is\_child[\,]$ require $O(\delta_x)$ space, where δ_x is the degree of x.
6. $x.loc\_net\_polarity$, integer, which can be positive, negative, or zero. Eventually, the value of this variable is $\sum_{y \in N(x)} x.polarity[y]$.
7. $x.net\_polarity$, integer. Eventually, $x.net\_polarity = \sum_{y \in T_x} y.loc\_net\_polarity$ for all x, where T_x is the subtree of T rooted at x, and $Root.net\_polarity = 0$,

We now list the functions of ROOT-UID, each of which (except *Root*) is defined for one process x, or for two processes x and $y \in N(x)$. Some functions have names which are capitalized versions of variable names. In each of those cases, the function returns the corrected value of the corresponding variable.

8. *Root*, a designated process. Let $Root.parent = \perp$ be fixed, and $Root.level = 0$ be fixed.

9. $Is\_Child\,(x, y) \equiv (y.parent = x) \wedge (y.level = 1 + x.level)$, Boolean, meaning that y is a child of x.

10. $Chldrn\,(x) = \{y \in N(x) : x.is\_child\,[y]\}$, the set of children of x.

11. $Family(x) = \{y \in N(x) : x.is\_child\,[y] \vee x.parent = y\}$

12. $False\_Root(x) \equiv (x \neq Root) \wedge ((x.parent = \perp) \vee (x.level \neq 1 + x.parent.level))$, Boolean.

13. $Is\_Root\,(x) \equiv (x = Root) \vee False\_Root(x)$, Boolean, namely x is a *root* of the forest T. At any step in the algorithm, T is the directed graph whose edges are all (x, y) such that $Is\_Child\,(y, x)$. Thus T has out-degree at most 1 at each process, and cannot contain a cycle, since the variable $x.level$ is decreasing along any directed path, and thus T is a forest.

14. $Level(x) = \begin{cases} 0 \text{ if } Is\_Root\,(x) \\ 1 + \min\{y.level : y \in N(x)\} \text{ otherwise} \end{cases}$

15. $Polarity(x, y) = \begin{cases} 1 \text{ if } y.id < x.id \\ -1 \text{ if } y.id > x.id \end{cases}$

16. $Loc\_Net\_Polarity(x) = \sum_{y \in N(x)} x.polarity\,[y]$, integer.

17. $Net\_Polarity(x) = x.loc\_net\_polarity + \sum_{y \in Chldrn\,(x)} y.net\_polarity$, integer.

18. $Nbr\_Ok\,(x, y) = \Big(x.polarity\,[y] = Polarity(x, y)\Big) \wedge \Big(x.is\_child\,[y] = Is\_Child\,(x, y)\Big) \wedge \Big(|x.level - y.level| \leq 1\Big)$, Boolean, meaning that y appears, to x, to have values consistent with legitimacy.

19. $Unrestricted\_Nbrs\,(x) = Family(x) \cup \{y \in N(x) : x.polarity\,[y] = 1\}$, the set of neighbors that can be read by x when x is *resting*.

20. $Visible\_Nbrs\,(x) = \begin{cases} Unrestricted\_Nbrs\,(x) \text{ if } x.color = 0 \\ N(x) \text{ otherwise} \end{cases}$
    the neighbors of x that x can read, given its current state.

21. $Visible\_Nbrs\_Ok\,(x) \equiv (\forall y \in Visible\_Nbrs\,(x))\, Nbr\_Ok\,(x, y)$, Boolean.

22. $Ok\,(x) \equiv Visible\_Nbrs\_Ok\,(x) \wedge \neg False\_Root(x) \wedge (x = Root \Rightarrow x.net\_polarity = 0)$ Boolean.
    The clause that $Root.net\_polarity = 0$ is explained in Section 3.6.

23. $Color\,(x) = \begin{cases} 0 \text{ if } Ok\,(x) \wedge \Big((x = Root) \vee (x.parent.color = 0)\Big) \wedge \\ \quad \forall y \in Chldrn\,(x)(y.color \neq 1) \\ 1 \text{ if } \neg Ok\,(x) \vee \exists y \in Chldrn\,(x)(y.color = 1) \\ 2 \text{ if } Ok\,(x) \wedge (x.parent.color \neq 0) \wedge \forall y \in Chldrn\,(x)(y.color \neq 1) \end{cases}$

Additional notation which is common to the rest of this paper includes

- T_x, the subtree of T rooted at x.
- $\delta_x = |N(x)|$, the *degree* of x.
- Tree edge, an undirected edge of G that connects some process x with $x.parent$.
- Cross edge, an undirected edge of G that is not a tree edge.
- $Cross(x) = N(x) \setminus Family(x)$, the *cross neighbors* of x.

3.3   Actions of ROOT-UID

We classify the actions of ROOT-UID as *easy* or *hard*. The guard of an easy action can be evaluated only by examining unrestricted neighbors, while evaluation of the guard of a hard action may require examining all neighbors. If a process x is resting, meaning that $x.color = 0$, then x cannot be enabled to execute a hard action, since one of the clauses of the guard of every hard action is that $x.color \neq 0$, *i.e.*, x is alert.

Figure 3.2:   Code of ROOT-UID for one process x and $y \in N(x)$.

Easy Actions:

3.2.1. $x.color \neq Color(x)$ $\longrightarrow x.color \leftarrow Color(x)$

3.2.2. $x.loc\_net\_polarity \neq Loc\_Net\_Polarity(x)$ $\longrightarrow x.loc\_net\_polarity$
$\leftarrow Loc\_Net\_Polarity(x)$

3.2.3. $x.net\_polarity \neq Net\_Polarity(x)$ $\longrightarrow x.net\_polarity \leftarrow Net\_Polarity(x)$

Hard Actions:

3.2.4. $x.color \neq 0 \ \wedge \ x.level \neq Level(x)$ $\longrightarrow x.level \leftarrow Level(x)$

3.2.5. $x.color \neq 0 \ \wedge \ x.level \neq 1 + x.parent.level \ \wedge \longrightarrow x.parent \leftarrow y$
$y \in N(x) \ \wedge \ 1 + y.level = x.level$

3.2.6. $x.color \neq 0 \ \wedge \ y \in N(x) \ \wedge$ $\longrightarrow x.polarity[y] \leftarrow Polarity(x,y)$
$x.polarity[y] \neq Polarity(x,y)$

3.2.7. $x.color \neq 0 \ \wedge \ y \in N(x) \ \wedge$ $\longrightarrow x.is\_child[y] \leftarrow Is\_Child(x,y)$
$x.is\_child[y] \neq Is\_Child(x,y)$

Figure 3.2 gives the actions of ROOT-UID. We define a configuration of ROOT-UID to be *legitimate* if T is a BFS tree, and if the variables $x.polarity[y]$, $x.is\_child[y]$, $x.loc\_net\_polarity$, and $x.net\_polarity$ have the correct values for all x and y, and if $x.color = 0$ for all x.

Construction of the BFS tree T is done by Actions 3.2.4 and 3.2.5, both of which are hard actions. The purpose of the three color control structure is to ensure that there is enough communication to enable that construction, and also to ensure that, once that construction is finished, the algorithm is $(m + n - 1)$-communication efficient. Action 3.2.6 is executed at most once for each ordered pair (x, y), since $Polarity(x, y)$ never changes. $\sum_{x \in G} \sum_{y \in N(x)} Polarity(x, y) = 0$, since $Polarity(x, y) + Polarity(y, x) = 0$ for every edge $\{x, y\}$. Actions 3.2.2 and 3.2.3 are both easy, and are bottom-up waves which cause $Root.net\_polarity$ to be set to $\sum_{x \in G} \sum_{y \in N(x)} x.polarity[y]$ which will be zero if all values of $x.polarity[y]$ are correct. If not, $Ok(Root) = \text{FALSE}$, and the three color control scheme causes all processes to become alert, ensuring that polarities become correct.

3.4   The Three Color Scheme

We now describe in detail how the three color scheme is used in ROOT-UID. We define the predicate $Enabled\_Hard(x)$ to mean that either x is enabled to execute a hard action, or that x is resting and would be enabled to execute a hard action if $x.color$ were changed. In order for the three color structure scheme to work, Property 1, given below, must hold.

Property 1. If the current configuration is illegitimate and legitimacy cannot be achieved by easy actions alone, then there exist processes x, y such that $\neg Ok(x) \wedge \widehat{Enabled\_Hard}(y)$ either holds or will hold at some later step, and x, y lie in the same component of T.

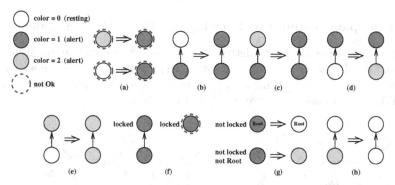

Figure 3.3: Some examples of color actions. In each case, the value of $x.color$ changes to match $Color(x)$.

We say that a process x is *locked* if $x.color = 1$, and if either $\neg Ok(x)$ or $y.color = 1$ for some $y \in Chldrn(x)$. Note that a locked process is not enabled to change its color. Suppose all processes have color 0, and $Ok(x) = $ FALSE for some x. Then x executes Action 3.2.1; $x.color \leftarrow 1$, and x becomes locked. In a convergecast wave, every ancestor of x, including *Root*, changes color to 1 and becomes locked. A broadcast wave, initiated by *Root* and by every other process of color 1, changes all processes to color 2 except for those which already have color 1 or 2.

The entire path between x and *Root* will remain in color 1, and locked, and all processes will remain alert as long as $\neg Ok(x)$ holds. When *Ok* holds for all processes, all color 1 processes except *Root* will change color to 2 in a bottom-up wave, after which *Root* will change color to 0. In a broadcast wave, all processes will then change color to 0. After possibly more executions of easy actions, legitimacy will be achieved.

Figure 3.3 shows the effect of Action 3.2.1 in various situations. 3.3(a) shows a convergecast color 1 wave initiated by a process x whose current color is either 0 or 2, where $Ok(x) = $ FALSE. Then $x.color \leftarrow 1$, starting the wave. The wave is propagated upward, as shown in 3.3(b) and 3.3(c). Figure 3.3(d) shows how a process in color 1 initiates a broadcast wave which alerts all processes below it. In this figure, a resting (color 0) process has a parent of color 1, and becomes alert by changing color to 2. Propagation of that wave downward is illustrated in 3.3(e).

The next three parts of Figure 3.3 deal with restoring the resting state. Figure 3.3(f) illustrates locked processes. Figure 3.3(g) shows propagation of the convergecast wave which eliminates color 1. If a color 1 process is unlocked, indicating that its original purpose has been fulfilled, it changes color to 2, unless it is *Root*, in which case it changes color to 0. Once *Root.color* = 0, a broadcast wave will change the color of all processes to 0. (Unless, of course, there is a new instance of a process where *Ok* does not hold.) The propagation of that wave is illustrated in Figure 3.3(h).

3.5   Example Computation Showing the Three Color Scheme

In Figure 3.4, each ID is the letter shown inside the circle representing the process. (The ID of *Root* is not relevant.) Parent pointers are shown as arrows, and cross edges as dashed lines. The figure shows an example computation of ROOT-UID on a network where $n = 7$ and $m = 7$.

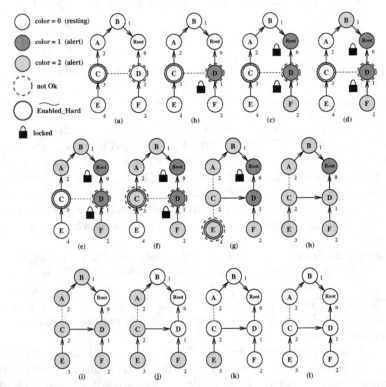

Figure 3.4: Example Computation of ROOT-UID. The links of the spanning tree T are shown as arrows, while other edges of G are shown as dashed lines. Values of *level* are shown in red. Initially, T is not a BFS tree. Figure 3.4 (l) shows a configuration in which T is correct and all processes are resting.

In the initial configuration, shown in Figure 3.4(a), all processes are resting. T is not correct, since C.*parent* should be D. The solid outer circle around C indicates that $Enabled\_Hard(C) = \text{TRUE}$, while the dashed circle indicates that

$Ok(\mathrm{D}) = $ FALSE. It may seem strange that action by C is required to "fix" the configuration, but that D, which is far from C in the tree, is the only process aware of this need. This situation is not unusual. In this case, since C < D, D can see that $C.level = 3$, which is inconsistent with its own level, which is 1. On the other hand, C does not see D, and hence does not notice the inconsistency.

In 3.4(b), D initiates a color 1 convergecast wave, which reaches *Root* in Figure 3.4(c). The resulting color 2 broadcast wave, which is initiated by both D and *Root*, eventually reaches all processes. As soon as the wave reaches C, in 3.4(f), that process executes a hard action, changing *C.parent* to D and *C.level* to 2. The configuration is still not legitimate, since now $E.level = 4$. In this case, there is a circle of each color around E, since $Ok(\mathrm{E}) = $ FALSE, and E is also enabled to execute a hard action, which it does in the next step, changing its level to 3. Meanwhile, since D is no longer locked in Figure 3.4(g), it changes its color to 2. If there were more color 1 processes between D and *Root*, their color would change from 1 to 2 in a bottom-up wave. *Root* changes color from 1 directly to 0 at 3.4(i), initiating a broadcast wave that causes all processes to change color to 0, as shown in Figure 3.4(l).

3.6   The Purpose of Polarity

The purpose of the variables $\left\{x.polarity[y]\right\}$ is to tell x whether y is a restricted neighbor, information that x must first obtain by reading y. But that presents a problem: what if both x and y are initialized in such a way that each believes the other to be a restricted neighbor? In that case, one of them has the wrong information, but neither knows it.

We solve that problem by using the *net polarity* of T, $\sum_{x \in T} \sum_{y \in N(x)} x.polarity[y]$. If all values of the polarity variables are correct, this net polarity will be zero, since each edge contributes 1 at one end and -1 at the other. The net polarity is computed bottom up, and the total value is stored at *Root.net_polarity*. If that value is not zero, there is an error, and *Root* will send a broadcast wave, alerting all processes. After sufficiently many actions, both easy and hard, T will be correct.

Figure 3.5 illustrates an example which shows the necessity of the net polarity variables. Suppose the variables *loc_net_polarity* and *net_polarity* as well as Actions 3.2.2 and 3.2.3 are removed from the definition of ROOT-UID, but all other variables and actions remain. In this case, there is a configuration, illustrated in Figure 3.5(a) which is both illegitimate and final, *i.e.*, a deadlock.

In Figure 3.5, each ID is the letter shown inside the circle representing the process, and $E = Root$. Parent pointers are shown as arrows, and cross edges as dashed lines. The "+" and "−" signs at the ends of each edge indicate polarity; if $y \in N(x)$, a plus sign indicates that $x.polarity[y] = 1$, meaning that x can read y when x is *resting*, and a minus sign indicates that $x.polarity[y] = -1$, meaning that x must be alert in order to read y. Note that $x.polarity[y] = 1 \Leftrightarrow x.id > y.id$ in a legitimate configuration. In 3.5(a), there are two places where the legitimacy fails: $D.polarity[B] = -1$ when it should be 1, and $B.parent = G$ when it should be D. Both processes are enclosed with solid circles, indicating that they would

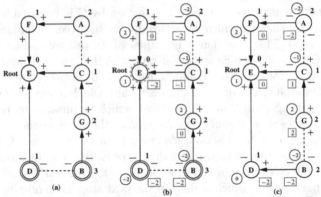

Figure 3.5: Example showing the necessity of net polarity variables. E = *Root*. Tree edges are black arrows, cross edges are dashed, and levels are shown as numerals. A "+" or "−" near x on the edge between x and y indicates that $x.polarity[y]$ is 1 or −1, respectively.

be enabled to execute a hard action if they became alert. B would be enabled to execute Action 3.2.5 if it were alert, while D would be enabled to execute Action 3.2.6 if it were alert. Neither action is enabled, since both processes are resting, *i.e.*, B.*color* = D.*color* = 0; in fact no process is enabled. The configuration is a deadlock.

In Figure 3.5(b), we illustrate the same configuration, but with the values of *loc_net_polarity* and *net_polarity* filled in. The value of $x.loc\_net\_polarity$ is shown as a numeral enclosed in a solid circle, while the value of $x.net\_polarity$ is shown as a numeral enclosed in a box. Note that E.*net_polarity* = −2, which means that $Ok(\text{E})$ = FALSE. since $Net\_Polarity(Root)$ must be zero in a legitimate configuration. (Conversely, if $Net\_Polarity(Root)$ were zero, it would indicate legitimacy, since if both end processes of an edge have positive polarity, that error would be detected.) The remaining steps of the computation are not shown, other than the final configuration at the end. E changes color to 1, triggering a broadcast wave in which all other processes change color to 2, which means they are alert and able to execute the needed actions, after which the configuration is legitimate, as shown in Figure 3.5(c). We list the sequence of changes below.

(a)	B.*parent* ← D	(b)	B.*level* ← 2	(c) D.*polarity*[B] ← 1
(d)	D.*loc_net_polarity* ← 0	(e)	D.*net_polarity* ← −2	(f) G.*net_polarity* ← 2
(g)	C.*net_polarity* ← 1	(h)	E.*net_polarity* ← 0	

3.7   Proof Sketches for ROOT-UID

Lemma 1. *The number of steps of any computation of* ROOT-UID *at which any hard action is executed is finite.*

Proof Sketch. A process can execute Action 3.2.6 at most once in a computation. We prove by induction on $\|x, Root\|$ that a process x can only execute Action

3.2.4 finitely many times; as a corollary, we have that x can only execute Action 3.2.5 finitely many times.   □

Lemma 2. *There is no infinite computation of* ROOT-UID *during which no hard action is executed.*

Proof Sketch. Suppose Γ is an infinite computation of ROOT-UID, during which no process executes a hard action. The shape of T does not change, and since Action 3.2.6 is not executed, the value of $x.polarity[y]$ for any x and y does not change. Therefore, no process can execute Action 3.2.2 more than once, and hence there is a last step at which any process executes 3.2.2. After the last action of Action 3.2.2, the values of $x.loc\_net\_polarity$ do not change, and thus no process can execute Action 3.2.3 more than once thereafter, hence there is a last step at which any process executes 3.2.3. After the last action of 3.2.3, the value of $Ok(x)$ does not change for any x. Thus, there is an infinite computation during which Action 3.2.1 is executed infinitely many times. Finally, we obtain a contradiction by defining a non-negative integral potential which decreases at any step at which Action 3.2.1 is executed, but no other action is executed. Since the potential cannot be less than zero, the computation is finite.   □

Lemma 3. *Any configuration of* ROOT-UID *that is final is also legitimate.*

Proof Sketch. Let γ be an illegitimate configuration of ROOT-UID. We break into cases, and show that in each case, there is some process that is enabled to execute an action, and thus γ is not final.   □

Theorem 1. ROOT-UID *is* $\Diamond$-$(m+n-1)$-*communication efficient and is silent and self-stabilizing under the unfair daemon, reaches a legitimate configuration within* $O(n)$ *rounds, and uses* $O(\log n + \delta_x)$ *space per process.*

Proof Sketch. The communication efficiency of ROOT-UID follows from Lemmas 1, 2, and 3.

In the initial configuration, the longest chain in T is no longer than $n - 1$. If the configuration is not final, every process will be alerted within $O(n)$ rounds. Thereafter, convergence will occur within $O(d)$ rounds in the same manner as in SIMPLE-TREE. Thus, ROOT-UID reaches a legitimate configuration within $O(n)$ rounds.

The value of $x.level$ takes $O(\log d)$ space, and x must retain $O(1)$ memory for each neighbor y to hold the value of $x.polarity[y]$. The value of $x.net\_polarity$ is an integer whose absolute value cannot exceed m, and no other variable requires more space. Thus, the per process space complexity of ROOT-UID is $O(\log n + \delta_x)$.   □

4   Conclusion

We have given a self-stabilizing and silent algorithm, ROOT-UID, for BFS construction in a network with the UID property which has a designated root process, which is $\Diamond$-$(m + n - 1)$-communication efficient, and which takes $O(n)$ rounds to reach legitimacy. The space complexity of ROOT-UID is $O(\log n + \delta_x)$ for each process x.

Acknowledgement. This work was supported in part by JSPS KAKENHI, Grant Numbers (B)26280022 and 24650012.

References

1. Aguilera, M.K., Delporte-Gallet, C., Fauconnier, H., Toueg, S.: Stable leader election. In: Welch, J. (ed.) DISC 2001. LNCS, vol. 2180, pp. 108–122. Springer, Heidelberg (2001)
2. Aguilera, M.K., Delporte-Gallet, C., Fauconnier, H., Toueg, S.: On implementing omega with weak reliability and synchrony assumptions. In: Proceedings of the 22rd ACM Symposium on Principles of Distributed Computing, pp. 306–314 (2003)
3. Aguilera, M.K., Delporte-Gallet, C., Fauconnier, H., Toueg, S.: Communication-efficient leader election and consensus with limited link synchrony. In: Proceedings of the 23rd ACM Symposium on Principles of Distributed Computing, pp. 328–337 (2004)
4. Biely, M., Widder, J.: Optimal message-driven implementations of omega with mute processes. ACM Transactions on Autonomous and Adaptive Systems 4(1), 4:1–4:22 (2009)
5. Datta, A.K., Larmore, L.L., Vemula, P.: Self-stabilizing leader election in optimal space under an arbitrary scheduler. Theoretical Computer Science 412(40), 5541–5561 (2011)
6. Delporte-Gallet, C., Devismes, S., Fauconnier, H.: Robust stabilizing leader election. In: Masuzawa, T., Tixeuil, S. (eds.) SSS 2007. LNCS, vol. 4838, pp. 219–233. Springer, Heidelberg (2007)
7. Devismes, S., Masuzawa, T., Tixeuil, S.: Communication efficiency in self-stabilizing silent protocols. In: Proceedings of the 29th International Conference on Distributed Computing Systems, pp. 474–481 (2009)
8. Dolev, S.: Self-stabilization. MIT Press (2000)
9. Dolev, S., Schiller, E.: Communication adaptive self-stabilizing group membership service. In: IEEE Transactions on Parallel and Distributed Systems, pp. 709–720 (2003)
10. Kutten, S., Zinenko, D.: Low communication self-stabilization through randomization. In: Lynch, N.A., Shvartsman, A.A. (eds.) DISC 2010. LNCS, vol. 6343, pp. 465–479. Springer, Heidelberg (2010)
11. Larrea, M., Fernandez, A., Arevalo, S.: Optimal implementation of the weakest failure detector for solving consensus. In: Proceedings of the 19th IEEE Symposium on Reliable Distributed Systems, pp. 52–59 (2000)
12. Masuzawa, T.: Silence is golden: Self-stabilizing protocols communication-efficient after convergence. In: Défago, X., Petit, F., Villain, V. (eds.) SSS 2011. LNCS, vol. 6976, pp. 1–3. Springer, Heidelberg (2011)
13. Masuzawa, T., Izumi, T., Katayama, Y., Wada, K.: Brief Announcement: communication-efficient self-stabilizing Protocols for spanning-tree construction. In: Abdelzaher, T., Raynal, M., Santoro, N. (eds.) OPODIS 2009. LNCS, vol. 5923, pp. 219–224. Springer, Heidelberg (2009)
14. Takimoto, T., Ooshita, F., Kakugawa, H., Masuzawa, T.: Communication-efficient self-stabilization in wireless networks. In: Richa, A.W., Scheideler, C. (eds.) SSS 2012. LNCS, vol. 7596, pp. 1–15. Springer, Heidelberg (2012)

Self-stabilizing Algorithms for Connected Vertex Cover and Clique Decomposition Problems

François Delbot[1], Christian Laforest[2], and Stephane Rovedakis[3],*

[1] Université Paris Ouest Nanterre / LIP6, CNRS UMR 7606. 4 Place Jussieu,
75252 Paris Cedex, France
francois.delbot@lip6.fr
[2] Université Blaise Pascal / LIMOS, CNRS UMR 6158, ISIMA. Campus Scientifique
des Cézeaux, 24 Avenue des Landais, 63173 Aubiere Cedex, France
christian.laforest@isima.fr
[3] Conservatoire National des Arts et Métiers / CEDRIC, EA 4629. 292 rue
Saint-Martin, F-75141 Paris Cedex 03, France
stephane.rovedakis@cnam.fr

Abstract. In many wireless networks, there is no fixed physical backbone nor centralized network management. The nodes of such a network have to self-organize in order to maintain a virtual backbone used to route messages. Moreover, any node of the network can be *a priori* at the origin of a malicious attack. Thus, in one hand the backbone must be fault-tolerant and in other hand it can be useful to monitor all network communications to identify an attack as soon as possible. We are interested in the minimum *Connected Vertex Cover* problem, a generalization of the classical minimum Vertex Cover problem, which allows to obtain a connected backbone. Recently, Delbot *et al.* [11] proposed a new centralized algorithm with a constant approximation ratio of 2 for this problem. In this paper, we propose a distributed and self-stabilizing version of their algorithm with the same approximation guarantee. To the best knowledge of the authors, it is the first distributed and fault-tolerant algorithm for this problem. The approach followed to solve the considered problem is based on the construction of a connected minimal clique partition. Therefore, we also design the first distributed self-stabilizing algorithm for this problem, which is of independent interest.

Keywords: Distributed algorithms, Self-stabilization, Connected Vertex Cover, Connected Minimal Clique Partition.

1 Introduction

In many wireless networks, there is no fixed physical backbone nor centralized network management. In such networks, the nodes need to regularly flood control messages which leads to the "broadcast storm problem" [42]. Thus, the

* The author was supported in part by CNAM Project CONDOR.

M.K. Aguilera et al. (Eds.): OPODIS 2014, LNCS 8878, pp. 307–322, 2014.
© Springer International Publishing Switzerland 2014

nodes have to self-organize in order to maintain a virtual backbone, used to route messages in the network. Routing messages are only exchanged inside the backbone, instead of being broadcasted to the entire network. To this end, the backbone must be connected. The construction and the maintenance of a virtual backbone is often realized by constructing a Connected Dominating Set. A *Connected Dominating Set (CDS)* of a graph $G = (V, E)$ is a set of nodes $S \subseteq V$ such that $G[S]$ (the graph induced by S in G) is connected and each node in $V - S$ has at least one neighbor in S. Nodes from S are responsible of routing the messages in the network, whereas nodes in $V - S$ communicate by exchanging messages through neighbors in S. In order to minimize the network resources consumption, the size of the backbone (and thus of the CDS) is minimized. This problem is NP-hard [19] and has been extensively studied due to its importance for communications in wireless networks. Many algorithms have been proposed in a centralized setting (e.g., see [4] for a survey). In addition to message routing, there is the problem of network security. Indeed, a faulty node infected by a virus or an unscrupulous user can be at the origin of flooding or a malicious attack. Thus, it is necessary to monitor all network communications to identify these situations, as soon as possible, in order to isolate this node. A CDS S will not support this feature since two nodes in $V - S$ can be neighbors, i.e., $V - S$ is not always an independent set.

In order to monitor all network communications, we can consider the Vertex Cover problem. A *vertex cover* of a graph $G = (V, E)$ is a set of nodes $S \subseteq V$ such that each edge $e = uv$ is *covered* by S, i.e., $u \in S$ or $v \in S$ (or both). A vertex cover is *optimal* if it's size is minimum. This is a classical NP-complete problem [19] that can be approximated with a ratio of 2. However, if a vertex cover allows to monitor all network communications, it is not always connected and cannot be used as a backbone. A *Connected Vertex Cover* S of G is a vertex cover of G with the additional property that $G[S]$ (the graph induced by S in G) is connected. Similarly, an *optimal* connected vertex cover is one of minimum size and the associated problem is also NP-complete. Not a lot of work has been done on this problem (see [18,44]). More recently, Delbot *et al.* in [11] proposed another (centralized) 2-approximation algorithm based on connected clique partitions of G (presented in the following sections).

In practice, it is more convenient to use distributed and fault-tolerant algorithms, instead of centralized algorithms due to the communications cost to obtain the network topology. *Self-stabilization* introduced first by Dijkstra in [13, 15] is one of the most versatile techniques to ensure a distributed system to recover a correct behaviour. A distributed algorithm is self-stabilizing if after faults and attacks hit the system and place it in some arbitrary global state, the system recovers from this catastrophic situation without external (e.g., human) intervention in finite time. Many self-stabilizing algorithms have been proposed to solve a lot of graph optimization problems, e.g., Guellati and Kheddouci [21] give a survey for several problems related to independence, domination, coloring and matching problems in graphs. For the minimal CDS problem, Jain and Gupta [27] design the first self-stabilizing algorithm for this problem. Drabkin *et al.* [17] gave then two self-stabilizing approaches for the construction of CDS. The first approach connects the nodes of a maximal independent set in order to obtain a CDS, while the second approach uses a dominating set instead of a maximal independent set. More recently, Kamei *et al.* [30–32] considered the minimum CDS problem. They proposed several self-stabilizing algorithms with a constant approximation ratio and useful properties during algorithms convergence.

However, as explained above a CDS does not meet all the desired properties. This is why we study the minimum connected vertex cover from a distributed and self-stabilizing point of view.

Contributions. We consider the minimum *Connected Vertex Cover* problem in a distributed system subject to transient faults. In this paper, we propose a distributed and self-stabilizing version of the algorithm given recently by Delbot *et al.* [11] for this problem while guaranteeing the same approximation ratio of 2. To the best of our knowledge, it is the first distributed and fault-tolerant algorithm for this problem. Moreover, our approach is based on the construction of a *Connected Minimal Clique Partition*. As discussed in Section 4, this allows us to attain a higher level of parallelism in comparison with a self-stabilizing approach which follows the sequential construction given by Savage [44]. Furthermore, we also design the first distributed self-stabilizing algorithm for the Connected Minimal Clique Partition problem, which is of independent interest.

In the following section, we describe the model assumed in this paper. In Section 3, we consider first the Connected Minimal Clique Partition problem. We give a state of the art related to the graph decomposition problem, then we present our self-stabilizing algorithm for this problem. Section 4 is devoted to the Connected Vertex Cover problem. We introduce first related works associated with this problem, then we give the self-stabilizing connected vertex cover algorithm that we propose. The last section concludes the paper and present several perspectives. Due to space constraints the correctness proofs are not included in the present paper, but they are available in [12].

2 Model

Notations. We consider a network as an undirected connected graph $G = (V, E)$ where V is a set of nodes (or *processors*) and E is the set of *bidirectional asynchronous communication links*. We state that n is the size of G ($|V| = n$) and m is the number of edges ($|E| = m$). We assume that the graph $G = (V, E)$ is a simple connected graph. In the network, p and q are neighbors if and only if a communication link (p,q) exists (i.e., $(p,q) \in E$). Every processor p can distinguish all its links. To simplify the presentation, we refer to a link (p,q) of a processor p by the *label* q. We assume that the labels of p, stored in the set $Neig_p$, are locally ordered by $\prec_p$. We also assume that $Neig_p$ is a constant input from the system. $Diam$ and Δ are respectively the diameter and the maximum degree of the network (i.e., the maximal value among the local degrees of the processors). Each node $p \in V$ has a unique identifier in the network, noted ID_p.

Programs. In our model, protocols are *uniform*, i.e., each processor executes the same program. We consider the local shared memory model of computation[1]. The program of every processor consists in a set of *variables* and an *ordered finite set of actions* inducing a *priority*. This priority follows the order of appearance of the actions into the text of the protocol. A processor can write to its own variable only, and read its own variables and that of its neighbors. Each action is constituted as follows: $< label > :: < guard > \rightarrow < statement >$. The guard of an action in the program of p is a boolean expression involving variables of p and its neighbors. The statement of an action of p updates one or more variables of p. An action can be executed only if its guard is satisfied. The *state* of a processor is defined by the value of its variables. The *state* of a system is the product of the states of all processors. We will refer to the state of a processor and the system as a (*local*) *state* and (*global*) *configuration*, respectively. We note $\mathcal{C}$ the set of all possible configuration of the system. Let $\gamma \in \mathcal{C}$ and A an action of p ($p \in V$). A is said to be *enabled* at p in γ if and only if the guard of A is satisfied by p in γ. Processor p is said to be *enabled* in γ if and only if at least one action is enabled at p in γ. When several actions are enabled simultaneously at a processor p: only the enabled action of highest priority can be activated.

Let a distributed protocol P be a collection of binary transition relations denoted by $\mapsto$, on $\mathcal{C}$. A *computation* of a protocol P is a *maximal* sequence of configurations $e = (\gamma_0, \gamma_1, ..., \gamma_i, \gamma_{i+1}, ...)$ such that, $\forall i \geq 0$, $\gamma_i \mapsto \gamma_{i+1}$ (called a *step*) if γ_{i+1} exists, else γ_i is a terminal configuration. *Maximality* means that the sequence is either finite (and no action of P is enabled in the terminal configuration) or infinite. All computations considered here are assumed to be maximal. $\mathcal{E}$ is the set of all possible computations of P.

As we already said, each execution is decomposed into steps. We consider the *composite atomicity model*. That is, each step consists of four sequential phases atomically executed: (*i*) every processor evaluates its guards, (*ii*) a *daemon* (also called *scheduler*) selects a non-empty subset of enabled processors, (*iii*) each selected processor computes its new state (invisibly to neighbors) by executing its enabled action of highest priority, and (*iv*) each selected processor overwrites its state with the new one. When the four phases are done, the next step begins.

A *daemon* can be defined in terms of *fairness* and *distributivity*. In this paper, we use the notion of *weak fairness*: if a processor p is continuously enabled then p will be eventually chosen by the daemon to execute an action. Concerning the *distributivity*, we assume that the daemon is *distributed* meaning that, at each step, if one or more processors are enabled, then the daemon chooses at least one of these processors to execute an action.

We consider that any processor p executed a *disabling action* in the computation step $\gamma_i \mapsto \gamma_{i+1}$ if p was *enabled* in γ_i and not enabled in γ_{i+1}, but did not execute any protocol action in $\gamma_i \mapsto \gamma_{i+1}$. The disabling action represents

---

[1] To execute the proposed algorithms in wireless networks, we can use the fined grained communication atomicity model [5] with the transformers for shared memory model protocols to act in message passing systems suggested in [15, 16].

the following situation: at least one neighbor of p changes its state in $\gamma_i \mapsto \gamma_{i+1}$, and this change effectively made the guard of all actions of p false in γ_{i+1}.

To compute the time complexity, we use the definition of (asynchronous) *round*. This definition captures the execution rate of the slowest processor in any computation. Given a computation e ($e \in \mathcal{E}$), the *first round* of e (let us call it e') is the minimal prefix of e containing the execution of one action (an action of the protocol or a disabling action) of every enabled processor from the initial configuration. Let e'' be the suffix of e starting from the last configuration of e'. The *second round* of e is the first round of e'', and so on.

3   Connected Minimal Clique Partition Problem

In this section, we consider a first problem whose aim is the partitioning of the input graph into subgraphs of maximal size in a distributed fashion, while maintaining a connectivity constraint between some subgraphs. More particularly, the goal is to decompose an input undirected graph $G = (V, E)$ into a set of cliques of maximal size such that all cliques of size at least two are connected. The connectivity constraint can be used for communication facilities. In the following, we define more formally the Connected Minimal Clique Partition problem.

Consider any undirected graph $G = (V, E)$. A *clique* is a complete subgraph of G and we call *trivial* any clique that contains only one node.

Definition 1 (Connected Minimal Clique Partition). *A clique partition of G is a partition of the set V into disjoint cliques $C_1, \ldots, C_k$ (i.e., $\bigcup_{i=1}^{k} C_i = V$ and if $i \neq j, C_i \cap C_j = \emptyset$). A clique partition is called* minimal *if for all $i \neq j$ the graph induced by $C_i \cup C_j$ is not a clique. A minimal clique partition $C_1, \ldots, C_k$ is* connected *iff for any pair of nodes $u, v \in C$ there is a path between u and v in the graph induced by C, with C the union of the nodes of all the non trivial cliques of $C_1, \ldots, C_k$.*

Since we consider that faults can arise in the system, we give in Specification 1 the conditions that a self-stabilizing algorithm solving the Connected Minimal Clique partition problem have to satisfy.

Specification 1 (Self-stabilizing Connected Minimal Clique Partition)
Let $\mathcal{C}$ be the set of all possible configurations of the system. An algorithm $\mathcal{A}_{CMCP}$ solving the problem of constructing a stabilizing connected minimal clique partition satisfies the following conditions:

1) Algorithm $\mathcal{A}$ reaches a set of terminal configurations $\mathcal{T} \subseteq \mathcal{C}$ in finite time, and 2) Every configuration $\gamma \in \mathcal{T}$ satisfies Definition 1.

3.1   Related Works

The decomposition of an input graph into patterns or partitions has been extensively studied in the literature, and also in the self-stabilizing context. Most of graph partitioning problems are NP-complete. For the graph decomposition into patterns, Ishii and Kakugawa [26] proposed a self-stabilizing algorithm for the construction of cliques in a connected graph with unique nodes identifier. Each node has to compute the largest set of cliques of same maximum size it can belong to in the graph. A set of cliques is constructed in $O(n^4)$ computation steps assuming an unfair central daemon. Moreover, the authors show that there exists no self-stabilizing algorithm in arbitrary anonymous graphs for this problem. Neggazi *et al.* [40] considered the problem of decomposing a graph into a maximal set of disjoint triangles. They give the first self-stabilizing algorithm for this problem whose convergence time is $O(n^4)$ steps under an unfair central daemon with unique nodes identifier. Neggazi *et al.* [41] studied later the uniform star decomposition problem, i.e., the goal is to divide the graph into a maximum set of disjoint stars of p leaf nodes. This is a generalization of the maximum matching problem which is a NP-complete. The aim is to construct a maximum set of independent edges of the graph, thus a 1-star decomposition is equivalent to a maximum matching. The authors proposed a self-stabilizing algorithm constructing a maximal p-star decomposition of the input graph in $O(\frac{n}{p+1})$ asynchronous rounds and a (exponential) bounded number of steps under an unfair distributed daemon with unique nodes identifier.

A well studied problem related with graph decomposition is the maximum matching problem. Many works address the maximal matching problem which is polynomial. The first self-stabilizing algorithm for this problem has been proposed by Hsu *et al.* [24]. The algorithm converges in $O(n^4)$ steps under a central daemon. Hedetniemi *et al.* [23] showed later that the algorithm proposed by Hsu *et al.* has a better convergence time of $2m + n$ steps under a central daemon. Goddar *et al.* [20] considered the construction of a maximal matching in ad-hoc networks and give a solution which stabilizes in $n+1$ rounds under a synchronous distributed daemon. Manne *et al.* [37] have shown that there exists no self-stabilizing algorithm for this problem under a synchronous distributed daemon in arbitrary anonymous networks. So, they proposed an elegant algorithm which converges in $O(n)$ rounds and $O(m)$ steps under an unfair distributed daemon in arbitrary networks with unique nodes identifier. Recently, several works considered the maximum matching problem to find an optimal or an approximated solution. Hadid *et al.* [22] give an algorithm which constructs an optimal solution in $O(Diam)$ rounds under a weakly fair distributed daemon only in bipartite graphs. Manne *et al.* [38] presented a self-stabilizing algorithm constructing a $\frac{2}{3}$-approximated maximum matching in general graphs within $O(n^2)$ rounds and a (exponential) bounded number of steps under an unfair distributed daemon. Manne *et al.* [36] proposed the first self-stabilizing algorithm for the maximum weighted matching problem achieving an approximation ratio of 2 in a (exponential) bounded number of steps under a central daemon and a distributed daemon. Turau *et al.* [47] gave a new analysis of the algorithm of Manne *et al.* [36]. They showed that this algorithm converges in $O(nm)$ steps under a central daemon and an unfair distributed daemon.

More recently, self-stabilizing works investigated the graph decomposition into disjoint paths. Al-Azemi *et al.* [1] studied the decomposition of the graph in two edge-disjoint paths in general graphs, while Neggazi *et al.* [39] considered the problem of dividing the graph in maximal disjoint paths of length two.

Finally, the cluster partitioning of the input graph has been extensively studied. Belkouch *et al.* [3] proposed an algorithm to divide a graph of order k^2 into k partitions of size k. The algorithm is based on spanning tree constructions of height h and converges in $O(h)$ rounds under a weakly fair distributed daemon. Johnen *et al.* [29] studied the weighted clustering problem and introduced the notion of robustness allowing to reach quickly (after one round) a cluster partition. A cluster partition is then preserved during the convergence to a partition satisfying the cluster head's weight. Bein *et al.* [2] design a self-stabilizing clustering algorithm dividing the network into non-overlapping clusters of depth two, while Caron *et al.* [6] considered the k-clustering problem in which each node is at most at distance k from its cluster head. Recently, Datta *et al.* [10] design a self-stabilizing k-clustering algorithm guaranteeing an approximation ratio in unit disk graphs.

All the works presented above concern the decomposition of the graph using different patterns. However, none of them allow to construct a disjoint maximal clique partition of the graph. Note that Ishii and Kakugawa [26] compute a set of maximal cliques which are not necessary disjoint. Moreover, the *non trivial* cliques (with at least two nodes) of the partition must be connected.

In [11], Delbot *et al.* studied the decomposition of an input graph in cliques while satisfying a connectivity property. They propose a centralized algorithm for the Connected Minimal Clique Partition problem (see Definition 1). The proposed algorithm constructs iteratively a set S of maximal cliques. At the beginning of the algorithm, S is empty and a node $u_1 \in V$ is randomly (with equiprobability) selected. A first maximal clique C_1 containing u_1 is added to S and all the nodes of C_1 are marked in G. Then for any iteration i, any non marked node $u_i \in V, 1 \leq i \leq k$, neighbor of at least one marked node of G is randomly (with equiprobability) selected. As for the first clique, a new maximal clique containing u_i is greedily built among non marked nodes of G. This procedure is executed iteratively while there is a non marked node in G. As mentioned in [11], every *trivial* clique (clique of size one) in the constructed set S is neighbor of no other trivial clique. Otherwise, it could be possible to merge two trivial cliques in order to obtain a clique of size two. So, the set of trivial cliques of any minimal partition computed by this algorithm induces an independent set of G.

3.2   Self-stabilizing Construction

In this section, we present our self-stabilizing algorithm $\mathcal{SS} - \mathcal{CMCP}$ which is based on the approach proposed by Delbot *et al.* [11] (see description in the precedent subsection). A formal description of $\mathcal{SS} - \mathcal{CMCP}$ is given in Algorithm 1. To design a distributed version of this approach, we consider here a designated node in the network called the *root* node r, and distances (in hops) from r given in input at each node p noted $dist_p$. These distance values can be obtained by computing a BFS tree rooted at r. Several self-stabilizing BFS algorithms can be used, e.g., [9, 16, 25, 28]. As described below, this allows to define an order on the construction of a clique partition of the graph.

In the proposed algorithm, the construction of maximal cliques is performed starting from the root r and following the distances in the graph. Indeed, the pair *(distance, node identifier)* allows to define a construction priority for the cliques. First of all, each node exchanges its neighbors set in its neighborhood, allowing for each node to know its 2-hops neighborhood. The 2-hops neighborhood is used by each node to identify its neighbors which can belong to its maximal clique. For each node p, we define by *candidate leaders* the set of neighbors q of p such that the pair $(dist_q, \mathrm{ID}_q)$ is lexicographically smaller than $(dist_p, \mathrm{ID}_p)$.

In Algorithm $\mathcal{SS} - \mathcal{CMCP}$, each node p can construct its maximal clique by selecting in a greedily manner a set of neighbors $S \subseteq Neig_p$ such that (i) for any $q \in S$ we have $(dist_p, \mathrm{ID}_p) < (dist_q, \mathrm{ID}_q)$ and (ii) $S \cup \{p\}$ is a complete subgraph. A node p is called a *local leader* if it has not been selected by one of its candidate leaders, otherwise p is no more a local leader and clears out its set S. Each node selected by one of its candidate leaders has to accept only the selection of its candidate leader q of smallest pair $(dist_q, \mathrm{ID}_q)$. Finally, any local leader p which has initiated the construction of its maximal clique considers in its clique only the selected neighbors which have accepted p's selection.

The proposed algorithm maintains a connectivity property between non trivial cliques of the constructed partition. This is a consequence of the construction order of the maximal cliques, which follows the distances in the network from r. Indeed, every non trivial clique C_i (that does not contain the root node r) is adjacent to at least another non trivial clique C_j, such that $dist_{l_j} < dist_{l_i}$ with l_k the local leader of the clique C_k. Otherwise, l_i has been selected to belong to the maximal clique of another local leader l_g such that $(dis_{l_g}, \mathrm{ID}_g) < (dist_{l_i}, \mathrm{ID}_i)$, and the maximal clique C_i would have been removed. In fact, the algorithm constructs a specific clique partition among the possible partitions computed by the centralized approach proposed in [11].

Detailed description. In the following, we give more details on the proposed algorithm $\mathcal{SS} - \mathcal{CMCP}$. Our algorithm is composed of four rules executed by every node and five variables maintained at each node $p \in V$:

- N_p: this variable contains the set of neighbors of p which allows to each node to be informed of its 2-hops neighborhood,
- d_p: this variable is used to exchange the value of $dist_p$ with p's neighbors,
- S_p: this variable is used by p to indicate in its neighborhood the nodes selected by p (if p is a local leader),
- C_p: this variable contains the set of nodes which belong to the maximal clique of p (if p is a local leader),
- $lead_p$: this variable stores the local leader in the neighborhood of p.

As explained above, each node stores in variable N_p the set of its 1-hop neighborhood, this is done using Rule N-action of the algorithm which is executed in case we have $N_p \neq Neig_p$. Rule N-action allows also to correct variable d_p such that $d_p = dist_p$. The information stored in this variable is used by each node in p's neighborhood for the computation of the maximal cliques.

For each node, the set of candidate leaders is given by Macro $LNeig_p$, and among this set of nodes the Macro $SNeig_p$ indicates the neighbors which have selected p for the construction of their own maximal clique. Every node p which is not selected by a candidate leader does not satisfy Predicate $Selected(p)$ and can execute $C1$-action to start the construction of its own maximal clique. The procedure $Clique\_temp()$ selects in a greedily manner the neighbors which form with p a complete subgraph. By executing $C1$-action, a node p stores its identifier in its variable $lead_p$ to become a local leader and notifies with its variable S_p the neighbors it has selected using Procedure $Clique\_temp()$. $C1$-action can be executed by a node p only if S_p does not contain the correct set of selected neighbors, i.e., we have $S_p \neq Clique\_temp()$.

Then, each node p selected by a candidate leader (i.e., which satisfies Predicate $Selected(p)$) can execute $C2$-action to accept the selection of its candidate leader q of smallest pair $(dist_q, \mathrm{ID}_q)$. In this case, we say that q has been elected as the local leader of p. This is given by Macro $Leader_p$ and stored in the variable $lead_p$. $C2$-action is only executed if the variable $lead_p$ does not store the correct local leader for p, i.e., we have $lead_p \neq Leader_p$.

Finally, $C3$-action allows to each local leader p to establish the set of neighbors q which are contained in its maximal clique. This set is stored in variable C_p and is given by Macro $Clique(p)$ considering only the neighbors q of p which have accepted the selection of p (i.e., $lead_q = \mathrm{ID}_p$). This last rule is executed only by local leaders which are not selected to belong to another clique (i.e., $Selected(p)$ is not satisfied) and have not computed the correct set of neighbors contained in their maximal clique (i.e., $S_p = Clique\_temp()$ and $C_p \neq Clique_p$).

Example of an execution. We illustrate with an example given in Figure 1 how the proposed algorithm $SS - CMCP$ constructs a Connected Minimal Clique Partition. In this example, we consider a particular execution following the distances in the graph and we give only the correct cliques which are constructed by the algorithm. We consider the topology given in Figure 1(a). First of all, each node exchanges its neighbors set using N-action. The root node r cannot be selected by one of its neighbors, so by executing $C1$-action it becomes a local leader (i.e., $lead_r = \mathrm{ID}_r$) and selects among its neighbors the nodes to include in its maximal clique, i.e., by indicating in its variable S_r the nodes 1, 2 and 5. Then, nodes 1, 2 and 5 detect that they have been selected by r (their unique possible candidate leader) and in response they accept r's selection using $C2$-action. The node r executes $C3$-action to construct its maximal clique by adding in its variable C_r the nodes which have accepted r's selection, i.e., nodes 1, 2 and 5, as illustrated in Figure 1(b). Next, the nodes 3, 4 and 6 become local leaders since they are not selected to belong to a clique. They execute $C1$-action to select among their neighbors of equal or higher distance those which form a complete subgraph (including themselves), i.e., neighbors 10 and 15 for node 3, neighbor 7 for node 4 and neighbor 9 for 6. The selected neighbors execute $C2$-action to accept the selection of their single candidate leader. We remind that in case of

Algorithm 1. Self-Stabilizing Connected Minimal Clique Partition $\forall p \in V$

Inputs: $Neig_p$: set of (locally) ordered neighbors of p;
   ID_p: unique identifier of p;
   $dist_p$: distance between p and the root (leader node);
Variables:
   N_p: variable used to exchange the neighbors set $Neig_p$ in p's neighborhood, $N_p \subseteq Neig_p$;
   d_p: variable used to exchange the distance $dist_p$ in p's neighborhood, $d_p \in \mathbb{N}$;
   S_p: variable used by p to select neighbors for the construction of its maximal clique, $S_p \subseteq Neig_p$;
   C_p: variable used to store the set of neighbors belonging to the maximal clique of p, $C_p \subseteq Neig_p$;
   $lead_p$: variable used to store the local leader of p, $lead_p \in Neig_p$;

. . . .

Macros:
   $Clique_p = \{q \in S_p : lead_q = ID_p\}$
   $LNeig_p = \{q \in Neig_p : d_q < d_p \vee (d_q = d_p \wedge ID_q < ID_p)\}$
   $SNeig_p = \{q \in LNeig_p : p \in S_q\}$
   $Leader_p = \begin{cases} \bot & \text{If } SNeig_p = \emptyset \\ \min\{q \in SNeig_p : (\forall s \in SNeig_p : d_q \leq d_s)\} & \text{Otherwise} \end{cases}$

. . . .

Predicate: $Selected(p) \equiv SNeig_p \neq \emptyset$

. . . .

Procedure:
Clique_temp()
1: $S := \{p\}$;
2: **for all** $q \in (Neig_p - LNeig_p)$ **do**
3:    **if** $S \subseteq N_q$ **then**
4:       $S := S \cup \{q\}$;
5:    **end if**
6: **end for**
7: **return** S;

. . . .

Actions:
   N-action :: $N_p \neq Neig_p \vee d_p \neq dist_p$ $\rightarrow N_p := Neig_p; d_p := dist_p$;
   $C1$-action :: $\neg Selected(p) \wedge S_p \neq Clique\_temp()$ $\rightarrow S_p := Clique\_temp(); lead_p := ID_p$;
   $C2$-action :: $Selected(p) \wedge lead_p \neq Leader_p$ $\rightarrow lead_p := Leader_p; S_p := \emptyset; C_p := \emptyset$;
   $C3$-action :: $\neg Selected(p) \wedge S_p = Clique\_temp()$
      $\wedge C_p \neq Clique_p$ $\rightarrow C_p := Clique_p$;

a selection from multiple candidate leaders a selected node accepts the selection of the candidate leader x of smallest pair $(dist_x, ID_x)$ given by Macro *Leader*. Then, the local leaders 3, 4 and 6 execute $C3$-action to construct respectively their maximal clique as illustrated in Figure 1(c). In the same way, nodes 8 and 12 become local leaders and select respectively no neighbor and neighbors 11 and 14 to join their clique. The neighbors selected by node 12 accept its selection and node 12 constructs its maximal clique, while node 8 constructs a trivial clique as illustrated in Figure 1(d). Finally, node 13 becomes a local leader in Figure 1(e).

Definition 2 (Correct clique). *Given a clique partition $C_1, \ldots, C_k$ of a graph $G = (V, E)$, a clique $C_i, 1 \leq i \leq k$, is correct iff the following conditions are satisfied:*

1. *There is a single local leader $p_i \in V$ in C_i;*
2. *p_i has selected a subset $S_{p_i} \subseteq Neig_{p_i}$ of its neighbors such that $\forall q \in S_{p_i}$, $(dist_{p_i}, ID_{p_i}) < (dist_q, ID_q)$ and $p_i \cup S_{p_i}$ forms a maximal clique;*
3. *Every node q selected by p_i has accepted p_i's selection iff p_i is the local leader with the smallest pair $(dist, ID)$ in q's neighborhood;*
4. *Every node selected by p_i which has accepted the selection of p_i belongs to the clique C_i maintained by p_i.*

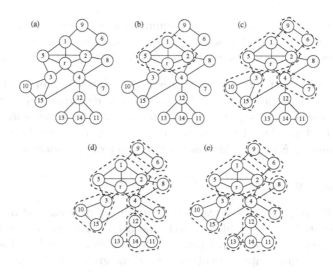

Fig. 1. Execution of Algorithm $SS - CMCP$

Definition 3 (Legitimate configuration). *Let C be the set of all possible configuration of the system. A configuration $\gamma \in C$ is legitimate for Algorithm $SS - CMCP$ iff every constructed clique in γ satisfies Definition 2.*

Lemma 1. *Starting from an arbitrary configuration, the fair composition of Algorithms A_{BFS} and $SS - CMCP$ reach a configuration satisfying Definition 3 in at most $O(T_{BFS} + \min(n_c \times Diam, n))$ (asynchronous) rounds, with T_{BFS} the round complexity of self-stabilizing BFS algorithm A_{BFS}, and n_c the maximum number of cliques at any distance from r in G.*

Theorem 1. *Algorithm $SS - CMCP$ is a self-stabilizing algorithm for Specification 1 under a weakly fair distributed daemon.*

4   Self-stabilizing Connected Vertex Cover

We define below the Connected Vertex Cover problem.

Definition 4 (2-approximation Connected Vertex Cover). *Let $G = (V, E)$ be any undirected graph. A vertex cover S of the graph G is connected iff for any pair of node $u, v \in S$ there is a path between u and v in the graph induced by S. Moreover, S is a 2-approximation Connected Vertex Cover, i.e., we have $|S| \leq 2|CVC^*|$ with CVC^* an optimal solution for the Connected Vertex Cover.*

In [11], Delbot *et al.* presented a centralized optimization algorithm to solve the minimum Connected Vertex Cover problem which uses a solution obtained for the Connected Minimal Clique Partition problem (see Definition 1). Given a Connected Minimal Clique Partition S, the authors have shown in [11] that we can construct a solution S' for the minimum Connected Vertex Cover with an approximation ratio of 2 by selecting in S' all the cliques in S which are not *trivial*, i.e., by selecting all the cliques composed of at least two nodes.

Specification 2 (Self-stabilizing Connected Vertex Cover) *Let C be the set of all possible configurations of the system. An algorithm $\mathcal{A}_{CVC}$ solving the problem of constructing a stabilizing connected vertex cover satisfies the following conditions: 1) Algorithm $\mathcal{A}$ reaches a set of terminal configurations $\mathcal{T} \subseteq C$ in finite time, and 2) Every configuration $\gamma \in \mathcal{T}$ satisfies Definition 4.*

4.1   Related Works

The Vertex Cover problem is a classical optimization problem and many works have been devoted to this problem or to its variations. This problem is known to be APX-complete and not approximable within a factor of $10\sqrt{5} - 21 \approx 1.36067$ [14]. Some very simple approximation algorithms gives a tight approximation ratio of 2 [19,44,48]. Despite a lot of works, no algorithm whose approximation ratio is bounded by a constant less than 2 has been found and it is conjectured that there is no smaller *constant* ratio unless $P = NP$ [34]. Karakostas [33] proposed an algorithm with ratio of $2 - \Theta(\frac{1}{\sqrt{\log n}})$.

From a self-stabilizing point of view, Kiniwa [35] proposed the first self-stabilizing algorithm for this problem which constructs a 2-approximation vertex cover in general networks with unique nodes identifier and under a fair distributed daemon. This algorithm is based on a maximal matching construction which allows to obtain a 2-approximation vertex cover by selecting the extremities of the matching edges. Turau *et al.* [46] considered the same problem in anonymous networks and gave a 3-approximation algorithm under a distributed daemon. Since it is impossible to construct a maximal matching in an anonymous network, this algorithm establishes first a bicolored graph of the network allowing then to construct a maximal matching to obtain a vertex cover. Turau [45] designed a self-stabilizing algorithm for the vertex cover problem with approximation ratio of 2 in anonymous networks under an unfair distributed daemon. This algorithm uses the one proposed in [46] executed several times on sub-parts of the graph to improve the quality of the constructed solution.

For the Connected Vertex Cover problem, Savage [44] proposed a (centralized) algorithm achieving an approximation ratio of 2 in general graphs. It is based on the construction of a Depth First Search tree T and selecting in the solution the nodes with at least a child in T. In 2010 Escoffier *et al.* [18] proved that the problem is NP-complete, even in bipartite graphs (whereas it is polynomial to construct a vertex cover in bipartite graphs). This problem is polynomial in chordal graphs and can be approximated with better ratio than 2 in several restricted graphs classes.

To our knowledge, there exists no self-stabilizing algorithm for the Connected vertex cover problem. However, the approach proposed by Savage [44] can be used to design a self-stabilizing algorithm. Indeed, any self-stabilizing algorithm performing a depth first search traversal of the graph (e.g., see [7,8,43]) used with a modified version of the algorithm described in this section can be used to select the appropriate set of nodes in the solution. However, this does not enable to obtain the best complexity in terms of time. Although a low memory complexity of $O(\log(\Delta))$ bits per node is reached, this approach has a time complexity of $\Theta(n)$ rounds. Indeed, a low level of parallelism is achieved because of the DFS traversal. In contrast, our self-stabilizing algorithm is based on the algorithm presented in the previous section. Our solution has a better time complexity of $O(\min(n_c \times Diam, n))$ rounds because of the parallel construction of cliques, but a higher memory complexity of $O(\Delta \log(n))$ bits per node is necessary.

4.2 Self-stabilizing Construction

In this subsection, we present our self-stabilizing Connected Vertex Cover algorithm called $\mathcal{SS} - \mathcal{CVC}$ which follows the approach given in [11]. A solution to the Connected Vertex Cover problem contains all the non trivial cliques of a Connected Minimal Clique Partition. We give in this section a self-stabilizing algorithm allowing to select the nodes of all the non trivial cliques, a formal description is given in Algorithm 2. So, Algorithm $\mathcal{SS} - \mathcal{CVC}$ is defined as a fair composition [15] of Algorithms 1 and 2 which are executed at each node $p \in V$.

Algorithm 2 takes in input at each node p the local leader of p and the set of nodes belonging to the maximal clique of p given by Algorithm 1 (i.e., variables $lead_p$ and C_p of Algorithm 1) in case p is a local leader. Moreover, in Algorithm 2 each node maintains a single boolean variable In_p. Any node p belongs to the Connected Vertex Cover if and only if (1) either it is a local leader and its maximal clique is not trivial (i.e., $lead_p = ID_p$ and $|C_p| > 1$), or (2) it is contained in a maximal clique constructed by a neighbor which is the local leader of p (i.e., $lead_p \neq ID_p$). Predicate $InVC(p)$ is satisfied at each node p if p is part of the Connected Vertex Cover. Therefore, Algorithm 2 is composed of a single rule executed by each node $p \in V$ to correct the value of Variable In_p in order that In_p equals the value of Predicate $InVC(p)$. So, a solution to the Connected Vertex Cover problem contains every node p such that $In_p = true$.

Algorithm 2.   Self-Stabilizing Connected Vertex Cover algorithm $\forall p \in V$

Inputs: ID_p: unique identifier of p;
     $lead_p$: leader of p computed by Algorithm 1;
     C_p: maximal clique of p computed by Algorithm 1;
Variable: $In_p \in \{true, false\}$;

....

Predicate: $InVC(p) \equiv (lead_p \neq ID_p \vee |C_p| > 1)$

....

Action: $VC\text{-}action :: In_p \neq InVC(p) \rightarrow In_p := InVC(p)$;

Definition 5 (Legitimate configuration). *A configuration $\gamma \in \mathcal{C}$ is legitimate for Algorithm $\mathcal{SS} - \mathcal{CVC}$ iff for every node $p \in V$ we have $In_p = InVC(p)$.*

Lemma 2. *Starting from any configuration, the fair composition of Algorithms $\mathcal{A_{BFS}}$ and $\mathcal{SS} - \mathcal{CVC}$ reach a configuration satisfying Definition 5 in at most*

$O(T_{\mathcal{BFS}} + \min(n_c \times Diam, n) + 1)$ *(asynchronous) rounds, with $T_{\mathcal{BFS}}$ the round complexity of self-stabilizing BFS algorithm $\mathcal{A_{BFS}}$, and n_c the maximum number of cliques at any distance from r in G. $O(\Delta \log(n))$ bits of memory are necessary at each node.*

Theorem 2. *Algorithm $\mathcal{SS} - \mathcal{CVC}$ is a self-stabilizing algorithm for Specification 2 under a weakly fair distributed daemon.*

5   Conclusion

In this paper, we give the first distributed and self-stabilizing algorithm for the minimum Connected Vertex Cover problem with a constant approximation ratio of 2. Moreover, to solve this problem we propose also a self-stabilizing algorithm for the construction of a Connected Minimal Clique partition of the graph. There are two natural perspectives to this work. First, our distributed self-stabilizing clique partition construction need a designated root node. This allows to ensure the connectivity property for the clique partition. This hypothesis is not used in the centralized algorithm. It could be interesting to design a distributed algorithm which does not need this hypothesis while guaranteeing the connectivity property. Second, the self-stabilizing algorithm we propose for the minimum Connected Vertex Cover problem achieves a better time complexity than a self-stabilizing solution based on Savage's approach, but at the price of a higher memory complexity. Therefore, a natural question is to investigate the existence of a self-stabilizing algorithm with a low time and memory complexity.

References

1. Al-Azemi, F.M., Karaata, M.H.: Brief announcement: A stabilizing algorithm for finding two edge-disjoint paths in arbitrary graphs. In: Défago, X., Petit, F., Villain, V. (eds.) SSS 2011. LNCS, vol. 6976, pp. 433–434. Springer, Heidelberg (2011)
2. Bein, D., Datta, A.K., Jagganagari, C.R., Villain, V.: A self-stabilizing link-cluster algorithm in mobile ad hoc networks. In: 8th Int. Symp. on Parallel Architectures, Algorithms, and Networks, pp. 436–441 (2005)
3. Belkouch, F., Bui, M., Chen, L., Datta, A.K.: Self-stabilizing deterministic network decomposition. J. Parallel Distrib. Comput. 62(4), 696–714 (2002)
4. Blum, J., Ding, M., Thaeler, A., Cheng, X.: Connected Dominating Set in Sensor Networks and MANETs. Springer US (2005)
5. Burman, J., Kutten, S.: Time optimal asynchronous self-stabilizing spanning tree. In: Pelc, A. (ed.) DISC 2007. LNCS, vol. 4731, pp. 92–107. Springer, Heidelberg (2007)
6. Caron, E., Datta, A.K., Depardon, B., Larmore, L.L.: A self-stabilizing k-clustering algorithm for weighted graphs. J. Par. Distrib. Comput. 70(11), 1159–1173 (2010)
7. Collin, Z., Dolev, S.: Self-stabilizing depth-first search. Information Processing Letters 49(6), 297–301 (1994)
8. Cournier, A., Devismes, S., Petit, F., Villain, V.: Snap-stabilizing depth-first search on arbitrary networks. The Computer Journal 49(3), 268–280 (2006)
9. Cournier, A., Rovedakis, S., Villain, V.: The first fully polynomial stabilizing algorithm for BFS tree construction. In: Fernàndez Anta, A., Lipari, G., Roy, M. (eds.) OPODIS 2011. LNCS, vol. 7109, pp. 159–174. Springer, Heidelberg (2011)

10. Datta, A.K., Larmore, L.L., Devismes, S., Heurtefeux, K., Rivierre, Y.: Competitive self-stabilizing k-clustering. In: IEEE 32nd Int. Conference on Distributed Computing Systems, pp. 476–485 (2012)
11. Delbot, F., Laforest, C., Phan, R.: New approximation algorithms for the vertex cover problem. In: Lecroq, T., Mouchard, L. (eds.) IWOCA 2013. LNCS, vol. 8288, pp. 438–442. Springer, Heidelberg (2013)
12. Delbot, F., Laforest, C., Rovedakis, S.: Self-stabilizing algorithms for connected vertex cover and clique decomposition problems. Tech. rep., HAL (July 2014), https://hal.archives-ouvertes.fr/hal-01053491
13. Dijkstra, E.: Self-stabilizing systems in spite of distributed control. Commun. ACM 17(11), 643–644 (1974)
14. Dinur, I., Safra, S.: On the hardness of approximating minimum vertex cover. Annals of mathematics 162(1), 439–485 (2005)
15. Dolev, S.: Self-Stabilization. MIT Press (2000)
16. Dolev, S., Israeli, A., Moran, S.: Self-stabilization of dynamic systems assuming only read/write atomicity. Distributed Computing 7(1), 3–16 (1993)
17. Drabkin, V., Friedman, R., Gradinariu, M.: Self-stabilizing wireless connected overlays. In: Shvartsman, M.M.A.A. (ed.) OPODIS 2006. LNCS, vol. 4305, pp. 425–439. Springer, Heidelberg (2006)
18. Escoffier, B., Gourvès, L., Monnot, J.: Complexity and approximation results for the connected vertex cover problem in graphs and hypergraphs. J. Discrete Algorithms 8(1), 36–49 (2010)
19. Garey, M., Johnson, D.: Computers and Intractability. Freeman and Co., New York (1979)
20. Goddard, W., Hedetniemi, S.T., Jacobs, D.P., Srimani, P.K.: Self-stabilizing protocols for maximal matching and maximal independent sets for ad hoc networks. In: 17th Int. Parallel and Distributed Processing Symp., p. 162 (2003)
21. Guellati, N., Kheddouci, H.: A survey on self-stabilizing algorithms for independence, domination, coloring, and matching in graphs. J. Parallel Distrib. Comput. 70(4), 406–415 (2010)
22. Hadid, R., Karaata, M.H.: Stabilizing maximum matching in bipartite networks. Computing 84(1-2), 121–138 (2009)
23. Hedetniemi, S.T., Jacobs, D.P., Srimani, P.K.: Maximal matching stabilizes in time o(m). Inf. Process. Lett. 80(5), 221–223 (2001)
24. Hsu, S.C., Huang, S.T.: A self-stabilizing algorithm for maximal matching. Inf. Process. Lett. 43(2), 77–81 (1992)
25. Huang, S.T., Chen, N.S.: A self-stabilizing algorithm for constructing breadth-first trees. Information Processing Letters 41(2), 109–117 (1992)
26. Ishii, H., Kakugawa, H.: A self-stabilizing algorithm for finding cliques in distributed systems. In: 21st Symp. on Reliable Distributed Systems, pp. 390–395. IEEE Computer Society (2002)
27. Jain, A., Gupta, A.: A distributed self-stabilizing algorithm for finding a connected dominating set in a graph. In: 6th Int. Conference on Parallel and Distributed Computing, Applications and Technologies, pp. 615–619. IEEE Comp. Soc. (2005)
28. Johnen, C.: Memory-efficient self-stabilizing algorithm to construct bfs spanning trees. In: 3rd Workshop on Self-stabilizing Systems, pp. 125–140 (1997)
29. Johnen, C., Nguyen, L.H.: Robust self-stabilizing weight-based clustering algorithm. Theor. Comput. Sci. 410(6-7), 581–594 (2009)

30. Kamei, S., Izumi, T., Yamauchi, Y.: An asynchronous self-stabilizing approximation for the minimum connected dominating set with safe convergence in unit disk graphs. In: Higashino, T., Katayama, Y., Masuzawa, T., Potop-Butucaru, M., Yamashita, M. (eds.) SSS 2013. LNCS, vol. 8255, pp. 251–265. Springer, Heidelberg (2013)

31. Kamei, S., Kakugawa, H.: A self-stabilizing distributed approximation algorithm for the minimum connected dominating set. Int. J. Found. Comput. Sci. 21(3), 459–476 (2010)

32. Kamei, S., Kakugawa, H.: A self-stabilizing 6-approximation for the minimum connected dominating set with safe convergence in unit disk graphs. Theor. Comput. Sci. 428, 80–90 (2012)

33. Karakostas, G.: A better approximation ratio for the vertex cover problem. In: Int. Colloquium on Automata, Languages and Programming, pp. 1043–1050 (2005)

34. Khot, S., Regev, O.: Vertex cover might be hard to approximate to within $2 - \epsilon$. Journal of Computer and System Sciences 74(3), 335–349 (2008)

35. Kiniwa, J.: Approximation of self-stabilizing vertex cover less than 2. In: Tixeuil, S., Herman, T. (eds.) SSS 2005. LNCS, vol. 3764, pp. 171–182. Springer, Heidelberg (2005)

36. Manne, F., Mjelde, M.: A self-stabilizing weighted matching algorithm. In: Masuzawa, T., Tixeuil, S. (eds.) SSS 2007. LNCS, vol. 4838, pp. 383–393. Springer, Heidelberg (2007)

37. Manne, F., Mjelde, M., Pilard, L., Tixeuil, S.: A new self-stabilizing maximal matching algorithm. Theor. Comput. Sci. 410(14), 1336–1345 (2009)

38. Manne, F., Mjelde, M., Pilard, L., Tixeuil, S.: A self-stabilizing 2/3-approximation algorithm for the maximum matching problem. Theor. Comput. Sci. 412(40), 5515–5526 (2011)

39. Neggazi, B., Haddad, M., Kheddouci, H.: Self-stabilizing algorithm for maximal graph decomposition into disjoint paths of fixed length. In: 4th Workshop on Theoretical Aspects of Dynamic Distributed Systems, pp. 15–19. ACM (2012)

40. Neggazi, B., Haddad, M., Kheddouci, H.: Self-stabilizing algorithm for maximal graph partitioning into triangles. In: Richa, A.W., Scheideler, C. (eds.) SSS 2012. LNCS, vol. 7596, pp. 31–42. Springer, Heidelberg (2012)

41. Neggazi, B., Turau, V., Haddad, M., Kheddouci, H.: A self-stabilizing algorithm for maximal p-star decomposition of general graphs. In: Higashino, T., Katayama, Y., Masuzawa, T., Potop-Butucaru, M., Yamashita, M. (eds.) SSS 2013. LNCS, vol. 8255, pp. 74–85. Springer, Heidelberg (2013)

42. Ni, S.Y., Tseng, Y.C., Chen, Y.S., Sheu, J.P.: The broadcast storm problem in a mobile ad hoc network. In: 5th Annual ACM/IEEE Int. Conference on Mobile Computing and Networking, pp. 151–162 (1999)

43. Petit, F., Villain, V.: Optimal snap-stabilizing depth-first token circulation in tree networks. Journal of Parallel and Distributed Computing 67(1), 1–12 (2007)

44. Savage, C.D.: Depth-first search and the vertex cover problem. Information Processing Letters 14(5), 233–237 (1982)

45. Turau, V.: Self-stabilizing vertex cover in anonymous networks with optimal approximation ratio. Parallel Processing Letters 20(2), 173–186 (2010)

46. Turau, V., Hauck, B.: A fault-containing self-stabilizing (3-2/(delta+1))-approximation algorithm for vertex cover in anonymous networks. Theoretical Computer Science 412(33), 4361–4371 (2011)

47. Turau, V., Hauck, B.: A new analysis of a self-stabilizing maximum weight matching algorithm with approximation ratio 2. Theor. Comp. Sci. 412(40), 5527–5540 (2011)

48. Vazirani, V.V.: Approximation algorithms. Springer-Verlag New York, Inc., New York (2001)

Fast and Compact Distributed Verification and Self-stabilization of a DFS Tree

Shay Kutten and Chhaya Trehan

Faculty of Industrial Engineering and Management, Technion, Haifa, Israel
kutten@ie.technion.ac.il, chhaya.dhingra@gmail.com

Abstract. We present algorithms for distributed verification and silent-stabilization of a DFS(Depth First Search) spanning tree of a connected network. Computing and maintaining such a DFS tree is an important task, e.g., for constructing efficient routing schemes. Our algorithm improves upon previous work in various ways. Comparable previous work has space and time complexities of $O(n \log \Delta)$ bits per node and $O(nD)$ respectively, where Δ is the highest degree of a node, n is the number of nodes and D is the diameter of the network. In contrast, our algorithm has a space complexity of $O(\log n)$ bits per node, which is optimal for silent-stabilizing spanning trees and runs in $O(n)$ time. In addition, our solution is modular since it utilizes the distributed verification algorithm as an independent subtask of the overall solution. It is possible to use the verification algorithm as a stand alone task or as a subtask in another algorithm. To demonstrate the simplicity of constructing efficient DFS algorithms using the modular approach, we also present a (non-silent) self-stabilizing DFS token circulation algorithm for general networks based on our silent-stabilizing DFS tree. The complexities of this token circulation algorithm are comparable to the known ones.

Keywords: Fault Tolerance, Self-* Solutions, Silent-Stabilization, DFS, Spanning Trees.

1 Introduction

A clear separation is common between the notions of computing and verification in sequential systems. A similar separation in the context of distributed systems has been emerging. Distributed verification of global properties like minimum spanning trees have been devised [21].

An area of distributed systems that can greatly benefit from this separation is that of self-stabilization. Self-stabilization is the ability of a system to recover from transient faults. A self-stabilizing distributed system can be started in any arbitrary configuration and must eventually converge to a desired legal behavior. Self-stabilizing algorithms can run a distributed verification algorithm repeatedly to detect the occurrence of faults in the system and take the necessary action for convergence to a legal behavior. This is the approach we take here in devising a silent-stabilizing DFS algorithm. The concept of first detecting a fault and

M.K. Aguilera et al. (Eds.): OPODIS 2014, LNCS 8878, pp. 323–338, 2014.

then taking the corrective measures for self-stabilization was first introduced by [20], [1] and [3]. The approach taken by Katz and Perry in [20] is that of global detection of faults by a leader node that periodically takes the *snapshots* of the global state of the network and resets the system if a fault is detected. Afek, Kutten and Yung [1], and Awebuch et al. [3] on the other hand, suggested that the faults in the global state of a system could sometimes be detected by local means - i.e., by having each node check the states of all its neighbors. Göös and Suomela further formalized the idea of local detection of faults in [16]. Korman, Kutten and Peleg [23] introduced the concept of *proof labeling schemes*. A *proof labeling scheme* works by assigning a *label* to every node in the input network. The collection of labels assigned to the nodes acts as a locally checkable *distributed proof* that the global state of the network satisfies a specific global predicate. A *proof labeling scheme* consists of a pair of algorithms $(\mathcal{M}, \mathcal{V})$, where $\mathcal{M}$ is a *marker* algorithm that generates a label for every node and $\mathcal{V}$ is a *verifier* algorithm that checks the *labels* of neighboring nodes. In this paper, we present a *proof labeling scheme* for detecting faults in the distributed representation of a DFS spanning tree. For self-stabilization, the DFS tree is computed afresh and new labels are assigned to the nodes by the marker on detection of faults.

1.1   Additional Related Work

Dijkstra introduced the concept of Self-stabilization [10] in distributed systems. Self-stabilization deals with the *faults* that entail an arbitrary corruption of the state of a system. These faults are rather severe in nature but do not occur very frequently in reality [31].

Table 1 summarizes the known complexity results for self-stabilizing DFS algorithms. Collin and Dolev presented a silent-stabilizing DFS tree algorithm in [6]. Their algorithm works by having each node store its path to the root node in the DFS tree. Since the path of a node to the root in a DFS tree can be as long as n, the number of nodes in the network, the space complexity of their algorithm is $O(n \log \Delta)$ per node, where Δ is the highest degree of a node in the network. The time complexity of their algorithm under the *contention* time model is $(nD\Delta)$. We drop the multiplicative factor of Δ from their time complexity here for the sake of comparison with all the other algorithms that do not count their time under the *contention* model. Cournier et al. presented a snap-stabilizing DFS *wave* protocol in [7] which snap stabilizes with a space complexity of $O(n \log n)$.

Considerable work has been invested in developing self-stabilizing depth-first token circulation algorithms with multiple successive papers improving each other. All of these algorithms also generate a DFS tree in every token circulation round, however these algorithms are not silent. Self-stabilizing depth-first token circulation on arbitrary rooted networks was first considered by Huang and Chen in [17]. Their algorithm stabilizes in $O(nD)$ time with a space complexity of $O(\log n)$ bits per node. Subsequently several self-stabilizing DFS token circulation algorithms [9,19,18,26] were devised. All these papers worked on improving the space complexity of [17] from $O(\log n)$ to a function of Δ, the highest

Table 1. Comparing self-stabilizing DFS algorithms

Algorithm	Space	Stabilization Time	Remarks
[6]	$O(n \log \Delta)$	$O(nD)$	Silent
[7]	$O(n \log n)$	0	Snap Stabilizing *first* DFS wave, needs Unique IDs
[8]	$O(\log n)$	0	Snap Stabilizing Wave takes $O(n^2)$ rounds
[17]	$O(\log n)$	$O(nD)$	Token Circulation, not silent
[25]	$O(\log n)$	$O(n)$	Token Circulation, not silent
[9]	$O(\log \Delta)$	$O(nD)$	Token Circulation, not silent
[19]	$O(\log \Delta)$	$O(nD)$	Token Circulation, not silent Requires neighbor of neighbor info
[18]	$O(\Delta)$	$O(nD)$	Token Circulation, not silent
[26]	$O(\log \Delta)$	$O(nD)$	Token Circulation, not silent
OUR RESULTS	$O(\log n)$	$O(n)$	**Two algorithms: Silent and token circulation; both with the same complexity**

degree of a node in the network. The time complexity of all of the above token circulation algorithms [17,19,18,26] is $O(nD)$ rounds, which is much more than the time it takes for one token circulation cycle on a given network. Petit improved the stabilization time complexity of depth-first token circulation to $O(n)$ in [25] with a space complexity of $O(\log n)$ bits per node. Petit and Villain [28] presented the first self-stabilizing depth-first token circulation algorithm that works in asynchronous message passing systems.

1.2 Our Contribution

The main contribution of the current paper is a silent self-stabilizing DFS spanning tree algorithm. The space complexity of our algorithm is $O(\log n)$ bits per node. The only other *silent-stabilizing* DFS tree algorithm [6] has a space complexity of $O(n \log \Delta)$. Dolev et al. [12] established a lower bound of $O(\log n)$ bits per node on the memory requirement of silent-stabilizing spanning tree algorithms. Thus, ours is the first memory optimal silent-stabilizing DFS spanning tree algorithm. The silent-stabilizing DFS construction algorithm is designed in a modular way consisting of separate modules for fault detection and correction. The distributed verification module of this algorithm can be considered a contribution in itself.

Composing self-stabilizing primitives using fair combination of protocols is a well-known technique(see e.g. [13,30]) to ensure that the resulting protocol is self-stabilizing. We use this approach of protocol combination to design a self-stabilizing depth-first token circulation algorithm which uses our silent-stabilizing DFS tree as a module of the overall algorithm. The space and time complexities of our token circulation algorithm are as good as the previously published work on *fast* self-stabilizing depth-first token circulation [25].

1.3   Outline of the paper

In the next section (Section 2), we describe the model of distributed systems considered in this paper. That section also includes some basic definitions and notations. Section 3 addresses the distributed verification algorithm which acts as the *Verifier* $\mathcal{V}$ of the proof labeling scheme. The *Marker* $\mathcal{M}$ of the *proof labeling scheme* is presented in Section 4. Section 5 describes the technique used to make the algorithm self-stabilizing. Section 6 presents the correctness proofs and performance analysis. Section 7 describes a token circulation scheme based on the new silent-stabilizing DFS spanning tree.

2   Preliminaries

A distributed system is represented by a connected undirected graph $G(V, E)$ without self-loops and parallel edges, where each node $v \in V$ represents a processor in the network and each edge $e \in E$ corresponds to a communication link between its incident nodes. Processors communicate by writing into their own shared registers and reading from the shared registers of the neighboring processors. The network is assumed to be *asynchronous*. We do not require processors to have unique identifiers. We do assume the existence of a distinguished processor, called the root of the network. Each node $v \in V$ orders its edges by some arbitrary ordering α_v as in [6]. For an edge (u, v), let $\alpha_u(v)$ denote the index of the edge (u, v) in α_u.

As opposed to Collin and Dolev [6], We use the (rather common) ideal time complexity which assumes that a node reads all of its neighbors in at most one time unit. Our results translate easily to an alternative, stricter, *contention* time complexity used by Collin and Dolev in [6], where a node can access only one neighbor in one time unit. The time cost of such a translation is a multiplicative factor of $O(\Delta)$, the maximum degree of a node (it is not assumed that Δ is known to nodes). As is commonly assumed in the case of self-stabilization, each node has only some bounded number of memory bits available to be used. Here, this amount of memory is $O(\log n)$.

Self-stabilization and silent-stabilization: A distributed algorithm is self-stabilizing if it can be started in any arbitrary global state and once started, the algorithm converges to a legal state by itself and stays in the legal state unless additional faults occur [11]. A self-stabilizing algorithm is *silent* if starting from an arbitrary state it converges to a legal global state after which the values stored in the communication registers do not change, see e.g. [12]. While some problems like token circulation are non-silent by nature, many *input/output* algorithms allow a silent solution.

Spanning Tree: Distributed Representation: A spanning tree T of a connected, undirected graph $G(V, E)$ is a tree composed of all the nodes and some of the edges of G. A spanning tree T of some graph G is represented in a distributed manner by having each node locally mark some of its incident edges such that the collection of marked edges of all the nodes forms a spanning tree of G. Actually, it is enough that each node marks its edge leading to its parent on the tree in a local variable.

DFS Tree and the *first* DFS Tree of a Graph: A DFS Tree of a connected, undirected graph $G(V, E)$ is the spanning tree generated by a depth first search traversal of G. In a DFS traversal, starting from a specified node called the root, all the nodes of the graph are visited one at a time, exploring as far as possible before backtracking, see e.g. [15]. The *first* DFS traversal is the one that acts as follows: whenever a node v has a set of unexplored edges to choose from, the chosen edge is the edge with the smallest port number in the port ordering α_v. The tree thus generated is called the *first DFS* tree [6]. While a connected, undirected graph can have more than one DFS spanning trees, it can have only one *first DFS* spanning tree.

Lexicographic Ordering. A simple path from the root of a graph G to some node $v \in V$ can be represented as a string starting with a $\perp$ followed by a sequence of the port numbers of the outgoing edges on the path [6]. Given such a string representation of a path, a lexicographic operator $\prec$ can be defined to compare multiple paths of a given node v from the root, where $\perp$ is considered the minimum character. In the *first* DFS tree of a graph, the path leading from the root to some node $v \in V$ is the lexicographically smallest (w.r.t. $\prec$) among all the simple paths from the root to v [6].

DFS Intervals. In a DFS traversal, it is common to assign to each node an interval (in, out) corresponding to the discovery and finish time of exploration of that node. The discovery time or in is the time at which a node is discovered for the first time. The discovery time of a node $v \in V$ is denoted as in_v. The finish time of node v denoted by out_v is the time at which a node has finished exploring all its neighbors. These intervals have the property that given any two intervals (in, out) and (in', out'), either one includes the other or they are totally disjoint. Assuming without loss of generality that $in < in'$, we can write this formally as: either $(in < in' < out' < out)$ or $(in < out < in' < out')$ [15]. In other words, the DFS intervals induce a partial order on the nodes of a graph.

2.1 Notation

We define the following notation to be used throughout:

- $\eta(v)$ denotes the set of neighbors of v in G. $\forall v \in V$ $(\eta(v) = \{u | u \in V \wedge (u, v) \in E)\})$.
- $interval_v$ denotes the (in, out) *label* of v.
- in_v denotes the in *label* of v and out_v denotes the out *label* of v.
- Relational operator $\subset$ between two intervals (in, out) and (in', out') indicates the inclusion of of the first interval in the second one. For example: $(in, out) \subset (in', out')$ indicates that (in, out) is included in (in', out').
- Relational operator $\supset$ is defined similarly.

3   DFS Verification: *Verifier* $\mathcal{V}$

Given a graph $G(V, E)$ and the distributed representation of a spanning tree T of G, the DFS verification algorithm is required to verify that T is the *first* DFS tree of G. The *Verifier* $\mathcal{V}$ takes as input a connected graph $G(V, E)$ where each node $v \in V$ bears an (in_v, out_v) label in addition to v's parent on T. Note that $\mathcal{V}$ takes (in, out) labels of nodes as input and is not concerned with how they are generated.

We assume that each node can read the labels of all its neighbors in addition to its own label and state. A node cannot look at the state of any of its neighbors, however. Each node $v \in V$ periodically reads the labels of all its neighbors and locally computes the following additional information from its own state and label as well as the labels of its neighbors.

3.1   Intermediate Computations

Each node computes the following *macros* to be used for verification.

1. There are zero or more neighbors of v whose interval includes v's interval. Let us call the set of all such nodes the *neighboring ancestors* of v and denote this set by by $anc_l(v)$.

$$anc_l(v) = \{w | w \in \eta(v) \text{ and } interval_w \supset interval_v\}$$

2. The parent of v as perceived by the labels : $parent_l(v) = w | w \in anc_l(v) \wedge \forall u \in anc_l(v) \ (u \neq w \rightarrow interval_w \subset interval_u)$.

3. There are zero or more neighbors of v whose interval is included in v's interval, let us call the set of all such nodes the *neighboring descendants* of v and denote this set by $desc_l(v)$.

$$desc_l(v) = \{w | w \in \eta(v) \text{ and } interval_w \subset interval_v\}$$

4. A *child neighbor* of v is a neighboring descendant of v whose interval is not included in the interval of any other neighboring descendant of v.

$$child_l(v) = u | u \in desc_l(v) \wedge \neg \exists u' \in desc_l(v)(u' \neq u \wedge interval_{u'} \supset interval_u)$$

5. $children_l(v) \subseteq desc_l(v)$ is the set of all *child neighbors* of v.

The subscript l in $anc_l(v)$ above denotes that the set $anc_l(v)$ is computed by the node v by just looking at the labels of v and those of v's neighbors. The same holds for all the other *macros* defined above. It is worth pointing out that all these are intermediate computations and the data they generate need not be stored on the node.

The verification is performed by having each node compute a set of predicates. If T is indeed the *first* DFS tree of G and the labels on all the nodes are proper (i.e. they are as if they were generated by an actual *first* DFS Traversal of the input graph); then the verifier **accepts** continuously on every node until a fault occurs. If a fault occurs either due to the corruption of the state of some nodes or due to some nodes having incorrect labels, at least one node **rejects**. The node that rejects is called a detecting node. The verifier self-stabilizes trivially since it runs periodically.

3.2 Local Interval Predicates

Let $parent_v$ denote the local variable used to store the parent of v in T. Following is the set of local predicates that each node has to compute:

3.2.1 Predicates for the Root Node r

1. $parent_r = null$.
2. $anc_l(r) = \phi$.

3.2.2 Predicates for a Non-root Node v

1. $parent_v \neq null$.
2. $anc_l(v) \neq \phi$.
3. $parent_v = parent_l(v)$. The parent of v on T denoted by $parent_v$ is the same as v's parent as computed by v from the labels of v and its neighbors.
4. $interval_v \subset interval_{parent_v}$.
5. $\forall u \in anc_l(v)$ such that $u \neq parent_v$ $(interval_{parent_v} \subset interval_u)$.

3.2.3 Predicates for Every Node(root as well as a non-root) v

1. $out_v > in_v$.
2. There is no neighbor of v such that its interval is totally disjoint with v. Formally
   $\forall u \in \eta(v)$ $(interval_u \subset interval_v \vee interval_u \supset interval_v)$.
3. if $|children_l(v)| = 0$ then $out_v = in_v + 1$.
4. if $|children_l(v)| > 0$ and let $childrenD_l(v)$ denote the list of children of v sorted in ascending order of their in labels and $firstChild_l(v)$ and $lastChild_l(v)$ be the first and last members of $childrenD_l(v)$ then $in_{firstChild_l} = in_v + 1 \wedge out_v = out_{lastChild_l} + 1$.
5. if $|children_l(v)| > 1$ and let $childrenP_l(v)$ denote the list of children of v sorted in the ascending order of their port numbers in v, then $childrenD_l(v)$ and $childrenP_l(v)$ sort the members of $children_l(v)$ in the same order.
6. Let u and $w \in desc_l(v)$, $u \neq w$, such that $u \in children_l(v)$ and $w \notin children_l(v)$ and $in_u < in_w$ then $\alpha_v(u) < \alpha_v(w)$.
7. $\forall(u, w) \in childrenD_l(v)$ such that u and w are adjacent in $childrenD_l(v)$ and $in_u < in_w$, then $in_w = out_u + 1$

Remark 1. The only predicates that deal with the order in which the neighbors of a node are explored are 5 and 6 of Section 3.2.3. Omitting these two Predicates leaves us with a set of predicates sufficient to verify that T is *some* DFS tree(may not be same as the initial input to the verifier) of G. If an algorithm that uses the verifier as a subtask is not concerned about the order, it can simply drop these predicates.

4   Generating the Labels: *Marker* $\mathcal{M}$

A natural method for assigning the (in, out) labels is to perform an actual DFS traversal of the network starting from the root. The required labels can be generated by augmenting some known DFS tree construction algorithm (e.g. [4], [2], [5]) by adding new variables for the labels and specific actions for updating these label variables. We assume that the DFS construction algorithm of Awerbuch [2] can be easily translated to shared memory and the resulting algorithm can be easily augmented with actions to update the *in* and *out* labels. Note that translating [2] to shared memory is trivial and it decreases the memory from $O(\Delta)$ to $O(log\Delta)$, if it changes memory at all, since a node does not need to store the *VISITED* message(the message broadcasted by a node to all its neighbors when it is visited for the first time, See [2]) of a neighbor, instead it can read the shared register of the neighbor.The pseudo code of the marker will appear in the full paper.

5   The Silent-Stabilizing DFS Construction Algorithm

We have constructed a *proof labeling scheme* $(\mathcal{M}, \mathcal{V})$ with a non-stabilizing marker $\mathcal{M}$ that takes as input a connected graph G and assigns (in, out) labels to every node in G. It also has a verifier $\mathcal{V}$ that takes as input a labeled (with (in, out) intervals) distributed data structure and verifies whether the input structure is the *first* DFS tree. The proofs for the correctness and the performance of $(\mathcal{M}, \mathcal{V})$ are presented in Section 6. In the meanwhile, we use them here assuming they are correct.

A simple way to stabilize any input/output algorithm is to run the algorithm repeatedly to maintain the correct output along with a self-stabilizing synchronizer [3]. This however would not be a silent algorithm. Still, let us use this approach to generate a non-silent self-stabilizing algorithm as an exercise, before presenting the silent one. Awerbuch and Varghese, in their seminal paper [3], present a transformer algorithm for converting a non-stabilizing input/output algorithm into its self-stabilizing version. Following theorem is taken from the paper of Awerbuch and Varghese [3]:

Theorem 1. *Given a non-stabilizing distributed algorithm Π to compute an input/output relation with a space complexity of S_Π and a time complexity of T_Π. The Resynchronizer compiler produces a self-stabilizing version of Π whose time complexity is $O(T_\Pi + \hat{D})$ and whose space complexity is same as that of Π, where $\hat{D}$ is an upper bound on the diameter of the network.*

Informally, the transformer that Awerbuch and Varghese developed to prove the above theorem is a self-stabilizing synchronizer. The transformer takes as input a non-stabilizing input/output algorithm Π whose running time and space requirement are T_Π and S_Π respectively. Another input it takes is $\hat{D}$ which is an *upper bound* on the actual diameter D of the network. Given these inputs, the transformer performs Π for T_Π(recall that the transformer is a synchronizer and transforms the network to be synchronous). Then it retains the results, performs Π again and compares the new results to the old ones. If they are the same, the old results are retained. if they differ, then some faults occurred, the new results are retained. This is repeated forever.

Since we do not assume the knowledge of n (required for input : $T_{\mathcal{M}}$) or $\hat{D}$, we use a slightly modified version of theorem 1 here, that appeared in [22]. The modified Awerbuch Varghese theorem presented in [22] is as follows:

Theorem 2. *Given a non-stabilizing distributed algorithm Π to compute an input/output relation with a space complexity of S_Π and a time complexity of T_Π. The enhanced Resynchronizer compiler produces a self-stabilizing version of Π whose time complexity is $O(T_\Pi+n)$ for asynchronous networks and $O(T_\Pi+D)$ for synchronous networks with a space complexity of $O(S_\Pi + log n)$.*

Informally, Korman et al. used a better synchronizer plus a simple self-stabilizing algorithm that computes n and D to prove the above theorem. To obtain a non-silent self-stabilizing DFS construction algorithm, we just plug the marker $\mathcal{M}$ of Section 4 into theorem 2 and obtain the following corollary.

Corollary 1. *There exists a non-silent self-stabilizing DFS construction algorithm that can operate in a dynamic asynchronous network, with a time complexity of $O(T_{\mathcal{M}} + n)$ and a space complexity of $O(S_{\mathcal{M}} + \log n)$.*

5.1   Achieving Silent-Stabilization

Before going into the details of achieving silence, let us go over how the self-stabilizing synchronizer of the enhanced transformer of theorem 2 helps coordinate repeated executions of the marker in the algorithm of corollary 1. A synchronizer simulates a synchronous protocol in an asynchronous network by using a pulse count at each node which is updated in increments of 1 subject to certain rules. A node u executes the ith step of the algorithm when pulse count at u, $pulse_u$ is equal to i. The synchronizer maintains the *invariant* that the pulse count of a node u differs from any of its neighbors by at most one. Since the synchronizer module is self-stabilizing, all the nodes may be initialized to an arbitrary pulse count and thus the network may not be synchronized in the beginning. The stabilization time of the synchronizer module of the enhanced transformer is $O(n)$, thus starting from any arbitrary set of pulse counters, the network is guaranteed to be synchronized after $O(n)$ time. The enhanced transformer waits for *sufficient* time for the nodes to get synchronized and then starts the execution of the algorithm to be stabilized, in our case, the marker $\mathcal{M}$. If T_e denotes the pulse count at which all the nodes are synchronized, the nodes run the marker from T_e to $T_e + T_{\mathcal{M}}$. Due to an allowed difference of at most 1 between pulse counts of neighboring nodes, the maximum difference between the pulse counts of any two nodes is D, the diameter of the network. Thus any node with a pulse count of $T_e + T_{\mathcal{M}}$ has to wait a maximum of D pulses to be sure that all the nodes in the network have written their output [3]. The node with a pulse count of $T_e + T_{\mathcal{M}} + D$ wraps around its pulse count to 0 which destroys the synchronization. Essentially the first node(s) to *wrap around* invoke the *reset* module of the transformer which brings the nodes back in sync for the next execution of the marker. To make the algorithm silent-stabilizing, we execute the marker(along with the synchronizer) only once in the beginning to generate the labels. The silence is achieved by turning the synchronizer off after all the nodes have finished executing the marker. As explained above, the nodes can easily detect when the marker has finished by looking at their respective pulse counts. When a node reaches a pulse count of $T_e + T_{\mathcal{M}} + D$, it stops updating its pulse count, thus turning the synchronizer off. When all the nodes in the neighborhood of a node have reached $T_e + T_{\mathcal{M}} + D$, it turns on the verifier $\mathcal{V}$. Since $\mathcal{V}$ can detect a fault in exactly one pulse, if one occurs, we can manage without running a synchronizer during the verification. The verifier keeps running repeatedly until a fault occurs. If a node v detects a fault, it invokes the synchronizer of the enhanced transformer again by dropping v's pulse count to 0. Again, as in case of non-silent algorithm, this invokes a reset which resynchronizes the network and subsequently invokes the marker again. Note that the nodes need not know the $T_{\mathcal{M}}$ a priori. The running time of $\mathcal{M}$ is a function of n, the number of nodes which can be computed in a self-stabilizing manner by the module of the enhanced transformer responsible for computing n.

Observation 1 *The only communication that takes place at each node during verification is the reading of the shared registers of the neighbors. The computations performed during verification do not affect the contents of the shared registers at all, thus ensuring silence as defined in [12].*

Thus we obtain a silent-stabilizing DFS construction algorithm. The following theorem summarizes our result:

Theorem 3. *The proof labeling scheme $(\mathcal{M}, \mathcal{V})$ for a DFS tree implies a silent-stabilizing DFS construction algorithm, that runs in $O(T_{\mathcal{M}}+n)$ time with a space complexity of $O(S_{\mathcal{M}} + S_{\mathcal{V}} + \log n)$.*

6   Correctness and Performance Analysis

In this section, we establish the correctness of our algorithm. The proofs follow easily from the known properties of a DFS tree and the predicates of the verifier. Given a labeled (with (in, out) labels) graph $G(V, E)$ and the distributed representation of a spanning subgraph T of G, the following lemmas holds on G, if the local interval predicates (Section 3.2) hold true at every node of G:

Lemma 1. *T is a spanning tree of G.*

Proof. In order to prove that a graph is a tree, it is sufficient to prove that it has no cycles and its number of edges is $n - 1$, where n is the number of nodes in this graph [15]. For the subgraph T of G to have a cycle, one of the ancestors of some node $v \in V$ has to mark v as its parent. However, this leads to a contradiction by predicate 4 of Section: 3.2.2 which requires that the interval of a node be included in the interval of its parent. Applying predicate 4 to v and v's ancestors, implies that for an ancestor u of v which points to v as its parent, $interval(v) \subset interval(u) \wedge interval(u) \wedge interval(v)$, a contradiction. The parent pointer of each node $v \in V$ except the root comprises of a single incident edge of v and the parent pointer of the root is $null$, therefore there are exactly n nodes and $n - 1$ edges in T.

Observation 2 *The macros defined in Section 3.1 extract (periodically) a perceived tree T_l from the (in, out) labels of the nodes in G.*

While input tree T is encoded only by the collection of the parent pointers of the nodes, T_l is extracted by having each node compute its perceived parent, denoted by $parent_l$ as well as its perceived children, denoted by the set $children_l$ on T_l.

Lemma 2. *For any node $v \in V$, the set of children of v in T is same as the set of perceived children of v in T_l.*

Proof. The predicate 3 of section 3.2.2, ensures that the parent pointer $parent_v$ of a node v on the input tree T is the same as v's perceived parent $parent_l(v)$ on T_l. The set of children of a node v on T is implicitly implied by the parent pointers of v's children. Hence, it is sufficient to prove that the set of perceived children of v on T_l is the same as those implied by the perceived parent pointers of perceived children of v, i.e., the collection of perceived parents is consistent with the collection of perceived children on T_l. In what follows, we prove that if a node v has a node p as its perceived parent ($parent_l(v) = p$), then $v \in children_l(p)$. Assume, for contradiction, that the above does not hold. Note that, by the definition of a perceived parent and simple inductive arguments, p has the *narrowest* interval of any node whose interval includes $interval_v$, i.e., the interval of p does not include the interval of any other node whose interval includes $interval_v$. Having $v \notin children_l(p) \wedge parent_l(v) = p$ implies that there is a node $x \in \eta(p)$ with $interval_x \supset interval_v$ and moreover $interval_p \supset interval_x$. This implies that p can not be the parent of v. In a similar way, one can prove that if $c \in children_l(v)$ then v is the perceived parent of c.

Following lemma 2, in the discussion that follows, $children_l(v)$ implies the children of v in T and vice versa.

Lemma 3. *For any two children u, w of a node v in T, the intervals of all the nodes in the subtree of u in T are disjoint from the intervals of all the nodes in the subtree of w in T.*

Proof. The set $childrenD_l(v)$ is the set $children_l(v)$ sorted in the ascending order of the in labels of the nodes $\in children_l(v)$ as defined in Section 3.2.3. Let us assume, without loss of generality, that $in_w > in_u$. Consider a node $u' \in \eta(v)$ such that u' is adjacent to u and appears after u in $childrenD_l(v)$(possibly $u' = w$). Applying predicate 7 of Section 3.2.3 to u and u', $in_{u'} = out_u + 1$. By predicate 1 of Section 3.2.3, $out_{u'} > in_{u'}$. Thus neither of the two intervals, $interval(u)$ and $interval(u')$, includes the other, i.e. they are totally disjoint.

Applying predicate 4 of Section: 3.2.2 inductively, it is easy to see that the intervals of all the descendants of u in T are included in u's own interval. Similarly, the intervals of all the descendants of u' are included in u''s interval . Therefore, intervals of all the descendants of u are disjoint from the intervals of u' and all its descendants. By inductively applying the above argument to every adjacent pair of nodes in $childrenD_l(v)$ starting from u' to w, it is easy to show that the subtrees of any two children of a node have disjoint intervals.

Lemma 4. *For any two children u, w of some node v in T, every simple path in G from some node in the subtree of u to any node in the subtree of w in $T goes through either v or v's ancestors.*

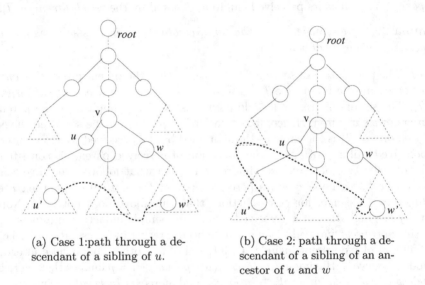

(a) Case 1:path through a descendant of a sibling of u.

(b) Case 2: path through a descendant of a sibling of an ancestor of u and w

Fig. 1. Figure for proof of lemma 4

Proof. Let u' be some node in the subtree of u and w' be some node in the subtree of w. Let us assume, by way of contradiction, that there is a simple path P in G between u' and w' that does not go through v or v's ancestors. There are two possibilities:

- P goes through a descendant of a sibling of u (possibly w).
- or, it goes through a descendant of a sibling of an ancestor (possibly v) of u and w.

Both these cases require an edge to exist in G that connects a pair of nodes in two sibling subtrees, known as a *cross* edge [15]. By lemma 3, the intervals of all the nodes in the subtree of some node x are disjoint from the intervals of all the nodes in the subtree of a sibling of x. Thus, the existence of any such edge in G is ruled out by predicate 2 of Section 3.2.3.

Observation 3 *The proof of Lemma 4 shows that there are no cross edges in the input tree T which implies that T is a DFS(not necessarily the first DFS) tree of G.*

Theorem 4. *If a graph $G(V, E)$ has every node $v \in V$ labeled with its (in, out) interval and interval assignments are such that all the local interval predicates (Section 3.2) hold true at every node, then the spanning tree T encoded in a distributed manner in the states of all the nodes of G is the first DFS tree of G.*

Proof. The problem of finding the *first* DFS Tree of a graph can be thought of as the one of selecting the lexicographically smallest simple path of every node $v \in V$ out of all the simple paths from the root to v, see [6]. Let P_v^T denote the path leading from the root to some node v in T. We now prove that for any node $v \in V$, P_v^T is the lexicographically smallest among all the simple paths from the root to v in G. By way of contradiction, let us assume that there is another simple path P_v^{Alt} from the root to v which is smaller than P_v^T. Let us assume, w.l.o.g., that P_v^T and P_v^{Alt} are the same up-to(and including) some node v_m, the mth node of the common prefix. Let v_{m+1}^T and v_{m+1}^{Alt} denote the $(m+1)^{th}$ node of P_v^T and P_v^{Alt} respectively.

Observation 4 *For P_v^{Alt} to be lexicographically smaller than P_v^T, the edge index (as defined in Section 2) $\alpha_{v_m}(v_{m+1}^{Alt})$ must be smaller than the corresponding index $\alpha_{v_m}(v_{m+1}^T)$.*

There are three possibilities for P_v^{Alt} based on how v_{m+1}^{Alt} is related to v_m :

1. v_{m+1}^{Alt} *is an ancestor of v_m*: This case is ruled out since any such path will not be a simple path.
2. v_{m+1}^{Alt} *is a child of v_m*: v_{m+1}^{Alt} and v_{m+1}^T are both children of v_m. According to lemma 4, there is no simple path from v_{m+1}^{Alt} to any node in the subtree of v_{m+1}^T that does not go through v_m or any of its ancestors. Since v_{m+1}^T falls on P_v^T, v belongs to the subtree of v_{m+1}^T in T. Thus, there is no simple path connecting v_{m+1}^{Alt} to v that does not go through v_m or its ancestors. The path from v_{m+1}^{Alt} to v that goes through either v_m or any of its ancestors would not be a simple path as in case 1. Therefore, this case is also ruled out.
3. v_{m+1}^{Alt} *is a descendant which is not a child of v_m*: This case can be further subdivided into two sub cases:
    (a) v_{m+1}^{Alt} *is also a descendant of V_{m+1}^T in addition to being a descendant of v_m*: This implies that $in_{v_{m+1}^{alt}} > in_{v_{m+1}^T}$. Also, v_{m+1}^T is a child of v_m. This leads to a contradiction due to local interval predicate 6 (Section 3.2.3) which requires that the edge index of the edge (v_m, v_{m+1}^T) be smaller than the edge index of the edge (v_m, v_{m+1}^{Alt}) in $alpha_{v_m}$.
    (b) v_{m+1}^{Alt} *is a proper descendant of of v_m, but not a descendant of v_{m+1}^T* : This case is similar to that of 2.

Theorem 5. *The verifier $\mathcal{V}$ described in section 3 runs in one time unit and requires $O(\log n)$ bits of memory per node.*

Proof. The running time of $\mathcal{V}$ follows from the fact that each node needs to look only at the labels of its immediate neighbors in order to compute its predicates. Every node shares its (in, out) labels with its neighbors. The maximum value of a label is $2n$ which can be encoded using $O(\log n)$ bits.

The following theorem establishes the correctness and performance of the marker $\mathcal{M}$:

Theorem 6. *There exists a marker that constructs the* first *DFS tree and assigns* (in, out) *labels to all the nodes of the input graph* $G(V, E)$ *in time* $O(n)$ *using* $O(\log n)$ *bits of memory per node.*

Proof. As described in Section 4, it is easy to design a marker that adds new actions[1] to a standard DFS tree construction algorithm for computing the *in* and *out* labels. The standard DFS tree construction algorithm in shared memory model, without any actions for computing the (in, out) labels has a space complexity of $O(\log \Delta)$ bits per node. The variables for updating the (in, out) labels require $O(\log n)$ bits per node. Therefore the overall space complexity of such a marker is $O(logn)$.

The actions for computing the labels do not change the values of any of the variables of the original algorithm. Also, these actions do not change the algorithm's flow of control. The addition of these actions cannot violate the correctness of the construction algorithm, nor change its time complexity of $O(n)$.

It is easy to modify the algorithm such that a node v always picks the unvisited neighbor with the smallest port number. This ensures that the output of the algorithm is the *first* DFS tree of the input graph.

7   Self-stabilizing DFS token circulation

The silent-stabilizing DFS tree of Section 5.1 can be combined with a self-stabilizing mutual exclusion algorithm for tree networks to obtain a self-stabilizing token circulation scheme for general networks with a specified root. Self-stabilizing mutual exclusion algorithms that circulate a token in the DFS order on a tree network can be found in [14,24,27]. Petit and Villain presented a space optimal *snap-stabilizing* DFS token circulation algorithm for tree networks in [29] with a *waiting time* (See [29] for a definition of waiting time) of $O(n)$. We can combine our silent-stabilizing DFS tree with the snap stabilizing DFS token circulation protocol of [29] using the fair composition method [13] to obtain a DFS token circulation for general networks. The space complexity of [29] is $O(\log \Delta)$ and that of our silent-stabilizing DFS tree is $O(\log n)$. Therefore the space complexity of the resulting self-stabilizing DFS token circulation algorithm is $O(\log n)$.

Acknowledgements. This research was supported in part by a grant from ISF and Technion TASP center.

---

[1] Actually, these are just common actions of various versions of non-distributed DFS.

References

1. Afek, Y., Kutten, S., Yung, M.: The local detection paradigm and its applications to self-stabilization
2. Awerbuch, B.: A new distributed depth-first-search algorithm. Information Processing Letters 20(3), 147–150 (1985)
3. Awerbuch, B., Varghese, G.: Distributed program checking: A paradigm for building self-stabilizing distributed protocols (extended abstract). In: Proceedings of the 32nd Annual Symposium on Foundations of Computer Science, SFCS, 1991, pp. 258–267. IEEE Computer Society, Washington, DC (1991)
4. Chlamtac, I., Kutten, S.: Tree-based broadcasting in multihop radio networks. IEEE Transactions on Computers 100(10), 1209–1223 (1987)
5. Cidon, I.: Yet another distributed depth-first-search algorithm. Inf. Process. Lett. 26(6), 301–305 (1988)
6. Collin, Z., Dolev, S.: Self-stabilizing depth-first search. Information Processing Letters 49(6), 297–301 (1994)
7. Cournier, A., Devismes, S., Petit, F., Villain, V.: Snap-stabilizing depth-first search on arbitrary networks. Comput. J., 268–280 (2006)
8. Cournier, A., Devismes, S., Villain, V.: A snap-stabilizing DFS with a lower space requirement. In: Tixeuil, S., Herman, T. (eds.) SSS 2005. LNCS, vol. 3764, pp. 33–47. Springer, Heidelberg (2005)
9. Datta, A.K., Johnen, C., Petit, F., Villain, V.: Self-stabilizing depth-first token circulation in arbitrary rooted networks. Distrib. Comput. 13(4), 207–218 (2000)
10. Dijkstra, E.W.: Self-stabilizing systems in spite of distributed control. Commun. ACM 17(11), 643–644 (1974)
11. Dolev, S.: Self-stabilization. MIT Press (2000)
12. Dolev, S., Gouda, M.G., Schneider, M.: Memory requirements for silent stabilization. Acta Informatica 36(6), 447–462 (1999)
13. Dolev, S., Israeli, A., Moran, S.: Self-stabilization of dynamic systems assuming only read/write atomicity. Distrib. Comput. 7(1), 3–16 (1993)
14. Dolev, S., Israeli, A., Moran, S.: Self-stabilization of dynamic systems assuming only read/write atomicity. Distrib. Comput. 7(1), 3–16 (1993)
15. Even, S.: Graph Algorithms. W. H. Freeman & Co., New York (1979)
16. Göös, M., Suomela Locally, J.: checkable proofs. In: Proceedings of the 30th Annual ACM Symposium on Principles of Distributed Computing, PODC 2011, San Jose, CA, USA, June 6-8, pp. 159–168 (2011)
17. Huang, S.-T., Chen, N.-S.: Self-stabilizing depth-first token circulation on networks. Distributed Computing 7(1), 61–66 (1993)
18. Johnen, C., Alari, G., Beauquier, J., Datta, A.K.: Self-stabilizing depth-first token passing on rooted networks. In: Mavronicolas, M. (ed.) WDAG 1997. LNCS, vol. 1320, pp. 260–274. Springer, Heidelberg (1997)
19. Johnen, C., Beauquier, J.: Space-efficient, distributed and self-stabilizing depth-first token circulation. In: In Proceedings of the Second Workshop on Self-Stabilizing Systems, pp. 4–1 (1995)
20. Katz, S., Perry, K.J.: Self-stabilizing extensions for meassage-passing systems. Distributed Computing 7(1), 17–26 (1993)
21. Korman, A., Kutten, S.: Distributed verification of minimum spanning trees. In: Proceedings of the Twenty-fifth Annual ACM Symposium on Principles of Distributed Computing, pp. 26–34. ACM, New York (2006)

22. Korman, A., Kutten, S., Masuzawa, T.: Fast and compact self stabilizing verification, computation, and fault detection of an mst. In: Proceedings of the 30th Annual ACM SIGACT-SIGOPS Symposium on Principles of Distributed Computing, PODC 2011, pp. 311–320. ACM, New York (2011)

23. Korman, A., Kutten, S., Peleg, D.: Proof labeling schemes. Distributed Computing 22(4), 215–233 (2010)

24. Petit, F.: Highly space-efficient self-stabilizing depth-first token circulation for trees. In: Lengauer, C., Griebl, M., Gorlatch, S. (eds.) Euro-Par 1997. LNCS, vol. 1300, pp. 47647–47649. Springer, Heidelberg (1997)

25. Petit, F.: Fast self-stabilizing depth-first token circulation. In: Datta, A.K., Herman, T. (eds.) WSS 2001. LNCS, vol. 2194, pp. 200–215. Springer, Heidelberg (2001)

26. Petit, F., Villain, V.: Color optimal self-stabilizing depth-first token circulation. In: ISPAN, pp. 317–323. IEEE Computer Society (1997)

27. Petit, F., Villain, V.: Optimality and self-stabilization in rooted tree networks. Parallel Processing Letters 10(01), 3–14 (2000)

28. Petit, F., Villain, V.: Self-stabilizing depth-first token circulation in asynchronous message-passing systems. Computers and Artificial Intelligence 19(5) (2000)

29. Petit, F., Villain, V.: Optimal snap-stabilizing depth-first token circulation in tree networks. Journal of Parallel and Distributed Computing 67(1), 1–12 (2007)

30. Stomp, F.A.: Structured design of self-stabilizing programs. In: Proceedings of the 2nd Israel Symposium on the Theory and Computing Systems, pp. 167–176 (June 1993)

31. Varghese, G., Jayaram, M.: The fault span of crash failures. Journal of the ACM 47(2), 244–293 (2000)

Loosely-Stabilizing Leader Election on Arbitrary Graphs in Population Protocols

Yuichi Sudo[1,2], Fukuhito Ooshita[2],
Hirotsugu Kakugawa[2], and Toshimitsu Masuzawa[2]

[1] NTT Secure Platform Laboratories
3-9-11 Midori-cho, Musashino, Tokyo, 180-8585, Japan
sudo.yuichi@lab.ntt.co.jp
[2] Graduate School of Information Science and Technology, Osaka University
1-5 Yamadaoka, Suita, Osaka, 565-0871, Japan
{y-sudou,f-oosita,kakugawa,masuzawa}@ist.osaka-u.ac.jp

Abstract. In the population protocol model Angluin et al. proposed in 2004, there exists no self-stabilizing protocol that solves leader election on complete graphs without knowing the exact number of nodes. To circumvent the impossibility, we previously introduced the concept of *loose-stabilization*, which relaxes the closure requirement of self-stabilization. A loosely-stabilizing protocol guarantees that starting from any initial configuration a system reaches a loosely-safe configuration, and after that, the system keeps its specification (e.g. the unique leader) not forever, but for a sufficiently long time. Our previous work presented a loosely-stabilizing protocol that solves the leader election on complete graphs using only the upper bound N of n, not the exact value of n. We take this work one step further in this paper: We propose two loosely-stabilizing protocols that solve leader election for *arbitrary graphs*. One is a deterministic protocol that uses the identifiers of nodes while the other is a probabilistic protocol that works on anonymous networks. Given the upper bounds N and Δ of the number of nodes and the maximum degree of nodes respectively, both protocols keep a unique leader for $\Omega(Ne^N)$ expected steps after entering a loosely-safe configuration. The former enters a loosely-safe configuration within $O(m\Delta N \log n)$ expected steps while the latter does within $O(m\Delta^2 N^3 \log N)$ expected steps where m is the number of edges of the graph.

Keywords: Loose-stabilization, Population protocols, Leader election.

1 Introduction

The *population protocol* (PP) model, which was presented by Angluin et al.[1], represents wireless sensor networks of mobile sensing devices that cannot control their movement. Two devices (say *agents*) communicate with each other only when they come sufficiently close to each other (we call this event an *interaction*). One example represented by this model is a flock of birds where each bird is equipped with a sensing device with a small transmission range; each device

M.K. Aguilera et al. (Eds.): OPODIS 2014, LNCS 8878, pp. 339–354, 2014.
© Springer International Publishing Switzerland 2014

can communicate with another device only when the corresponding birds come sufficiently close to each other. This unique but meaningful model has attracted broad attention, and there have been numerous studies involving it.

Self-stabilizing leader election (SS-LE) requires that starting from any configuration, a system (say *population*) reaches a safe-configuration in which a unique leader is elected, and after that, the population has the unique leader forever. Self-stabilizing leader election is important in the PP model because (i) many population protocols in the literature work on the assumption that a unique leader exists [1,2,3], and (ii) self-stabilization tolerates any finite number of transient faults and this property suits systems consisting of numerous cheap and unreliable nodes. (Such systems are the original motivation of the PP model.) However, there exists strict impossibility of SS-LE in the PP model: no protocol can solve SS-LE for complete graphs, arbitrary graphs, trees, lines, degree-bounded graphs and so on unless the exact size of the graph (the number of agents n) is available [3]. This impossibility holds even if we strengthen the PP model by assigning unique identifies to agents, allowing agents to use random numbers, introducing memory of communication links (mediated population protocols [10]), or allowing more than two agents (k agents) to interact at the same time (the PP_k model [5]).

Accordingly, many studies of SS-LE took either one of the following two approaches. One approach is to accept the assumption that the exact value of n is available and focus on the space complexity of the protocol. Cai et al. [6] proved that n states of each agent is necessary and sufficient to solve SS-LE for a complete graph of n agents. Mizoguchi et al.[12] and Xu et al.[14] improved the space-complexity by adopting the mediated population protocol model and the PP_k model respectively. The other approach is to use *oracles*, a kind of failure detectors. Fischer and Jiang [8] took this approach for the first time. They introduced oracle Ω? that informs all agents whether at least one leader exists or not and proposed two protocols that solve SS-LE for rings and complete graphs by using Ω?. Beauquier et al.[4] presented an SS-LE protocol for arbitrary graphs that uses two copies of Ω?. Canepa et al.[7] proposed two SS-LE protocols that use Ω? and consume only 1 bit of each agent: one is a deterministic protocol for trees and the other is a probabilistic protocol for arbitrary graphs although the position of the leader is not static and moves among the agents.

Our previous work [13] took another approach to solve SS-LE. We introduced the concept of loose-stabilization, which relaxes the closure requirement of self-stabilization: we allow protocols to deviate from the specification after following it for a sufficiently long time. Concretely, starting from any initial configuration, the population must reach a loosely-safe configuration within a relatively short time; after that, the specification of the problem (the unique leader) must be kept for a sufficiently long time, though not forever. We then proposed a loosely-stabilizing protocol that solves leader election on complete graphs using only an upper bound N of n, not using the exact value of n. Starting from any configuration, the protocol enters a loosely-safe configuration within $O(nN \log n)$

expected steps. After that, the unique leader is kept for $\Omega(Ne^N)$ expected steps. Since the specification is kept for an exponentially long time, we can say this loosely-stabilizing protocol is practically equivalent to a self-stabilizing leader election protocol. Furthermore, this protocol works on any complete graph whose size is no more than N while protocols using the exact value of n work only on the complete graph of size n.

Some works on population protocols assume the probabilistic distribution regarding the interactions of agents: any interaction occurs uniformly at random [1,2,13]. This assumption have been used partly for evaluating the time complexity of protocols. We also adopt this assumption because the measure of time is crucial in the concept of loose-stabilization.

1.1 Our Contribution

In this paper, we consider loosely-stabilizing leader election for *arbitrary undirected graphs*. We adopt two settings: the population with agent-identifiers as in [9] [1] and the population in which agents can use random numbers for state-transition as in [7]. As mentioned above, no self-stabilizing protocol can solve SS-LE for arbitrary graphs, even in these settings, unless the exact value of n is available. For each setting, we propose two protocols P_{ID} and P_{RD} respectively. To elect the unique leader, we take "the minimum ID selection" approach for P_{ID} utilizing the identifiers of agents while we take a novel approach we call "virus war mechanism" for P_{RD} utilizing random numbers.

Given upper bounds N of n and Δ of the maximum degree of nodes, both protocols keep the unique leader for $\Omega(Ne^N)$ expected steps after entering a loosely-safe configuration. Protocol P_{ID} enters a loosely-safe configuration within $O(mN\Delta \log n)$ expected steps while P_{RD} does within $O(mN^3\Delta^2 \log N)$ expected steps where m is the number of edges of the graph. Both protocols consume only $O(\log N)$ bits of each agent's memory. We can say this space complexity is small because even space optimal self-stabilizing protocols that use exact value of n consume $O(\log n)$ bits of each agent[6,12]. For simplicity, our protocols are presented for undirected graphs. However, they work on *directed* graphs with slight modification which is discussed in the conclusion.

Angluin et al.[1] proves that for any population protocol P working on complete graphs, there exists a protocol that simulates P on any arbitrary graph. However, this simulation can be achieved assuming that all the agents have the common initial states at the start of the execution. Since we cannot assume the specific initial states (This is the essence of self-stabilization), we cannot translate our previous loosely-stabilizing algorithm[13] for complete graphs to a loosely-stabilizing algorithm that works for arbitrary graphs.

---

[1] Strictly speaking, our model with identifiers is stronger than the model in [9]. We use identifiers to compare their values while Guerraoui et al.[9] only allow equality-test of identifiers and prohibited any other calculation of identifiers such as value-comparing.

2 Preliminaries

This section defines the model we consider for this paper. The model includes both agent-identifiers and random numbers while protocols P_{ID} and P_{RD} use only one of them. In what follows, we denote set $\{z \in \mathbb{N} \mid x \leq z \leq y\}$ by $[x, y]$.

A *population* is a simple and weakly-connected directed graph $G(V, E, \text{id})$ where V ($|V| \geq 2$) is a set of *agents*, $E \subseteq V \times V$ is a set of directed edges and id defines unique identifiers of agents. Each edge represents a possible *interactions* (or communication between two agents): If $(u, v) \in E$, agents u and v can interact with each other where u serves as an *initiator* and v serves as a *responder*. Each agent v has the unique identifier $\text{id}(v) \in I$ ($I = [0, \text{id}_{\max}]$, $\text{id}_{\max} \in O(n^c)$ for constant c). We say that G is undirected if it satisfies $(u, v) \in E \Leftrightarrow (v, u) \in E$. We define $n = |V|$ and $m = |E|$.

A *protocol* $P(Q, Y, I, R, T, O)$ consists of a finite set Q of states, a finite set Y of output symbols, a set of possible identifiers I, a range of random numbers $R \subset \mathbb{N}$, transition function $T : (Q \times I) \times (Q \times I) \times R \to Q \times Q$, and output function $O : (Q \times I) \to Y$. When an interaction between two agents occurs, T determines the next states of the two agents based on the current states of the agents, identifiers of the two agents, and a random number $r \in R$ generated at each interaction. The *output of an agent* is determined by O: the output of agent v with state $q \in Q$ is $O(q, \text{id}(v))$. We assume that the set of possible identifiers I is a given parameter and not subject to protocol design.

A *configuration* is a mapping $C : V \to Q$ that specifies the states of all the agents. We denote the set of all configurations of protocol P by $\mathcal{C}_{\text{all}}(P)$. We say that configuration C changes to C' by interaction $e = (u, v)$ and integer $r \in R$, denoted by $C \xrightarrow{e,r} C'$, if we have $(C'(u), C'(v)) = T(C(u), \text{id}(u), C(v), \text{id}(v), r)$ and $C'(w) = C(w)$ for all $w \in V \setminus \{u, v\}$. A scheduler determines which interaction occurs at each time. In this paper, we consider a uniformly random scheduler $\Gamma = \Gamma_0, \Gamma_1, \ldots$: each $\Gamma_t \in E$ is a random variable such that $\Pr(\Gamma_t = (u, v)) = 1/m$ for any $t \geq 0$ and any $(u, v) \in E$. We also define the random number sequence as $\Lambda = R_1, R_2, \ldots$: each number $R_t \in R$ is a random variable such that $\Pr(R_t = r) = 1/|R|$ for any $t \geq 0$ and $r \in R$. Given an initial configuration C_0, Γ, and Λ, the *execution* of protocol P is defined as $\Xi_P(C_0, \Gamma, \Lambda) = C_0, C_1, \ldots$ such that $C_t \xrightarrow{\Gamma_t, R_t} C_{t+1}$ for all $t \geq 0$. We denote $\Xi_P(C_0, \Gamma, \Lambda)$ simply by $\Xi_P(C_0)$ when no misunderstanding can arise.

The leader election problem requires that every agent should output L or F which means "leader" or "follower" respectively. We say that a finite or infinite sequence of configurations $\xi = C_0, C_1, \ldots$ preserves a unique leader, denoted by $\xi \in LE$, if there exists $v \in V$ such that $O(C_t(v), \text{id}(v)) = L$ and $O(C_t(u), \text{id}(u)) = F$ for any $t \geq 0$ and $u \in V \setminus \{v\}$. For $\xi = C_0, C_1, \ldots$, the holding time of the leader $\text{HT}(\xi, LE)$ is defined as the maximum $t \in \mathbb{N}$ that satisfies $(C_0, C_1, \ldots, C_{t-1}) \in LE$. We define $\text{HT}(\xi, LE) = 0$ if $C_0 \notin LE$. We denote $\mathbf{E}[\text{HT}(\Xi_P(C), LE)]$ by $\text{EHT}_P(C, LE)$. Intuitively, $\text{EHT}_P(C, LE)$ is the expected number of interactions for which the population keeps the unique leader after protocol P starts from configuration C. For configuration sequence

$\xi = C_0, C_1, \ldots$ and a set of configurations $\mathcal{C}$, we define convergence time $\text{CT}(\xi, \mathcal{C})$ as the minimum $t \in \mathbb{N}$ that satisfies $C_t \in \mathcal{C}$. We define $\text{CT}(\xi, \mathcal{C}) = |\xi|$ if $C_t \notin \mathcal{C}$ for any $t \geq 0$, where $|\xi|$ is the length of ξ. We denote $\mathbf{E}[\text{CT}(\Xi_P(C), \mathcal{C})]$ by $\text{ECT}_P(C, \mathcal{C})$. Intuitively, $\text{ECT}_P(C, \mathcal{C})$ is the expected number of interactions by which the population enters a configuration in $\mathcal{C}$ after P starts from C.

Definition *Protocol* $P(Q, Y, I, R, T, O)$ *is an* (α, β)*-loosely-stabilizing leader election protocol if there exists set* $\mathcal{S}$ *of configurations satisfying two inequalities* $\max_{C \in \mathcal{C}_{\text{all}}(P)} \text{ECT}_P(C, \mathcal{S}) \leq \alpha$ *and* $\min_{C \in \mathcal{S}} \text{EHT}_P(C, LE) \geq \beta$.

2.1   Chernoff Bounds

In this section, we quote the three variants of Chernoff bounds [11] used in several proofs of this paper.

Lemma 1 (from Eq. (4.2) in [11]). *The following inequality holds for any binomial random variable* X:

$$\Pr(X \geq 2\mathbf{E}[X]) \leq e^{-\mathbf{E}[X]/3}.$$

Lemma 2 (from Eq. (4.5) in [11]). *The following inequality holds for any binomial random variable* X:

$$\Pr(X \leq \mathbf{E}[X]/2) \leq e^{-\mathbf{E}[X]/8}.$$

Lemma 3 (from Eq. (4.5) in [11]). *The following inequality holds for any binomial random variable* X:

$$\Pr(X \leq \mathbf{E}[X]/4) \leq e^{-9\mathbf{E}[X]/32}.$$

3   Leader Election with Identifiers

This section presents loosely-stabilizing leader election protocol P_{ID}, which works on arbitrary undirected graphs with unique identifiers of agents (Protocol 1). In the protocol description, we regard a state of agents as a collection of *variables* (e.g. `timer`), and denote a transition function as pseudo code that updates variables of initiator x and responder y. We denote the value of variable `var` of agent $v \in V$ by $v.\text{var}$. We also denote the value of `var` in state $q \in Q$ by $q.\text{var}$.

This protocol elects the agent with the minimum identifier, denoted by v_{min}, as the leader. Each agent v tries to find the minimum identifier and stores it on $v.\text{lid}$. At interaction, two agents x and y compare their `lid` and store the smaller value on their `lid` (Lines 3 and 6), by which the smallest identifier $\text{id}(v_{\text{min}})$ eventually spreads to all the agents. Then, after some point, v_{min} is the unique leader because output function O makes only agents v satisfying $\text{id}(v) = v.\text{lid}$ output L and other agents output F.

Protocol 1. Leader Election with Identifiers P_{ID}

Variables of each agent:

   $\mathtt{lid} \in I$, $\mathtt{timer} \in [0, t_{\max}]$

Output function O:

   if $v.\mathtt{lid} = \mathrm{id}(v)$ holds, then the output of agent v is L; Otherwise, F.

Interaction between initiator x and responder y:

1: **if** $x.\mathtt{lid} > \mathrm{id}(x)$ **then** $x.\mathtt{lid} \leftarrow \mathrm{id}(x)$ **endif**
2: **if** $x.\mathtt{lid} < y.\mathtt{lid}$ **then**
3:    $y.\mathtt{lid} \leftarrow x.\mathtt{lid}$
4:    $x.\mathtt{timer} \leftarrow y.\mathtt{timer} \leftarrow \max(x.\mathtt{timer} - 1,\ 0)$
5: **else if** $x.\mathtt{lid} > y.\mathtt{lid}$ **then**
6:    $x.\mathtt{lid} \leftarrow y.\mathtt{lid}$
7:    $x.\mathtt{timer} \leftarrow y.\mathtt{timer} \leftarrow \max(y.\mathtt{timer} - 1,\ 0)$
8: **else**   // $x.\mathtt{lid} = y.\mathtt{lid}$ at this time
9:    $x.\mathtt{timer} \leftarrow y.\mathtt{timer} \leftarrow \max(x.\mathtt{timer} - 1,\ y.\mathtt{timer} - 1,\ 0)$
10: **end if**
11: **if** $\mathrm{id}(x) = x.\mathtt{lid}$ **or** $\mathrm{id}(y) = y.\mathtt{lid}$ **then**   // a leader resets timers
12:    $x.\mathtt{timer} \leftarrow y.\mathtt{timer} \leftarrow t_{\max}$
13: **else if** $x.\mathtt{timer} = 0$ **then**   // a new leader is created at timeout
14:    $x.\mathtt{lid} \leftarrow y.\mathtt{lid} \leftarrow \min(\mathrm{id}(x),\ \mathrm{id}(y))$
15:    $x.\mathtt{timer} \leftarrow y.\mathtt{timer} \leftarrow t_{\max}$
16: **end if**

However, in the initial configuration, some agents may have false identifiers (or the integers that are not identifiers of any agent in the population) on $\mathtt{lid}$. A false identifier may spread to the population instead of $\mathrm{id}(v_{\min})$ if it is smaller than $\mathrm{id}(v_{\min})$. We define $\mathrm{ID} = \{\mathrm{id}(v) \mid v \in V\}$, which is the correct identifiers set (Note that $\mathrm{ID} \subseteq I$). Protocol P_{ID} removes false identifiers $i \notin \mathrm{ID}$ from $\mathtt{lid}$ of all the agents by the *timeout mechanism*. Specifically, if $x.\mathtt{lid} \neq y.\mathtt{lid}$, we take the timer value of the agent with smaller $\mathtt{lid}$, decrease it by one, and substitute the decreased value into $x.\mathtt{lid}$ and $y.\mathtt{lid}$ (Lines 4 and 7). If $x.\mathtt{lid} = y.\mathtt{lid}$, we take the larger value of $x.\mathtt{timer}$ and $y.\mathtt{timer}$, decrease it by one, and substitute the decreased value into $x.\mathtt{lid}$ and $y.\mathtt{lid}$ (Line 9). We call this event *larger value propagation*. If x or y is a leader, both timers are reset to $t_{\max}$ (Line 12). We call this event *timer reset*. When a timer becomes zero, agents x and y suspect that there exists no leader in the population. In this case, they elect the one with a smaller identifier as a leader by substituting $\min(\mathrm{id}(x), \mathrm{id}(y))$ into $x.\mathtt{lid}$ and $y.\mathtt{lid}$ (Line 14). We call this event *timeout*. Agents with false identifiers never experience timer reset; thus, their timers keep on decreasing. Hence, timeout eventually occurs and their $\mathtt{lid}$s satisfy $\mathtt{lid} \in \mathrm{ID}$. This mechanism rarely ruins the stability of the unique leader because agents with $\mathtt{lid} \in \mathrm{ID}$ keep high value timers because of timer reset and lager value propagation.

Complexity Analysis The upper bound $t_{\max}$ of variable $\mathtt{timer}$ is the only parameter of P_{ID}, which affects the correctness and complexities of the protocol. We assume $t_{\max} \geq 8\delta \max(d,\ 2 + \log n)$ where δ is the maximum degree of the agents and

d is the diameter of population G. (Note that δ is an even number because G is undirected.) We prove the following equations under this assumption:

$$\max_{C \in \mathcal{C}_{all}} \mathrm{ECT}_{P_{ID}}(C, \mathcal{S}_{id}) = O(m\delta\tau \log n), \tag{1}$$

$$\min_{C \in \mathcal{S}_{id}} \mathrm{EHT}_{P_{ID}}(C, LE) = \Omega(\tau e^\tau), \tag{2}$$

where $\tau = t_{max}/(8\delta)$ and $\mathcal{S}_{id}$ is the set of configurations in which $v.\mathtt{lid} = id(v_{min})$ and $v.\mathtt{timer} > t_{max}/2$ hold for all $v \in V$ and $v_{min}.\mathtt{timer} = t_{max}$ holds. When upper bounds N of n and Δ of δ are available and we assign $t_{max} = 8N\Delta$, protocol P_{ID} is an $(O(m\Delta N \log n), \Omega(Ne^N))$-loosely-stabilizing leader election protocol.

First, we analyze the expected holding time. Let $C_0 \in \mathcal{S}_{id}$ and $\Xi_{P_{ID}}(C_0) = C_0, C_1, \ldots$. To prove (2), it suffices to show that both $C_0, \ldots, C_{2m\tau} \in LE$ and $C_{2m\tau} \in \mathcal{S}_{id}$ hold with probability at least $p_{suc} = 1 - O(ne^{-\tau})$. Then, we have $\min_{C_0 \in \mathcal{S}_{id}} \mathrm{EHT}_{P_{ID}}(C_0, LE) \geq 2m\tau/(1 - p_{suc}) = \Omega(\tau e^\tau)$.

Lemma 4. *The probability that every $v \in V$ joins only less than $t_{max}/2$ interactions among $\Gamma_0, \ldots, \Gamma_{2m\tau-1}$ is at least $1 - ne^{-\tau}$.*

Proof. For any $v \in V$ and $t \geq 0$, v joins interaction Γ_t with probability at most δ/m. Thus, the number of interactions v joins during the $2m\tau$ interactions is bounded by binomial random variable $X \sim B(2m\tau, \delta/m)$. Applying a variant of Chernoff bound (Lemma 1), we have

$$\begin{aligned}
\Pr(X \geq t_{max}/2) &= \Pr(X \geq 2\mathbf{E}[X]) &&\because t_{max} = 8\delta\tau \\
&\leq e^{-\mathbf{E}[X]/3} \\
&= e^{-2\delta\tau/3} &&\text{(By Chernoff Bound of Lemma 1)} \\
&\leq e^{-\tau}. &&\because \delta \geq 2
\end{aligned}$$

Summing up the probabilities for all $v \in V$ gives the lemma. □

Lemma 5. *Let $C_0 \in \mathcal{L}_{lid}$ and $\Xi_{P_{ID}}(C_0) = C_0, C_1, \ldots$. Then, we have the following inequality:*

$$\Pr(\forall v \in V, \ C_{2m\tau}(v).\mathtt{timer} > t_{max}/2) \geq 1 - 2ne^{-\tau}.$$

Proof. It suffices to show $\Pr(C_{2m\tau}(v).\mathtt{timer} > t_{max}/2) \geq 1 - 2e^{-\tau}$ for any agent $v \in V$. We denote the shortest path from v_{min} to v by $(v_0, v_1, \ldots, v_k)$ where $v_0 = v_{min}$, $v_k = v$, $0 \leq k \leq d$ and $(v_{i-1}, v_i) \in E$ for all $i = 1, \ldots, k$. For any $t \in [0, 2m\tau]$, we define $v_{head}(t)$ as v_l with maximum $l \in [1, k]$ such that there exist $t_1, t_2, \ldots, t_l$ satisfying $0 \leq t_1 < t_2 < \cdots < t_l < t$ and $\Gamma_{t_i} \in \{(v_{i-1}, v_i), (v_i, v_{i-1})\}$ for $i = 1, 2, \ldots, l$. We define $v_{head}(t) = v_0$ if such l does not exist. Intuitively, $v_{head}(t)$ is the head of the agents in path $(v_0, v_1, \ldots, v_k)$ to which a large timer value is propagated from v_{min}. (Remember that v_{min} resets the timers to t_{max}.) We define $J(t)$ as the number of integers $i \in [0, t]$ such that $v_{head}(i)$ joins interaction Γ_i. Intuitively, $J(t)$ is the number of interactions that the head agent joins among $\Gamma_0, \ldots, \Gamma_t$. Obviously, we have $C_t(v_{\text{head}}(t)).\text{timer} \geq t_{\max} - J(t)$ for any $t \geq 0$.

In what follows, we prove $\Pr(v_{\text{head}}(2m\tau) = v) \geq 1 - e^{-\tau}$ and $\Pr(J(2m\tau) < t_{\max}/2) \geq 1 - e^{-\tau}$, which give $\Pr(C_{2m\tau}(v).\text{timer} > t_{\max}/2) \geq 1 - 2e^{-\tau}$. For any $i \in [1, k]$, a pair v_{i-1} and v_i interacts with probability $2/m$ at each interaction. Hence, we can say each interaction makes v_{head} forward with probability $2/m$. Therefore, by letting Z be a binomial random variable such that $Z \sim B(2m\tau, 2/m)$, we have

$$
\begin{aligned}
\Pr(v_{\text{head}}(t) = v) &= 1 - \Pr(Z < k) \\
&\geq 1 - \Pr(Z < d) \\
&\geq 1 - \Pr\left(Z < \frac{1}{4} \cdot \mathbf{E}[Z]\right) \qquad \because d \leq \tau = \frac{1}{4} \cdot \mathbf{E}[Z] \\
&\geq 1 - e^{-9\mathbf{E}[Z]/32} \qquad \text{(By Chernoff bound of Lemma 3)} \\
&> 1 - e^{-\tau}.
\end{aligned}
$$

The probability that $v_{\text{head}}(t)$ joins interaction Γ_t is at most δ/m regardless of t. Hence, by letting Z' be a binomial random variable such that $Z' \sim B(2m\tau, \delta/m)$, we have

$$
\begin{aligned}
\Pr(J(2m\tau) < t_{\max}/2) &> 1 - \Pr(Z' \geq t_{\max}/2) \\
&= 1 - \Pr(Z' \geq 2\mathbf{E}[Z']) \\
&> 1 - e^{-\mathbf{E}[Z']/3} \qquad \text{(By Chernoff bound of Lemma 1)} \\
&= 1 - e^{-2\delta\tau/3} \\
&> 1 - e^{-\tau}. \qquad \because \delta \geq 2
\end{aligned}
$$

Thus, we have shown $\Pr(C_{2m\tau}(v).\text{timer} > t_{\max}/2) \geq 1 - 2e^{-\tau}$.   □

Lemma 6. $\min_{C \in \mathcal{S}_{\text{id}}} \text{EHT}_{P_{\text{ID}}}(C, LE) = \Omega(\tau e^{\tau})$.

Proof. We have $C_0, \ldots, C_{2m\tau} \in LE$ and $C_{2m\tau} \in \mathcal{S}_{\text{id}}$ if $C_0 \in \mathcal{S}_{\text{id}}$ holds, no timeout happens, and any agent interacts at most $t_{\max}/2$ times during $2m\tau$ interactions. Hence, probability p_{suc} discussed in the beginning of this section is at least $1 - 3ne^{-\tau}$ by Lemmas 4 and 5, which leads to the lemma.   □

Next, we analyze the expected convergence time. To prove (1), we define two sets of configurations: $\mathcal{C}_{\text{lid}} = \{C \in \mathcal{C}_{\text{all}}(P_{\text{ID}}) \mid \forall v \in V, C(v).\text{lid} \in ID\}$ and $\mathcal{L}_{\text{lid}} = \mathcal{C}_{\text{lid}} \cap \{C \in \mathcal{C}_{\text{all}}(P_{\text{ID}}) \mid C(v_{\min}).\text{lid} = \text{id}(v_{\min}) \wedge C(v_{\min}).\text{timer} = t_{\max}\}$.

Lemma 7. $\max_{C \in \mathcal{C}_{\text{all}}(P_{\text{ID}})} \text{ECT}_{P_{\text{ID}}}(C, \mathcal{C}_{\text{lid}}) = O(m\delta\tau \log n)$.

Proof. Let z be the maximum value of $v.\text{timer}$ such that $v.\text{lid} \notin ID$. This z decreases by one every time all interactions of E occur. Thus, it takes at most $\frac{m}{m} + \frac{m}{m-1} + \ldots \frac{m}{1} \leq m(1 + \log m)$ expected steps to decrease z by one. Hence, $\max_{C \in \mathcal{C}_{\text{all}}(P_{\text{ID}})} \text{ECT}_{P_{\text{ID}}}(C, \mathcal{C}_{\text{lid}}) \leq t_{\max} m(1 + \log m) = O(m\delta\tau \log n)$.   □

Lemma 8. $\max_{C \in \mathcal{C}_{lid}} \text{ECT}_{P_{ID}}(C, \mathcal{L}_{lid}) = O(m)$.

Proof. We have $v_{\min}.\texttt{lid} = \text{id}(v_{\min})$ and $v_{\min}.\texttt{timer} = t_{\max}$ just after $v_{\min}$ interacts in any configuration of $\mathcal{C}_{lid}$. This takes $O(m)$ expected interactions.   □

Lemma 9. $\max_{C \in \mathcal{L}_{lid}} \text{ECT}_{P_{ID}}(C, \mathcal{S}_{id}) = O(m\tau)$.

Proof Sketch. Let $C_0 \in \mathcal{L}_{lid}$ and $\Xi_{P_{ID}}(C_0) = C_0, C_1, \ldots$. By similar argument to Lemmas 4 and 5, we can prove $\Pr(C_{2m\tau} \in \mathcal{S}_{id}) > 1 - 2ne^{-\tau}$. Since $C \in \mathcal{L}_{lid}$ cannot change to $D \notin \mathcal{L}_{lid}$, we have $\max_{C \in \mathcal{L}_{lid}} \text{ECT}_{P_{ID}}(C, \mathcal{S}_{id}) \leq 2m\tau + 3ne^{-\tau} \cdot \max_{C \in \mathcal{L}_{lid}} \text{ECT}_{P_{ID}}(C, \mathcal{S}_{id})$. Solving this inequality gives the lemma.   □

The following lemma immediately follows from Lemmas 7, 8, and 9.

Lemma 10. $\max_{C \in \mathcal{C}_{all}(P_{ID})} \text{ECT}_{P_{ID}}(C, \mathcal{S}_{id}) = O(m\delta\tau \log n)$.

Lemmas 6 and 10 gives the following theorem.

Theorem 1 *Protocol P_{ID} is a $(O(m\delta\tau \log n), \Omega(\tau e^\tau))$ loosely-stabilizing leader election protocol for arbitrary graphs when $t_{\max} \geq 8\delta \max(d, 2 + \log n)$.*

Therefore, given upper bound N and Δ of n and δ respectively, we get a $(O(m\Delta N \log n), \Omega(N e^N))$ loosely-stabilizing leader election protocol for arbitrary graphs by assigning $t_{\max} = 8N\Delta$.

4   Leader Election with Random Numbers

This section presents loosely-stabilizing leader election protocol P_{RD}. It works on arbitrary undirected anonymous graphs with a random number generated at each interaction (Protocol 2). Random numbers are used in Line 11: When the protocol enters Line 11, the code is executed with probability $p = 1/|R|$. This is implemented as the code is executed only when a specific number is generated. For example, $p = 0.01$ if we assign $R = [0, 99]$ and treat 0 as a specific number.

Each agent has binary variable $\texttt{DoA} \in \{\text{DEAD}, \text{ALIVE}\}$ and three timers $\texttt{timer}_L$, $\texttt{timer}_V$ and $\texttt{timer}_S$. The output function defines leaders based on $\texttt{DoA}$: agent v is a leader if v is alive (or $v.\texttt{DoA} = \text{ALIVE}$), and a follower if v is dead (or $v.\texttt{DoA} = \text{DEAD}$). Protocol P_{RD} consists of a timeout mechanism (Lines 1-7) and a virus-war mechanism (Lines 8-14). By using $\texttt{timer}_L$, the timeout mechanism creates a leader when it is suspected that no leader exists. By using $\texttt{timer}_V$ and $\texttt{timer}_S$, the virus-war mechanism reduces the number of leaders.

The timeout mechanism is almost the same as P_{ID}. By the timer reset and the larger value propagation, timeout eventually occurs when no leader exists, and all agents keep high timer values with high probability when one ore more leaders exist. At timeout, a dead agent becomes a leader (Line 5).

In the virus-war mechanism, each leader tries to kill other leaders by viruses and become the unique leader. We say that agent v has a virus if $v.\texttt{timer}_V > 0$, and v wears a (head) shield if $v.\texttt{timer}_S > 0$. A leader creates a new virus with probability p when it interacts as an initiator (Line 11). When creating a virus,

---

Protocol 2. Leader Election with Random Numbers P_{RD}

Variables of each agent:

DoA $\in$ {DEAD, ALIVE}, $\mathtt{timer_L} \in [0, t_{max}]$, $\mathtt{timer_V} \in [0, t_{virus}]$, $\mathtt{timer_S} \in [0, t_{shld}]$

Output function O:

if $v.\mathtt{DoA} = $ ALIVE holds, then the output of agent v is L, otherwise F.

Interaction between initiator x and responder y:

1: $x.\mathtt{timer_L} \leftarrow y.\mathtt{timer_L} \leftarrow \max(x.\mathtt{timer_L} - 1,\ y.\mathtt{timer_L} - 1,\ 0)$
2: **if** $x.\mathtt{DoA} = $ ALIVE **or** $y.\mathtt{DoA} = $ ALIVE **then**
3:    $x.\mathtt{timer_L} \leftarrow y.\mathtt{timer_L} \leftarrow t_{max}$ // a leader resets timer
4: **else if** $x.\mathtt{timer_L} = 0$ **then**   // a new leader is created at timeout
5:    $x.\mathtt{DoA} \leftarrow$ ALIVE
6:    $x.\mathtt{timer_L} \leftarrow y.\mathtt{timer_L} \leftarrow t_{max}$
7: **end if**
8: $x.\mathtt{timer_V} \leftarrow y.\mathtt{timer_V} \leftarrow \max(x.\mathtt{timer_V} - 1,\ y.\mathtt{timer_V} - 1,\ 0)$
9: $x.\mathtt{timer_S} \leftarrow \max(0, x.\mathtt{timer_S} - 1)$
10: **if** $x.\mathtt{DoA} = $ ALIVE **then**
11:    Execute ($x.\mathtt{timer_V} \leftarrow t_{virus}$, $x.\mathtt{timer_S} \leftarrow t_{shld}$) with probability p
      // An alive initiator creates a new virus and a new shield with probability p.
12: **end if**
13: **if** $x.\mathtt{timer_V} > 0$ **and** $x.\mathtt{timer_S} = 0$ **then** $x.\mathtt{DoA} \leftarrow$ DEAD **endif**
14: **if** $y.\mathtt{timer_V} > 0$ **and** $y.\mathtt{timer_S} = 0$ **then** $y.\mathtt{DoA} \leftarrow$ DEAD **endif**

---

the agent wears a shield so as not to be killed by the new virus (Line 11). A virus spreads among agents by interactions (Line 8), and an agent is killed when it has a virus without a shield (Lines 13-14). A virus has TTL (time to live), which is memorized on $\mathtt{timer_V}$ and decreased by one at each interaction of its owner (line 8). When $\mathtt{timer_V}$ becomes zero, the virus vanishes and looses the ability to kill agents. A shield also has TTL, which is memorized on $\mathtt{timer_S}$ and decreased by one at each interaction of its owner (Line 9). When $\mathtt{timer_S}$ becomes zero, the shield vanishes and looses the ability to protect its owner from viruses.

The virus-war mechanism correctly works if p is sufficiently small and t_{shld} is sufficiently greater than t_{virus}. Consider the case multiple leaders exist. Since p is small, all viruses and shields eventually vanishes. After that, some agent eventually creates a new virus and shield. The created virus kills all other agents unless some of them also create a new virus and shield before the virus reaches them. Since p is sufficiently small, the probability of the latter is small. Thus, the unique leader is elected within a relatively short time. Even after that, the unique leader keeps on creating new viruses, each of which may kill the leader. However, the leader is not killed for an extremely long time: since $t_{shld} \gg t_{virus}$, the leader's shield rarely vanishes before all viruses vanish from the population.

Complexity Analysis We have four parameters in P_{RD}: three upper bounds t_{max}, t_{virus}, and t_{shld} of the timers, and probability p. We assume $t_{virus} = t_{max}/2$, $t_{max} \geq 8\delta \max(d,\ 2 + \log(13n\delta\lceil \log n \rceil))$, $t_{shld} \geq 2\delta t_{max}\lceil \log n \rceil$ and $p \leq$

$(4mt_{\text{shld}})^{-1}$. We prove the following equations under this assumption:

$$\max_{C\in\mathcal{C}_{\text{all}}} \text{ECT}_{P_{\text{RD}}}(C,\mathcal{S}_{\text{RD}}) = O(mp^{-1}), \tag{3}$$

$$\min_{C\in\mathcal{S}_{\text{RD}}} \text{EHT}_{P_{\text{RD}}}(C,LE) = \Omega(\tau e^{\tau}), \tag{4}$$

where $\tau = t_{\max}/(8\delta)$ and $\mathcal{S}_{\text{RD}}$ is the set of configurations we define later. When upper bounds N and Δ are available and we assign $t_{\max} = 8N\Delta$, $t_{\text{shld}} = 2\Delta t_{\max}\lceil\log N\rceil$ and $p = (4N^2 t_{\text{shld}})^{-1}$ (i.e., $R = [0, 4N^2 t_{\text{shld}} - 1]$), then P_{RD} is an $(O(m\Delta^2 N^3 \log N), \Omega(Ne^N))$-loosely-stabilizing leader election protocol.

Before proving equations (3) and (4), we define five sets of configurations:

$$\mathcal{G}_{\text{half}} = \{C \in \mathcal{C}_{\text{all}}(P_{\text{RD}}) \mid \exists v \in V,\ C(v).\text{DoA} = \text{ALIVE} \wedge C(v).\text{timer}_S > t_{\text{shld}}/2\},$$

$$\mathcal{V}_{\text{clean}} = \{C \in \mathcal{C}_{\text{all}}(P_{\text{RD}}) \mid \forall v \in V,\ C(v).\text{timer}_V = 0\},$$

$$\mathcal{L}_{\text{half}} = \{C \in \mathcal{C}_{\text{all}}(P_{\text{RD}}) \mid \#_L(C) \geq 1 \wedge \forall v \in V,\ C(v).\text{timer}_L > t_{\max}/2\},$$

$$\mathcal{L}_{\text{one}} = \{C \in \mathcal{C}_{\text{all}}(P_{\text{RD}}) \mid \#_L(C) = 1\},$$

$$\mathcal{S}_{\text{RD}} = (\mathcal{G}_{\text{half}} \cup \mathcal{V}_{\text{clean}}) \cap \mathcal{L}_{\text{half}} \cap \mathcal{L}_{\text{one}},$$

where $\#_L(C)$ denotes the number of leaders in configuration C. Note that $\mathcal{G}_{\text{half}}$ requires that not all agents but at least one leader has timer_S more than $t_{\text{shld}}/2$.

First, we analyze the expected holding time. Let $C_0 \in \mathcal{S}_{\text{RD}}$ and $\Xi_{P_{\text{RD}}}(C_0) = C_0, C_1, \ldots$. To prove (4), it suffices to show that both $C_0, \ldots, C_{8m\delta\tau\lceil\log n\rceil} \in LE$ and $C_{8m\delta\tau\lceil\log n\rceil} \in \mathcal{S}_{\text{RD}}$ hold with probability no less than $p_{\text{suc}} = 1 - O(n\delta\log n \cdot e^{-\tau})$. Then, $\min_{C_0\in\mathcal{S}_{\text{RD}}} \text{EHT}_{P_{\text{RD}}}(C_0, LE) \geq 8m\delta\tau\lceil\log n\rceil\tau/(1 - p_{\text{suc}}) = \Omega(\tau e^{\tau})$.

We define two predicates PROP_i and HALF_i for any $i \geq 0$: $\text{PROP}_i = 1$ if $C_{2m\tau(i+1)}(v).\text{timer}_L > t_i - t_{\max}/2$ for all $v \in V$, otherwise $\text{PROP}_i = 0$, where $t_i = \max_{v\in V} C_{2m\tau i}(v)$; $\text{HALF}_i = 1$ if every agent joins only less than $t_{\max}/2$ interactions among $\Gamma_{2m\tau i}, \ldots, \Gamma_{2m\tau(i+1)-1}$, otherwise $\text{HALF}_i = 0$. Intuitively, $\text{PROP}_i = 1$ means the maximum value of timer_L propagates to all the agents well during the $2m\tau$ interactions, and $\text{HALF}_i = 1$ means every agent does not interact so much during the $2m\tau$ interactions.

Lemma 11. *Let $C_0 \in \mathcal{S}_{\text{RD}}$ and $\Xi_{P_{\text{RD}}}(C_0) = C_0, C_1, \ldots$. Then, we have both $C_0, \ldots, C_{8m\delta\tau\lceil\log n\rceil} \in LE$ and $C_{8m\delta\tau\lceil\log n\rceil} \in \mathcal{S}_{\text{RD}}$ if the following conditions hold:*

(A) $\#_L(C_t) \geq 1$ for all $t = 0, \ldots, 8m\delta\tau\lceil\log n\rceil$,
(B) $C_{8m\delta\tau\lceil\log n\rceil} \in \mathcal{G}_{\text{half}} \cup \mathcal{V}_{\text{clean}}$,
(C) $\text{PROP}_i = 1$ for all $i = 0, \ldots, 4\delta\lceil\log n\rceil - 1$, and
(D) $\text{HALF}_i = 1$ for all $i = 0, \ldots, 4\delta\lceil\log n\rceil - 1$.

Proof. We have $C_{2m\tau i}(v).\text{timer}_L > t_{\max}/2$ for any $i \in [0, 4\delta\lceil\log n\rceil]$ from (A) and (C). Since no agent interacts more than $t_{\max}/2$ times among each $2m\tau$ interactions (i.e. (D)), timeout does not occur at any interaction $\Gamma_0, \ldots, \Gamma_{8m\delta\tau\lceil\log n\rceil-1}$, by which we obtain $C_0, \ldots, C_{8m\delta\tau\lceil\log n\rceil} \in LE$. We also obtain $C_{8m\delta\tau\lceil\log n\rceil} \in \mathcal{L}_{\text{half}} \cap \mathcal{L}_{\text{one}} \cap (\mathcal{G}_{\text{half}} \cup \mathcal{V}_{\text{clean}}) = \mathcal{S}_{\text{RD}}$ from above discussion and (B). $\square$

Lemma 12. *The probability that every agent joins only less than $t_{\text{shld}}/2$ interactions as an initiator among $\Gamma_0, \ldots, \Gamma_{8m\delta\tau\lceil\log n\rceil-1}$ is at least $1 - ne^{-\delta\tau}$.*

Proof. For any $v \in V$ and $t \geq 0$, v joins interaction Γ_t as an initiator with probability at most $\delta/(2m)$ since v has at most $\delta/2$ outgoing edges. Thus, the number of interactions v joins as an initiator during the $8m\delta\tau\lceil \log n \rceil$ interactions is bounded by binomial random variable $X \sim B(8m\delta\tau\lceil \log n \rceil, \delta/(2m))$. We have

$$
\begin{aligned}
\Pr(X \geq t_{\text{shld}}/2) &\leq \Pr(X \geq 8\delta^2\tau\lceil \log n \rceil) \qquad \because t_{\text{shld}} \geq 16\delta^2\tau\lceil \log n \rceil \\
&= \Pr(X \geq 2\mathbf{E}[X]) \\
&\leq e^{-\mathbf{E}[X]/3} \qquad \text{(By Chernoff Bound of Lemma 2)} \\
&= e^{-4\delta^2\tau\lceil \log n \rceil/3} \\
&= e^{-\delta\tau}.
\end{aligned}
$$

Summing up these probabilities gives the lemma. $\qquad\square$

Lemma 13. Let $C_0 \in \mathcal{S}_{\text{RD}}$ and $\Xi_{P_{\text{RD}}}(C_0) = C_0, C_1, \ldots$. Then, we have $\Pr(\forall t \in [0, 8m\delta\tau\lceil \log n \rceil - 1], \#_L(C_t) \geq 1) \geq 1 - ne^{-\delta\tau}$.

Proof. By Lemma 12, it suffices to show that $\#_L(C_t) \geq 1$ holds for all $t \in [0, 8m\delta\tau\lceil \log n \rceil]$ when we assume every agent joins only less than $t_{\text{shld}}/2$ interactions as an initiator among $\Gamma_0, \ldots, \Gamma_{8m\delta\tau\lceil \log n \rceil - 1}$. Since $C_0 \in \mathcal{S}_{\text{RD}}$, we have $C_0 \in \mathcal{G}_{\text{half}} \cup \mathcal{V}_{\text{clean}}$. If $C_0 \in \mathcal{G}_{\text{half}}$, there exists a leader v such that $C_0(v).\texttt{timer}_\texttt{S} > t_{\text{shld}}/2$. By the assumption, v decrease its $\texttt{timer}_\texttt{S}$ by at most $t_{\text{shld}}/2$; thus, v is never killed and remains a leader in $C_0, \ldots, C_{8m\delta\tau\lceil \log n \rceil}$. If $C_0 \in \mathcal{V}_{\text{clean}}$, no leader is killed before a new virus is created. Even if some leader u creates a new virus at interaction Γ_t $(0 \leq t < 8m\delta\tau\lceil \log n \rceil)$, u wears a new shield at the same time. Hence, u remains a leader in $C_t, \ldots, C_{8m\delta\tau\lceil \log n \rceil}$ by the assumption. $\qquad\square$

We define the first round time $\text{RT}_\Gamma(1)$ as the minimum t satisfying $\forall e \in E, 0 \leq \exists t' \leq t, \Gamma_{t'} = e$. For any $i \geq 2$, we define the i-th round time $\text{RT}_\Gamma(i)$ as the minimum t satisfying $\forall e \in E, \text{RT}_\Gamma(i - 1) < \exists t' \leq t, \Gamma_{t'} = e$. Lemma 15 bounds $\text{RT}_\Gamma(i)$ from above with high probability. To prove the lemma, we firstly prove Lemma 14.

Lemma 14. Let $v_1, v_2, \ldots, v_l$ be any l $(l < n)$ agents in V. There exists at least $2l$ edges of E that are incident to at least one of the l agents.

Proof. Since $l < n$, there exists agent $r \in V$ that differs from any $v_1, v_2, \ldots, v_l$. Since G is strongly connected, there exists a rooted spanning tree T on G where r is the root agent of T. Then, every v_i $(i \in [1, k])$ has two edges between v_i and the parent agent of v_i in T. (Remind that G is undirected, that is, $(u, v) \in E \Leftrightarrow (v, u) \in E$ for any $u, v \in V$.) These edges are mutually exclusive. Thus, we have $2l$ edges of E that are incident to at least one of the l agents. $\qquad\square$

Lemma 15. $\Pr(\text{RT}_\Gamma(i) < 2im\lceil \log n \rceil) \geq 1 - ne^{-i/4}$ holds for any $i \geq 1$.

Proof. Each round j $(j \geq 1)$ finishes when every agent $v \in V$ interacts in round j. Consider the case that k $(k \geq 1)$ agents have not yet interacted in round j and only $n-k$ agents have interacted in round j. We call the former uninvolved agents and the latter involved agents. If $k < n$, one of the k uninvolved agents joins the next interaction and becomes an involved agent with probability more than $2k/m$ by Lemma 14. If $k = n$, some uninvolved agent joins the next interaction with probability 1. Let $X_{j,k}$ ($j \geq 1$, $k \geq 1$) be the random variable that corresponds to the number of trials to the first success in which the success probability of each trial is $2k/m$. From the above discussion, we obtain

$$\Pr(\mathrm{RT}_\Gamma(i) \geq 2im\lceil \log n\rceil) \leq \Pr\left(\sum_{j=1}^{i}\left(1 + \sum_{k=1}^{n-1}X_{j,k}\right) \geq 2im\lceil\log n\rceil\right)$$

$$\leq \Pr\left(\sum_{k=1}^{n-1}\sum_{j=1}^{i}X_{j,k} \geq 2im\lceil\log n\rceil - i\right). \tag{5}$$

For binomial random variable $Y_k \sim B(\lceil\frac{im}{k}\rceil, \frac{2k}{m})$, we have $\Pr(\sum_{j=1}^{i}X_{j,k} > \frac{im}{k}) \leq \Pr(\sum_{j=1}^{i}X_{j,k} \geq \lceil\frac{im}{k}\rceil) \leq \Pr(Y_k \leq i)$. Hence, we have

$$\Pr\left(\sum_{j=1}^{i}X_{j,k} > \frac{im}{k}\right) \leq \Pr(Y_k \leq i)$$

$$\leq \Pr\left(Y_k \leq \frac{1}{2}\cdot \mathbf{E}[Y_K]\right) \tag{6}$$

$$\leq e^{-\mathbf{E}[Y_k]/8} \qquad \text{(By Chernoff Bound of Lemma 2)}$$

$$\leq e^{-i/4}.$$

From Inequalities (5) and (6), we have

$$\Pr(\mathrm{RT}_\Gamma(i) \geq 2im\lceil\log n\rceil) \leq \Pr\left(\sum_{k=1}^{n-1}\sum_{j=1}^{i}X_{j,k} \geq 2im\lceil\log n\rceil - i\right)$$

$$\leq \Pr\left(\sum_{k=1}^{n-1}\sum_{j=1}^{i}X_{j,k} > \sum_{k=1}^{n-1}\frac{im}{k}\right)$$

$$\leq \sum_{k=1}^{n-1}\Pr\left(\sum_{j=1}^{i}X_{j,k} > \frac{im}{k}\right)$$

$$\leq ne^{-i/4},$$

where $\sum_{k=1}^{n-1}\frac{im}{k} \leq im(1 + \log n) - i < 2im\lceil\log n\rceil - i$ is used for the second inequality. Thus, $\Pr(\mathrm{RT}_\Gamma(i) < 2im\lceil\log n\rceil) \geq 1 - ne^{-i/4}$ holds. $\square$

Lemma 16. *Let $C_0 \in \mathcal{S}_{\mathrm{RD}}$ and $\Xi_{P_{\mathrm{RD}}}(C_0) = C_0, C_1, \ldots$. Then, we have $\Pr(C_{8m\delta\tau\lceil\log n\rceil} \in \mathcal{G}_{\mathrm{half}} \cup \mathcal{V}_{\mathrm{clean}}) \geq 1 - 2ne^{-\delta\tau}$.*

Proof.   Assume that $\mathrm{RT}_\Gamma(t_\text{virus}) < 8m\delta\tau\lceil\log n\rceil$ holds and every agent joins only less than $t_\text{shld}/2$ interactions as an initiator among $\Gamma_0,\ldots,\Gamma_{8m\delta\tau\lceil\log n\rceil-1}$. These assumptions lead to $C_{8m\delta\tau\lceil\log n\rceil} \in \mathcal{G}_\text{half} \cup \mathcal{V}_\text{clean}$ as follows. If a new virus is not created among $\Gamma_0,\ldots,\Gamma_{8m\delta\tau\lceil\log n\rceil-1}$, then all viruses in the initial configuration vanish during the period since each round decreases the maximum value of $\mathtt{timer_V}$ by at least one. Thus, $C_{8m\delta\tau\lceil\log n\rceil} \in \mathcal{V}_\text{clean}$ holds. If some agent v creates a new virus at Γ_t, then v wears a new shield at the same time. Thus, $C_{t+1}(v).\mathtt{timer_S} = t_\text{shld}$. Since v interacts as an initiator only less than $t_\text{shld}/2$ times among $\Gamma_{t+1},\ldots,\Gamma_{8m\delta\tau\lceil\log n\rceil-1}$, we have $C_{8m\delta\tau\lceil\log n\rceil}(v).\mathtt{timer_S} > t_\text{shld}/2$, which means $C_{8m\delta\tau\lceil\log n\rceil} \in \mathcal{G}_\text{half}$. By $t_\text{virus} = 4\delta\tau$ and Lemmas 12 and 15, the probability that the two assumptions hold is at least $1 - 2ne^{-\delta\tau}$. $\square$

Lemma 17. $\Pr(\mathrm{PROP}_i = 1) \geq 1 - 2ne^{-\tau}$ *for any* $i \geq 0$.

Proof.   The same argument as the proof of Lemma 5 gives the lemma. $\square$

Lemma 18. $\Pr(\mathrm{HALF}_i = 1) \geq 1 - ne^{-\tau}$ *for any* $i \geq 0$.

Proof.   Each interaction is independent. Thus, Lemma 4 gives the lemma. $\square$

Lemma 19. $\min_{C \in \mathcal{S}_\mathrm{RD}} \mathrm{EHT}_{P_\mathrm{RD}}(C, LE) = \Omega(\tau e^\tau)$.

Proof.   Probability p_suc, discussed in the beginning of this section, is at least $1 - 3ne^{-\delta\tau} - 4\delta\lceil\log n\rceil \cdot 3ne^{-\tau} \geq 1 - 13n\delta\lceil\log n\rceil e^{-\tau}$ by Lemmas 11, 13, 16, 17 and 18, which leads to the lemma. $\square$

Next, we analyze the expected convergence time. We define two sets of configurations: $\mathcal{N}_\text{oVG} = \{C \in \mathcal{C}_\text{all}(P_\mathrm{RD}) \mid \forall v \in V, \; C(v).\mathtt{timer_V} = C(v).\mathtt{timer_S} = 0\}$ and $\mathcal{L} = \{C \in \mathcal{C}_\text{all}(P_\mathrm{RD}) \mid \#_L(C) \geq 1\}$.

Lemma 20. $\max_{C \in \mathcal{C}_\text{all}(P_\mathrm{RD})} \mathrm{ECT}_{P_\mathrm{RD}}(C, \mathcal{S}_\mathrm{RD}) = O(mp^{-1})$.

Proof Sketch.   Probability p, with which a leader creates a virus at each interaction, is sufficiently small ($p < 1/(4mt_\text{shld})$). Thus, the probability that all viruses and shields vanish (i.e. the population enters a configuration of $\mathcal{N}_\text{oVG}$) within $2mt_\text{shld}$ interactions is at least $1 - (2mt_\text{shld} \cdot p + O(ne^{-\tau})) > 1/2 - O(ne^{-\tau})$. Even if the reached configuration of $\mathcal{N}_\text{oVG}$ does not have any leader, the timeout mechanism creates a leader, and the population enters a configuration of $\mathcal{N}_\text{oVG} \cap \mathcal{L}$. This takes less than $16m\delta\tau\lceil\log n\rceil$ interactions with probability $1 - O(ne^{-\tau})$. After the population enters into $\mathcal{N}_\text{oVG} \cap \mathcal{L}$, additional $\lceil m/p\rceil$ interactions create a new virus with probability $1 - e^{-2}$. Let v be a leader that creates the virus. Since v wears a new shield at the same time, v is not killed and remains a leader during the next $2m\tau$ interactions with probability $1 - O(e^{-\tau})$. On the other hand, the virus spreads to all the agents within these $2m\tau$ interactions with probability $1 - O(ne^{-\tau})$, killing all the agents other than v. A leader other than v may create a new virus during the $2m\tau$ interactions, and survives with a shield. However, this probability is at most $2m\tau \cdot p \leq 1/4$. Hence, v becomes the unique leader within the $2m\tau$ interactions with probability $3/4 - O(ne^{-\tau})$. After the $2m\tau$ interactions, all the agents have $\mathtt{timer_L} > t_{\max}/2$ with probability $1 - O(ne^{-\tau})$ by the larger value propagation, and $v.\mathtt{timer_S} > t_{\mathrm{shld}}/2$ holds with probability $1 - O(ne^{-\tau})$. Hence, the population enters a configuration of $\mathcal{L}_{\mathrm{one}} \cap \mathcal{L}_{\mathrm{half}} \cap \mathcal{G}_{\mathrm{half}} \subset \mathcal{S}_{\mathrm{RD}}$ within the $2m\tau$ interactions with probability $3/4 - O(ne^{-\tau})$. As a result, starting from any configuration, the population enters into $\mathcal{S}_{\mathrm{RD}}$ within $O(mp^{-1})$ interactions with probability $1/4 - e^{-2} - O(ne^{-\tau}) > 0.11 - o(1)$, which gives the lemma. □

Lemmas 19 and 20 gives the following theorem.

Theorem 2 *Protocol* P_{RD} *is a* $(O(mp^{-1}), \Omega(\tau e^{\tau}))$ *loosely-stabilizing leader election protocol for arbitrary graphs when* $t_{\max} \geq 8\delta \max(d, 2 + \log(13n\delta\lceil \log n \rceil))$, $t_{\mathrm{virus}} = t_{\max}/2$, $t_{\mathrm{shld}} \geq 2\delta t_{\max}\lceil \log n \rceil$ *and* $p \leq (4mt_{\mathrm{shld}})^{-1}$.

Therefore, given upper bound N and Δ of n and δ respectively, we get a $(O(m\Delta^2 N^3 \log N), \Omega(Ne^N))$ loosely-stabilizing leader election protocol for arbitrary graphs by assigning $t_{\max} = 8N\Delta$, $t_{\mathrm{virus}} = t_{\max}/2$, $t_{\mathrm{shld}} = 2\Delta t_{\max}\lceil \log N \rceil$ and $p = (4N^2 t_{\mathrm{shld}})^{-1}$.

5 Conclusion

We have presented two loosely-stabilizing leader election protocols for arbitrary undirected graphs in the PP model: one works with agent-identifiers and the other works with random numbers. Both protocols keep a unique leader for an exponentially long expected time after entering a loosely-safe configuration. The protocols use only upper bounds N of n and Δ of δ while any self-stabilizing leader election protocol needs the exact knowledge of n. The restriction of the protocols to *undirected* graph is only for simplicity of protocol description and complexity analysis. The proposed protocols also work on arbitrary *directed* graphs with slight modification: it is only necessary that a responder also executes some actions of an initiator (Line 1 of Protocol 1 and Lines 10-12 of Protocol 2). Our future work is to develop a loosely-stabilizing leader election protocol without agent-identifiers or random numbers for arbitrary graphs. We will also tackle with loosely-stabilizing leader election for some classes of graphs (e.g. rings and trees). We are also interested in the empirical evaluation of the holding time of loosely-stabilizing protocols. Since our probabilistic evaluation of the holding time in this paper is not tight, the actual holding time of the protocols should be much longer. By simulation experiments, we will empirically evaluate the actual holding time (and convergence time) for various network sizes and graph topologies.

Acknowledgements. This work was supported by JSPS KAKENHI Grant Numbers 24500039, 24650012, 25104516, 26280022, and 26330084.

References

1. Angluin, D., Aspnes, J., Diamadi, Z., Fischer, M.J., Peralta, R.: Computation in networks of passively mobile finite-state sensors. Distributed Computing 18(4), 235–253 (2006)

2. Angluin, D., Aspnes, J., Eisenstat, D.: Fast computation by population protocols with a leader. In: Dolev, S. (ed.) DISC 2006. LNCS, vol. 4167, pp. 61–75. Springer, Heidelberg (2006)
3. Angluin, D., Aspnes, J., Fischer, M.J., Jiang, H.: Self-stabilizing population protocols. ACM Transactions on Autonomous and Adaptive Systems 3(4), 13 (2008)
4. Beauquier, J., Blanchard, P., Burman, J.: Self-stabilizing leader election in population protocols over arbitrary communication graphs. In: Baldoni, R., Nisse, N., van Steen, M. (eds.) OPODIS 2013. LNCS, vol. 8304, pp. 38–52. Springer, Heidelberg (2013)
5. Beauquier, J., Burman, J., Rosaz, L., Rozoy, B.: Non-deterministic population protocols. In: Baldoni, R., Flocchini, P., Binoy, R. (eds.) OPODIS 2012. LNCS, vol. 7702, pp. 61–75. Springer, Heidelberg (2012)
6. Cai, S., Izumi, T., Wada, K.: How to prove impossibility under global fairness: On space complexity of self-stabilizing leader election on a population protocol model. Theory of Computing Systems 50(3), 433–445 (2012)
7. Canepa, D., Potop-Butucaru, M.G.: Stabilizing leader election in population protocols (2007), http://hal.inria.fr/inria-00166632
8. Fischer, M., Jiang, H.: Self-stabilizing leader election in networks of finite-state anonymous agents. In: Shvartsman, A. (ed.) OPODIS 2006. LNCS, vol. 4305, pp. 395–409. Springer, Heidelberg (2006)
9. Guerraoui, R., Ruppert, E.: Even small birds are unique: Population protocols with identifiers. Rapport de Recherche CSE-2007-04, Department of Computer Science and Engineering, York University, York, ON, Canada (2007)
10. Michail, O., Chatzigiannakis, I., Spirakis, P.G.: Mediated population protocols. Theoretical Computer Science 412(22), 2434–2450 (2011)
11. Mitzenmacher, M., Upfal, E.: Probability and Computing: Randomized Algorithms and Probabilistic Analysis. Cambridge University Press (2005)
12. Mizoguchi, R., Ono, H., Kijima, S., Yamashita, M.: On space complexity of self-stabilizing leader election in mediated population protocol. Distributed Computing 25(6), 451–460 (2012)
13. Sudo, Y., Nakamura, J., Yamauchi, Y., Ooshita, F., Kakugawa, H., Masuzawa, T.: Loosely-stabilizing leader election in a population protocol model. Theoretical Computer Science 444, 100–112 (2012)
14. Xu, X., Yamauchi, Y., Kijima, S., Yamashita, M.: Space complexity of self-stabilizing leader election in population protocol based on k-interaction. In: Higashino, T., Katayama, Y., Masuzawa, T., Potop-Butucaru, M., Yamashita, M. (eds.) SSS 2013. LNCS, vol. 8255, pp. 86–97. Springer, Heidelberg (2013)

LCD: Local Combining on Demand[*]

Dana Drachsler-Cohen and Erez Petrank

Computer Science Department, Technion, Israel
{ddana,erez}@cs.technion.ac.il

Abstract. Combining methods are highly effective for implementing concurrent queues and stacks. These data structures induce a heavy competition on one or two contention points. However, it was not known whether combining methods could be made effective for parallel scalable data structures that do not have a small number of contention points. In this paper, we introduce *local combining on-demand*, a new combining method for highly parallel data structures. The main idea is to apply combining locally for resources on which threads contend. We demonstrate the use of local combining on-demand on the common linked-list data structure. Measurements show that the obtained linked-list induces a low overhead when contention is low and outperforms other known implementations by up to 40% when contention is high.

Keywords: Concurrent Data-Structures, Multiprocessors, Synchronization.

1 Introduction

In the era of multi-core architectures, there is a growing need for scalable concurrent data structures, which are fundamental building-blocks in a wide range of algorithms. A common approach to design scalable concurrent algorithms is to let each thread execute as independently as possible of other threads, while making its own progress as fast as possible. This approach is often highly effective, especially when contention is low. When resources become contended it is beneficial to consider *combining* techniques in which threads help each other to complete operations.

Combining techniques entail overhead and thus one may expect to integrate them only in highly-contended data structures. Much work follows this guideline and focuses on techniques designated for data structures which have few contention points, and thus are contention-prone [1,5,6,7,12,13,18].

We present *local combining on-demand* (LCD), a combining technique for data structures with unbounded number of contention points. We show combining contributes even to such data structures for which contention occurs for short periods.

For such data structures, applying one of the general combining techniques (e.g., [7,12,13]) often results in a high overhead. General techniques apply combining globally and involve all threads accessing the data structure. *LCD* is applied *locally*, namely only in contended sections, and does not introduce any overhead to threads accessing other sections. This is achieved by applying *LCD* independently for each resource. This approach is beneficial for data structures with an unbounded number of contention points which typically observe contention on small sections.

---

[*] This work was supported by the Israeli Science Foundation grant No. 275/14.

M.K. Aguilera et al. (Eds.): OPODIS 2014, LNCS 8878, pp. 355–371, 2014.

In addition, *LCD* applies combining *on-demand* when contention is observed and it applies the proper combining routine. Examples for combining routines are executing operations on behalf of other threads, eliminating complementary operations without affecting the data structure, and notifying threads waiting to lock a resource which has become irrelevant to their operations due to concurrent updates. Thereby, *LCD* reduces the number of accesses to the data structure and the overall waiting time of threads.

We demonstrate the *LCD* methodology by incorporating it into a fundamental data structure, the *linked-list*, for which various concurrent implementations were suggested (e.g., [8,10,11,20]). The linked-list is a simple data structure that enables us to present the main challenges and solutions of an *LCD* design and evaluating its efficiency on standard workloads. We introduce the *LCD-list* which is an extension of the *lazy-list*. The lazy-list [11] is a lock-based implementation which is arguably the most efficient and scalable implementation of the linked-list on most workloads. We implemented the *LCD*-list in Java and we show it improves the lazy-list performance. While the application of *LCD* to data structures whose operations acquire multiple locks is not trivial, we believe it can be beneficial for other concurrent data structures as well.

We consider one implementation choice as an additional contribution of this work. We integrated *LCD* into the *Java reentrant lock*. A thread acquiring this lock first attempts to obtain it using a single CAS operation. If the attempt succeeds, it usually implies that contention is low. If this attempt fails, the thread is added to a waiting queue. *LCD* is triggered only for threads added to this queue, and thus redundant combining overhead is avoided. Furthermore, *LCD* leverages the lock queue to detect the operations to combine. Utilizing one queue for serving threads waiting for the lock and detecting threads for combining reduces space and maintenance overhead. Measurements show that performance is improved when integrating *LCD* within the Java lock.

The main contributions of this paper are:

- A novel combining methodology adequate for data structures with an unbounded number of contention points which is triggered locally on contended sections only.
- An application of the methodology to the linked-list data structure.
- Implementation and evaluation of the *LCD* methodology and its integration into the Java lock. Results show that they perform well especially when contention is high and introduce negligible overhead under little or no contention.
- Implementation and evaluation of the *LCD*-list. Results show that it outperforms the lazy-list and other linked-list implementations on most workloads.

Related Work. The technique of combining operations first appears in *combining trees* [21]. In combining trees each hot-spot is associated with a tree whose leaves are pre-assigned to threads. Threads traverse upwards in the tree to gain exclusive access to the hot-spot which is assigned to threads that reached the root. If during the traversal two threads access concurrently the same tree node, one thread collects the other's operations, and the other thread ceases its traversal and waits until its operations are completed. Several enhancements of this technique have been presented, such as adaptive combining tree [18], barriers implementations [9,16], and counting networks [19].

A different approach lets threads waiting to acquire a global lock to append their operation details to a list of requests [17]. This list is collected and executed by threads that acquire the lock. The *flat combining* technique [12] enhances this method by eliminating the hot spot caused by threads contending on appending requests to the list. Flat combining was implemented for stacks, queues, and priority queues and showed excellent performance. Later work showed that its application to skip-lists did not improve performance [3]. Various extensions were suggested to improve performance. Hendler et al. [13] extend the method to support multiple locks and delegation of requests from one thread to another. Fatourou and Kallimanis [7] used one queue to implement the lock and maintain the request list for combining. These approaches were evaluated for data structures with a small number of contention points, whereas *LCD* is effective (and was evaluated) for data structures with an unbounded number of contention points. Flat combining was also extended for skip-lists by allowing combiners to access non-intersected sections concurrently [3]. In this technique, combiners are pre-assigned to static sections on which they operate exclusively. In contrast, our approach reduces overhead by triggering combining locally and dynamically only when required.

In addition, *LCD* differs from previous work in its integration of the combining structure into the Java lock. Previous work either maintained designated data structures for combining (e.g., [12]) or presented new locks supporting combining (e.g., [5,7]).

Paper Organization. The rest of the paper is organized as follows. Section 2 provides background. Section 3 overviews the *LCD* methodology and its application to the linked-list. Section 4 describes the algorithm details. Section 5 discusses the *LCD*-list correctness. Section 6 reports performance evaluation, and Section 7 concludes.

2   Background

Here, we provide the background on the lazy-list algorithm [11], which we extend in this work, and the Java reentrant lock, in which we embed our combining technique.

2.1   The Lazy-List

The lazy-list algorithm is a concurrent sorted linked-list implementation. It consists of nodes, each storing: (i) a data object and its unique key, (ii) a lock, (iii) a *marked* flag, and (iv) a pointer to the next node in the list. The *marked* flag signifies whether the node has been *logically* removed from the list, even if it has not been *physically* unlinked from it. For simplicity, the data object is ignored in the sequel.

Throughout the execution, the list contains two sentinel nodes, denoted by *head* and *tail*, where *head*'s key is $-\infty$ and *tail*'s key is ∞. Initially, *head*'s next node is *tail*.

The list supports three operations:

- `insert(k)` – inserts k if it is not in the list; returns `true` if it inserted k, and `false` if not.
- `remove(k)` – removes k if it is in the list; returns `true` if it removed k, and `false` if not.
- `contains(k)` – checks if k is in the list; returns `true` if so, and `false` otherwise.

Algorithm 1. Search(n, k)
1 prev = n
2 curr = prev.next
3 **while** *curr.key* < *k* **do**
4   prev = curr; curr = curr.next
5 **return** *prev, curr*

Algorithm 2. Contains(k)
1 prev, curr = search(head, k)
2 **if** *curr.marked* **then return** *false*
3 **return** *curr.key* == *k*

All operations begin with a traversal along the list to find the correct location for the operation invocation (Alg. 1). This location is captured by two consecutive nodes, *prev* and *curr*, such that k is greater than *prev*'s key and smaller or equal to *curr*'s key. After the correct location is reached, the operations proceed differently.

Algorithm 3. Insert(k)	**Algorithm 4.** Remove(k)
1 prev = head	1 prev = head
2 **while** *true* **do**	2 **while** *true* **do**
3 prev, curr = search(prev, k)	3 prev, curr = search(prev, k)
4 prev.lock()	4 prev.lock()
5 next = prev.next	5 next = prev.next
6 **if** *!prev.marked && next.key $\geq$ k* **then**	6 **if** *!prev.marked && next.key $\geq$ k* **then**
7 **if** *next.key == k* **then**	7 **if** *next.key != k* **then**
8 prev.unlock()	8 prev.unlock()
9 **return** *false*	9 **return** *false*
10 new = Node(k)	10 next.lock()
11 new.next = next	11 next.marked = true
12 prev.next = new	12 prev.next = next.next
13 prev.unlock()	13 next.unlock(); prev.unlock()
14 **return** *true*	14 **return** *true*
15 prev.unlock()	15 prev.unlock()
16 **if** *prev.marked* **then** prev = head	16 **if** *prev.marked* **then** prev = head

In `contains(k)` (Alg. 2), *curr*'s *marked* is examined and if it is `true`, `false` is returned (since *curr* was removed). If *marked* is `false`, *curr*'s key is examined and if it is k, `true` is returned; otherwise, `false` is returned. Note no locks are acquired.

In `insert(k)` and `remove(k)` (Alg. 3 and Alg. 4) *prev* is locked, its successor (which may be different from *curr*) is stored in *next*, and then the location is checked to meet two conditions: (i) *prev*'s *marked* is `false` (it was not removed), and (ii) *next*'s key is greater or equal to k . If condition (i) fails, the lock is released and the operation restarts. If condition (ii) fails, the lock is released and the traversal for the correct location resumes from *prev*. If both conditions are met, `insert` and `remove` take place.

In `insert(k)`, *next*'s key is examined and if it is k, `false` is returned. Otherwise, a new node with key k is inserted between *prev* and *next* and `true` is returned.

In `remove(k)`, *next*'s key is examined and if it is greater than k, then k is not in the list and `false` is returned. Otherwise, *next*'s key is k and the removal begins. First, *next*'s lock is acquired, then its *marked* is set to `true` (the *logical* removal), and finally *prev*'s next node is set to *next*'s next node (the *physical* removal) and `true` is returned.

This description is a slightly optimized version of the lazy-list, in which unnecessary lock acquires and restarts were removed. In the original lazy-list, `insert` and `remove` lock both *prev* and *curr*, check that their *marked* flags are `false` and that *prev* points to *curr*. If any condition fails, the operation restarts. Here, *curr* is not locked, instead, *next* (*prev*'s successor after locking *prev*) is examined. While holding *prev*'s lock, its successor cannot be concurrently changed, not by adding nodes between *prev* and *next*, nor by removing *next*. Thus, reading *next*'s key is safe and there is no need to lock it.

The presented description also avoids restarts, namely, a new traversal from the head of the list. In the original lazy-list, a restart is triggered if any of the conditions following the lock acquisition has failed. Here, restarts occur only if condition (i) fails (i.e., *prev* is marked as removed). If only condition (ii) fails, the traversal resumes from *prev*. This is safe, since if only condition (ii) fails, then the key of *prev*'s successor is smaller than k, and thus the correct location must appear after *prev*.

2.2 The Java Reentrant Lock

The Java reentrant lock, available at Java's
concurrent package, is a variant of the
CLH lock [4]. For simplicity, the below
description omits some details.

Algorithm 5. Lock()
1 **if** CAS(*owner, null, thread*) **then return**
2 enqueue(thread)
3 **while** *true* **do**
4   **if** *thread == next(head)* **then**
5   **if** *CAS(owner, null, thread)* **then**
6   head = thread
7   **return**
8   sleep()

The Java lock is a semi-honest lock
which provides fair access but allows op-
portunistic attempts to acquire it unfairly.
To provide fair access, a queue of pending threads is maintained using a doubly-
linked list. The lock's main fields are
head, *tail*, and *owner*. The *head* and *tail*

Algorithm 6. Unlock()
1 owner = null
2 wake up next(head)

point to the queue head and tail. The
owner points to the thread holding the
lock or to null if the lock is free.

The lock supports the lock and unlock operations. The lock operation (Alg. 5)
first attempts to acquire the lock unfairly by updating *owner* to the current thread via
the CAS operation[1]. This may succeed only if *owner* is null, namely no other thread
holds it. If the CAS operation fails, the thread is added to the end of the queue. It is then
allowed to attempt acquiring the lock only when the queue head is its predecessor (i.e.,
the thread is the second in line). In this case, attempting to acquire the lock may fail if its
predecessor has not released the lock yet, or if another thread has acquired it unfairly. If
the acquisition fails, the thread yields and attempts again later. After acquiring the lock,
the queue head is updated and the operation terminates.

The unlock operation (Alg. 6) sets *owner* to null and notifies the successor of the
queue head it can acquire the lock.

3 Overview

In this section, we provide an overview of the local combining on-demand methodology.

LCD is executed independently for each lock, thus allowing multiple combining
threads to execute concurrently (one for each lock). *LCD* is executed by threads that
acquired the lock fairly, and thus waited for permission to lock in the lock queue. In
LCD, the permission to lock is granted to the *newest* thread in the queue (and not to
the oldest one, as in the Java lock). When a thread acquires the lock fairly, it becomes a
combiner and collects the operation *requests* of threads preceding it in the queue. The
combiner examines the collected requests. Requests not requiring this lock are removed
(and their owners are notified), and identical or complementary requests are eliminated.
Then, the remaining requests are executed. Finally, the owner threads of the combined
requests are reported. Fig. 1 illustrates a flow example of *LCD* in the lazy-list.

LCD is embedded in the Java lock but requires cooperation from the data structure
operations. The lock is responsible for the combining logic, e.g., picking the next combiner and collecting requests. The data structure operations are responsible for searching and updating the data structure, and reporting to the owners of combined operations.

---

[1] CAS(*field, old, new*) is an atomic operation that updates *field* to *new* if *field* stores *old*. If *field*
is updated, the operation succeeded and true is returned; otherwise false is returned.

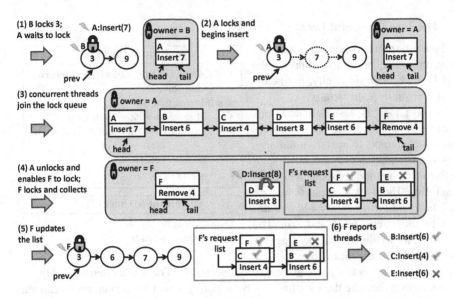

Fig. 1. *LCD* in the lazy-list. (1) *A* invokes `insert(7)`, attempts to lock 3, observes *B* has acquired it, and waits in the lock queue, (2) *B* unlocks and *A* begins to insert 7, (3) concurrent threads join the lock queue, (4) *A* unlocks and wakes the queue tail, *F*; *F* locks and collects requests: *D* is notified to search for another lock for its `insert(8)` request, *C*'s `insert(4)` and *F*'s `remove(4)` are grouped and marked as successful, and *B*'s and *E*'s `insert(6)` are grouped and *E* is marked as failed. (5) *F* inserts 6, and marks *B*'s request as successful, and finally, (6) *F* reports *B*, *C*, and *E*, and terminates.

The *LCD* Lock. The Java lock is extended with *LCD* via `LCDlock` and `LCDunlock`. The `LCDlock` operation begins with an attempt to lock unfairly. If it fails, the thread joins the lock queue, and waits for permission to lock or for a notification that its request was combined. If it was granted permission and acquired the lock, it becomes a combiner and collects requests of threads preceding it in the lock queue. During the collection, the requests are examined. Identical requests are grouped so eventually only one may update the data structure (e.g., *B*'s and *E*'s requests in Fig. 1). Complementary requests are grouped and completed without affecting the data structure (e.g., *C*'s and *F*'s requests in Fig. 1). Threads whose requests require a different lock are notified to search again. This case may arise since threads choose a lock after observing a certain state of the data structure, however, this state may change before they acquire the lock (e.g., *D*'s request in Fig. 1). If the combiner itself requires a different lock, it delegates the collected requests and the combiner role to another thread that requires this lock.

The `LCDunlock` operation releases the lock, and if it is invoked by a combiner, a new combiner is chosen, and it is chosen to be the last thread that joined the lock queue.

Data Structure Adaptation. The data structure operations are adapted to cope with the `LCDlock` results. If the lock was acquired unfairly, the operation proceeds without any *LCD* overhead. If the lock was acquired fairly, the operation executes the collected requests and reports to the request owners. If the request was combined, the operation terminates; and if the lock was unsuitable, the operation searches for the correct lock.

Operations acquiring several locks require a special care. *LCD* is applied independently for each lock and thus such operations need to be split into sub-operations that each require one lock for executing its updates safely. We denote such sub-operations as *single-lock operations*. Operations are split into single-lock operations as follows.

First, all possible executions are represented as series of steps of the following form:

1 . execute s_0, acquire lock l_1, and execute s_1,
2 . acquire lock l_2 and execute s_2,

   ...

k .   acquire lock l_k, execute s_k, and release l_k,

   ...

2'.   execute s_2' and release l_2,
1'. execute s_1', release l_1, and execute s_0'.

Where s_0, s_0' are sequences of computational operations and for $i > 0$, s_i, s_i' are sequences of computational or update operations which require the lock l_i.

Next, we divide such execution into k phases, where the ith phase is:

   (i) execute step i,
   (ii) invoke LCDlock(l_{i+1}) to initiate the next phase and upon its completion,
   (iii) execute step i'.

Each phase assumes previous steps have acquired locks and completed successfully.

Finally, phases requiring the same lock are grouped into a single-lock operation, and some criteria are used to decide which phase to execute.

The benefit of single-lock operations is that different combiners may help separately and independently. Note that combiners may execute several consecutive single-lock operations provided they gained access to the required locks.

We exemplify this technique on the remove(k) operation of the lazy-list.

A successful removal execution requires the locks of two nodes: *prev* (the node preceding k) and *next* (the node storing k), and it can be split into the following steps:

1 . locate *prev* (s_0), lock it (l_1), verify it is not marked and that *next*'s key is k (s_1),
2 .   lock *next* (l_2), set its *marked* flag to true (s_2), and release the lock (l_2),
1'. update *prev*'s next to *next*'s next (s_1'), release *prev*'s lock (l_1), and return true (s_0').

An unsuccessful removal execution requires *prev*'s lock and consists of two steps:

1 . locate *prev*, lock it, verify it is not marked and that *next*'s key is greater than k,
1". release *prev*'s lock and return false.

A restart removal execution requires *prev*'s lock and consists of two steps:

1 . locate *prev*, lock it, verify it is marked or that *next*'s key is smaller than k,
1'''. release *prev*'s lock and repeat step 1.

Thus, remove(k) is split into two single-lock operations:

* *remove* – execute step 1. Then, if *prev* is not marked and *next*'s key is k, initiate step 2 and upon its completion execute step 1'; if *prev* is not marked and *next*'s key is greater than k, execute step 1"; else, execute step 1'''.
* *mark* – execute step 2.

This separation allows *remove* and *mark* to be executed by different threads.

4   Implementation Details

In this section we present the implementation details and provide pseudo-code. We first describe the request object, next the *LCD* lock, and finally the *LCD* linked-list.

4.1   The Request Object

To enable *LCD*, each thread is equipped with a unique *request* object which remains throughout its execution. The request object (Alg. 7) stores all information required for the combining. Op and key contain the operation details. Op may be *insert*, *remove*, or *mark*. Result stores the operation outcome after it is completed, and initially it is set to null. ThreadID stores the owner thread ID to notify upon events, e.g., operation completion.

Algorithm 7. Request
1 Operation op
2 Key key
3 Boolean result
4 Thread threadID
5 State state
6 Request head
7 Request next
8 Set combined

The request status is stored in state and may be either *none*, *locked*, *pending*, *completed*, or *search*. *None* is the initial state. *Locked* indicates that the request owner acquired the lock fairly and acts as a combiner (if the lock is acquired unfairly, the request fields are ignored). *Pending* indicates that a combiner has collected this request, and *completed* indicates that a combiner has executed it. *Search* signals that the request owner is waiting for a lock unsuitable for its request.

The head field points to the request list of the request owner. The owner is responsible for executing these requests. Initially, each thread is responsible only to its own request and thus its request list is of size one. The request list may become longer if its owner thread becomes a combiner and adds requests to its list. The request list may become empty if a combiner thread collects this request. An invariant of the execution is that uncompleted requests always belong to a single request list. As a result, each uncompleted request has at most one successor in the request list it belongs to, and this successor is stored in next. Thus, the request list can be implemented as a linked-list whose elements are connected via the next fields. When registering a new operation to a request, the head is set to this request and next is set to null.

The combined field stores requests which were eliminated or combined by this request. These requests' owners are reported after this request completes its execution.

4.2   The LCD Lock

The *LCD* lock applies combining via the LCDlock and LCDunlock operations.

The LCDlock (Alg. 8) attempts to acquire the lock and returns false if the lock was acquired unfairly and true otherwise (i.e., true indicates that local combining was initiated). LCDlock begins with an attempt to acquire the lock unfairly by attempting to set the owner to be the invoking thread. If it fails, fair locking begins by initializing the given request, registering the operation details, and adding the thread to the lock queue. An attempt to acquire the lock occurs only when the thread is the queue head's successor. If the lock is acquired, the queue head is advanced to point to the thread that acquired the lock, the request state is set to *locked*, and requests are collected.

If the lock is not acquired, the state is examined. If it is *pending*, the thread yields until state is updated. If state is *none*, the thread yields and attempts to acquire the lock later. Otherwise, if state is *completed*, *search*, or *locked* (in case a combiner thread delegated it the lock), LCDlock returns since the lock is not required anymore.

The `LCDunlock` operation (Alg. 9) releases the lock, and picks a new combiner if it is invoked by a combiner thread and there are threads waiting in the lock queue. The new combiner is chosen to be the current queue tail. To allow it acquire the lock, the queue head is set to be its predecessor in the lock queue. Requests of threads preceding the new combiner in the queue join the `reqList`, and they will collected by the new combiner after it acquires the lock.

After choosing the new combiner (if required), the lock is released by clearing the `owner` field and notifying the queue head successor.

Collecting Requests. Combiners collect requests via the `collect` operation (Alg. 10). `Collect` begins by checking whether the combiner needs the current lock and if so, the requests will be collected to its request list (lines 2-4). If the combiner does

---

Algorithm 8. LCDlock(req, op, key)

1 **if** *CAS(owner, null, thread)* **then return** *false*
2 req.state = none; res.result = null
3 req.head = req; res.next = null
4 req.op = op; req.key = key
5 enqueue(thread)
6 **while** *true* **do**
7   **if** *thread == next(head)* **then**
8   **if** *CAS(owner, null, thread)* **then**
9   head = thread
10   req.state = locked
11   collect(req)
12   **return** *true*
13   **while** *req.state == pending* **do** sleep()
14   **if** *req.state != none* **then return** *true*
15   sleep()

Algorithm 9. LCDunlock(req)

1 reqList = new List()
2 **if** *acquired fairly && tail != head* **then**
3   **for** *t = next(head); t != tail; t = next(t)* **do**
4   reqList.add(t.request)
5   head = prev(tail)
6 owner = null
7 wake next(head)

---

not need this lock, another thread will be selected and requests will be added to its request list. The request whose request list is extended with the collected requests is stored in `dest`. To check whether the combiner needs this lock, its request key is compared to the key of the locked list node's successor. If the request key is smaller or equal to the successor's key, the combiner needs this lock and thus its request is stored in `dest`. Otherwise, the combiner's `state` is set to *search* (instead of *locked*) indicating that after collecting requests, it passes the lock and the combiner role to another thread, and searches for the lock required for its own request.

Next, the combiner collects the requests from the `reqList` (constructed by the `LCDunlock` operation) (lines 5-14). At each iteration, one request, denoted by r, is processed. First, its `state` is set to *pending*. Then, if `dest` is `null` and r requires this lock, `dest` is set to r (lines 7-8). Otherwise, r's request list is added to `dest`'s list via `addReqs` (line 11). Then, if r's `result` was not `null` before calling `addReqs`, then r's `owner` is a combiner that completed its own request and proceeded executing other requests. Thus, after collecting its requests, r completes its execution (lines 12-14).

After all requests were collected, `dest` is examined. If it is the combiner's request, the operation terminates. If `dest` is `null`, the lock is released as no thread requires it (line 15). If `dest` is a different request, the lock is delegated to `dest`'s owner by updating the lock's `owner` and `dest`'s `state` (lines 16-19). The lock queue head remains unchanged since its task is to signal to `LCDunlock` the starting point of waiting threads.

Algorithm 10. Collect(req)	**Algorithm 11.** AddReqs(dst, src, k)
1 k = ownerNode.next.key	1 d = p = dst
2 dest = null	2 **for** s = src; s != null; s = s.next **do**
3 **if** *req.head.key ≤ k* **then** dest = req	3 **if** s.key > k **then**
4 **else** req.state = search	4 s.head = s
5 **for** *r in reqList* **do**	5 s.state = search
6 r.state = pending	6 notify s.theadID
7 **if** *dest == null && r.head.key ≤ k* **then**	7 **return**
8 dest = r	8 **while** d.key < s.key **do**
9 **else**	9 p = d
10 rResult = r.result	10 d = d.next
11 addReqs(dest.head, rHead, k)	11 **if** d.key > s.key **then**
12 **if** *rResult != null* **then**	12 s.next = d
13 r.locked = completed	13 p.next = s
14 notify r.threadID	14 **else**
15 **if** *dest == null* **then** LCDunlock(req)	15 **if** d.op == s.op || d.result **then**
16 **if** *dest != req && dest != null* **then**	16 s.result = false
17 owner = dest.thread	17 **else**
18 dest.state = locked	18 d.result = s.result = true
19 notify dest.threadID	19 d.combined.add(s)

The addReqs operation (Alg. 11) is invoked after a list node was locked, and it transfers requests from a source list, src, to a destination list, dst, provided that the requests in src require this lock. To verify this, addReqs receives the key of the list node's successor, k, and only requests whose keys are no greater than k are transferred. The src and dst lists are sorted by the operation keys in an ascending order.

The addReqs operations iterates the src list and at each iteration examines one request, denoted by s. If s has a key greater than k, s's state is set to *search* and s's head is set to s. Thereby, s becomes responsible to all requests succeeding it in src, whose keys are also greater than k as src is sorted (lines 3-7).

If s's key is not greater than k, the combiner looks for two requests in dst, p and d, such that s's key is greater than p's key and not greater than d's key (lines 8-10)[2]. If d's key is greater than s's key, s is inserted between p and d (lines 11-13). Otherwise, if the keys are equal, elimination is applied (lines 14-19). If d and s are identical operations, s is eliminated by setting its result to false. This is correct since if the combiner were to execute d and s, then s would fail. To illustrate, consider two requests of insert(4): the first may succeed if 4 is not in the list, but the second will observe 4 and fail.

If d and s are complementary operations (i.e., an insert-remove pair) their result fields are set to true. If the combiner were to execute them, it could choose an ordering that would not affect the data structure. To illustrate, assume d and s are insert(4) and remove(4). If 4 is not in the list, the combiner can execute d and then s; otherwise it can execute s and then d. Such elimination can be applied once for each request and thus if d is discovered to be eliminated, then s is eliminated by the request that eliminated d (and is identical to s), and thus s's result is set to false.

After eliminating s, it is added to d's combined set (line 19). After d's operation is completed, the state of these requests is set to completed. This is required for

---

[2] For simplicity's sake, we omit the special treatment required if s becomes dst's head or tail.

Algorithm 12. LCDRemove(k,req)	**Algorithm 13.** Combine(req, n)
1 prev = head	1 **while** *true* **do**
2 **while** *true* **do**	2 **if** *req.state == completed* **then**
3 prev, curr = search(prev, k)	3 **return** *complete(req)*
4 **if** *prev.LCDlock(req, remove, k)* **then**	4 **if** *req.state == locked* **then**
5 **return** *combine(req, prev)*	5 dne = execute(req, req.head, n)
6 next = prev.next	6 n.LCDunlock(req)
7 **if** *!prev.marked && next.key $\geq$ k* **then**	7 **if** *dne* **then**
8 **if** *next.key != k* **then**	8 **return** complete(req)
9 prev.LCDunlock(req)	9 **if** *n.marked* **then**
10 **return** *false*	10 n = list.head
11 **if** *next.LCDlock(req, mark, k)* **then**	11 curr = n.next
12 combine(req, next)	12 **while** *curr.key < req.head.key* **do**
13 **else**	13 n = curr; curr = curr.next
14 next.marked = true	14 n.LCDlock(req, req.op, req.key)
15 next.LCDunlock(req)	

Algorithm 14. Complete(req)

16 prev.next = next.next	1 **for** *r in req.combined* **do**
17 prev.LCDunlock(req)	2 r.state = completed
18 **return** *true*	3 notify r.threadID
19 prev.LCDunlock(req)	4 **return** *req.result*
20 **if** *prev.marked* **then** prev = head	

operations eliminated by identical operations, since their results are valid only after the eliminating requests were executed. In addition, it reduces the number of updates to the list. For example, delaying the completion of the complementary requests insert(4) and remove(4) enables to eliminate additional insert(4) with the first insert(4).

4.3 LCD-List

The extended lazy-list insert and remove operations, denoted by LCDInsert and LCDRemove, access locks via LCDlock and LCDunlock (Alg. 12 shows LCDRemove which extends remove, and LCDInsert extends insert similarly). If LCDlock returns false, *LCD* was not initiated and the operation proceeds as in the lazy-list. If LCDlock returns true, then the invoking thread is either a combiner or its request was combined, and in any case, the operation is completed via the combine operation.

The combine operation (Alg. 13) receives the thread request, req, and the node which LCDlock attempted to lock, n. It begins by reading req's state which may be *completed*, *locked*, or *search*. If state is *completed*, req was combined, and the operation completes (via the complete operation). If state is *locked*, execute is called and n's lock is released. If execute returned true, then all requests were executed and combine completes. If there are remaining requests or if state is *search*, a new n node is located and attempted to be locked, and the loop begins again (lines 9-14).

The complete operation (Alg. 14) receives a request, reports threads whose requests were combined or eliminated by this request, and returns the request result.

The execute operation (Alg. 15) receives the combiner request, req, the head of its request list, r, and a locked node, n. It executes all requests requiring n and returns true if all the requests in the list were executed. It begins by finding the first request not eliminated in the request list, namely, a request whose result field is null (lines 1-2). If all requests were eliminated, it returns true.

If a non-eliminated request was found, two conditions are checked similarly to the lazy-list (line 7). If any condition fails, `execute` terminates (lines 31-36). Note that if the second condition fails, namely, `r`'s key is greater than `next`'s key, then also all requests succeeding `r` require a different lock since the list is sorted by the operation keys.

If both conditions are met, `r` begins execution. First, it is added to `req`'s `combined` set (line 8) so it will be reported after `req` is completed[3]. Then, `r`'s operation type is examined.

If `r`'s operation is *insert* (lines 9-19), and it is a successful insert, a new node is created and added to a temporary sublist consisting of all nodes created by insertion requests. The sublist is connected to the list when `execute` terminates and only then these requests' `result` fields are set to `true` (lines 31-34).

If `r`'s operation is *remove* (lines 20-26), and it is a successful removal, the `removeNode` operation is invoked. Note that the sublist cannot contain `r`'s key, as complementary requests are eliminated.

If `r`'s operation is *mark*, it is delayed until all other requests are completed (lines 27, 35). If `r` would have been served immediately by marking n, then no subsequent request could have been executed as the check in line 7 would fail. `Execute` may encounter at most one *mark* request, since a thread initiating a *mark* request locks the node preceding n until it is completed.

After serving `r`, the next request to execute is searched for (lines 28-30).

The `removeNode` operation (Alg. 16) receives two consecutive nodes, p and n, and it removes n from the list. It begins by invoking `LCDlock` to lock n with a fresh request, `frq` (to avoid overriding

---

Algorithm 15. Execute(req, r, n)

```
1  while r != null && r.result != null do
2      complete(r); r = r.next
3  if r == null then return true
4  insReqs = new List()
5  subHead = null; subTail = null
6  next = n.next
7  while !n.marked && r.key ≤ next.key do
8      req.combined.add(r)
9      if r.op == Insert then
10         if r.key == next.key then
11             r.result = false
12         else
13             new = Node(r.key)
14             if subHead == null then
15                 subHead = new
16             else
17                 subTail.next = new
18             subTail = new
19             insReqs.add(r)
20     if r.op == Remove then
21         if r.key < next.key then
22             r.result = false
23         else
24             removeNode(req, n, next)
25             next = n.next
26             r.result = true
27     if r.op == Mark then mark = r
28     while r != null && r.result != null do
29         complete(r); r = r.next
30     if r == null then break
31 if subHead != null then
32     subTail.next = n.next
33     n.next = subHead
34 for r in insReqs do r.result = true
35 if mark != null then n.marked = true
36 return r == null
```

Algorithm 16. RemoveNode(req, p, n)

```
1  n.LCDlock(frq, mark, n.key)
2  if frq.state == locked then
3      n.marked = true
4      n.LCDunlock(frq)
5      k = n.next.key
6      addReqs(req.head, frq.head.next, k)
7  p.next = n.next
```

---

[3] This is a simplification, requests are actually notified sooner.

req fields). Then, frq's state is examined, and it may be either *locked* or *completed* (it cannot be *search*, as p is locked and thus n's lock must be the required lock). If state is *completed* then p's next is set to n's next. Otherwise, if state is locked, n's marked flag is set to true and n's lock is released. Then, collected requests are transferred from frq to req, starting from the second request in frq's request list (the first one is frq since the list is sorted). Finally, p's next is set to n's next.

5   Correctness

Here, we provide the linearization points [15] of the *LCD*-list operations.

Unsuccessful inserts and removes (which were not eliminated) are linearized when discovered to be unsuccessful in LCDInsert (similarly to the discovery in insert, line 7), LCDRemove (line 8), or execute (lines 10, 21).

A successful insert is linearized in LCDInsert when prev is set to point to the new node (similarly to the update in insert, line 12), or in execute when the sublist containing the new node is appended to the linked-list (line 33). This linearization point linearizes "atomically" all successful inserts whose nodes share the same sublist. That is, these inserts are linearized at a single point, and no other operation is linearized between them. The linearization order between the inserts is by their keys.

A successful remove is linearized when the node's marked flag is set to true, which is either in LCDRemove (line 14), execute (line 35), or removeNode (line 3).

Eliminated operations are grouped into disjoint sets as follows. Two complementary operations which were eliminated (addReqs, line 18), belong to the same set. Operations that were eliminated by an identical operation or by a complementary eliminated operation (addReqs, line 16) belong to the set of that operation. The operations of each disjoint set are linearized together "atomically" in one of the following points. If the operation set contains two complementary operations, then the linearization point is upon the first update to a state field of any of these operations (complete, line 2). If the operation set does not contain complementary operations, then it consists of identical operations and one of them is executed in execute. In this case, the linearization point of the operation set is the linearization point of the executed operation.

The linearization point ordering between operations belonging to the same operation set is as follows. If the operation set contains two complementary operations, i.e., insert(k) and remove(k), then the ordering is one of the following. If k is present in the linked-list, then first appears the remove(k) whose result is true. Next, all the other remove(k) appear in some order. Then, the insert(k) whose result is true appears; and finally, all the other insert(k) appear in some order. If k is not present in the linked-list, the ordering is similar only that inserts are linearized before removes. If the operation set contains only identical operations then the order is first the operation that was executed (in execute) and then the other operations in some order.

The linearization point of the contains operation is set similarly to the original lazy-list algorithm. A successful contains is linearized when the marked flag is found false. An unsuccessful contains has two possible linearization points. The first case is when k is indeed not in the linked-list and curr's key is found to be greater than k, or curr is found to be marked. Here, the linearization point is when curr is examined. The second case is similar to the first case only that k is concurrently inserted. Since contains returns false, it is linearized just before the concurrent insert. Note that this linearization point occurs after contains begins, since otherwise contains would have found k. Also note that the linearization points of contains are not affected by inserts which were eliminated by complementary removes, since these operations do not affect the list and are linearized "atomically" with their complementary removes.

6   Performance Evaluation

We implemented the *LCD*-list in Java and ran experiments on an AMD Opteron Processor 6376 with 128GB RAM and 64 cores: four processors with sixteen cores each. We used Ubuntu 12.04 LTS and OpenJDK Runtime version 1.7.0_65 using the HotSpot 64-bit Server VM (build 24.45-b08, mixed mode).

We compared the *LCD*-list to the following list implementations:

- The Lazy-List – the optimized lazy-list algorithm as presented in Section 2.1.
- The Flat Combining List [12] – this list is protected by a global lock and the lock owner executes operations of contending threads. We used the authors' code [2].
- The Lock Free List [10] – the lock-free list by Harris. We used the code provided with the book "The Art of Multiprocessor Programming" [14].

We also consider two variations of the *LCD*-list to evaluate two of its main features:

- *LCD* without elimination – the *LCD*-list without eliminating operations.
- *LCD* without integration – the *LCD*-list without the Java lock integration.

We ran two workloads:

(I) 0% contains, 50% insert, and 50% remove.

(II) 60% contains, 20% insert, and 20% remove.

We ran five-second trials, where each thread reported the number of operations it completed. We report the total throughput, namely the total number of completed operations. The number of threads is 2^i where i varies between 0-8.

During the trial, each thread randomly chooses a type of operation according to the workload distribution and then randomly chooses the key for that operation from a given range. The examined range sizes were: 128, 512, and 1024. Before each trial, the list was prefilled to a size of half of the key range. Every experiment was run 8 times and the arithmetic average is reported along with the 95% confidence interval. Each batch of 8 trials was run in its own JVM instance, and a warm-up phase was run before to avoid HotSpot effects. Threads were not bound to processors in our experiments.

Table 1 reports the results. The experimental results show that *LCD* improves the lazy-list performance on most workloads, mostly by 5%-15% and up to 40%. Under low contention, the *LCD*-list performance is similar to the lazy-list and its overhead is negligible. The lock-free list performance is at best under heavy contention and it decreases significantly under lower contention. The flat combining performs poorly as it blocks all concurrent accesses to the list even if they do not contend.

The *LCD* features appear to improve the *LCD* performance. The performance of the *LCD*-list without the operation elimination is mostly similar to the *LCD*-list, but as contention increases and more operations can be eliminated its performance is reduced. The *LCD*-list without the Java lock integration performs better than the *LCD*-list under low contention and performs poorly under heavy contention. The integration into the Java lock introduces overhead since volatile fields of the Java lock are updated (e.g., the queue head and tail), and this overhead is significant under low contention. However, the lock integration becomes very beneficial as contention increases.

Table 1. Throughput of linked-list implementations (64 h/w threads)

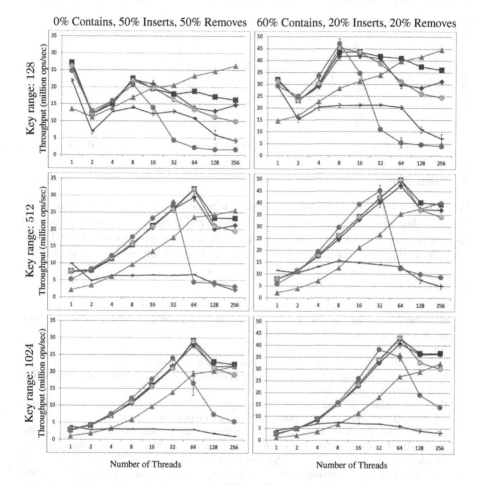

We next study how often *LCD* is applied in the *LCD*-list. There are four avenues for an operation to be executed: (i) a fast execution without *LCD*, (ii) a slow execution when the operation owner is a combiner, (iii) operation elimination, or (iv) combining.

Table 2 reports the percentage of operations completed in each *LCD* avenue ((ii)–(iv)). The summation of each column is the percentage of operations completed in an *LCD* avenue. The two workloads yielded similar results, thus we present only the results of the second workload. The results show that as contention increases, more operations use *LCD* (up to 22% of the operations). As the number of threads increases, many operations are eliminated. The number of combined operations is low on all workloads.

Table 2. Distribution of *LCD* avenue types

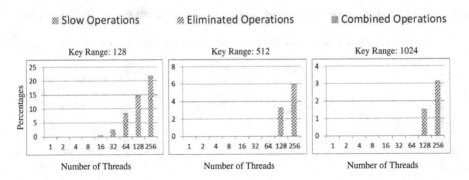

7 Summary

We presented the *local combining on-demand* methodology (*LCD*), a combining technique for data structures with an unbounded number of contention points. We designed and implemented the *LCD*-list, an extension of the lazy-list that provides *LCD*. Evaluation shows that the *LCD*-list outperforms the lazy-list typically by 7% and up to 40%.

References

1. Bar-Nissan, G., Hendler, D., Suissa, A.: A dynamic elimination-combining stack algorithm. In: Fernàndez Anta, A., Lipari, G., Roy, M. (eds.) OPODIS 2011. LNCS, vol. 7109, pp. 544–561. Springer, Heidelberg (2011)
2. Bronson, N.: The flat combining project,
   `http://mcg.cs.tau.ac.il/projects/flat-combining`
3. Budovsky, V.: Combining techniques application for tree search structures, m.sc. thesis. Tel-Aviv University, Israel (2010)
4. Craig, T.: Building fifo and priority-queuing spin locks from atomic swap. Technical Report TR 93-02-02, University of Washington, Dept. of Computer Science (1993)
5. Dice, D., Marathe, V.J., Shavit, N.: Flat-combining numa locks. In: Proceedings of the Twenty-third Annual ACM Symposium on Parallelism in Algorithms and Architectures, SPAA 2011, San Jose, California, USA, pp. 65–74. ACM, New York (2011)
6. Dice, D., Marathe, V.J., Shavit, N.: Lock cohorting: A general technique for designing numa locks. In: Proceedings of the 17th ACM SIGPLAN Symposium on Principles and Practice of Parallel Programming, PPOPP 2012, New Orleans, Louisiana, USA, pp. 247–256. ACM, New York (2012)
7. Fatourou, P., Kallimanis, N.D.: Revisiting the Combining Synchronization Technique. In: Proceedings of the 17th ACM SIGPLAN Symposium on Principles and Practice of Parallel Programming, PPoPP 2012, New Orleans, Louisiana, USA, pp. 257–266. ACM, New York (2012)
8. Fomitchev, M., Ruppert, E.: Lock-free Linked Lists and Skip Lists. In: Proceedings of the Twenty-third Annual ACM Symposium on Principles of Distributed Computing, PODC 2004, St. John's, Newfoundland, Canada, pp. 50–59. ACM, New York (2004)
9. Gupta, R., Hill, C.R.: A Scalable Implementation of Barrier Synchronization Using an Adaptive Combining Tree. International Journal of Parallel Programming 18(3), 161–180 (1990)
10. Harris, T.L.: A pragmatic implementation of non-blocking linked-lists. In: Welch, J. (ed.) DISC 2001. LNCS, vol. 2180, pp. 300–314. Springer, Heidelberg (2001)

11. Heller, S., Herlihy, M.P., Luchangco, V., Moir, M., Scherer III, W.N., Shavit, N.N.: A lazy concurrent list-based set algorithm. In: Anderson, J.H., Prencipe, G., Wattenhofer, R. (eds.) OPODIS 2005. LNCS, vol. 3974, pp. 3–16. Springer, Heidelberg (2006)

12. Hendler, D., Incze, I., Shavit, N., Tzafrir, M.: Flat Combining and the Synchronization-parallelism Tradeoff. In: Proceedings of the Twenty-second Annual ACM Symposium on Parallelism in Algorithms and Architectures, SPAA 2010, Thira, Santorini, Greece, pp. 355–364. ACM, New York (2010)

13. Hendler, D., Incze, I., Shavit, N., Tzafrir, M.: Scalable flat-combining based synchronous queues. In: Lynch, N.A., Shvartsman, A.A. (eds.) DISC 2010. LNCS, vol. 6343, pp. 79–93. Springer, Heidelberg (2010)

14. Herlihy, M., Shavit, N.: The art of multiprocessor programming. Morgan Kaufmann Publishers Inc., San Francisco (2008)

15. Herlihy, M.P., Wing, J.M.: Linearizability: A correctness condition for concurrent objects. ACM Trans. Program. Lang. Syst. 12, 463–492 (1990)

16. Mellor-Crummey, J.M., Scott, M.L.: Algorithms for Scalable Synchronization on Shared-memory Multiprocessors. ACM Trans. Comput. Syst. 9(1), 21–65 (1991)

17. Oyama, Y., Taura, K., Yonezawa, A.: Executing parallel programs with synchronization bottlenecks efficiently. In: Proceedings of International Workshop on Parallel and Distributed Computing for Symbolic and Irregular Applications, PDSIA 1999, pp. 182–204 (July 1999)

18. Shavit, N., Zemach, A.: Combining Funnels: A Dynamic Approach to Software Combining. J. Parallel Distrib. Comput. 60(11), 1355–1387 (2000)

19. Shavit, N., Zemach, A.: Diffracting Trees. ACM Trans. Comput. Syst. 14(4), 385–428 (1996)

20. Valois, J.D.: Lock-free Linked Lists Using Compare-and-swap. In: Proceedings of the Fourteenth Annual ACM Symposium on Principles of Distributed Computing, PODC 1995, Ottowa, Ontario, Canada, pp. 214–222. ACM, New York (1995)

21. Yew, P.-C., Tzeng, N.-F., Lawrie, D.H.: Distributing Hot-Spot Addressing in Large-Scale Multiprocessors. IEEE Trans. Comput. 36(4), 388–395 (1987)

ParMarkSplit: A Parallel Mark-Split Garbage Collector Based on a Lock-Free Skip-List

Nhan Nguyen[1] and Philippas Tsigas[1,*] and Håkan Sundell[2]

[1] Chalmers University of Technology, Gothenburg, Sweden
{nhann,tsigas}@chalmers.se
[2] University of Borås, Borås, Sweden
Hakan.Sundell@hb.se

Abstract. Mark-split is a garbage collection algorithm that combines advantages of both the mark-sweep and the copying collection algorithms. In this paper, we present a parallel mark-split garbage collector (GC). Our parallel design introduces and makes use of an efficient concurrency control mechanism for handling the list of free memory intervals. This mechanism is based on a lock-free skip-list design which supports an extended set of operations. Beside basic operations, it can perform a composite one that can search and remove and also insert two elements atomically. We have implemented the parallel mark-split GC in OpenJDK's HotSpot virtual machine. We experimentally evaluate our collector and compare it with the default concurrent mark-sweep GC in HotSpot, using the DaCapo benchmarks, on two contemporary multiprocessor systems; one has 12 Intel Nehalem cores with HyperThreading and the other has 48 AMD Bulldozer cores. The evaluation shows that our parallel mark-split keeps the characteristics of the sequential mark-split, that it performs better than the concurrent mark-sweep in applications that have low live/garbage ratio, and have live objects locating contiguously, therefore being marked consecutively. Our parallel mark-split performs significantly better than a trivial parallelization based on locks in terms of both collection time and scalability.

Keywords: garbage collector, concurrent programming, mark-split, mark-sweep, parallel garbage collection, lock-free data structures.

1 Introduction

Garbage collection (GC) is a form of automatic memory management to reclaim memory occupied by objects that are no longer used. Being introduced in 1960 [1], GC has evolved to become an important feature offered by many modern programming languages. Mark-sweep [1], copying [2], and their derivations are among the algorithms that have been extensively studied in the literature;

---

* The research leading to these results has been partially supported by the European Union Seventh Framework Programme (FP7/2007-2013) through the EXCESS Project (www.excess-project.eu) under grant agreement 611183.

M.K. Aguilera et al. (Eds.): OPODIS 2014, LNCS 8878, pp. 372–387, 2014.
© Springer International Publishing Switzerland 2014

and their pros and cons have been identified in a range of scenarios. The mark phase in mark-sweep has a time complexity proportional to the amount of live data, while the sweep phase has one proportional to the size of the heap. Mark-sweep can be improved by executing the sweep phase concurrently with the execution of the mutator, which has been suspended while marking. This technique is referred to as *lazy sweeping* [3]. Mark-region [4] improves the mark-sweep by dividing the heap in several regions and compacts objects to one end of the regions, and can thus reduce memory fragmentation. Garbage-First[5], which also works in per-region manner, marks objects and then evacuates them from current regions to new ones so that current regions can be reclaimed as a whole. Differing from the mark-sweep collectors, copying ones need time proportional to the amount of live data. However, they waste half of the space reserved for the need of the collectors, and move objects during collection. Copying collectors perform better than mark-sweep ones when the amount of live data is small compared to the size of the heap. This is the case where mark-sweep is penalized by the complexity of its sweep phase.

Sagonas and Wilhelmsson [6] introduced a GC technique called mark-split that can combine advantages of mark-sweep and copying collection. Mark-split evolves from mark-sweep but removes the sweep phase. Instead, the list of free spaces is built during marking, and can thus be used for allocation when the mark phase completes. Mark-split starts by creating the list of free intervals containing only a big free interval spanning the whole collected space. Then it proceeds to the mark phase. For each unmarked live object, it marks the object and calls a special *split* operation to exclude the marked space from the free intervals. The *split* operation which takes an object as an argument splits a free interval containing that object into two smaller free intervals, one to the left and the other to the right of the object. When the mark phase completes, the list of free intervals contains only free memory, thus can be used for new allocation.

Mark-split removes the sweep phase from mark-sweep, and thus achieves a time complexity proportional to the size of the live data set. However, this comes with an overhead cost of maintaining a set of free memory intervals. The number of free intervals is much smaller than the number of live objects because some live objects reside adjacent to each other. It seems beneficial, in certain situations, to avoid the sweep phase at the cost of this overhead, which depends on the distribution of live objects and also highly on the data structure selected to store the free intervals. The data structure should preferably provide search for an interval at sub-linear cost, e.g. binary search trees, splay trees, or skip-lists. The original mark-split uses a sequential balanced search tree [6], which might hurt its performance.

While mark-split is comparable to mark-sweep and even outperforms it in some situations, to the best of our knowledge, this is the first effort to design a mark-split collector for multi-core systems. Our contribution is to parallelize mark-split based on a highly concurrent data structure to handle the free intervals. We consider using lock-free data structures for their many advantages such as providing high performance, progress guarantees and immunity to deadlocks and livelocks [7]. However, previous implementations of concurrent data structures that supported the basic operations couldn't be used directly to parallelize mark-split, as they were not powerful enough to build a list of free intervals in mark-split. This was because mark-split frequently performs a combined operation of multiple basic operations. First it finds the correct interval and then performs *split* on it. This latter operation is also a combination of two operations; i) remove one interval and possibly ii) add two intervals. Concurrent environments require that *split* operations must perform all those actions in an atomic step, and thus the concurrency control is a challenge for the data structure to be used. A lock-free skip-list such as the one introduced in [8] can satisfy the performance but not the capability requirements of mark-split. We therefore extend it with a novel concurrency control to handle the free intervals and use the new skip-list to parallelize mark-split.

The rest of this paper is organized as follows. Section 2 introduces our extended skip-list algorithm to meet the requirements for parallelization of mark-split. The implementation of a parallel mark-split algorithm with the design of a lazy-splitting mechanism are presented in Section 3. Section 4 shows our evaluation of the GC inside HotSpot, along with result discussions before section 5 concludes the paper.

2   Concurrent Skip-List with Extended Functionality

We present a skip-list with extended functionality offering significant extensions over the original lock-free skip-list in [8]. A skip-list is a search data structure which stores elements in different layers of ordered linked lists with different densities to achieve tree-like behaviour. The original skip-list [8] can insert a new element, search for or remove an exiting element, but not a combination of those in one atomic operation. The use of recursion in that skip-list also made its memory management complicated and not efficient. Our extensions of the new skip-list are significant both when it comes to operations that it supports and in the algorithmic design. The new *replace2* operation gives the ability to atomically replace a node with one or two new

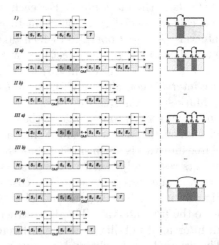

Fig. 1. Multiple-step process for marking and deleting blocks simultaneously with inserting new nodes, thus fulfilling the corresponding (to the right) abstract operations on the free-list

nodes; making the skip-list usable in the context of mark-split. Regarding the performance, we redesigned the data structure to make use of hazard pointers[9] for memory reclamation purposes and thread-local-storage.

The *split* operation described in the mark-split algorithm operates on an abstract free-list representing a set of free intervals. A free interval can be represented by a node in a skip-list, where key represents the start address S of the interval and the corresponding value represents its end address E. As used in [8], the skip-list is basically made out of a singly-linked list with the nodes ordered by their keys. To allow probabilistic logarithmic expected time complexity for searching a particular node, nodes are inserted with a varying height such that several auxiliary lists are created with several layers of decreasing density with increasing height. For modifications to the abstract state of the free-list, only changes on the lowest layer's linked list are representative, i.e., changes are first performed atomically on the lowest layer and then modifications of the other layers can be performed concurrently with other operations. All necessary additional steps of the operation are eventually completed by making use of a suitably designed helping scheme. The helping scheme is designed to allow a concurrent operation to help another on-going operation when the former want to access the data that the latter is processing. A node in the skip-list can be defined to be *present* as soon as it is inserted on the lowest layer (i.e., there is another present node with a next pointer on the lowest level pointing to it) and *deleted* whenever the next pointer on the lowest layer for the corresponding node is marked (e.g. bit 0 set to 1). Atomic changes to the state of each node being present or deleted can be made using the Compare-And-Swap (*CAS*) primitive[1].

The *split* operation can result in four distinct changes on the abstract free-list. Each of these four changes must be possible to perform atomically with respect to each other. The possible changes are to either change S or E of an interval, replace the interval with two new intervals, or remove the interval altogether. To facilitate the representation of these abstract changes in the skip-list, an important observation is that it is possible to extend the skip-list to actually allow atomic deletion *and* insertion. The *CAS* primitive has the capability to both mark the next pointer and change it in the same operation. Thus, it is possible to atomically replace a node in the skip-list with one or more new nodes. The way that this modified skip-list is made to represent the abstract changes on the free-list, is shown in Fig. 1.

-**Step I** illustrates how a free-list containing the intervals $\langle S_1, E_1 \rangle$ and $\langle S_2, E_2 \rangle$ can be represented with two corresponding nodes in the skip-list.

- **In Step IIa,** the interval $\langle S_2, E_2 \rangle$ is split into two intervals $\langle S_3, E_3 \rangle$ and $\langle S_4, E_4 \rangle$, where $S_3 = S_2$ and $E_4 = E_2$. By means of a *CAS*, the pointer on the lowest level of node $[S_2, E_2]$ is atomically marked and made to point to the new node $[S_3, E_3]$ which is already pointing to the new node $[S_4, E_4]$. The deleted node is then removed (also part of the helping scheme) in **step IIb**, with the *CAS* operating on the previous node's corresponding next pointer. The remaining layers are then handled in a similar manner.

---

[1] CAS, a synchronization primitive available in most modern processors, compares the content of a memory word to a given value and, only if they are the same, modifies the content of that word to a given new value.

- **In Step IIIa,** the interval $\langle S_3, E_3 \rangle$ is modified to become $\langle S_5, E_5 \rangle$ where either $S_5 = S_3$ or $E_5 = E_3$. By means of a *CAS*, the pointer on the lowest level of node $[S_3, E_3]$ is atomically marked and made to point to the new node $[S_5, E_5]$. The deleted node is then finally removed (also part of the helping scheme) in **step IIIb,** with the *CAS* operating on the previous node's corresponding next pointer. The remaining layers are then handled in a similar manner.

- **In Step IVa,** the interval $\langle S_5, E_5 \rangle$ is removed altogether from the free-list. By means of a *CAS*, the pointer on the lowest level of node $[S_5, E_5]$ is atomically marked. The deleted node is then finally removed (part of the helping scheme) in **step IVb,** with the *CAS* operating on the previous node's corresponding next pointer. The remaining layers are then handled in a similar manner.

The lock-free property is fulfilled by properly designing the helping scheme so that whenever an attempt made to perform a *CAS* for the a-part of the steps fails, the helping scheme makes sure that the b-part is being performed before attempting the a-part again.

2.1   Implementation

The implementation of the extended skip-list is described in Figs. 2 and 3 . The operation *split* removes a given interval (i.e., the *start* and *end* memory addresses of the live object) from the list of free intervals represented by the skip-list. The node that contains the given interval is searched for, with the search starting from the *head* node at the highest level. As the search is done in the skip-list level by level downwards, the previous node on each level is stored in the thread-local-storage *savedNodes* array. These remembered previous nodes are later used when deciding to either replace or remove the found node, according to the rules described in Section 2. If the found node, represented by *node*, is concurrently modified, the corresponding replace or remove attempts will fail, and the whole *split* operation is repeated.

Operation *replace2* describes how *node* can be atomically replaced by two new nodes *node1* and *node2*. First the next pointer of *node* on the lowest level is atomically modified using *CAS*, to both contain the deletion mark (represented by the pointer value of 1) and instead point to *node1*. Thereafter, *node* is fully removed from the skip-list, and then *node1* and *node2* are inserted together, starting from level 1 and going upwards. During this insertion, *node1* or *node2* can have been concurrently deleted, in which case the insertion is aborted and helping is applied to make sure the deleted node is fully removed. Before actually starting modifying next pointers of previous nodes, the deletion mark is propagated upwards on all levels of the next pointer of *node* using *CAS* operations. This step is also required to be done by all concurrent operations that apply helping. The next step is then to modify the next pointer of all previous nodes of *node* such that they should instead point to the next node of *node*, starting with the highest level of the next pointers of *node* and going downwards. This is done by using *CAS* to atomically update the next pointer of the previous node, possibly given by *savedNodes*[i], from originally pointing to *node* to instead

```
1 void split(void *start,void *end)
2   do
3       Node *node, *prev = head;
4       for(i = MAX_HEIGHT; i >= 0; i--)
5         for(;;)
6             node = prev.next[i];
7             if(node & 1)
8                 Go backwards in path using savedNodes[i + 1] or higher and help prev if needed
9             if(node matches interval) break;
10            prev = node;
11          savedNodes[i] = prev;
12        bool keepLeft = (start-node.start) ≥ T;
13        bool keepRight = (node.end-end) ≥ T;
14        int height = log2random(1, MAX_HEIGHT);
15        if(keepLeft && keepRight)
16            ok = replace2(node, new Node(node.start,start,height), new Node
                  (end, node.end,height));
17        else if(keepLeft)
18            ok = replace1(node, new Node(node.start,start,height));
19        else if(keepRight)
20            ok = replace1(node, new Node(end,node.end,height));
21        else
22            ok = remove(node);
23      while(!ok);

24 bool replace2(Node *node, Node *node1, Node *node2)
25    Connect all next[] of node1 to node2
26    do
27        Node *next = node.next[0];
28        if(next & 1) return false;
29        node2.next = next;
30        ok = CAS(&node.next[0], next, (node1 | 1));
31      while(!ok);
32      do_remove(node);
33      for(i=1; i<node1.height; i++)
34        do
35            Node *prev = savedNodes[i];
36            Node *next = prev.next[i];
37            If prev is deleted or not the previous node according to node1, update savedNodes[i]
                  while applying helping if necessary, and repeat
38            node2.next[i] = next;
39            ok = CAS(&prev.next[i], next, node1);
40          while(!ok);
41        If node1 or node2 has been marked for deletion, perform helping if needed and exit
                  for-loop
42      return true;

43 bool replace1(Node *node, Node *node1)
44    do
45        Node *next = node.next[0];
46        if(next & 1) return false;
47        node1.next = next;
48        ok=CAS(&node.next[0], next, (node1 | 1));
49      while(!ok);
50      do_remove(node);
51      for(i=1; i<node1.height; i++)
52        do
53            Node *prev = savedNodes[i], *next = prev.next[i];
54            If prev is deleted or not the previous node according to node1, then update
                  savedNodes[i] while applying helping if necessary, and repeat
55            node1.next[i] = next;
56            ok = CAS(&prev.next[i], next, node1);
57          while(!ok);
58        if node1 has been marked for deletion, perform helping if necessary and exit for-loop
59      return true;
```

Fig. 2. Operations of the skip-list

```
60 struct Node
61    void * start , end;
62    int height;
63    Node* next[height];
64 static Node *head = new Node(-∞, -∞, MAX_HEIGHT);
65 static Node *tail = new Node(∞, ∞, MAX_HEIGHT);
66 thread static savedNodes[MAX_HEIGHT];

68 void do_remove(Node *node)
69    Mark node.next[x] on all levels x using CAS
70    for(i = node.height-1; i >= 0; i--;)
71       Node *prev = savedNodes[i], *next = node.next[i] & (~1);
72       bool ok=CAS(&prev.next[i], node, next);
73       if(!ok)
74          Update savedNodes[i] to be the previous node of node and perform helping if
                necessary of deleted nodes in the path, and repeat. If previous node cannot be
                found, perform next lap in the for-loop

75 bool remove(Node *node)
76    do
77       Node *next = node.next[0];
78       if (next & 1) return false;
79       ok = CAS(&node.next[0], next, (next | 1));
80    while (!ok);
81    do_remove(node);
82    return true;
```

Fig. 3. Data structures, auxiliary *do_remove*, and *remove* operation of the skip-list

point to the next node. As concurrent helping can have been applied, it is important to notify this state when trying to update a possibly outdated pointer in $savedNodes[i]$.

Operation *replace1*, which replaces the free interval with another interval, follows the similar logic as *replace2* but only one new node, *node1*, atomically replaces *node*. *Remove* operation deletes a *node* as follow. First, the next pointer of *node* on lowest level is atomically modified using *CAS*, to contain the deletion mark. Thereafter, *node* is fully removed from the skip-list by *do_remove*.

For internal memory management, hazard pointers [9] are preferably used. Each hazard pointer represents a memory address that can be set by an individual thread in order to signal that the corresponding object is currently in use and should not be reclaimed. The thread-local-storage *savedNodes* can then be implemented by a corresponding number of hazard pointers. To also allow the search part of *split* to safely pass through (i.e., de-reference) next pointers that are marked, without applying helping, the same hand-over trick as used in [10] can be applied.

2.2   Correctness

We now sketch (because of space constraints) the proof of correctness for the linearizability and lock-free criteria.

Lemma 1. *The implementation of the split operation, described in Fig.2, is linearizable with respect to other concurrent split operations.*

Proof: Linearizability is demonstrated by giving the respective linearizability points for the corresponding executions of the *split* operations in four cases:

Case 1 - split into two intervals: A *split* operation that results in this case takes effect at the successful *CAS* in line 30. Before the *CAS* takes effect, the nodes *node*1 and *node*2 cannot be reached by the search part of any concurrent *split* invocation, and *node* is not marked for deletion. After the *CAS* takes effect, the nodes *node*1 and *node*2 can clearly be reached by the search part of a concurrent *split*, as *node* is now referring to *node*1 as being the next node, and *node* has been logically deleted.

Case 2 - keep the left interval: a *split* that results in this case takes effect at the successful *CAS* in line 48 . Before the *CAS* takes effect, the node *node*1 (containing the left interval) cannot be reached by the search part of any concurrent *split*, and *node* is not marked for deletion. After the *CAS* takes effect, the node *node*1 can clearly be reached by the search part of a concurrent *split* as *node* is referring to *node*1 as being the next node, and *node* has been logically deleted.

Case 3 - keep the right interval: a *split* that results in this case takes effect at the successful *CAS* in line 48 . Same arguments holds as for Case 2.

Case 4 - remove the interval: A *split* that results in this case takes effect at the successful *CAS* in line 79 . Before the *CAS* takes effect, *node* is not marked for deletion. After the *CAS* takes effect, *node* has been logically deleted, which will be noted by any concurrent *split* operations that will fail to modify *node*, as the *CAS* in lines 30 and 48 requires the mark to not be set of the next pointer.
□

Lemma 2. *The implementation of the split operation, described in Fig.2, is lock-free.*

Proof: The lock-free property of the *split* operation is maintained if a not finite execution of a loop for one invocation of the operation, is a result of a progress of another concurrent invocation. Assuming that the searched interval exists, the lines 6-10 are indefinitely repeated due to concurrent deletions. These deletions are due to successful concurrent *CAS* in lines 79, 30, and 48, all resulting in progress for the corresponding invocations. The lines 3-23 are repeated due to failed *replace2*, *replace1*, or *remove* functions. These functions fail in lines 28, 46, or 78, due to concurrent deletion of *node*. These deletions are due to successful concurrent *CAS* in lines 30, 48 and 79, all resulting in progress for the corresponding invocations. The lines 35-40 can indefinitely repeat due to concurrent deletions or insertions, which is progress for the corresponding invocations. Same arguments can be applied for the loops in lines 53-57 and lines 77-80.
□

3   Parallel Mark-Split

We are first presenting the design of a lazy-splitting mechanism for our parallel mark-split algorithm, and then the main implementation, a.k.a ParMarkSplit.

3.1   Lazy Splitting

We design a *lazy-splitting* mechanism to improve the efficiency of the splitting part. Originally, whenever a live object is marked, an interval is split to exclude the space occupied by the marked object from the free intervals. We called this design *aggressive splitting*. Splitting for every marked object is inefficient in multi-threaded environment as it causes high contention at the shared data structure. We observe that: marking threads often consecutively mark objects that locate adjacent. The number of those adjacent marked objects is observed about 10% to 61% of the total number of live objects in applications in the DaCapo benchmarks. It is possible to perform splitting one time for adjacent objects that are marked consecutively, instead of splitting for each individual marked objects. We design a mechanism to do so, called *lazy-splitting*.

The lazy-splitting mechanism works as follows. Each marking thread maintains a memory range of adjacent objects recently marked but not yet "split". When it marks a new object, the object's memory is coalesced to the range if they are adjacent. Otherwise, it performs *split* for the range and the range is set to the object's memory. At the end of marking, *split* is called for the remained range. The lazy-splitting mechanism reduces the number of accesses by marking threads to the list of free intervals compared to aggressive splitting at a cost of maintaining not-yet-split interval locally at each marking thread. The lazy-splitting benefits the parallel mark-split algorithm when the performance gain by the reduction of the number of calls to *split* can cover the cost: $(N - M).C_1 > N.C_2$, where N is the total number of live objects; M is the total number of *split* operations that the lazy-splitting performs; C_1 is the cost of a *split* operation and C_2 is the cost to add a marked object to the not-yet-split interval. It is reasonable to assume that, for specific application and platform, these costs are constants. Therefore, whether the lazy-splitting mechanism benefits ParMarkSplit collector mainly depends on the $(N - M)/N$ ratio. An auto switch mechanism for determining when to use lazy-splitting is easy to design by using a threshold t to decide when to use lazy-splitting. Based on the evaluation results, we recommend $t = 10\%$. By default, lazy-splitting is applied as it is observed to benefit the parallel mark-split GC. But, lazy-splitting is not going to be applied when the GC finds, while collecting, that $(N - M)/N < t$.

3.2 Implementation

A parallel version of mark-split can be achieved by performing the following modifications to the concurrent mark-sweep collector (CMS) [11]

- When GC starts, empty the skip-list, then add one interval of the entire region to it.
- When a thread marks an object during the mark phases: If *aggressive splitting* is used, the thread calls *split* to remove the occupied space from the skip-list. If *lazy-splitting* is used, the thread book-keeps the object for the lazy-splitting mechanism.
- At the end of the Remark phase (i.e., the mutator is still suspended), convert the list of free intervals to the format of the allocator's free list. Remove the Sweep phase.

The correctness of the algorithm in the presence of interleaving among concurrent operations can be achieved thanks to the design of the extended skip-list which allows *split* to be performed atomically and in a lock-free manner. The lock-free property of the skip-list, when the number of objects to be marked is finite, guarantees the termination of all executed *split*, and therefore, the mark phases.

We implement our parallel mark-split collector as a collector for the 64-bit OpenJDK 7's HotSpot virtual machine - an open source implementation of the Java SE Platform contributed and supported by Oracle. The collector is named ParMarkSplit. The HotSpot uses a generational heap layout which divides its memory space into two parts: young and old generations. The young generation is to contain recently allocated objects, while objects that have been lived for a while are placed in the old generation. ParMarkSplit serves as a collector for the old generation, similar to CMS. One implementation issue of ParMarkSplit based on the CMS is that CMS is dedicated to work for the old generation in a generational heap. This brings difficulty for a plain comparison of the two algorithms in which they are used to collect a whole heap. Disabling the generational option in HotSpot so that the collectors work on the whole heap requires thorough modifications of the memory management that would have touched the HotSpot intensively. We find that it is more practical to maintain the generational heap layout, similar to other known commercial JVM, which also allows the comparison of the collectors in an industrial standard environment.

4   Evaluation

We are presenting an experimental evaluation of our parallel mark-split collector and comparing it with other collectors in the HotSpot, using the DaCapo benchmarks. Then we discuss the memory overhead and characterize applications that can benefit from ParMarkSplit. We opted to compare ParMarkSplit, with lazy-splitting (PMS) and without it (PMS1), to existing HotSpot's CMS as it was implemented based on CMS. Our evaluation also includes a lock-based parallel mark-split (PMS_Lock) which uses a binary search tree that relies on a single mutex lock to synchronizes concurrent accesses to store free intervals.

The collectors were evaluated in two scenarios. In the first scenario, GCs were configured to work in a stop-the-world mode where the mutator was suspended during collection. This setting allowed us to exclude the synchronization cost between the old GCs and the mutator. Such an execution provides a better look at the performance of the design itself. In the second scenario, the GCs were evaluated in the concurrent mode where the mutator was not suspended during collection. This is the scenario that CMS was designed for.

The DaCapo suite [12] was used for benchmarking. DaCapo contains a set of open-source, general-purpose JVM benchmarks, and is representative of real-world Java applications. We ran the benchmarks and reported results from representative applications which have rich memory behaviors, as tested by Gidra L. et.al [13] and Kalibera T. et al. [14]: *lusearch, avrora, sunflow, tomcat* and *xalan*. We also found that the Dacapo benchmarks use much less memory than our available memory and do not produce much garbage in the old generation. Too big heap might not trigger any old generation collection, though the young generation collection could be triggered often. In order to focus on garbage collection of the old generation, the heap sizes were chosen to be close to the benchmark's working set size. They were 50 megabytes (MB) for aurora, 400MB for xalan and 100MB for the others. Corresponding flags are set to allow the GCs working in multi-threaded mode. The other flags were left on default values.

The experiments were run on two contemporary NUMA multiprocessor platforms running Ubuntu Linux with kernel 3.0.0. One has two Intel Nehalem 6-core processors running at 2.4GHz with HyperThreading, 48GB of RAM, and support up to 24 concurrent hardware threads. The other has four AMD Bulldozer 12-core processors at 2.6GHz, 64GB of RAM and supports up to 48 concurrent hardware threads. In each experiment, we iterated a benchmark six times so that the old generation's collection can be triggered for several cycles.

4.1   Stop-the-world Scenario

In the stop-the-world scenario, we evaluate the lazy-splitting mechanism and the garbage collection time of the evaluated GCs in five applications of the DaCapo benchmarks. We varied the number of threads that collect garbage (GC threads). As we observed that the performance of the evaluated GCs does not change significantly above 15 threads (due to the known poor scalability of CMS), we report the results up to this number of GC threads.

We first evaluate the lazy-splitting mechanism by comparing the number of *split* operations performed by ParMarkSplit in each collection cycles before and after adopting lazy-splitting mechanism. In general, the lazy-splitting mechanism helps ParMarkSplit reducing the number of *split* operations to be performed. In avrora and sunflow, lazy-splitting can reduce this number by around 50%. But in xalan applications, the reduction is only about $4 - 6\%$. The reason may be that live objects marked by the GC in xalan interleave with garbage. Therefore lazy-splitting can not reduce the number of calls to *split* as much as in other applications. We expect that the lazy-splitting mechanism benefits PMS, in term of collection time, the most in avrora and sunflow.

The benefits of lazy-splitting are reflected in the performance of the ParMark-Split collector. Fig. 4 presents the collection time of different GCs in the HotSpot in our Intel and AMD systems. In four out of five benchmarks on the Intel one, lazy-splitting helps reducing the collection time of ParMarkSplit, especially in avrora and sunflow. Only in xalan, the improvement of lazy-splitting are not clear as the gained performance is not enough to pay-off for the overhead cost. Comparing to PMS_Lock, the ParMarkSplit implementations perform significantly better in all applications. The performance of ParMarkSplit compared to CMS, however, are mixture of good and bad results. There are two applications, avrora and sunflow, in which ParMarkSplit works better than CMS. In others, CMS works better. In seeking for the reason of this result, we notice that avrora and sunflow have higher ratios of adjacent live objects over the total number of live objects compared to the other DaCapo applications. These applications can benefit ParMarkSplit from the caching effect as the GC accessing the same intervals for a short time and help ParMarkSplit to work more efficient. We analyze this observation further in section 4.4.

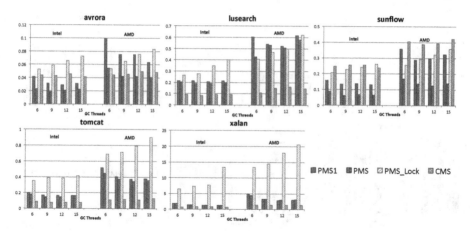

Fig. 4. Garbage collection time (sec) in the stop-the-world scenario

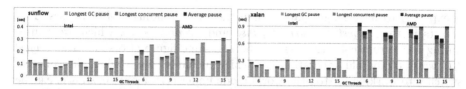

Fig. 5. Pause time when the GCs work concurrently with the mutator. Data columns at each label, from left to right: PMS1, PMS, PMS_Lock and CMS. *Longest concurrent (or GC) pause* when GC works in concurrent mode (or includes pauses when the GC switches to stop-the-world); *Average pause:* average of all the pauses by the old GC.

Another observation is the scalability of different GCs in Fig. 4. We can see that the PMS_Lock performs worse as the number of GC threads increases. This result is not surprised as the lock protecting the skip-list becomes a hot contention point when many GC threads concurrently access it. Meanwhile, the ParMarkSplit collectors, with and without lazy-splitting, as well as CMS are, at least, not scaling down its performance as the number of GC threads increases.

Considering the tested hardware platforms, we found that ParMarkSplit performs better on the Intel than on the AMD. One possible reason can be that the AMD system has NUMA architecture with four nodes which results in higher cost for accessing the shared skip-list.

4.2 Concurrent Scenario

In the concurrent scenario, GCs collect garbage concurrently with the mutator, i.e., the scenario that CMS was built for. We evaluated the pause times of our GC during the collection, in addition to the benchmark's execution times.

We measured the pause time at different number of GC threads. CMS suspends applications during the initial mark and remark phase. ParMarkSplit, which derives from CMS and adds the splitting work to these phases,

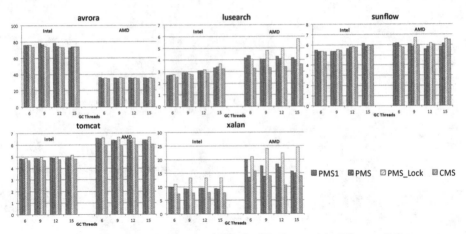

Fig. 6. Benchmark time (sec) for the HotSpot with different GCs

is expected to have longer pauses than the corresponding CMS's pauses. This reflects in the *longest concurrent pause* and *average pause*, which are pauses during concurrent collection, in Fig. 5. The same figure also shows the *longest GC pause* of the old generation GC which includes pauses when the collector switches to working in stop-the-world mode under certain circumstance, e.g the old generation is full during concurrent collection. Due to the lack of space, we include only the results of sunflow and xalan applications, representing for applications which may or may not benefit from ParMarkSplit. We can observe that both average and longest concurrent pauses of ParMarkSplit are longer than those of CMS, as expected from the design. In current HotSpot, the initial mark phase runs single-threaded while the remark phase, though can run multi-threaded, has many parts running sequentially as well. As these two phases run mostly sequentially, the pause time in ParMarkSplit, which uses lock-free synchronization based on compare-and-swap operation, are penalized dramatically. We can expect that when these two phases are fully parallelized in the HotSpot, pause time of ParMarkSplit will be improved significantly, at least proportionally to the speedup of the lock-free skip-list. Regarding the garbage collection pause time, we also notice that the longest GC pause time does not follow the trends of the longest concurrent pause time across the applications. In sunflow, the ParMarkSplit with or without lazy-splitting usually achieves shorter longest GC pauses than both the lock-based one and CMS. However, in xalan, the ParMarkSplit collectors have shorter longest GC pauses than the lock-based one, but longer than CMS. This observation can be drawn from both the AMD and Intel platforms. There are also different in term of absolute values between the two architecture. The AMD system usually has longer pauses than the Intel one.

Regarding the relation between the application's response time and the GC's pause time, it is noticeable that GC pause time is not necessarily the same as the application response time, which means how long it takes an application to responds to a request by users or by other applications. Even though pause time is an indicator for the maximum application response time in the worst case,

Table 1. The estimated size of the skip-list, and of the bitmap of Printesiz's technique

	avrora	lusearch	sunflow	tomcat	xalan
	Number of nodes (thousands) / Size (MB)				
Intel	2.0/ 0.3	14.4/ 2.1	4.9/ 0.7	49.1/ 7.1	55.0/ 7.9
AMD	2.2/ 0.3	16.6/ 2.3	4.3/ 0.7	46.4/ 6.7	57.6/ 8.3
	Estimated size of bitmap (MB)				
Bitmap	0.78	1.56	1.56	1.56	6.25

the contribution of the GC's pause time to the mean application response time
is less and less important in systems with heavy loads, as studied by Persson M.
and Cummins H. from IBM [15].

ParMarkSplit brings the split part to the mark phase, but it also removes the
sweep phase. Does this change reflect in the overall throughput of the applications? Fig. 6 plots the execution time of the benchmarks at different numbers
of GC threads. In some specific cases of lusearch and xalan applications, CMS
performs better than PMS. In sunflow, however, PMS performs slightly better
than CMS. Excepts for those cases, we did not observe significant differences
in the benchmark's execution time between PMS and CMS. Comparing to the
lock-based parallel mark-split, the benchmark times of ParMarkSplit are lower
in most cases. ParMarkSplit has also shown that it works better than CMS in
sunflow, both in terms of pause time and throughput. We will analyze the reason
that ParMarkSplit works well in certain applications in section 4.4.

4.3   Memory Usage

We can estimate the memory overhead used to store free intervals based on the
memory used by the skip-list. Each free interval is stored as a skip-list's node,
which occupies approximately 18 memory words; two for the start and the end of
the free interval, one for the node's level in the skip-list, and at most max_level
pointers pointing to the next nodes in the linked list at each level of the skip-
list. During its construction, the skip-list decides max_level so that 2^{max_level} is
approximately its average size. As our estimated average number of free intervals
is 32000, max_level is set statically to 15. The estimated memory used by the
skip-list in a 64-bit system is presented in Table 1.

We observe that avrora and sunflow have the lowest number of free intervals
among the benchmarks. This is because their marked live objects often reside
next to each other as discussed above. The memory overhead in avrora and sunflow is less than 1% over the heap size (0.3/50MB and 0.7/100MB, respectively),
which is negligible. This cost is higher in applications where the number of free
intervals are high; approximately 2% in lusearch and xalan, and 7% in tomcat,
where the heap sizes are 100MB, 400MB and 100MB respectively. The size of
the skip-list is usually small in applications where their live objects often reside
adjacent to each other, making the memory overhead become negligible. Compared to the memory overhead of Printezis's technique, which uses a bitmap to

skip over contiguous unmarked objects while sweeping [11], ParMarkSplit uses less memory in avrora and sunflow, but more in other benchmarks.

The fragmentation behavior of ParMarkSplit is similar to that of CMS, as it is expected by design. It is possible to check the fragmentation level during or after a collection cycle by checking the size of the skip-list. When the heap is considered too fragmented, a compaction algorithm can be applied in a similar way as it is applied in the CMS garbage collector in HotSpot.

4.4 Characterization of Applications That Benefit from ParMarkSplit

We try to characterize the applications in which ParMarkSplit performs better than CMS so that the system can adaptively select the best GC based on these characteristics. We have observed from the experimental results that ParMarkSplit outperforms CMS in the sunflow and avrora applications, and not in tomcat and xalan. As ParMarkSplit performance is highly dependent on its most frequent operation, i.e., *split*, it usually performs better in applications where the number of live/garbage ratios are low. Analysis on those applications shows that sunflow and avrora have live/garbage ratios as low as about 15% and 20%, respectively. Tomcat have higher ratio; 40% on average. We can speculate that ParMarkSplit maintains the property of its sequential counter-part that it performs better in applications which have low live/garbage ratio. However, this property could not be applied to explain ParMarkSplit's performance in other applications with the same characteristics. Xalan have similar live/garbage ratios as sunflow and avrora, but ParMarkSplit does not perform well in them. We need to distinct the formers from the latters to better characterize the application that clearly benefit from ParMarkSplit.

We observed that our lazy-splitting design brings significant performance gains to ParMarkSplit in applications where it already performs better than CMS, i.e., sunflow and avrora. The benefit of the design in xalan is however not as much. The main characteristic differentiating the two groups is the ratio of the number of adjacent marked objects over that of total marked objects. This ratio is high in sunflow and avrora; and lower in xalan. When the ratio is high, doing splitting interval operation in ParMarkSplit benefits in two ways. First one is a cache benefit when a free interval that is previously split can be cached and reused in the next splitting. Second benefit is that only one *split* operation is required for a set of adjacent objects. As ParMarkSplit brings more such advantages to sunflow and avrora than to xalan, it performs better than CMS in the former applications but not in the latters in our experimental evaluation. All above observations regarding the characterization of applications that benefit from ParMarkSplit are consistent across the two evaluated hardware platforms.

To conclude, ParMarkSplit has been shown to perform better than CMS in applications where the ratio of the number of live objects to that of garbage objects is low and live objects often reside adjacent to each other. ParMarkSplit can be used as a complement to other garbage collection mechanisms to target applications with such characteristics.

5   Conclusion

We present a parallel design of the mark-split garbage collector, called ParMark-Split. To the best of our knowledge, this is the first parallel mark-split design. The design is based on a lock-free data structure that extends the functionality of a skip-list to meet the requirements of the mark-split algorithm augmented with a lazy-splitting design. A complete implementation of the ParMarkSplit collector was developed and integrated in the OpenJDK HotSpot Java virtual machine. We evaluated its behavior experimentally and compared it with the default concurrent mark-sweep garbage collector present in HotSpot, using the DaCapo benchmarks. The experiments were performed on two multiprocessor systems of different architectures; Intel's Nehalem and AMD's Bulldozer. The results are encouraging in applications where the ratio of the number of live objects to that of garbage objects is low and live objects often reside adjacent to each other. We believe that ParMarkSplit can add weight to other garbage collection mechanisms when used for applications with such characteristics.

References

1. McCarthy, J.: Recursive functions of symbolic expressions and their computation by machine, part i. Commun. ACM 3, 184–195 (1960)
2. Cheney, C.J.: A nonrecursive list compacting algorithm. Commun. ACM 13, 677–678 (1970)
3. Hughes, R.J.M.: A semi-incremental garbage collection algorithm. Software: Practice and Experience 12(11), 1081–1082 (1982)
4. Blackburn, S.M., McKinley, K.S.: Immix: A mark-region garbage collector with space efficiency, fast collection, and mutator performance. SIGPLAN Not. 43(6), 22–32 (2008)
5. Detlefs, D., Flood, C., Heller, S., Printezis, T.: Garbage-first garbage collection. In: Proceedings of the 4th ISMM, pp. 37–48. ACM (2004)
6. Sagonas, K., Wilhelmsson, J.: Mark and split. In: Proceedings of the 5th International Symposium on Memory Management, ISMM 2006, pp. 29–39. ACM (2006)
7. Herlihy, M., Shavit, N.: The Art of Multiprocessor Programming. Morgan Kaufmann (2008)
8. Sundell, H., Tsigas, P.: Fast and lock-free concurrent priority queues for multi-thread systems. J. Parallel Distrib. Comput. 65(5), 609–627 (2005)
9. Michael, M.M.: Hazard pointers: Safe memory reclamation for lock-free objects. IEEE Transactions on Parallel and Distributed Systems 15(8) (August 2004)
10. Sundell, H., Gidenstam, A., Papatriantafilou, M., Tsigas, P.: A Lock-Free Algorithm for Concurrent Bags. In: Proceedings of the 23rd ACM SPAA. ACM (2011)
11. Printezis, T., Detlefs, D.: A generational mostly-concurrent garbage collector. SIGPLAN Not. 36, 143–154 (2000)
12. Blackburn, S.M., et al.: The dacapo benchmarks: Java benchmarking development and analysis. SIGPLAN Not. 41, 169–190 (2006)
13. Gidra, L., Thomas, G., Sopena, J., Shapiro, M.: Assessing the scalability of garbage collectors on many cores. In: Proceedings of the 6th PLOS Workshop. ACM (2011)
14. Kalibera, T., et al.: A black-box approach to understanding concurrency in dacapo. In: The UK MM-NET Workshop on Memory Management (April 2012)
15. Persson, M., Cummins, H.: Java technology, ibm style: Garbage collection policies. IBM DeveloperWorks (May 2006)

Practical Concurrent Unrolled Linked Lists Using Lazy Synchronization

Kenneth Platz, Neeraj Mittal*, and Subbarayan Venkatesan

The University of Texas at Dallas,
Richardson, TX 75080
{kplatz,neerajm,venky}@utdallas.edu

Abstract. Linked lists and other list-based sets are one of the most ubiquitous data structures in computer science. They are useful in their own right and are frequently used as building blocks in other data structures. A linked list can be "unrolled" to combine multiple keys in each node; this improves storage density and overall performance. This organization also allows an operation to skip over nodes which cannot contain a key of interest. This paper introduces a new high-performance concurrent unrolled linked list with a lazy synchronization strategy that allows wait-free read operations. Most write operations under this strategy can complete by locking a *single* node. Experiments show up to a 300% improvement over other concurrent list-based sets.

Keywords: concurrent data structures, lazy synchronization, linked lists.

1 Introduction

In recent years, processor manufacturers have shifted their development focus away from increasing clock speeds and single-threaded performance. The rising prevalence of multi-core and multi-processor systems adds additional import to the quest for high-performance data structures that permit concurrent reads and writes while maintaining correct behavior.

Concurrent data structures can synchronize via several methods; the most common techniques involve locking. An exclusive lock can be used to control access to some portion of a data structure. When a thread attempts to access a portion of a data structure, it must first acquire one or more locks. Multiple techniques exist offering varying degrees of performance. The performance of a technique depends upon both the number of locks which must be acquired during an operation and the *granularity* of each lock (the fraction of the data structure protected by each lock).

Other algorithms use atomic read-modify-write instructions in lieu of locks. These instructions, such as compare-and-swap (CAS) or load-linked/store conditional (LL/SC), can be used to provide lock-free or wait-free synchronization [1]. Efficient lock-free and wait-free algorithms are inherently more complex

* This work was supported, in part, by the National Science Foundation (NSF) under grant number CNS-1115733.

M.K. Aguilera et al. (Eds.): OPODIS 2014, LNCS 8878, pp. 388–403, 2014.

than lock-based algorithms; they are harder to design, analyze, implement, and debug.

The linked list is one of the most ubiquitous data structures in computer science. It implements the standard set operations: *insert, remove,* and *lookup.* A linked list is typically implemented via a sequence of *nodes,* each of which contains a *key,* possibly a *data* element, and a pointer to the *next* node in the sequence. Linked lists are of particular interest because many other data structures (such as graphs and hash tables) use linked lists as "black box" subroutines [2].

Linked lists have been well-studied from a concurrency perspective. Several efficient lock-based algorithms exist. The simplest algorithm consists of a single lock which protects all accesses to the list, but this does not allow for any true concurrency. Improvements have been seen with fine-grained locking, where each node contains its own lock. These fine-grained algorithms typically scan the list for a node of interest and acquire the lock on that node (and possibly other nodes). Two algorithms that use this technique include an "optimistic" algorithm by Herlihy and Shavit [3] and a "lazy" algorithm by Heller [4].

Linked lists, while extremely useful, do have several disadvantages. One major disadvantage to a linked list is that any operation must, on average, traverse half the nodes in the list. Each step in this traversal must dereference that node's next pointer and access a memory location that may be far removed from the prior node. This access pattern makes poor use of the memory hierarchy found in today's systems.

Several attempts have been made to increase the efficiency of linked lists by combining multiple keys into a single node. These "unrolled" lists, first described by Shao et al [5], improve performance in two ways. First, unrolling reduces the number of pointers which must be followed to find an item. Second, this groups multiple successive elements in sequential memory locations and better conforms to the principle of spatial locality [6, 7].

More recently Braginsky and Petrank developed a "chunked" lock-free linked list [8]. Their algorithm improves the locality of memory accesses by storing a sequential subset of key/data pairs within a contiguous block of memory. As time elapses and elements are inserted and removed from the list, their algorithm splits full chunks and combines sparsely populated ones. An operation can quickly locate the appropriate chunk, and searches within a chunk exhibit favorable spatial locality.

Our Contributions: We present a new lock-based data structure for an unrolled linked list based upon Heller's lazy synchronization wherein the majority of operations complete by locking a *single* node. We allow our data structure to contain up to K key/data pairs per node; this improves both the storage density and locality of reference within a node [6]. Using the algorithms we present, we can traverse this data structure in $O(n/K + K)$ operations, where n is the number of key/data pairs stored in the list. We also sketch a proof of correctness, using linearizability and deadlock-freedom as our safety and liveness properties, respectively.

The data structure we present is straightforward to implement and exhibits excellent throughput. Our analysis shows that it (i) exhibits high degrees of *spatial and temporal locality* by accessing sequential memory locations; (ii) responds extremely well to common *compiler optimizations*; and (iii) increases *cache efficiency* by eliminating extraneous pointers. In performance testing our implementation provides up to 300% higher throughput than the list presented by Braginsky and Petrank [8]; the improvement over other concurrent lists is even higher.

Roadmap: The rest of the paper is as follows. Section 2 describes prior work related to this paper. Section 3 briefly describes our system model. Section 4 describes our data structure, the algorithms to implement standard set operations, and provides a proof of correctness. Section 5 describes our experiments and analyzes the results. Section 6 consists of our conclusions and suggestions for further work.

2   Related Work

Linked lists have been extensively studied in terms of concurrency; A number of lock-free and lock-based algorithms for linked lists exist, including lock-free algorithms by Valois [9], Michael [10], and Harris [11]. In this paper, we present a lock-based algorithm which permits wait-free reads.

The data structure presented here is modeled after a list by Heller which uses a "lazy" locking strategy [4]. This implementation stores all keys in sorted order; a *scan* identifies the first key greater than or equal to the target key. The scan returns a *window* of two nodes: a node of interest and its immediate predecessor. A *lookup* operation returns TRUE if the key of the current node matches the key in question and FALSE otherwise. An *insert* or *remove* operation obtains a *window* from a scan operation and locks the predecessor and current nodes (in that order). The thread must next perform a *validate*; another thread may modify this section of the list before we acquire the locks. An *insert* then splices a new node into the list while a *remove* removes the node from the list.

Braginsky and Petrank recently developed a "chunked" lock-free linked list [8] which stores multiple sequential elements within the same memory block. This chunked list maintains chunk sizes within a specified minimum and maximum by splitting overfull chunks and merging underfull neighboring chunks. A merge or split requires "freezing" the chunk(s) in question to prevent further operations on a chunk. The operation must then *stabilize* the chunk to quiesce all pending operations. Multiple threads can help with the freeze and stabilize operations.

3   System Model

Our data structure implements a list-based set that supports three operations. A *lookup* operation accepts a *key* as an argument and returns either a *data* element indicating success or NIL indicating failure. An *insert* operation accepts a *key* and *data* element as arguments, returning TRUE if the operation successfully inserted the key/data pair or FALSE to indicate failure (due to the key already existing in the list)[1]. A *remove* operation accepts a *key* and returns TRUE for success or FALSE if the element was not found.

Our algorithms use exclusive locks for coordination between threads. Many locks exist which provide different performance characteristics and progress guarantees. We assume a "black-box" lock which provides the guarantees of *deadlock-freedom* and *mutual exclusion*. For the sake of brevity, we rely on the Resource Acquisition Is Initialization (RAII) [12] idiom when acquiring these locks. We assume that acquiring a lock involves creating object with local scope which releases the lock when destroyed. In C++11, this is implemented with the `std::lock_guard` class. Other languages have similar constructs, either language-provided or user-specified.

4   An Unrolled Linked List Using Lazy Synchronization

4.1   Algorithm Overview

Our unrolled linked list maintains a singly-linked list of nodes and stores keys in partially sorted order. Each node contains (i) an array of *key/data* pairs, (ii) a *next* pointer to the next node in the list, (iii) a *count* of the number of elements in the node, (iv) an exclusive *lock*, and (v) a *marked* flag indicating a node's logical removal. In this data structure *lock* protects access to the *next* pointer which allows most operations to complete while holding a single lock. We define the parameter K to indicate the maximum number of key/data pairs per node and the *anchor key* as the first key in a node. We further define parameters MINFULL and MAXMERGE. MINFULL is the minimum number of keys before we attempt to merge nodes; MAXMERGE is the maximum number of keys we allow in a merged node.

The data structure keeps track of the *head* pointer which points to the first element in the list. We maintain two invariants: (i) the anchor key of each node is strictly less than the anchor key of its successors and (ii) all (non-anchor) keys in a node are strictly greater than their anchor key. We do not impose any ordering among keys within a node; attempting to keep keys in sorted order would penalize write performance and complicate wait-free lookups. The layout of the data structure is depicted in Fig. 1.

We define two sentinel values of $-\infty$ and $+\infty$. We initialize the list with three sentinel nodes with anchor keys of $-\infty$, $-\infty$, and $+\infty$. The sentinel value of $\top$ indicates a key slot that is unused.

Each operation scans the list to find the appropriate node upon which to operate and returns that node and its predecessor. An insert replaces a sentinel key with our new key/data pair and returns TRUE if successful or FALSE if the element is already in the list. A remove replaces a key with a sentinel key,

---

[1] Sets do not permit multiple entries for a given key. Another option is to replace the existing data element with the new element.

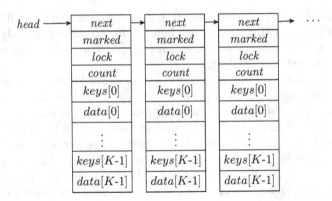

Fig. 1. Layout of the unrolled linked list

returning TRUE for success or FALSE for failure (i.e. the element was not found in the list). A lookup returns either the *data* element associated with the key or NIL if the key is not in the list.

4.2   Algorithm Detail

The first step in any operation on the list involves a *scan* (Alg. 1). We maintain three pointers during this scan, *prev*, *curr*, and *succ*. We scan through the list until the *succ* contains an anchor key greater than our key of interest. Once *succ* meets this criteria, *scan* returns the pair (*prev*, *curr*).

```
1  Function scan(item) : (node,node)
2      prev ← head
3      curr ← prev.next
4      succ ← curr.next
5      while succ.key > item do
6          prev ← curr
7          curr ← succ
8          succ ← succ.next
9      return (prev,curr)
```

Alg. 1. Scan

The *lookup* function starts with a *scan* of the list (Alg. 2). Once we have our *curr* node, we perform a single pass through its keys looking for *item*. At each slot, we read the key/data pair *atomically* (line 13). We can either select key and data elements that collectively fit within a machine word or use atomic snapshots [13, 14]. All of our tested implementations use the former technique[2] [8].

---

[2] Specifically we store a 32-bit key and data element in a 64-bit word.

If we encounter *key* during our scan, we return the associated *data* element. If we reach the end of the keys without finding *item*, it may still be present; a concurrent *remove* operation may relocate *item* to the anchor key of a node. In this case, we must re-read the anchor key/data pair. If *item* is present, we return its associated *data* element; otherwise, we return NIL.

```
10 Function lookup(item) : Data
11     (prev, curr) ← scan(item)
12     for i ← 0 to K − 1 do
13         (key, data) ← (curr.keys[i], curr.data[i])
14         if key = item then return data
15     (key, data) ← (curr.keys[0], curr.data[0])
16     if key = item then return data
17     return NIL
```

Alg. 2. Lookup

Our *insert* and *remove* functions depend upon a *validate* function (Alg. 3) similar to Heller's [4]. We must perform this validation because another thread may still manipulate *prev* or *curr* until we acquire the lock on *prev*. We validate by checking that neither *prev* nor *curr* are marked for removal, *prev.next* still points to *curr*, and our target key is not less than *curr*'s anchor key[3].

```
18 Function validate(prev, curr, item) : boolean
19     return ¬prev.marked ∧ ¬curr.marked
           ∧ prev.next = curr ∧ curr.keys[0] ≤ item
```

Alg. 3. Validate

The *insert* operation (Alg. 4), performs a *scan* to locate an appropriate insertion point, locks *prev*, performs a *validate*. If the validation fails, the operation must return to the head of the list and scan again. Once a validation succeeds, it checks *curr* for three conditions. If *curr* already contains *item* it leaves the node unchanged and returns FALSE. If there is at least one empty slot in *curr* (denoted by the sentinel ⊤) it atomically replaces the sentinel key and its associated data element with the new key/data pair, increments *count*, and returns TRUE. If there are no available slots, we must split the node.

To split a node (Alg. 5), we first lock *curr*. This will not require another validation since no other thread can modify *prev.next*. Next we allocate two new nodes, *new1* and *new2*. We copy all of the key/data pairs from *curr* to *new1*, sort them[4], and then copy the upper half to *new2*. Finally, we replace the upper half of *new1*'s keys with ⊤.

---

[3] A concurrent removal of *curr*'s anchor key may result in this violation.

[4] While there is a $O(n)$ algorithm to determine the median and partition a set of values, in real-world situations, an efficient sorting algorithm is faster [2].

```
20 Function insert(key,data) : boolean
21     while true do
22         (prev,curr) ← scan(item)
23         prev.lock()
24         if ¬validate(prev,curr,item) then
25             continue                    /* Return to head and re-scan */
26         if curr.contains(item) then return FALSE
27         slot ← first location of ⊤ in curr
28         if slot is defined then
29             (curr.keys[slot], curr.data[slot]) ← (key, data)
30             curr.count ← curr.count + 1
31         else
32             curr.lock()
33             (new1, new2) ← split(curr)
34             if key < new2's anchor key then
35                 (new1.keys⌈K/2⌉, new1.data⌈K/2⌉) ← (key, data)
36                 new1.count ← new1.count + 1
37             else
38                 (new2.keys⌊K/2⌋, new2.data⌊K/2⌋) ← (key, data)
39                 new2.count ← new2.count + 1
40             curr.marked ← TRUE
41             prev.next ← new1
42         return TRUE
```

Alg. 4. Insert

```
43 Function split(node) : (Node, Node)
44     Allocate two new nodes, new1 and new2
45     Copy all key/data pairs from node to new1
46     Sort all key/data pairs in new1 ascending by key
47     Copy the upper ⌊K/2⌋ key/data pairs from new1 to new2
48     Replace the upper ⌊K/2⌋ keys in new1 with ⊤
49     new1.next ← new2
50     new2.next ← node.next
51     new1.count ← ⌈K/2⌉, new2.count ← ⌊K/2⌋
52     return (new1, new2)
```

Alg. 5. Split

```
53 Function remove(item) : boolean
54 │   while true do
55 │   │   (prev, curr) ← scan(item)
56 │   │   prev.lock()
57 │   │   if ¬validate(prev, curr ,item) then
58 │   │   │   continue
59 │   │   slot = curr.contains(item)
60 │   │   if slot is not defined then
61 │   │   │   return FALSE
62 │   │   if slot = 0 then
63 │   │   │   min ← location of next smallest key in curr
64 │   │   │   (curr.keys[0], curr.data[0]) ← (curr.keys[min], curr.data[min]
65 │   │   │   curr.keys[min] ← ⊤
66 │   │   else
67 │   │   │   curr.keys[slot] ← ⊤
68 │   │   curr.count ← curr.count − 1
69 │   │   if curr.count <MINFULL then
70 │   │   │   curr.lock()
71 │   │   │   succ ← curr.next
72 │   │   │   if succ.keys[0] = +∞ then
73 │   │   │   │   return TRUE
74 │   │   │   if curr.count = 0 then
75 │   │   │   │   curr.marked ← TRUE
76 │   │   │   │   prev.next ← succ
77 │   │   │   │   return TRUE
78 │   │   │   succ.lock()
79 │   │   │   if curr.count + succ.count < MAXMERGE then
80 │   │   │   │   merge(curr, succ)
81 │   │   │   else
82 │   │   │   │   (new1, new2) ← rebalance(curr, succ)
83 │   │   │   │   prev.next ← new1
84 │   │   return TRUE
```

Alg. 6. Remove

Removing an element operates in a similar manner (Alg. 6). We perform a *scan* to locate the (*prev, curr*) pair, lock *prev*, and *validate*. If this succeeds, we attempt to locate *item* in *curr.keys*. If it is not present, we return FALSE. If *item* is present but not the anchor key, we replace *item* with the sentinel ⊤. If *item* is the anchor key, we locate *min*, the location of our next smallest key. We then replace the anchor key/data pair with the *min* key/data pair and replace the *min* key with ⊤. At this time, we also decrement the node's *count*. If our node now has fewer than MINFULL keys, some additional checking is required. Specifically, we neither merge with the tail (line 72) nor an empty node (line 74)[5]. Otherwise, we either merge with our successor node (Alg. 7) or create two new nodes and partition the key/data pairs equally among them (Alg. 8).

```
85 Function merge(curr, succ)
86 |   Copy valid key/data pairs from succ to curr
87 |   succ.marked ← TRUE
88 |   curr.next ← succ.next
```

Alg. 7. Merge

```
89 Function rebalance(curr, succ) : (Node, Node)
90  |   Create two new nodes new1 and new2
91  |   Copy valid key/data pairs from curr and succ to new1
92  |   newcount ← curr.count + succ.count
93  |   Sort all key/data pairs in new1 by ascending key value
94  |   Copy the upper ⌊newcount/2⌋ key/data pairs from new1 to new2
95  |   Replace the upper ⌊newcount/2⌋ keys in new1 with ⊤
96  |   new1.count ← ⌈newcount/2⌉, new2.count ← ⌊newcount/2⌋
97  |   new2.next ← succ.next, new1.next ← new2
98  |   curr.marked ← TRUE
99  |   succ.marked ← TRUE
100 |   return (new1, new2)
```

Alg. 8. Rebalance

Optimization. We can further modify the above algorithms to keep all valid keys at the head of the node; this requires minor changes to *remove*. Instead of replacing the affected key with ⊤, we would replace that key with the last valid key in the node and replace the last valid key with ⊤ (this is symmetric to removing the anchor key). This effectively can cause a valid key to move forward within a node; therefore, a *lookup* would need to scan from right-to-left to correctly identify whether the key is present.

4.3   Correctness Proof

Here we sketch a proof that our algorithm is correct; we provide more rigorous proof in the accompanying technical report [15]. In our proof, we use deadlock-freedom as the liveness property and linearizability as the safety property. We assume that garbage nodes are never reclaimed (all memory accesses are safe). We also assume that our key space is finite; any traversal of the list will terminate.

---

[5] This can happen if *succ* is the tail.

We will make use of the following terms: a *write* operation shall consist of an insert or a remove, an *active* node is a node currently reachable from the head of the list, and a *passive* node is a node that is no longer active. We similarly define *lookup-hit* and *lookup-miss* operations. We can further treat failed *insert* operations as lookup-hits, and failed *remove* operations can be treated as lookup-misses.

At any moment in time, one or more threads may hold locks on nodes. We can order these threads in head-to-tail order according to the lock(s) they hold. Since our key space is finite, our list is of finite length; therefore, one thread will hold the rightmost lock. Since a thread always acquires locks from head-to-tail, this thread will always be able to progress.

We can also define a linearization point for every operation.

- A successful *insert* operation either linearizes to when the key/data is written or (for a split operation) when the *prev.next* pointer is updated.
- A successful *remove* operation can linearize to the point where ⊤ is written
- Any *lookup* operation which operates on a *passive* node can be linearized to the point at which the node becomes passive.
- A *lookup-hit* operating on an *active* node can be linearized at the point it reads the key/data pair.
- A *lookup-miss* operating on an *active* node has two subcases. If the key is not present when the thread starts scanning, we can linearize to the instant the scan begins. If the key is present at that point, a successful *remove* operation must have removed it. We can therefore linearize the *lookup-miss* immediately to immediately follow the point of the *remove*.

5   Experimental Evaluation

5.1   Experiment Setup

We completed our experiments on a 2-processor AMD Opteron 6180SE system with a clock speed of 2.5GHz, 24 total execution cores, and 64GB of memory running Linux (kernel 2.6.43). All of our evaluation code was written in C++ and compiled using gcc-4.8.3 using the same set of optimizations (-O3 -funroll-loops -march=native). We evaluated the following list implementations:

1. **Lazy:** The lazy linked list by Heller [4].
2. **LockFree:** A lock-free linked list by Harris [11] and Michael [10, 16].
3. **Chunked:** The chunked linked list by Braginsky and Petrank [8][6].
4. **Unrolled:** The unrolled linked list described in this paper.

Each implementation used hazard pointers for garbage collection. For our initial experiments we tested node sizes ranging from 8 to 512 keys per node, key ranges from 1,024 to 1 million, thread counts ranging from 1 to 48, and multiple workload mixes. Based on our initial observations, we feel the following parameters accurately represent the performance of our and other list implementations[7]:

---

[6] Source code was obtained with permission from Braginsky and Petrank.

[7] Additional experimental data is available in the companion technical report [15].

1. **Node Size:** For the chunked and unrolled lists, we evaluated the performance with K of 8 and 64 keys per node, MINFULL of $K/4$ and MAXMERGE of $3K/4$.
2. **Workload Distribution:** We evaluated performance against three representative workloads: *write-dominant* with no lookups, 50% inserts, 50% removes; *balanced* with 70% lookups, 20% inserts, 10% removes; and *read-biased* with 90% lookups, 9% inserts, 1% removes.
3. **Degree of Concurrency:** We evaluated the performance with 1, 2, 4, 8, 12, 16, 20, and 24 threads.
4. **Key Range:** Keys were allowed to range from 0 to 5,000 (inclusive)

Each experiment was conducted by initially creating a list with 2,500 entries. We then spawned the specified number of threads and ran them concurrently. Each thread executed as many operations as possible using the specified mix of operations, and we recorded the total number of operations. Each experiment was repeated until we achieved a 95% confidence interval less than 10% of the mean. All results are reported in operations per microsecond.

5.2   Experimental Results

Fig. 2 depicts the results of our experiments. The graphs on the left depict results for 8 keys per node while those on the right show 64 keys per node. From top to bottom we display results for the 0/50/50, 70/20/10, and 90/9/1 workloads, respectively. These results show that the relative performance of each algorithm remains consistent for every workload and thread count. Specifically, we can rank them fastest to slowest: our unrolled algorithm, Braginsky and Petrank's chunked algorithm, Harris and Michael's lock-free algorithm, and Heller's lazy algorithm. The relative throughput at 24 threads is shown in Table 1.

Table 1. Relative throughput at 24 threads with respect to the Lazy algorithm

			$K = 8$		$K = 64$	
Workload	Lazy	Lock-Free	Chunked	Unrolled	Chunked	Unrolled
0/50/50	100	112	143	291	311	1012
70/20/10	100	108	177	435	665	1745
90/9/1	100	114	160	406	1086	2211

Intuitively we can divide the four algorithms into two groups. The lazy and lock-free lists operate as a single phase; each step along the list visits each entry in turn. The chunked list and unrolled list take a two phase approach. The first phase skips over multiple entries while seeking the correct node, and the second phase scans sequential entries in that node.

When we consider the performance differential between our data structure and the chunked list, we should consider the following points:

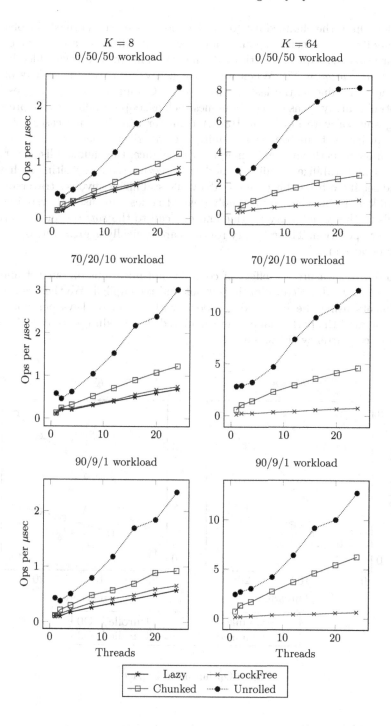

Fig. 2. Experimental Results

1. A lookup in the chunked list (once the node has been identified) involves repeatedly dereferencing a pointer and accessing a different area of the chunk. Our list scans sequentially through an array. This provides the added benefit of *spatial locality* [17]. When we access a key/data pair, it is likely on the same cache line as the last-accessed pair.[8] Compilers can also aggressively optimize array scans using techniques such as loop unrolling, cache prefetching and software pipelining [18]. A compiler can (in some cases) also use vector instructions to perform multiple comparisons concurrently.

2. In order to perform a split, merge, or rebalance, the chunked list must first freeze and stabilize the affected node(s). Freezing requires visiting each entry and setting a freeze bit (using CAS) while stabilizing involves traversing the chunk and removing any partially-deleted nodes. Our list only requires two calls to the *copy* library routine and one call to the *sort* routine to perform either operation. These library routines are typically aggressively optimized for performance.

In order to measure the effect of compiler optimizations on our list and the chunked list, we disabled all optimizations and recompiled. We then re-ran our experiments using the balanced workload with 8 and 64 keys per node. We then compared the performance and calculated the speedup percentage for each degree of concurrency (Figs. 3 and 4).

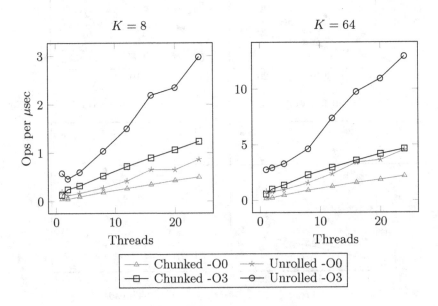

Fig. 3. Optimization Impact

---

[8] On the AMD architecture we tested, a key/data pair consumes 8 bytes, and the cache line stores 64 bytes. This allows 8 key/data pairs to share a cache line.

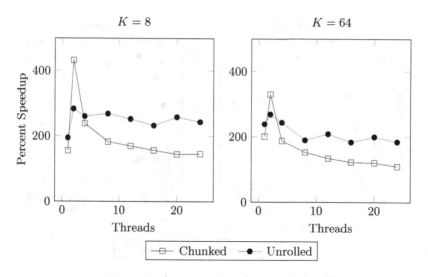

Fig. 4. Optimization Speedup

The results confirm our hypothesis. While the chunked list did see substantial improvements (150-200% in most cases), our list achieved a minimum of a 200% improvement for each thread count, with a maximum improvement of 330% at 2 threads and $K = 8$. The experiments did exhibit one pair of outlying data points. At 2 threads, the chunked list exhibited a 430% improvement for $K = 8$ and 330% improvement at $K = 64$.

Next we should consider how best to select the value of K. We expect to traverse $O(n/K)$ nodes to find the correct node; following that, expect to scan $O(K)$ keys. This results in $O(n/K + K)$ steps per operation. If we select $K = O(\sqrt{n})$, we should maximize the throughput for our algorithms. In order to evaluate this, we performed additional experiments. We executed the same tests as in Fig. 2 with concurrency of 12 and 24 threads, the 70/20/10 "balanced", a maximum key size of 5,000 (and therefore a bound on n), and varied the node size from 2 to 512 keys per node. The results are depicted in Fig. 5.

As expected, each algorithm exhibited peak performance near our predicted value of $\sqrt{n}$. Specifically, the "chunked" algorithm peaked out at 64 keys per node, while our unrolled algorithm continued to scale well up until 128 and 256 keys per node.

6   Conclusions and Future Work

Braginsky and Petrank described a means to reorganize a linked list to improve locality of memory access; in this paper we have improved upon their algorithms. By storing multiple keys in a node and skipping irrelevant nodes, we can improve performance within a constant factor over traditional linked lists. Storing the entries in an unsorted array allows us to sequentially scan these entries, a

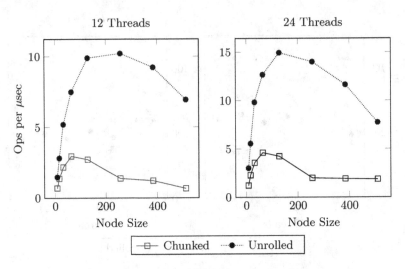

Fig. 5. Impact of Node Size on Performance

task which compilers can aggressively and effectively optimize. Our results are extremely encouraging and suggest that further research should be done in this area.

We envision three different ideas for further research. One possible improvement involves the use of a *group mutual exclusion* object to control access to a node [19]; this would permit multiple *insert* or multiple *remove* operations to operate on the same node concurrently. Second, we can develop a lock-free implementation of this object. We would expect either technique to provide an incremental performance improvement over what we have presented.

Additionally, we would like to explore the potential of unrolling other, more complex, data structures. We imagine that certain implementations of concurrent hash tables (such as those presented by Shalev and Shavit [20]) and concurrent skiplists (such as those by Herlihy [21] and Fraser [22]) would be amenable to this technique.

References

1. Herlihy, M., Shavit, N.: On the nature of progress. In: Fernàndez Anta, A., Lipari, G., Roy, M. (eds.) OPODIS 2011. LNCS, vol. 7109, pp. 313–328. Springer, Heidelberg (2011)
2. Cormen, T.H., Leiserson, C.E., Rivest, R.L., Stein, C.: Introduction to Algorithms, 3rd edn. MIT Press (2009)
3. Herlihy, M., Shavit, N.: The Art of Multiprocessor Programming, 1st edn. Elsevier, Inc., Waltham (2012)
4. Heller, S., Herlihy, M., Luchangco, V., Moir, M., Scherer III, W.N., Shavit, N.N.: A lazy concurrent list-based set algorithm. In: Anderson, J.H., Prencipe, G., Wattenhofer, R. (eds.) OPODIS 2005. LNCS, vol. 3974, pp. 3–16. Springer, Heidelberg (2006)

5. Shao, Z., Reppy, J.H., Appel, A.W.: Unrolling lists. In: ACM Conference on LISP and Functional Programming (LFP 1994), pp. 185–195. ACM, New York (1994)
6. Demaine, E.: Cache-oblivious algorithms and data structures. In: Lecture Notes from the EEF Summer School on Massive Data Sets (2002)
7. Patterson, D.A., Hennessy, J.L.: Computer Organization and Design: The Hardware/ Software Interface, 5th edn. Morgan Kaufmann Publishers Inc., San Francisco (2013)
8. Braginsky, A., Petrank, E.: Locality-conscious lock-free linked lists. In: Aguilera, M.K., Yu, H., Vaidya, N.H., Srinivasan, V., Choudhury, R.R. (eds.) ICDCN 2011. LNCS, vol. 6522, pp. 107–118. Springer, Heidelberg (2011)
9. Valois, J.D.: Lock-free linked lists using compare-and-swap. In: 14th ACM Symposium on Principles of Distributed Computing (PODC 1995), pp. 214–222. ACM, New York (1995)
10. Michael, M.M.: High performance dynamic lock-free hash tables and list-based sets. In: 14th ACM Symposium on Parallelism in Alorithms and Architectures (SPAA 2002), pp. 73–82. ACM, New York (2002)
11. Harris, T.L.: A pragmatic implementation of non-blocking linked-lists. In: Welch, J.L. (ed.) DISC 2001. LNCS, vol. 2180, pp. 300–314. Springer, Heidelberg (2001)
12. Stroustrup, B.: The Design and Evolution of C++. ACM Press/Addison-Wesley Publishing Co., New York (1994)
13. Afek, Y., et al.: Atomic snapshots of shared memory. J. ACM 40, 873–890 (1993)
14. Anderson, J.H.: Composite registers. In: Distributed Computing, pp. 15–30 (1993)
15. Platz, K., Mittal, N., Venkatesan, S.: Practical concurrent unrolled linked lists using lazy synchronization. Technical Report UTDCS-09-14, The University of Texas at Dallas Computer Science Department (2014)
16. Michael, M.: Hazard pointers: Safe memory reclamation for lock-free objects. IEEE Trans. Parallel Distrib. Syst. 15, 491–504 (2004)
17. Hennessy, J.L., Patterson, D.A.: Computer Architecture: A Quantitative Approach, 5th edn. Morgan Kaufmann Publishers Inc., San Francisco (2011)
18. Aho, A.V., Lam, M.S., Sethi, R., Ullman, J.D.: Compilers: Principles, Techniques, and Tools, 2nd edn. Addison-Wesley Longman Publishing Co., Inc., Boston (2006)
19. Joung, Y.J.: Asynchronous group mutual exclusion (extended abstract). In: 17th ACM Symposium on Principles of Distributed Computing (PODC 1998), pp. 51–60. ACM, New York (1998)
20. Shalev, O., Shavit, N.: Split-ordered lists: Lock-free extensible hash tables. J. ACM 53, 379–405 (2006)
21. Herlihy, M.P., Lev, Y., Luchangco, V., Shavit, N.N.: A simple optimistic skiplist algorithm. In: Prencipe, G., Zaks, S. (eds.) SIROCCO 2007. LNCS, vol. 4474, pp. 124–138. Springer, Heidelberg (2007)
22. Fraser, K.: Practical Lock-Freedom. PhD thesis, Kings College, University of Cambridge (2003)

Space- and Time-Efficient Long-Lived Test-And-Set Objects*

Zahra Aghazadeh and Philipp Woelfel

Department of Computer Science, University of Calgary
{zaghazad,woelfel}@ucalgary.ca

Abstract. We provide several space- and time-efficient implementations of randomized long-lived Test-And-Set (TAS) objects from registers, in the standard asynchronous shared memory system with n processes. Our main construction uses $O(n)$ registers, which is optimal, and TAS() and Reset() methods have expected step-complexity $O(\log \log n)$ against the oblivious adversary. Prior to this work, no long-lived TAS implementation from $O(n)$ registers was known, where all methods have sub-linear (expected) step complexity. Another construction achieves expected step-complexity $O(\log^* n)$ for TAS() against the oblivious adversary, constant worst-case step-complexity for Reset(), but requires $O(n^{1.5})$ registers. These results are obtained from general transformations of randomized one-time TAS implementations (e.g., [3, 11, 13]) into long-lived ones.

Keywords: test-and-set, long-lived, memory management, space efficiency.

1 Introduction

Test-And-Set (TAS) objects are standard synchronization primitives that have many algorithmic applications, for example in mutual exclusion and renaming algorithms [4–6, 8, 10, 19, 22]. A TAS object stores a bit, which is initially 0. The one-time version allows only one method TAS(), which sets a bit to 1, and returns the previous value of that bit. A long-lived TAS object also provides a Reset() method, which resets the bit to 0 (and has no return value). We say a process *wins* a TAS(), if that method call returns 0, otherwise it *loses*. Only the process that *wins* a TAS() method call is allowed to subsequently reset it [1].

One-time TAS objects and registers can be used to implement two-process consensus and vice versa [15, 20]. Hence, there is no deterministic wait-free linearizable implementation of TAS objects from registers. The famous randomized one-time TAS implementation of Afek, Gafni, Tromp, and Vitànyi [1] uses $O(n)$ registers, and the TAS() method takes $O(\log n)$ steps in expectation against a strong adaptive adversary. The long-lived variant provides a constant time

---

* This research was undertaken, in part, thanks to funding from the Canada Research Chairs program and from the Discovery Grants program of the Natural Sciences and Engineering Research Council of Canada (NSERC).

M.K. Aguilera et al. (Eds.): OPODIS 2014, LNCS 8878, pp. 404–419, 2014.
© Springer International Publishing Switzerland 2014

Reset() method but increases the expected step complexity of the TAS() method to $O(n)$.

Recently, a series of papers focused on improving the time- and space-complexities of one-time TAS implementations [3, 6, 11, 13]. In particular for weaker adversary models one-time TAS implementations with almost constant step complexity were devised. However, research on efficient implementations of long-lived TAS objects has trailed behind. In fact, prior to our work, no long-lived construction was known which achieves asymptotically optimal $O(n)$ space complexity, and is at the same time faster than Afek et al.'s (i.e., provides TAS() methods with sub-linear expected step complexity). On the other hand, it was recently shown that memory reclamation techniques can be used to obtain efficient long-lived TAS objects from a bounded number of one-time ones: The construction in [2] uses $O(n^2)$ one-time TAS objects and registers, preserves the asymptotic (expected) step complexity of TAS() method and provides a Reset() method with constant (worst-case) step complexity. However, even applied to the most space-efficient published randomized wait-free one-time TAS implementation [11], the long-lived implementation of [2] still results in an unreasonably high space complexity of $\Theta(n^{2.5})$ registers. This space requirement is significantly higher than the lower bound of $\Omega(n)$ ([9]).

It is a fundamental question whether there are inherent tradeoffs between the time and space efficiency of long-lived TAS implementations, or whether implementations exist that achieve similar low step complexities as the best known one-time TAS implementations, and at the same time have optimal (or nearly optimal) space complexity.

In this paper we show that the latter is the case. In particular, we construct an efficient randomized long-lived TAS implementation from $O(n)$ registers (which is optimal), where TAS() and Reset() methods have expected $O(\log \log n)$ step complexity against the oblivious adversary. Our construction applies a space-improved version of a recent memory reclamation technique [2] to a one-time TAS implementation with $O(\log \log n)$ expected step complexity [3, 13].

We also present two general transformations of one-time TAS objects into long-lived ones. Our base construction transforms any one-time TAS object implemented from m registers into a long-lived one that uses $O(n + m)$ registers while preserving the (asymptotic) step complexity of TAS() methods, and providing Reset() methods with $O(m)$ worst-case step complexity. For example, combined with a one-time algorithm in [11], this yields long-lived TAS objects with optimal space complexity $O(n)$, and where TAS() and Reset() methods have (expected) step complexity $O(\log^* n)$ respectively $O(\sqrt{n})$.[1] While in this case Reset() methods are significantly slower than TAS() methods, this may still be reasonable in applications where Reset()s may be executed less frequently than TAS() methods (e.g., mutual exclusion).

Our second transformation reduces the step complexity of Reset() methods to $O(1)$, but sacrifices space optimality by requiring $O(n \cdot m)$ registers.

---

[1] $\log^* n$ denotes the *iterated logarithm* of n, i.e., the number of times $\log_2$ must be applied iteratively until the resulting value is at most 1.

In particular, using again the one-time algorithm in [11] we obtain a long-lived TAS object from $O(n^{1.5})$ registers, where TAS() and Reset() methods have (expected) step complexity $O(\log^* n)$ respectively $O(1)$.

Other Related Work. A randomized two-process long-lived TAS implementation was given by Tromp and Vitànyi [24], which is used in almost all modern n-process TAS implementations. E.g., the randomized long-lived n-process implementation by Afek, Gafni, Tromp, and Vitànyi [1] is based on a tournament tree consisting of two-process TAS objects. Alistarh, Attiya, Gilbert, Giurgiu, and Guerraoui [6] proposed a one-time variant of that tournament tree in which the expected step complexity of the TAS() method is logarithmic in the contention, k (against the strong adaptive adversary). Their construction needs $O(n^3)$ registers. Randomized one-time TAS objects with sub-logarithmic expected step complexity are only known in weaker adversary models. Alistarh and Aspnes [3] propose a TAS() method that has $O(\log \log n)$ expected step complexity against the oblivious adversary, using $O(n^3)$ registers. Giakkoupis and Woelfel [13] reduce the space to $O(n)$ and provided an adaptive version of the algorithm with an expected step complexity of $O(\log \log k)$ against the oblivious adversary, where k is the contention. In fact, the double-logarithmic algorithms of [3] and [13] achieve the claimed step complexities even in slightly stronger read-write oblivious adversary. Giakkoupis and Woelfel also present an adaptive one-time algorithm which has $O(\log^* k)$ expected step complexity against the oblivious (and also a slightly stronger location oblivious) adversary, using $O(n)$ registers. Giakkoupis, Helmi, Higham, and Woelfel showed how to reduce the space complexity of this algorithm to $O(\sqrt{n})$ registers, while achieving an expected step complexity of $O(\log^* n)$ [11]. Note that the best known lower bound for the space complexity of one-time TAS implementations is $\Omega(\log n)$ [23], while from lower bounds for mutual exclusion [9] it follows that any long-lived TAS implementation needs at least a linear number of registers.

In [1], Afek et al. also discuss how their one-time TAS implementation can be transformed into a long-lived one which uses $O(n)$ registers, each using $O(n)$ bits. (To bound the size of the registers they use a modified version of the Sequential Time Stamps System by Israeli and Li [18].) The TAS() method has $O(n)$ expected step complexity against the adaptive adversary, and the Reset() method has constant step complexity. Hoepman [17] shows how one can implement a long-lived TAS object from $O(n)$ registers and $n+1$ one-time TAS objects, provided those TAS objects can be reset in a sequential execution. The resulting TAS() method has, up to a constant additive term, the same step complexity as the TAS() method on the one-time TAS object, and the Reset() method requires n steps in addition to the number of steps it takes to sequentially reset the one-time TAS object. In a more general approach, Aghazadeh, Golab, and Woelfel [2] show how to make any object *writeable*. Since Reset() methods are a special case of writes, the technique can be used to transform one-time TAS objects into long-lived ones. Specifically, any one-time TAS implemented from m registers can be transformed into a long-lived one that uses $O(m \cdot n^2)$

registers, while preserving the (expected) step complexity of the TAS() method, and providing constant time Reset().

2 Model and Definitions

We consider the standard asynchronous shared memory model with n processes with IDs $0, \ldots, n-1$. Processes communicate by executing operations on atomic shared multi-reader multi-writer registers. They can make random decisions using (private) coin flips as a source of randomness. A *schedule* determines the order in which processes take steps. In the case of randomized algorithms, the schedule may be determined by an *adversary*, in response to random choices made by processes. The *strong (adaptive)* adversary makes scheduling decisions based on the entire past execution history including the results of coin-flips. An adversary is *oblivious*, if the schedule is fixed in advance, and thus independent of any random decisions.

The sequence of steps executed by all processes is called *history* or equivalently *execution*. We assume w.l.o.g. that the invocation and the response of a method call coincide with the first respectively the last shared memory step of the method.

3 Base Algorithm

In this section, we describe our basic construction of a long-lived TAS object from an implementation of a one-time TAS object from registers. Later, more efficient constructions employ the techniques introduced here.

Theorem 1. *If there is an implementation of a one-time TAS object OT from m b-bit registers, then there is a a long-lived TAS object T with the following properties:*

1. *T uses $O(n + m)$ registers of length $\max\{\lceil \log(2n + m + 1)\rceil, b\}$ bits;*
2. *the TAS() method of T has asymptotically the same (expected) step complexity as the TAS() method of OT; and*
3. *the Reset() method of T has worst-case step complexity $O(m)$.*

For example, applying this theorem to the randomized one-time TAS implementation from $O(\sqrt{n})$ registers of [11], which has $O(\log^* n)$ expected step complexity, yields a long-lived TAS object implemented from $O(n)$ registers (which is optimal), where TAS() methods have $O(\log^* n)$ expected step complexity, and Reset() methods have worst-case step complexity $O(\sqrt{n})$. We will later show how the step complexity of Reset() methods can be improved to $O(\log \log n)$ (Section 4) respectively $O(1)$ (Section 5) by increasing the expected time-complexity of TAS() methods to $O(\log \log n)$ respectively the space requirement to $O(n^{1.5})$.

The proof of Theorem 1 is constructive, i.e., we show how any one-time TAS implementation OT can be transformed into a long-lived TAS implementation T satisfying the properties stated in the lemma.

We use an array $Ptr[0 \ldots m - 1]$ of $O(\log n)$-bit registers and an array $B[0 \ldots r - 1]$ of r registers, where r is determined below. Each element of Ptr stores an index j of an array entry of B, such that at each point no two entries of Ptr are the same. (Unlike [2], each process does not *own* a set of registers in B.) We say a register $B[j]$ is *in use*, if $Ptr[i] = j$, for some $i \in \{0, \ldots, m - 1\}$.

To motivate the construction, first assume that B is infinite ($r = \infty$). In order to execute a TAS() method call, a process follows the exact implementation of the TAS() method of object OT, except that when the process wants to executes an operation (read or write) on the i-th register of OT, it executes the same operation on the register whose index is in $Ptr[i]$. To reset TAS object T, a process goes through all entries of Ptr and replaces all indices in Ptr with indices of new registers. We say register $B[j]$ becomes *retired* when value j stored in an entry of Ptr is overwritten with some value j'.

However, the resulting implementation is not linearizable. A TAS() call which overlaps a Reset() can obtain an inconsistent view of the object while the resetter is replacing registers with new ones.

To avoid inconsistent views of registers obtained by TAS() calls during pending Reset() calls, we exploit that a process can only call Reset() right after it won a TAS() call. Thus, any TAS() call overlapping a Reset() method call can safely return 1, as long as it does not affect other TAS() calls (i.e., prevent them from winning), because a process has already won the TAS object. In addition, our implementation will ensure that each Reset() method call in any execution linearizes at its response. As a consequence, no Reset() call r_b can get invoked while a Reset() call r_a is still pending, as this would contradict linearizability of the prefix of the execution ending immediately before the invocation of r_b. Hence, no two Reset() calls can overlap.

All processes share a register Seq, which stores a sequence number, and is initially 0. For ease of exposition, we assume Seq can store unbounded values. (In Section 3.2, we discuss how to bound sequence numbers, using n additional single-bit registers, while maintaining the step complexity.) Register Seq is only modified in the Reset() method; in particular, it gets incremented twice during each Reset() call, once at the beginning and once at the end. We prove in Section 6, that Reset() method calls linearize when the value of Seq gets updated for the second time at the end of a Reset() call. Thus, Reset() method calls do not overlap. Hence, Seq stores an odd value if a Reset() call is pending, and stores a unique even value between every two Reset() calls.

A process that executes a TAS() method can identify whether its TAS() call overlaps a Reset() method call by checking the current value of Seq. More specifically, during a TAS() call, at the beginning of the method call and also after executing every shared memory step (i.e., reading from Ptr or reading/writing from registers of B), the process checks the value of Seq. If it reads an odd value from Seq or the value it reads from Seq has changed since the first time it read the value of Seq during the same TAS() call, then the process concludes that its TAS() overlaps a Reset() call and it returns 1 immediately. We can linearize such "failed" TAS() calls during the overlapping Reset() call; this puts them after the successful TAS() of the process that calls Reset() and before the linearization point of that Reset() (which occurs at its response).

Since the value of *Seq* is checked every other shared memory step, a process whose TAS() call overlaps a Reset() call executes at most one shared memory step on some register of B after the value of *Seq* is incremented at the beginning of the Reset() call. This prevents TAS() calls from making unnecessary steps and thus bounds their step complexity. Later, in Section 4, we use this technique to prevent a resetters from having to "clean up" too many registers; this allows us to improve the time-complexity of Reset() calls.

In the implementation described above, each Reset() puts into use m new registers, and thus the algorithm requires unbounded space. Various memory management techniques [7, 14, 16, 21] can be used to bound the space. However, if they are used directly without any additional tricks they increase the step complexity or they may even break wait-freedom. Moreover, for most memory management techniques, such as [7, 14, 16], stronger primitives such as CAS and FAA are required. Here we use a variant of a memory reclamation technique introduced by Aghazadeh, Golab, and Woelfel [2], which is time-efficient and uses only registers. But we exploit the fact that no two Reset() calls can overlap, to improve the space-complexity of that memory reclamation technique.

Correctness proofs are provided in Section 6.

3.1   Bounding the Space

Our idea for memory recycling builds on [2], but we modify that technique to reduce the space. This yields long-lived TAS implementations from one-time ones using $O(n + m)$ additional registers (where m is the number of registers used by the one-time implementation), while the direct application of the recycling technique from [2] would require $\Theta(n^2 + m)$ additional registers.

Figure 1 shows the implementation of the TAS() and Reset() methods. We also use a helper function recycle(). The array B in our implementation now has size $r = 2n + m + 1$. Processes share an array $Ann[0 \ldots n - 1]$ of registers, initially all equal to $\perp$. After a process p read an index b from an entry of *Ptr* (in line 11) during a TAS() call, and before it executes its operation on $B[b]$ (in line 17), it announces its access to this register by writing b into $Ann[p]$ (in line 14).

In order to reset the object, a process replaces the index stored in each entry $Ptr[j]$, $j \in \{0, \ldots, m - 1\}$, with an index $f \in \{0, \ldots, 2n + m\}$, such that $B[f]$ is not in use and is not announced. The fact that f is not announced guarantees that $B[f]$ will not be accessed until after it has been put back into use again and also that no TAS() methods that overlap with the Reset() will access $B[f]$. The resetter finds such an index f by executing a recycle() method in line 4 of the Reset() method. Then the process resets this register by writing $\perp$ in $B[f]$. Finally, it updates the value stored in $Ptr[j]$ to f (lines 5-6).

A naive implementation of the recycle() method requires processes to read the entire *Ann* and *Ptr* arrays in each recycle() operation. In order to reduce the step complexity of the recycle() method to $O(1)$, processes distribute

shared:
Array<register> $B[0\ldots 2n+m]$
Array<int> $Ptr[0\ldots m-1]=(0,\ldots,m-1)$
Array<int> $Ann[0\ldots n-1]=(\bot,\ldots,\bot)$
int $Seq=0$

Method Reset$_p$()

```
1  s := Seq + 1
2  Seq := s
3  for j = 0 to m − 1 do
4  |   f := recycle(Ptr[j])
5  |   B[f] := ⊥
6  |   Ptr[j] := f
7  s := Seq + 1
8  Seq := s
```

Method TAS$_p$()

```
9   seq := Seq
10  if seq.isOdd() then return 1
```
Execute the one-shot **TAS()** method, where every read or write operation
row() on a register i is replaced with the following code:
Read or write register i:
```
11  |   b := Ptr[i]
12  |   s := Seq
13  |   if s ≠ seq then return 1
14  |   Ann[p] := b
15  |   s := Seq
16  |   if s ≠ seq then return 1
17  |   B[b].row()
18  |   s := Seq
19  |   if s ≠ seq then return 1
```

Fig. 1. TAS() and Reset() methods of the long-lived **Test-And-Set**

the work of scanning Ann over many **recycle()** method calls, and avoid scanning Ptr altogether, as described below. Figure 2 shows the pseudo-code for the **recycle()** method.

We use several shared data-structures, but they are accessed only during **Reset()** method calls (which don't overlap), and thus sequentially. A set *Free* stores "free" indices in $\{0,\ldots,2n+m\}$, i.e., indices among which a process can choose an arbitrary one to return from a **recycle()** operation. The set data structure supports the operations **add(x)** which adds an element x to it, and **remove()** which removes and returns an arbitrary element from the set. Our algorithm ensures that **add(x)** is only called if x is not in the set, and that the set is never empty; therefore, sequential implementations with constant worst-case

```
shared:
Queue<int> RetQ[n] = (⊥, ..., ⊥)
Queue<int> AnnQ[n] = (⊥, ..., ⊥)
Set<int> Free = {m, · · · , 2n + m}
Array<Boolean> Use[0...2n + m]     // first m elements are 1, the rest 0
int Inx = 0
```

Method Recycle$_p$(int ℓ)

```
20  Use[ℓ] := 0
21  RetQ.enq(ℓ)
22  a := Ann[Inx]
23  AnnQ.enq(a)
24  Inx := (Inx + 1)  mod n
25  r₀ := AnnQ.deq()
26  r₁ := RetQ.deq()
27  for j = 0 to 1 do
28  │   if rⱼ ≠ ⊥ then
29  │   │   if ¬Use[rⱼ] ∧ ¬AnnQ.contains(rⱼ) ∧ ¬RetQ.contains(rⱼ)  then
30  │   │   └   Free.add(rⱼ)
31  f := Free.remove()
32  Use[f] := 1
33  return f
```

Fig. 2. Recycle() method of the long-lived **Test-And-Set**

step complexity exist (e.g., based on linked lists). A queue $AnnQ$ keeps track of the last n elements found in the announce array during the last n recycle() operations. Similarly, a queue $RetQ$ stores the last n indices that got retired. Initially, both queues contain n elements $\bot$, and the algorithm ensures that the length of each queue is n before and after each recycle() operation. We assume that each queue supports, in addition to the standard operations enq() and deq(), an operation contains(ℓ) which returns true if element ℓ is in the queue and false otherwise. Since the domain of elements stored in the queue is a bounded set of size $2n + m + 1$, a contains() method with constant step-complexity can be easily provided by using a register for each element of the domain that keeps track of the number of times the element occurs in the queue. Finally, a Boolean array $Use[0...2n + m]$ is used to indicate for each index $\ell \in \{0, ..., 2n + m\}$ whether the corresponding register $B[\ell]$ is in use or not.

Now consider a recycle(ℓ) call by a process p that is about to retire index ℓ. First, in line 20, the process resets the flag $Use[\ell]$ to indicate that index ℓ will not be in use anymore, once the Reset() method during which recycle() was called is completed. Then, in line 21, p enqueues ℓ into $RetQ$. After that, p reads one array entry $Ann[Inx]$, where Inx is incremented modulo n with every recycle() operation, and adds the index it read to $AnnQ$ (lines 22-24). Then p dequeues an element r_0 from $AnnQ$ and an element r_1 from $RetQ$ (in lines 25-26), and thus restores the length of both queues to n. Next, in lines 27-30, p checks each index r_j, $j \in \{0,1\}$, whether it is use (by testing the flag $Use[r_j]$), or whether it still appears in one of the queues. If none of those two conditions is met, the index is added to the set $Free$. Finally, process p removes an arbitrary element f from $Free$, sets the flag $Use[f]$ to indicate that f will be in use when the Reset() method terminates, and returns f (lines 31-33).

3.2   Bounding Sequence Numbers

We can bound the sequence numbers used in the base algorithm by using n additional one-bit registers. We increment Seq during a Reset() call modulo $2n$. This might cause an ABA problem: During a TAS() method a process p might fall asleep after reading some value seq from Seq, and then it wakes up after Seq has been incremented a multiple of $2n$ times, so that p later reads the same value seq from Seq again, even though Seq has changed. To deal with this, we use an array of shared bits $S[0 \ldots n-1]$, which are initially all 0. At the beginning of its TAS() call, just before process p reads the value stored in Seq for the first time, it sets $S[p]$ to 1. When a process executes a Reset(), it sets $S[C]$ to 0, where C is a counter incremented modulo n once during every Reset() call. Thus, each bit $S[c]$, $c \in \{0, \ldots, n-1\}$, is reset at least once while the value of Seq wraps around. Then during its TAS() call, when process p compares the value of Seq with the one it read initially at the beginning of its TAS() call, p also checks bit $S[p]$. If the value of Seq has wrapped around, that bit must be 0. On the other hand, if that bit is 0, then p's TAS() call must overlap some Reset() call. Therefore, even if p does not notice that Seq has changed, it can safely abort the TAS() call and return 1 if it reads $S[p] = 0$.

4   Space-Optimal and Fast Long-Lived Test-And-Set

In this section, we explain how we can improve the expected step complexity of the Reset() method, when we apply the base algorithm of Section 3 to a specific one-time TAS implementation.

Theorem 2. *There is an implementation of a long-lived TAS from $O(n)$ $O(\log n)$-bit registers, such that the TAS() and Reset() methods have $O(\log \log n)$ expected step complexity against the oblivious adversary.*[2]

This is the first implementation of a long-lived TAS object with $O(n)$ space complexity and sub linear step complexity for both, the TAS() and the Reset() methods. The space complexity is optimal due to the lower bound on the space requirement of mutual exclusion [9].

---

[2] The step complexity bound even holds for the stronger read-write oblivious adversary model used in [13].

We use the one-time TAS implementation by Giakkoupis and Woelfel [13] from $O(n)$ registers, which in turn is based on an algorithm by Alistarh and Aspnes [3]. The shared data structure used there can be viewed as having n special (randomized) objects $O_1, \ldots, O_n$ and n randomized 2-process TAS objects $T_1, \ldots, T_n$. Each object O_i, $1 \leq i \leq n$, is a combination of a randomized *sifter* (also called *group election object*), and a splitter. It supports an operation sift(), which each process can call at most once, and which returns one of *lose*, *win*, or *undecided*. (We say a process wins, loses, or is undecided at object O_i.) The implementation guarantees that not all processes lose O_i and if only one process calls O_i.sift(), then that process wins O_i. Moreover, the i-th object O_i is instantiated with some chosen parameter, which is a function f_i, and guarantees that if k processes call O_i.sift(), then in expectation at most $f_i(k) < k$ of them are undecided. Each object O_i is implemented from a constant number of registers and a sift() method call has constant worst-case step complexity.

We can imagine that the objects $O_1, \ldots, O_n$ form a path "down" and the 2-process TAS objects $T_1, \ldots, T_n$ form a path "up". A process p walks the path down, executing sift() operations on each object O_i that it visits. More precisely, after each sift() operation on an object O_i (starting with O_1) a process decides how to proceed based on the return value: If p is undecided at O_i, then it proceeds to object O_{i+1}; if it loses O_i, then it loses the entire TAS() method, and does not take any further steps; finally, if p wins O_i, then p switches to the path "up", i.e., it executes a TAS() method on the TAS object T_i. Whenever process p loses a TAS() call on some object T_i, it also immediately loses the implemented TAS() method, i.e., it returns 1. If it wins a TAS() call on some object T_i, then it continues to walk up the path by calling T_{i-1}.TAS() if $i > 1$, and if $i = 1$, then p wins the implemented TAS() and thus returns 0.

The functions f_i, $1 \leq i \leq n$, satisfy $f_i(k) = O(\sqrt{k})$, and therefore, in expectation only approximately $k^{1/2^i}$ processes reach object O_i. The smallest index i^*, such that no process reaches object $O(i^*)$ has expectation $E[i^*] = O(\log\log n)$. This yields the desired expected step complexity of $O(\log\log n)$.

Before transforming this one-time TAS implementation into a long-lived one, we first modify it slightly: We add n shared registers $F_1, \ldots, F_n$, which are initially $\bot$. During a TAS() method call, when a process wants to access O_i for the first time, it first has to write a non-$\bot$ value to F_i.

Now we transform the modified object into a long-lived one as described in Section 3. Recall that in the resulting Reset() method, a process first increments a sequence number Seq, and then it replaces all registers of the TAS() implementation with recycled ones. (In fact, it replaces indices stored in Ptr with indices of recycled registers, but this is conceptually the same.) At the end of the Reset() method, the process increases Seq again. Recall also that no process which executes a TAS() method, can execute more than one shared memory step once a concurrent resetter has incremented Seq. Hence, suppose that at the point when Seq gets incremented at the beginning of a Reset() method call, exactly registers $F_1, \ldots, F_\ell$ have non-$\bot$ values. Then it is not possible that any object O_j, T_j, or F_{j+1}, for $j > \ell$, will be accessed by any process before Seq

is incremented again at the end of that Reset() method call. (In fact, no such object O_j, T_j of F_{j+1}, for $j > \ell$, will be accessed by a pending TAS() method call that has already executed its first shared memory step.)

Hence, in the Reset() method it suffices to replace the registers used in $O_1, \ldots, O_\ell$, $T_1, \ldots, T_\ell$, and $F_1, \ldots, F_{\ell+1}$ with recycled ones. In particular, during a Reset() method the resetter can read registers $F_1, F_2, \ldots$, until it finds the first one, $F_{\ell+1}$, with value $\perp$. Then it only replaces those registers with recycled ones that need to be replaced, which can be done in $O(\ell)$ time. As proved in [13], the expected value of ℓ is $O(\log \log n)$. Each of $O_1, \ldots, O_n$ and $T_1, \ldots, T_n$ is implemented from a constant number of registers. Thus, we obtain a long-lived TAS object from $O(n)$ registers, where the TAS() and Reset() methods have $O(\log \log n)$ expected step complexity.

5 Long-Lived Test-And-Set with Constant Time Reset

In this section we show how to construct long-lived TAS objects from one-time ones, such that the Reset() method needs only constant time in the worst-case.

Hoepman [17] suggested the following straight-forward transformation of a one-time TAS object into an (inefficient) long-lived one. It is assumed that the one-time TAS object supports a safe_reset() method, which resets the TAS object, but must not overlap with any other method calls on that object. Processes share an array $T[0 \ldots n]$ of one-time TAS objects and one register Ptr, which stores an index $i \in \{0, \ldots, n\}$ to the one-time TAS object $T[i]$ that is in use. To execute a TAS(), a process reads index i from Ptr, announces that index, and reads Ptr again. If the value stored of Ptr changes between those two reads, then the process returns 1, and otherwise, it executes $T[i]$.TAS(). To reset the long-lived TAS object, a process reads all announced indices, and chooses an index $j \in \{0, \ldots, n\}$ that is not announced and such that the value of Ptr is not j. Then it resets the one-time TAS object $T[j]$ by executing safe_reset(). Finally, the process writes j into Ptr.

We can use this simple algorithm as the base algorithm, but instead of reading the entire announce array during each Reset(), we use the recycling technique proposed in Section 3.1, which has constant time complexity. The resulting algorithm still requires only $O(n)$ one-time TAS objects. In addition, we also employ a technique presented in [2] to reset any fixed set of registers in $O(1)$ steps, as long as that reset method does not overlap any other array accesses. For sake of completeness, we explain this simple technique here.

We use one additional shared register Ver, to store a version number. To execute a safe_reset() call, a process simply increments the version number stored in Ver. If some process wants to execute a TAS() call, it first reads the current version number from Ver. It then proceeds as in the original implementation, but whenever it writes some information to a register, it augments that information with the version number it read at the beginning of its TAS() call. Whenever the process reads information together with an augmented version number from a register, it checks whether that version number matches the one stored in *Ver*. If yes, it can use the information stored in the register, otherwise it assumes the register is in an initial state. To bound register *Ver*, we can increment the version number modulo k, where k is the number of registers used in the TAS implementation. In order to avoid the ABA problem, we "lazily" reset registers during a series of safe_reset() calls (in each safe_reset() we reset one register), and thus ensure that during any sequence of k consecutive safe_reset() executions, each register gets reset at least once.

As a result, we obtain a safe_reset() method with constant worst-case step-complexity for any one-time TAS object implemented from registers. This yields the following result.

Theorem 3. *If there is an implementation of a one-time TAS object OT from m ($\log n$)-bit registers, then there is a a long-lived TAS object T with the following properties:*

1. *T uses $O(n \cdot m)$ registers of length $O(\log n)$ bits;*
2. *the TAS() method of T has, up to a constant additive term, the same (expected) step complexity as the TAS() method of OT; and*
3. *the Reset() method of T has constant worst-case step complexity.*

Currently, the most space-efficient published randomized wait-free one-time TAS implementation uses $O(\sqrt{n})$ registers and has step complexity $O(\log^* n)$ [11]. Applying Theorem 3 to this construction yields the following.

Corollary 1. *There is an implementation of a long-lived TAS from $O(n^{3/2})$ $O(\log n)$-bit registers, such that against the oblivious adversary the TAS() method has $O(\log^* n)$ expected step complexity, and the Reset() method requires $O(1)$ steps in the worst case.*

A randomized wait-free one-time TAS object implemented from $O(\log n)$ registers and with $O(\log^* n)$ expected step complexity was recently announced [12], and thus yields a long-lived TAS object from $O(n \log n)$ registers and with the same time complexity as in Corollary 1. On the other hand, a lower bound of Styer and Peterson [23] shows that any deadlock-free one-time TAS object requires at least $\Omega(\log n)$ registers, so the space complexity of a long-lived TAS object obtained from a direct application of Theorem 3 cannot go below $\Omega(n \log n)$.

6   Correctness of the Base Algorithm

Consider the long-lived TAS implementation T in Figures 1 and 2 obtained from a randomized wait-free TAS implementation OT. We say a history H is *permissible on T*, if it can arise from a sequence of T.TAS() and T.Reset() method calls, where a process calls T.Reset() only immediately after it won a T.TAS() call.

Lemma 1. *If H is a permissible history on T, where no two Reset() calls overlap and all method calls complete, then H has a linearization S, such that all Reset() calls in H linearize at their responses.*

For a method call m, we denote the invocation of m by $\texttt{inv}(m)$, and the response of m by $\texttt{rsp}(m)$. As we mentioned before, we assume w.l.o.g. that $\texttt{inv}(m)$ and $\texttt{rsp}(m)$ coincide with the first respectively the last shared memory step of m.

Proof of Lemma 1. Suppose H contains $k \geq 0$ completed $\texttt{Reset}()$ method calls $r_1, \ldots, r_k$, with the same invocation order. Thus, since no two $\texttt{Reset}()$ method calls overlap in H, $\texttt{rsp}(r_i) < \texttt{inv}(r_{i+1})$, for all $i \in \{1, \ldots, k-1\}$. It is not hard to see that values $1, \ldots, 2k$ are written in the same order to Seq during history H. (A proof of this is omitted due to space restrictions.) Let t_0 be the starting point of H, t_i the point during H at which value i is written to Seq, for $i \in \{1, \ldots, 2k\}$, and t_{2k+1} the point at which H ends, or $t_{2k+1} = \infty$ if H is not finite. For $i \in \{0, \ldots, k\}$, let I_i denote time interval $[t_{2i}, t_{2i+1})$ in H. Also, let O_i be the set of all $\texttt{TAS}()$ method calls m in H, such that $\texttt{inv}(m) \in I_i$. Moreover, for $i \in \{1, \ldots, k\}$, let F_i be the set of all $\texttt{TAS}()$ method calls m in H, such that $\texttt{inv}(m) \in [t_{2i-1}, t_{2i})$. Since the values stored in array Ptr only change during interval (t_{2i-1}, t_{2i}) of a $\texttt{Reset}()$ method call r_i, we let R_i denote the set of all registers that are in use (pointed by Ptr) during time interval I_i, for $i \in \{0, \ldots, k\}$. Equivalently, for $i \in \{1, \ldots, k\}$, R_i is the set of all registers that are put into use during $\texttt{Reset}()$ call r_i, and for $i = 0$, R_i is the set of all registers that are put into use initially.

The remainder of the proof relies on the following claims, whose proofs are omitted due to space constraints.

Claim 1. *Consider some $\texttt{TAS}()$ method call m in H,*

a) *if $m \in F_i$, for some $i \in \{1, \ldots, k\}$, then method call m returns 1, without executing any operation on any register $B[j]$, for $j \in \{0, \ldots, 2n + m\}$; and*
b) *if $m \in O_i$, for some $i \in \{0, \ldots, k\}$, and $\texttt{rsp}(m) > t_{2i+1}$, then m returns 1.*

Claim 2. *During each time interval I_i, for $i \in \{0, \ldots, k\}$,*

(a) *only method calls in O_i access registers in R_i; and*
(b) *method calls in O_i access only registers in R_i.*

Let H_i be the history obtained from H, such that it contains only steps in H that occur during interval I_i, for $i \in \{0, \ldots, k\}$. Moreover, let $H_i|R_i$ be the history obtained from H_i, such that it contains only steps in H_i that are on registers of R_i and let $H_i|O_i$ be the history obtained from H_i, such that it contains only steps of method calls in O_i in H_i. Claim 2 immediately yields the following.

Corollary 2. *For any $i \in \{0, \ldots, k\}$, we have $H_i|R_i = H_i|O_i$.*

It can be shown (the proof is omitted due to lack of space) that for any $i \in \{0, \ldots, k\}$ the following is true.

Claim 3. *History $H_i|R_i$ has a linearization S_i of a one-time TAS object, such that S_i contains all method calls of O_i and the $\texttt{TAS}()$ call which wins the object responds before t_{2i+1}.*

We construct a sequential history S from H, and then we show that S is a linearization of H. Let S_i be the linearization of $H_i|R_i$, such that it contains all method calls in O_i, for $i \in \{0, \ldots, k\}$. By Claim 3, such a linearization exists. We add all method calls in S_0 to S in the same order in which they appear in S_0. Next, for $i \in \{1, \ldots, k\}$, we do the following. We append all method calls in F_i to S ordered by their response time, then `Reset()` call r_i, followed by all method calls in O_i, as they appear in S_i.

Now we prove that S is a linearization of H. From definitions of O_i and F_i, it is easy to see that $\left(\bigcup_{0 \leq i \leq k} O_i\right) \cup \left(\bigcup_{1 \leq i \leq k} F_i\right)$ contains all `TAS()` calls in H. Therefore, by the construction, sequential history S contains all `TAS()` and `Reset()` calls in H. Moreover, from the same definitions, it is easy to see that

$$
\begin{aligned}
\texttt{inv}(o_0) < \texttt{inv}(f_1) < \texttt{rsp}(r_1) < \texttt{inv}(o_1) < \texttt{inv}(f_2) < \texttt{rsp}(r_2) < \\
\texttt{inv}(o_2) < \cdots < \texttt{inv}(o_{k-1}) < \texttt{inv}(f_k) < \texttt{rsp}(r_k) < \texttt{inv}(o_k)
\end{aligned}
\tag{1}
$$

where $o_i \in O_i$ and $f_i \in F_i$, for $i \in \{1, \ldots, k\}$. Therefore, the partial happens-before order of operations as defined by H is preserved in S, i.e., if an operation o precedes o' in S, then $\texttt{inv}(o)$ precedes $\texttt{rsp}(o')$ in H. As each history S_i, for $i \in \{0, \ldots, k\}$, is a valid history on a one-time TAS object, the first `TAS()` method call in S_i returns 0, and all other method calls in S_i return 1. Moreover, by Claim 1(a) all method calls in F_i return 1. Therefore, the first `TAS()` method call in S and the first `TAS()` method call after each `Reset()` call return 0, and all other `TAS()` calls return 1. Thus, history S is a valid history on a long-lived TAS object. Hence, S is a linearization of H. Since all `TAS()` methods overlapping a `Reset()` call fail, they can be linearized before that `Reset()`. Thus, each `Reset()` call can be linearized at its response.   □

In the following we show that no two `Reset()` calls can overlap in any permissible history on T.

Lemma 2. *If H is a permissible history on T, where no two `Reset()` calls overlap, then H has a linearization S, where all complete `Reset()` calls in H linearize at their responses and a pending `Reset()` is the last method in S.*

Proof. Since no two `Reset()` calls overlap in H, there can be at most one pending `Reset()` call r. First, we let all pending `TAS()` method calls in H run to completion, and then the pending `Reset()` call r. (Since the implementation OT and thus also T is randomized wait-free, we can let any pending method calls run to completion in an arbitrary order.) Let H' denote the resulting execution. Then H' is a permissible history on T, where no two `Reset()` calls overlap and all method calls in H' complete. Thus, by Lemma 1, H' has a linearization S, such that all `Reset()` calls linearize at their responses. As the response of r is the last step in H, and r linearizes at its response, the last method in S is r. Moreover, history S is also a linearization of H, as H' is an extension of H which contains only method calls that are invoked in H.   □

Lemma 3. *In any permissible history on T, no two `Reset()` calls overlap.*

Proof. Suppose for the sake of contradiction that there are Reset() calls in a permissible history H that overlap. Let r_a and r_b be the Reset() calls that overlap other Reset() calls with the earliest respectively second earliest invocation, and let p_a and p_b be the processes executing r_a and r_b, respectively. Then r_a and r_b overlap each other and hence $p_a \neq p_b$. Let H' be the longest prefix of H in which no two Reset() calls overlap, i.e., H' ends just before the invocation of r_b. Then r_a is pending in history H'. By Lemma 2, H' has a linearization S, such that r_a is the last method in S.

Since in H process p_b invokes Reset() r_b immediately after the prefix H' of H completes, p_b's last TAS() call m_b in H' must return 0 in H'. Then p_b's last method call in S is also m_b, and in particular p_b does not call Reset() in S after m_b. By the sequential specification of TAS, the first Reset() call after m_b (which returns 0) in S must be by the same process p_b. However, m_b is p_b's last method call in S, and Reset() call r_a by process p_a appears after m_b in S—contradiction. □

Lemmas 2 and 3 imply that any permissible history on T is linearizable.

Corollary 3. *Implementation T is linearizable.*

7   Conclusion

We presented several implementations of time- and space-efficient long-lived Test-And-Set objects from registers in the standard shared memory model with n processes. As one of the results we obtain the first long-lived TAS implementation that has optimal (linear) space complexity, and at the same time yields sub-logarithmic (in fact, almost constant) expected step complexity for both, TAS() and Reset() methods (against the oblivious adversary). Our techniques heavily rely on the property that only the winner of a TAS() call can reset the object. We employ this property to prevent multiple Reset() calls from overlapping. It would be interesting to research whether similar techniques can be applied to obtain other long-lived objects from one-time ones, or whether restricted forms of write operations can be implemented more space efficiently than in [2].

Acknowledgments. The authors are grateful to Wojciech Golab for insightful discussions and helpful comments on earlier drafts of the paper. The authors also thank the anonymous reviewers for their valuable comments.

References

1. Afek, Y., Gafni, E., Tromp, J., Vitanyi, P.M.B.: Wait-free test-and-set. In: Segall, A., Zaks, S. (eds.) WDAG 1992. LNCS, vol. 647, pp. 85–94. Springer, Heidelberg (1992)
2. Aghazadeh, Z., Golab, W., Woelfel, P.: Making objects writable. In: Proc. of 33rd PODC, pp. 385–395 (2014)
3. Alistarh, D., Aspnes, J.: Sub-logarithmic test-and-set against a weak adversary. In: Peleg, D. (ed.) DISC 2011. LNCS, vol. 6950, pp. 97–109. Springer, Heidelberg (2011)

4. Alistarh, D., Aspnes, J., Censor-Hillel, K., Gilbert, S., Zadimoghaddam, M.: Optimal-time adaptive strong renaming, with applications to counting. In: Proc. of 30th PODC, pp. 239–248 (2011)
5. Alistarh, D., Aspnes, J., Gilbert, S., Guerraoui, R.: The complexity of renaming. In: Proc. of 52nd FOCS, pp. 718–727 (2011)
6. Alistarh, D., Attiya, H., Gilbert, S., Giurgiu, A., Guerraoui, R.: Fast randomized test-and-set and renaming. In: Lynch, N.A., Shvartsman, A.A. (eds.) DISC 2010. LNCS, vol. 6343, pp. 94–108. Springer, Heidelberg (2010)
7. Braginsky, A., Kogan, A., Petrank, E.: Drop the anchor: Lightweight memory management for non-blocking data structures. In: Proc. of 25th SPAA, pp. 33–42 (2013)
8. Buhrman, H., Panconesi, A., Silvestri, R., Vitányi, P.: On the importance of having an identity or, is consensus really universal? Distributed Computing 18(3), 167–176 (2006)
9. Burns, J., Lynch, N.: Bounds on shared memory for mutual exclusion. Information and Computation 107(2), 171–184 (1993)
10. Eberly, W., Higham, L., Warpechowska-Gruca, J.: Long-lived, fast, waitfree renaming with optimal name space and high throughput. In: Kutten, S. (ed.) DISC 1998. LNCS, vol. 1499, pp. 149–160. Springer, Heidelberg (1998)
11. Giakkoupis, G., Helmi, M., Higham, L., Woelfel, P.: An $O(\sqrt{n})$ space bound for obstruction-free leader election. In: Afek, Y. (ed.) DISC 2013. LNCS, vol. 8205, pp. 46–60. Springer, Heidelberg (2013)
12. Giakkoupis, G., Helmi, M., Higham, L., Woelfel, P.: Test-and-set in optimal space (2014) (manuscript submitted for publication)
13. Giakkoupis, G., Woelfel, P.: On the time and space complexity of randomized test-and-set. In: Proc. of 31th PODC, pp. 19–28 (2012)
14. Gidenstam, A., Papatriantafilou, M., Sundell, H., Tsigas, P.: Efficient and reliable lock-free memory reclamation based on reference counting. IEEE Transactions on Parallel and Distributed Systems 20(8), 1173–1187 (2009)
15. Herlihy, M.: Wait-free synchronization. ACM Transactions on Programming Languages and Systems 13(1), 124–149 (1991)
16. Herlihy, M., Luchangco, V., Moir, M.: The repeat offender problem: A mechanism for supporting dynamic-sized, lock-free data structures. In: Malkhi, D. (ed.) DISC 2002. LNCS, vol. 2508, pp. 339–353. Springer, Heidelberg (2002)
17. Hoepman, J.-H.: Long-lived test-and-set using bounded space. Technical report, University of Twente (1999)
18. Israeli, A., Li, M.: Bounded time-stamps. In: Proc. of 28th FOCS, pp. 371–382 (1987)
19. Kruskal, C., Rudolph, L., Snir, M.: Efficient synchronization on multiprocessors with shared memory. ACM Transactions on Programming Languages and Systems 10(4), 579–601 (1988)
20. Loui, M., Abu-Amara, H.: Memory requirements for agreement among unreliable asynchronous processes. Advances in Computing Research 4(163-183), 31 (1987)
21. Michael, M.: Hazard pointers: Safe memory reclamation for lock-free objects. IEEE Transactions on Parallel and Distributed Systems 15(6), 491–504 (2004)
22. Panconesi, A., Papatriantafilou, M., Tsigas, P., Vitányi, P.: Randomized naming using wait-free shared variables. Distributed Computing 11(3), 113–124 (1998)
23. Styer, E., Peterson, G.: Tight bounds for shared memory symmetric mutual exclusion problems. In: Proc. of 8th PODC, pp. 177–192 (1989)
24. Tromp, J., Vitányi, P.: Randomized two-process wait-free test-and-set. Distributed Computing 15(3), 127–135 (2002)

WFR-TM: Wait-Free Readers
without Sacrificing Speculation of Writers

Panagiota Fatourou[1], Eleni Kanellou[2],
Eleftherios Kosmas[1], and Md Forhad Rabbi[3]

[1] FORTH-ICS & University of Crete, Heraklion (Crete), Greece
{faturu,ekosmas}@csd.uoc.gr
[2] FORTH-ICS & IRISA, Université de Rennes 35042 Rennes Cedex, France
eleni.kanellou@irisa.fr
[3] FORTH-ICS
rabbi@ics.forth.gr

Abstract. Transactional Memory (TM) is a promising concurrent programming paradigm which employs *transactions* to achieve synchronization in accessing common data known as *transactional variables*. A transaction may either *commit* by making its updates to transactional variables visible or *abort* by discarding all its changes.

We introduce WFR-TM, a TM algorithm which attempts to combine the advantages of *pessimistic* and *optimistic* TM. In a pessimistic TM, no transaction ever aborts; however, *update* transactions are executed sequentially, thus decreasing the degree of achieved parallelism. In optimistic TM, transactions are executed concurrently and they commit if they have not encountered any conflict during their execution.

In WFR-TM, *read-only* transactions are *wait-free* and they never execute expensive synchronization operations, like CAS, Fetch&Increment, Swap, etc. This is achieved without sacrificing the parallelism between update transactions. Update transactions synchronize pessimistically with concurrently executed read-only transactions and they synchronize optimistically with each other.

Keywords: shared memory, software transactional memory, read-only, wait-free.

1   Introduction

The multi-core revolution that chip manufacturing has witnessed in the last decades has in turn created new challenges for software design and transactional memory (TM) [16] has emerged as one of the ways to face them. TM aims at facilitating concurrent programming by providing the programmer with the *transaction* abstraction. A transaction is a piece of code that is used to access data that become shared in an asynchronous, multiprocess shared memory system. Such data are referred to as *transactional variables* (or *t-variables*) and an underlying TM system is in charge of correctly carrying out the process synchronization. The TM system is thus in charge of executing a transaction as if

M.K. Aguilera et al. (Eds.): OPODIS 2014, LNCS 8878, pp. 420–436, 2014.

it had happened *atomically*. If this can be achieved, then the execution of the transaction is successful and the transaction *commits*. This means that all of its updates to t-variables will take effect. Otherwise, the execution of the transaction fails, in which case the transaction *aborts*, and none of its intended updates are realized. A transaction that has been aborted is usually re-executed as many times as needed to terminate successfully.

Most TM systems are *optimistic*: they execute transactions speculatively and they may pro-actively abort transactions if they "suspect" that their execution may jeopardize consistency. Unfortunately, this proactive behavior often leads to a big number of *spurious* aborts, i.e., transactions are aborted even in cases where they could commit without violating consistency. Research on TM [2,11,12,14,17,20] has given special attention on avoiding this, as it degrades performance.

Ideally, we would like to have TM systems in which all transactions terminate successfully within a finite number of steps. However, Bushkov *et al.* [4] proved that no TM algorithm can achieve this property. *Pessimistic* TM algorithms [1,18] use locks to achieve the execution of each transaction exactly once before committing it, i.e., a pessimistic TM never aborts any transaction but the use of locks results in transactions that may not terminate successfully within a finite number of steps. The way most existing pessimistic TM algorithms achieve to commit all transactions is by "pessimistically" imposing sequential execution to update transactions. This significantly restricts parallelism in many cases and therefore it also leads to performance degradations.

In this paper, we present WFR-TM, an algorithm which combines the advantages of both optimistic and pessimistic TM, while trying to avoid their drawbacks. WFR-TM ensures that *read-only* transactions (i.e., transactions that do not intend to update any t-variable) commit within a finite number of steps, i.e., they are wait-free; this property is also called *local progress* [4], in TM context. Moreover, a read-only transaction performs only two writes to shared memory which write to single-writer registers. Additionally to these desirable properties for the readers, WFR-TM allows also multiple *update* transactions (i.e., non read-only transactions) to execute in parallel. Thus, WFR-TM, in contrast to pessimistic TM algorithms, imposes less restriction on parallelism.

In WFR-TM, a read-only transaction T_r starts by announcing itself, so that an update transaction that wants to update a t-variable x after the announcement of T_r (and thus probably after T_r has read x), does so only after T_r has committed. So, before an update transaction T_w completes, it waits for all read-only transactions that have been initiated and not yet completed at some point of T_w's execution, to commit. We remark that it is not necessary to know in advance whether a transaction is read-only; any transaction is read-only when it is initiated and becomes an update transaction the first time it accesses a t-variable for write. Update transactions employ fine-grained locking for accessing t-variables, so that those of them that do not conflict can commit in parallel; a *conflict* occurs between two concurrent update transactions when they access the same t-variable and at least one of them writes it.

On the contrary, in pessimistic TM algorithms [1,18], the updaters use a single coarse-grain lock for accessing shared data. Popular lock-based TM implementations, which, like WFR-TM use fine-grained locking on each t-variable that they update, include [5,7,8,22]. However, in those algorithms, read-only transactions are not wait-free since they may be aborted spuriously.

In [9], a multi-version TM algorithm is introduced which supports wait-free read-only transactions by keeping a list for each t-variable, where each value that it has had is recorded; read-only transactions can find values for the t-variables that they read that are mutually consistent. In [20], a property, called *multi-version permissiveness* or *MV-permissiveness*, is introduced which requires that read-only transactions never abort. Multi-version MV-permissive TM algorithms are also presented in [19,20] enhanced with efficient garbage collection for obsolete versions of t-variables. WFR-TM ensures multi-version permissiveness while being single-version, i.e., it does not maintain multiple versions of t-variables. Thus, WFR-TM is more space efficient in comparison to multi-version algorithms. We remark that in WFR-TM read-only transactions not only never abort, but additionally, they always complete (by committing).

Attiya and Hillel present in [2] PermiSTM, a TM algorithm that ensures multi-version permissiveness without actually maintaining multiple versions of t-variables. Instead, transactions that read a t-variable x announce their presence by incrementing a dedicated read-counter linked to x; this is done by repeatedly executing CAS until it succeeds. So, if it executes concurrently with update transactions that read x, a read-only transaction may repeatedly fail to increment the read-counter of x. This means that read-only transactions in [2] are obstruction-free; *obstruction-freedom* does not ensure that a transaction completes unless the thread executing it runs solo for a sufficient number of steps after some point during the transaction's execution. PermiSTM pays this cost in order to ensure disjoint-access parallelism; roughly speaking, *disjoint-access parallelism* guarantees that transactions that do not conflict do not interfere with each other by accessing common base objects. It has been proved in [3] that in disjoint-access parallel TM implementations with wait-free read-only transactions, a read-only transaction that reads m t-variables has to perform non-trivial operations on at least $m - 1$ base objects; a *non-trivial operation* may change the status of the object on which it is applied. In WFR-TM, read-only transactions perform only two writes on base objects and no expensive synchronization operations at all. However, WFR-TM is not disjoint-access parallel.

Similarly to WFR-TM, PermiSTM supports parallelism among update transactions; update transactions are executed speculatively and they may abort. In PermiSTM, a write-transaction does not proceed in updating the t-variables until all read-only transactions that are accessing it are committed (after decrementing the read counter of the t-variable). Thus, update transactions writing to a t-variable may face a never-decrementing read-counter for this t-variable, leading them to run forever. WFR-TM avoids this by having update transactions waiting for the completion of only a limited number of read-only transactions.

Snooping into a transaction's write-set in order to read t-variable values has also been used in other algorithms, such as WSTM [10] and OSTM [10]. However, WFR-TM combines this with a waiting mechanism where update transactions let read-only transactions terminate, in order to guarantee that they are wait-free. Similar waiting techniques have also been used in [1,2].

2   Model

We consider an asynchronous shared memory system of n processes, which use transactions in order to synchronize in accessing t-variables. A transaction may contain multiple such accesses to different t-variables. In the following, we consider that the sequential code of each transaction accesses a bounded number of t-variables and those accesses can either be reads or writes. The t-variables read by a given transaction form the transaction's *read-set* and those written by it form the transaction's *write-set*. A transaction that is committed or aborted, is *completed*. An initiated transaction that has not completed, is *active*.

Although the code of a transaction is sequential, transactions executed by different processes may run concurrently. A TM algorithm uses a *shared representation* for each t-variable, which consists of the metadata that are necessary in order to monitor the state of the t-variable. A TM algorithm also implements certain *transactional* routines, which are used by processes to access a t-variable's shared representation. Commonly, those routines are the following: (1) BEGINTX, which initiates a transaction; (2) CREATETVAR, which creates the shared representation of a new t-variable; (3) READTVAR, which reads a t-variable; (4) WRITETVAR, which writes a t-variable; (5) COMMITTX, which is commonly invoked after all t-variable accesses of a transaction in order to attempt to effectuate the transaction's changes: if it finds that the execution of the transaction is correct, then the transaction commits, otherwise it aborts; (6) ABORTTX, which aborts a transaction intentionally. The *execution interval* of a transaction starts when BEGINTX is invoked for it and ends when either COMMITTX or ABORTTX return (routines CREATETVAR, READTVAR, and WRITETVAR may only be invoked during a transaction's execution interval).

A TM algorithm implements the above transactional routines by applying a sequence of operations on shared base objects provided by the system. WFR-TM uses *read/write* (R/W) *registers* and CAS *objects*. A R/W register R stores a value from some set and supports the operations $\text{read}(R)$, which returns the value of R, and $\text{write}(R, v)$, which writes the value v in R. A CAS object O stores a value from some set and supports the operations $\text{read}(O)$, which returns the value of O, and $\text{CAS}(O, u, v)$, which checks whether the value of O equals u and, if so, it sets the value of O to v and returns $\texttt{true}$, otherwise, the value of O does not change and $\texttt{false}$ is returned.

A *configuration* is a vector that consists of the state of each process and the state of each base object. It describes the system at some point in time. In an *initial configuration*, each process is in an initial state and each base object has an initial value. During a *step* an operation is applied on a base object by some process; each step may also contain local computation by that process. An *execution* α is a (finite or infinite) sequence $C_0, \phi_0, C_1, \phi_1, \ldots, \phi_{i-1}, C_i$ of alternating configurations (C_k) and steps (ϕ_k), starting from an initial configuration C_0, where the application of ϕ_k to configuration C_k results in configuration C_{k+1}, for each $0 \leq k < i$.

Correctness for a TM algorithm is defined through some *consistency condition*. A well-known one defined for TM is *opacity* [13]. Roughly speaking, opacity ensures that all transactions see consistent values for the t-variables in their read-sets. This is true not only for the committed transactions but also for the aborted and non-completed ones.

3   WFR-TM

Main Ideas. Each transaction starts by announcing itself into an appropriate element of an announce array. This array has size n, with one entry for each process, used by the corresponding process to announce its transactions. Update transactions execute speculatively and employ fine-grained locking to ensure consistency when updating t-variables. Specifically, each transaction T keeps track of the t-variables that it accesses by maintaining a read-set and a write-set. The read-set contains an entry for each t-variable that T reads, where the value read from the t-variable is stored. Similarly, for each t-variable T writes, the write-set contains an associated entry which stores the value that T wants to write to the t-variable. At commit time, T attempts to obtain the locks that are associated with each t-variable in its read-set and its write-set. It is easy to design a correct variant of WFR-TM that avoids locking the read-set (see details in paragraph describing the LOCKDATASET routine below).

In order to avoid deadlocks, the locks are acquired in ascending order based on the address of the t-variable. After acquiring the lock of some t-variable x in its write-set, T also maintains in the corresponding entry of its write-set, the value that x has at the time that it is locked by T. Once T acquires the locks, it enters its *updating phase*, where it actually updates the t-variables recorded in its write-set, and then enters its *waiting phase*, where it waits for active announced read-only transactions to commit. T finally releases all the acquired locks. We remark that WFR-TM guarantees that if T enters its updating phase, it will commit in a finite number of steps.

For each transaction T, WFR-TM maintains a record associated with it. The record for T contains T's status, a variable that represents the current state of T and can take the values *simulating, updating, waiting, committed* or *aborted*. Each transaction starts by speculatively executing its code during its simulating phase. An update transaction (that does not abort early) executes an updating phase and a waiting phase. This last phase is needed to ensure wait-freedom for read-only transactions. The record for T also contains the read-set and write-set of T, as well as a set called *beforeMe* of active transactions that will be linearized before T. This set is needed in order to ensure consistency of reads.

For each t-variable x, WFR-TM maintains a record containing the current value of x, its version which is a strictly increasing sequential number, and a pointer *owner* to some transaction's record which indicates whether x is locked. An update transaction T_w acquires the lock of x each time it successfully executes a CAS to identify itself as the owner of x; x is considered to be unlocked if either the owner field of its record is null or the status of the transaction that it points to is aborted or committed. T_w releases all the locks it has acquired by successfully changing its status to either committed or aborted (i.e. in one atomic step).

WFR-TM provides wait-freedom for any read-only transaction T by ensuring that T_r reads consistent values independently of whether the transactional variables that it accesses are locked, as follows. When a t-variable x is unlocked, T_r reads its value from x's record. Suppose that x is locked by some update transaction T_w at some point. We define an old value and a new value for x at that point. The *old value* for x is the value stored in x's record at the moment that it was locked by T_w, whereas the *new value* for x is the value that T_w wants to write to x. Notice that the old value of x is contained it its record until T_w writes the new value for it during its updating phase. Afterwards, the old value is recorded in the write-set of T_w.

During its initialization, each transaction T takes a *snapshot* of the announce array; this snapshot is a consistent view of the announced transactions together with their statuses. Using this snapshot, T decides whether it must read or ignore the values written by active update transactions. Specifically, T adds into the *beforeMe* set all those announced transactions whose status is waiting. If T reads from x and finds that it is locked by an update transaction T_w, then it checks if T_w is in T's *beforeMe* set. If this is so, T reads directly from the record of x, since T_w's status was waiting when it was recorded by T. This value is the new value of T_w. If T_w is not in T's *beforeMe* set, T decides which value to read based on the phase of T_w. If T_w is in its simulating phase, T returns the value found in x's record (and thus ignores the value that T_w wants to write since T_w has not yet started updating its t-variables). If T_w is in its updating phase, T reads the old value for x from T_w's write-set. This is necessary because in this case, T_w is in the process of updating the t-variables contained in its write-set, so some of them may contain the new values and some of them may still contain the old values; so, if for instance the read-set of T contains two t-variables x and y updated by T_w, and T reads both of them from their records, it may read the old value for x and the new value for y, which would be inconsistent. The same action is taken by T, if T_w is in its waiting phase, since similar consistency problems could appear if T has read other t-variables written by T_w while T_w was in earlier phases. This procedure ensures consistency of read-only transactions.

Before committing, each update transaction reads all entries of the announce array and *waits* for the completion of each announced read-only transaction that it encounters. By incorporating this *waiting mechanism*, WFR-TM ensures that if a read-only transaction T_r ignores the value written to a t-variable by an update transaction T_w, then T_w does not commit before T_r has committed. This is necessary to argue that at the time that T_r commits, it will not have read

```
1    typedef statval {SIMULATING, UPDATING, WAITING, COMMITTED, ABORTED}
```

```
                                                      12   type rnode
                      6    type txrec                 13      tvarrec *tvar
2    type tvarrec     7       uint pid                14      value val
3       value val     8       statval status          15      uint ver
4       uint ver      9       set of rnode elements rset
5       txrec *owner  10      set of wnode elements wset   16   type wnode
     Shared variable: 11      set of pointers to txrec     17      tvarrec *tvar
21   txrec *A[1..n]          elements beforeMe             18      value oldval
                                                           19      uint oldver
                                                           20      value newval
```

Fig. 1. Data structures of WFR-TM

an inconsistent set of values. It is also necessary for guaranteeing the progress properties of the algorithm.

For each t-variable x, there is a version that is associated to it whose value is unique for each value stored in x. An update transaction T_w performs its reads by executing the same actions described above for read-only transactions. Additionally, since the waiting mechanism is not employed between update transactions, in order to ensure opacity, T_w must validate its read-set whenever it reads a t-variable for the first time, as well as a final time before it starts its updating phase. Specifically, T_w validates the read-set by comparing the current version of each t-variable contained there in, against the version that T_w last read for this t-variable (which is contained in its read-set). T_w aborts if a mismatch is found for some t-variable. We remark that, T_w performs a final (indirect) validation by acquiring the lock of each t-variable contained in its read-set. If a version mismatch is found, the CAS used to acquire the lock of the corresponding t-variable, fails, and T_w aborts.

Data Structures. Figure 1 presents the data structures of WFR-TM. For each t-variable x, WFR-TM stores a CAS object of type tvarrec, containing: i) the value val of x, ii) the version ver of x, an unsigned integer, and iii) a pointer $owner$ to a txrec record. To implement WFR-TM with single-word CAS objects, indirection can be used (as in [15,23]).

For each transaction T, WFR-TM stores a record of type txrec that contains: 1) the identifier pid of the process that initiated T, 2) a three-bit variable $status$, storing the $status$ of T, 3) a set $rset$ of rnode elements, implementing the read-set of T, 4) a set $wset$ of wnode elements, implementing the write-set of T, 5) a set $beforeMe$ of pointers to elements of type txrec.

We remark that an element of type rnode contains: i) a pointer $tvar$ to the tvarrec record of x, ii) the value val of x read by T, and iii) an unsigned integer value ver representing the version of x read by T. Moreover, an element of type wnode contains: i) a pointer $tvar$ to the tvarrec record of x, ii) the (old) value $oldval$ of x at the current point in time, iii) an unsigned integer $oldver$ representing the (old) version of x at the current point in time, and iv) the value $newval$ that T will store into x.

Finally, A is the announce array maintained by WFR-TM. Initially, all entries of A are null and for each t-variable x, the fields of the **tvarrec** record of x have the following values: i) *val* contains an initial value, ii) *ver* is equal to 0, and iii) *own* points to a dummy **txrec** record whose *status* field is equal to COMMITTED.

Pseudocode Description. The pseudocode is provided in Figures 2 and 3.

BeginTx. When called by process p for transaction T, it creates (line 23) and initializes (lines 24 - 29) the **txrec** record of T, and then announces T in $A[p]$. Finally, it calls CHECKIFPERFORMED to initialize the *beforeMe* set of T (line 30). Each iteration of the while loop of CHECKIFPERFORMED, reads all elements of A (lines 34 - 35) and adds to T's *beforeMe* (line 37) new update transactions (i.e. those are not already in *beforeMe*) whose status is either waiting or committed (line 36). A new iteration will start if some transaction is added to *beforeMe* in the current iteration. This procedure guarantees that *beforeMe* contains a consistent snapshot at the beginning of the last execution of the **for** of line 34. We now explain why CHECKIFPERFORMED terminates within a finite number of steps. Any transaction T' that is announced after the announcement of T cannot commit before CHECKIFPERFORMED completes, given that even if T' reaches its commit phase, T' will consider T as a read-only transaction (since T has an empty write-set as long as it executes CHECKIFPERFORMED), so T' will wait for T to either terminate or become an update transaction. This ensures that only a limited number of new transactions can appear while CHECKIFPERFORMED is executed, which in turn ensures that CHECKIFPERFORMED returns in a finite number of steps.

ReadTvar. When called by T to read the value of some t-variable x, it first checks if there is an entry for x in the write-set or in the read-set of T. If this is the case, READTVAR returns the value from there (to ensure consistency). Otherwise, the value of x is determined in lines 47-50.

Initially, the value $\langle val, ver, owner \rangle$ of x's **tvarrec** record (line 47) and the status of x's *owner* (line 48) are read. Assume first that x is not locked. Then, the value for x that T returns is *val*, as read in line 47. Assume now that x is locked by a transaction T_w. If the status of T_w is either simulating or committed, then again the value for x that T returns is *val*. Otherwise, the status of T_w is either updating or waiting, and the first and third condition of line 49 evaluate to **true**. Recall that we consider that x has an *old value* and a *new value*, which are stored in T_w's write-set entry for x (specifically, in fields *oldval* and *newval* of this entry, respectively). If T_w is contained in T's *beforeMe* set, i.e. the second condition of line 49 evaluates to **true**, then T_w's update on x has already been performed before the beginning of T. Therefore, again the value for x that T should read is *val*. However, if T_w is not contained in T's *beforeMe* set, then T should not read T_w's update on x, i.e. the new value of x, and should instead read the old value of x; this value is read in line 50.

Then, after determining the value that has to be read, it is added together with its corresponding version in the read-set of T (line 51). In case T is an update transaction, then its read-set is validated by calling VALIDATE (line 52);

```
22  txrec *BeginTx() by process p:
23    txrec *newTx := new txrec
24    newTx → pid := p
25    newTx → status = SIMULATING
26    newTx → rset := empty set of rnode elements
27    newTx → wset := empty set of wnode elements
28    newTx → beforeMe := empty set of pointers to txrec elements

29    A[p] := newTx                          /* T announces itself */
30    CHECKIFPERFORMED(newTx)                /* initialize set beforeMe */
31    return (newTx)

32  CheckIfPerformed(txrec *newTx) by process p:
33    do
34      for i = 1 up to n, excluding p, do
35        tran := A[i]        /* if tran is an update transaction that has entered its waiting phase, ... */
36        if (tran ∉ newTx → beforeMe AND tran → wset ≠ ∅ AND
                  tran → status ∈ {WAITING, COMMITTED}) then
37          add tran in newTx → beforeMe          ... then add it once to beforeMe */
38      while a new element is added in newTx → beforeMe

39  tvarrec *CreateTvar(txrec *tx) by process p:
40    tvarrec newTvar := new tvarrec ⟨⊥, 0, tx⟩
41    return (newTvar)

42  ⟨boolean, value⟩ *ReadTvar(txrec *tx, tvarrec *tvar) by process p:
43    if an element el with el.tvar = tvar exists in tx → wset then
44      return ⟨true, el.newval⟩
45    if an element el with el.tvar = tvar exists in tx → rset then
46      return ⟨true, el.val⟩

47    ⟨val, ver, owner⟩ := *tvar
48    status := owner → status
                /* if tvar is write-locked by some transaction T that is not to be linearized before tx
                   and T is in its updating or waiting phase, then read the old value of tvar from T */
49    if (an element el with el.tvar = tvar ∈ owner → wset AND
          owner ∉ tx → beforeMe AND status ∈ {UPDATING, WAITING}) then
50      ⟨val, ver⟩ := ⟨el.oldval, el.oldver⟩

51    add ⟨tvar, val, ver⟩ in tx → rset

52    if (tx → wset ≠ ∅ AND VALIDATE(tx) = false) then  /* VALIDATE here ensures opacity */
53      tx → status = ABORTED
54      return ⟨false, ⊥⟩

55    return ⟨true, val⟩

56  WriteTvar(txrec *tx, tvarrec *tvar, value value) by process p:
57    if an element el with el.tvar = tvar exists in tx → wset then
58      update el.newval with value
59    else add ⟨tvar, ⊥, ⊥, value⟩ in tx → wset
```

Fig. 2. Pseudocode for BEGINTX, CHECKIFPERFORMED, CREATETVAR, READTVAR, and WRITETVAR of WFR-TM

```
60   boolean CommitTx(txrec *tx)by process p:
61     if (tx → wset = null) then                  /* if tx is read-only, then commit */
62       tx → status := COMMITTED
63       return true

64     if (LOCKDATASET(tx) = false) then            /* if locking of some t-variable fails, then abort */
65       tx → status := ABORTED
66       return false

67     tx → status := UPDATING                      /* tx enters updating phase */
68     for each element el in tx → wset do
69       CAS(el.tvar,*el.tvar, ⟨el.newval, el.tvar → ver + 1, tx⟩)

70     tx → status := WAITING                       /* tx enters waiting phase */
71     WAITREADERS(tx)                              /* tx waits announced read-only transactions */

72     tx → status := COMMITTED                     /* tx commits */
73     return true

74   boolean Validate(txrec *tx) by process p:
75     for each element el in tx → rset
76       ⟨val, ver, owner⟩ := *el.tvar
77       if (ver ≠ el.ver) then return false
78     return true

79   boolean LockDataSet(txrec *tx) by process p:
80     for each element el′ of tx → wset ∪ tx → rset, in ascending order (based
       on tvar field)
81       if ∃ an element el ∈ tx → rset with el.tvar = el′.tvar then
                   /* if tx has read the tvar before, use this old value for consistency */
           ⟨val, ver, owner⟩ := ⟨el.val, el.ver, el.tvar → owner⟩
                   /* otherwise, if the tvar was not read before, use the current value as old value */
82       else ⟨val, ver, owner⟩ := *(el′.tvar)

83       if (owner → status ∉ {COMMITTED, ABORTED})        /* el′.tvar is locked */
84         if ∃ an element el″ ∈ owner → wset with el″.tvar = el′.tvar then
85           return false         /* it if is write-locked, locking fails; otherwise, wait until it is unlocked */
86         else wait until owner → status ∈ {COMMITTED, ABORTED}
                   /* try to lock el.tvar, with l-cas */
87       if (CAS(el′.tvar, ⟨val, ver, owner⟩, ⟨val, ver, tx⟩) = false) then return false
                   /* if el′ is written by tx, then maintain the old value of el′.tvar */
88       if (el′ ∈ tx → wset) then update ⟨el′.oldval, el′.oldver⟩ with ⟨val, ver⟩

89     return true

90   void WaitReaders(txrec *tx) by process p:
91     for i = 0 up to n − 1 excluding p do
92       tran := A[i]
93       if (tran ≠ null AND tran → wset = null) then
94         wait until (tran → status = COMMITTED OR tran → wset ≠ null)
```

Fig. 3. Pseudocode for COMMITTX, VALIDATE, LOCKDATASET, and WAITREADERS of WFR-TM

VALIDATE (lines 75-78) returns **true** when no version of the elements in T's read-list has changed and **false** otherwise. We remark that this validation mechanism can also be implemented using a timestamping mechanism as that presented in TLII [5] or LSA [21], to boost performance.

WriteTvar. When called by T to update some t-variable x with value *val*, T first checks whether it has previously invoked WRITETVAR to modify x. If this is so, then there is already an element for x in T's write-set (line 57) and WRITETVAR updates the *newval* field of this element to *val* (line 58). Otherwise, a **wnode** element is added in T's write-set (line 59).

Recall that when T enters its updating phase, the *oldval* and *oldver* fields of x's **wnode** must contain the value and version, respectively, written by the transaction for which it holds that it had x in its write-set and is the last to commit before T acquired the lock of x (i.e., before x is updated by T). WFR-TM allows another transaction T' to snoop into T's write-set in order to read the old value of some t-variables. Therefore, T's write-set must offer a way to T' to read values that are mutually consistent. To achieve this, WRITETVAR sets the *oldval* and *oldver* fields of new **wnode** elements that are added in a write-set equal to $\perp$ (line 59). This is necessary for avoiding bad scenarios such as the following: Apart from x, assume that T wants to write also another t-variable y and let C be a configuration at which T has written x but not yet y. Thus, T has created a write set entry for x, but there is no such entry in T's write-set for y. T has also read (before C) the contents of x's **tvarrec** to store in the *oldval* and *oldver* fields of x's **wnode**. Now, let another transaction T'' lock and update both x and y, and commit. Then, T continues by writing y. So it places an entry in its write-set for y and reads the contents of y's **tvarrec** to store in the *oldval* and *oldver* fields of this entry. Then, T acquires the locks of both x and y. So, if T' snoops both x and y from T's write-set, it reads inconsistent values.

CommitTx. If T is a read-only transaction (its write-set is empty), COMMITTX changes T's status to committed and returns **true** (lines 61-63). If T is an update transaction, it attempts to acquire the required locks by calling LOCKDATASET (line 64). If it fails to acquire some lock (LOCKDATASET returns **false**), T is aborted (lines 64-66). Otherwise, all the required locks have been acquired (LOCKDATASET returns **true**). Then, T enters its updating phase (line 67) and updates the t-variables in its write-set (line 69). Notice that it also increments the version of each t-variable by one. Afterwards, T enters its waiting phase (line 70) and waits until all announced read-only transactions commit. This is done by calling WAITREADERS (line 71). WAITREADERS goes through the announce array A, and waits until each active read-only transaction (line 93) either commits or turns out to be an update transaction (line 94). Finally, T commits and COMMITTX returns **true** (lines 72-73).

LockDataSet. It is called by T to lock each t-variable in its read-set and write-set. Recall that deadlocks are avoided by acquiring the locks in (ascending) order (based on the *tvar* pointer contained in each **rnode** or **wnode** element). Initially, LOCKDATASET determines the value and version of each t-variable x that it wants to lock, as follows: If x exists in T's read-set, these values are taken from the corresponding read-set entry (line 81). Otherwise, they are read from x's tvarrec record.

LOCKDATASET tries to lock x using a CAS operation which stores a pointer to T's txrec record into the *owner* field of x's tvarrec record (line 87). Notice that this CAS also serves as a final validation of the value of x read by T (in case x is in T's read-set). LOCKDATASET returns true only if it successfully locks all the t-variables in T's read-set and write-set (line 89). If x is already locked by some transaction T' (lines 83 to 84), LOCKDATASET by T returns false (line 84). If x is locked by some transaction that does not intend to update it, LOCKDATASET waits until this transaction completes (line 86). Finally, recall that when LOCKDATASET is invoked, the contents of the *oldval* and *oldver* fields of x's element in T's write-set are $\perp$. In case x is locked, these fields are updated with the determined current values for x (line 88), so that if T enters its updating phase these fields are appropriately set in each element of T's write-set.

We remark that it is very easy to design a correct variant of WFR-TM that avoids locking the read-set. In this variant, the only difference would be that a final read-set validation must be performed explicitly since LOCKDATASET will not then lock the variables in T's read-set. Moreover, in this case, only the *owner* field of a tvarrec is required to be a CAS object, whereas the rest can be updated with writes.

Correctness. We now provide a sketch of the proof of correctness. The full proof is provided in [6]. Fix any execution α of WFR-TM and let T be any transaction in α; let α_T be the *execution interval* of T. Each transaction T is associated with a unique txrec record; the *status* of T is the value of the field *status* in this record. We abuse notation and we use the same notation to refer both to the name of some transaction and to its txrec record.

By the code, $T.status$ is initially SIMULATING. If T is a read-only transaction, its status chances from SIMULATING directly to COMMITTED. If T is an update transaction, then the code implies that its status will change either from SIMULATING to ABORTED, or from SIMULATING to UPDATING, and then from UPDATING to WAITING, and from WAITING to COMMITTED. As long as its status is SIMULATING, UPDATING, or WAITING, we say that T is in its *simulating, updating,* or *waiting phase*, respectively.

An update transaction T_w acquires the lock for x when it successfully executes the CAS of line 87. Notice that this CAS changes the *owner* field of the tvarrec of x to point to T_w. T_w releases this lock (as well as all other locks it has acquired) by changing its status to COMMITTED or ABORTED. We denote by α_{x,T_w} the execution interval of α_{T_w} during which T_w maintains the lock for x.

The code (lines 68 to 70) implies that before T_w enters its waiting phase, it has finished with the updating of the t-variables in its write-set. It is easy to argue that during α_{x,T_w} no transaction $T'_w \neq T_w$ can either lock or update x. This and the code (lines 40 and 69) imply that the version of each t-variable is strictly increasing. This holds since (i) at most one transaction may maintain the lock for x at each configuration and (ii) each transaction updates x only if it maintains the lock for x and when it does so it increments by one the version of x. So, each version of x is unique and it is written by a single transaction.

Lemma 1. *The following holds:*

1. *Consider any update transaction T_w that acquires the lock for some t-variable x. During α_{x,T_w}, the owner field of the* tvarrec *record of x contains a pointer to the* txrec *record of T_w.*
2. *The version of each t-variable x is strictly increasing.*

We *assign linearization points* to read-only transactions that commit in α and to update transactions that enter their waiting phase in α. If T is an update transaction that enters its waiting phase, we assign the linearization point for it at the configuration after the execution of line 70, which updates the status of T to WAITING. If T is a read-only transaction, let CR_T be the configuration at the beginning of the last execution of the for of line 34 in CHECKIFPERFORMED by T. Notice that this for iteration is executed in the last iteration of the do while loop of line 38 during this instance of CHECKIFPERFORMED. Also, notice that during this do while iteration no other element is added in T's *beforeMe* set. Thus, T's *beforeMe* set contains a consistent snapshot of the announced transactions. We assign the linearization point for T at CR_T.

By the way linearization points are assigned, the linearization point of each transaction is placed in its execution interval. Moreover, at each configuration C, there is a sequence l_C of transactions of α that have been linearized before or at C. The *read-set* of T is the set indicated by the *rset* field of T. We say that T *reads* the version v for some t-variable x if a triple of the form $\langle x, -, v \rangle$ is added to the read-set of T. We say that the version v read by T for some t-variable x is *consistent* at C, if it is the version written by the last transaction in l_C that updates x. The read-set of T is consistent at C if the version of each t-variable included in it is consistent.

We first argue about the consistency of read-only transactions.

Lemma 2. *Consider a read-only transaction T_r that completes in α and reads version $v \neq 0$ for some t-variable x. Let T_w be the update transaction that writes the value v in the version field of the* tvarrec *of x. Then, T_w is the last update transaction that is linearized before T_r.*

Proof (Sketch). To prove that T_w is the last update transaction writing to x that is linearized before the linearization point of T_r, we consider the following cases. Let C be the configuration at which the value v is read either on line 47 or on line 50. We prove that CR_{T_r} follows the configuration CW_{T_w}, at which the status of T_w changes to WAITING. Recall that T_w is linearized at CW_{T_w}.

Assume that x is unlocked at C. We prove that if some other update transaction writing x was linearized between T_w and T_r, then T_r would read a version larger than v for x which is a contradiction.

Assume now that T_w maintains the lock for x at C. By inspection of the code, C follows CR_{T_r}. Thus, C follows CW_{T_w}. By inspection of the code, it follows that

T_w acquires the lock for x before CW_{T_w}. We prove that T_w is contained in the *beforeMe* set of T_r. By inspection of the code, it follows that T_r adds T_w in its *beforeMe* set before CR_{T_r}, i.e. before the configuration at which T_r is linearized. Since T_w is still active at C, when this occurred the status of T_w was `WAITING`. Thus, Lemma 1 implies that T_w holds the lock for x in every configuration between CW_{T_w} and CR_{T_r}. Since any update transaction T'_w that updates x holds the lock of x at $CW_{T'_w}$, Lemma 1 implies that no other transaction is linearized between CW_{T_w} and CR_{T_r}.

Assume next that x is locked by some other transaction T'_w at C. Since C follows CW_{T_w}, T'_w acquires the lock for x after T_w. If T'_w is not the first to acquire the lock for x after T_w, then we argue that T_r reads a version larger than v for x, which is a contradiction. Thus, T'_w must be the first such update transaction. We consider again two cases. Assume first that T'_w has not yet updated x at C. Then, no update transaction writing x (including T'_w since its execution has not yet reached $CW_{T'_w}$) has been linearized between T_w and T_r. Assume finally that T'_w has updated x at C, then T_r reads v for the version of x through the *oldver* field of the corresponding write-set entry of T'_w (line 50); thus, T'_w is not contained in the *beforeMe* set of T_r (line 49). We use this to prove that T'_w is linearized after T_r.   □

Lemma 2 implies that the read-set of T_r is consistent at CR_{T_r}. We next argue for the consistency of each update transaction T_w.

Lemma 3. *The read-set of an update transaction T_w that enters its waiting phase is consistent at CW_{T_w}, where the linearization point of T_w is placed.*

Proof (Sketch). Until T_w performs its first write on a t-variable, it is considered as a read-only transaction (since its write-set is still empty). So, Lemma 2 implies that its read-set is consistent up to the point that it performs its first write. So, let's assume that T_w has performed its first write on a t-variable. Whenever it reads some t-variable, T_w successfully validates its read-set (line 52); so, the read-set of T_w is consistent up to the configuration preceding the invocation of the last instance of VALIDATE. Moreover, before T_w changes its status to waiting (line 70) it has successfully acquired locks. During this lock acquisition procedure, T_w has successfully validated its read-set in an implicit way through the l-cas that acquires the locks. By Lemma 1, in any configuration that T_w maintains these locks, no other transaction can update them. Since T_w locks the variables in its read-set, its read-set remains consistent. Recall that T_w releases the locks it has acquired when it changes its status to committed.   □

Theorem 1. WFR-TM *is an opaque* TM *algorithm.*

Progress. During BEGINTX, each read-only transaction T_r first announces itself and then calculates its *beforeMe* set using the `for` of line 34. This `for` is executed each time a new transaction is added to the *beforeMe* set of T_r. We remark that if T_r performs its announcement before an update transaction T_w enters its waiting phase, then T_w will wait (line 94) for T_r to commit. This implies that during the execution of T_r, each process may initiate at most one update transaction; so, the for of line 34 is executed at most n times. Thus, CHECKIF-PERFORMED has step complexity $O(n^2)$. Recall that m is the maximum number of t-variables accessed by any transaction. Since T_r may have to snoop into an update transaction's write-set (line 49), it executes each instance of READTVAR in $O(m)$ steps. Finally, by inspecting the code, it follows that T_r completes the execution of any other transactional routine after a constant number of steps.

Theorem 2. *Each read-only transaction executed by a nonfaulty process commits after $O(n^2 + m^2)$ steps, where m is the maximum number of t-variables accessed by any transaction.*

Consider now a set of concurrently executing update transactions. Since each update transaction acquires the locks for the t-variables it accesses in (ascending) order, it follows that at least one transaction in this set will be able to acquire all the locks it requires. This transaction will also commit since it waits (line 91) for at most $n - 1$ read-only transactions to complete and Theorem 2 implies that each read-only transaction commits after $O(n^2 + m^2)$ steps.

Theorem 3. *In an infinite execution in which no process crashes, infinitely many update transactions commit.*

4   Discussion

In this paper our effort was mainly directed towards introducing and combining TM techniques to design a new TM algorithm, which has some interesting theoretical properties. An experimental evaluation of WFR-TM is still to be done. An implementation of WFR-TM for this purpose could include several techniques to enhance its performance. An important such optimization is the use of a global clock [5,21] in order speed up the validation process through timestamping.

WFR-TM forces each update transaction to wait for each active read-only transaction it encounters, even if their read-sets share no t-variables with the update transaction's write-set. An efficient way of avoiding this type of scenario is an interesting open problem.

Acknowledgements. This work has been supported by the project "IRAKLI-TOS II - University of Crete" of the Operational Programme for Education and Lifelong Learning 2007 - 2013 (E.P.E.D.V.M.) of the NSRF (2007 - 2013), co-funded by the European Union (European Social Fund) and National Resources. It has also been supported by the European Commission under the 7th Framework Program through the TransForm (FP7-MC-ITN-238639) project and by the ARISTEIA Action of the Operational Programme Education and Lifelong Learning which is co-funded by the European Social Fund (ESF) and National Resources through the GreenVM project.

References

[1] Afek, Y., Matveev, A., Shavit, N.: Pessimistic software lock-elision. In: Aguilera, M.K. (ed.) DISC 2012. LNCS, vol. 7611, pp. 297–311. Springer, Heidelberg (2012)

[2] Attiya, H., Hillel, E.: A single-version stm that is multi-versioned permissive. Theory of Computing Systems 51(4), 425–446 (2012)

[3] Attiya, H., Hillel, E., Milani, A.: Inherent limitations on disjoint-access parallel implementations of transactional memory. In: Proceedings of the 21st Symposium on Parallelism in Algorithms and Architectures, SPAA 2009, pp. 69–78. ACM Press, New York (2009)

[4] Bushkov, V., Guerraoui, R., Kapalka, M.: On the liveness of transactional memory. In: Proceedings of the 31st ACM Symposium on Principles of Distributed Computing, PODC 2012, pp. 9–18. ACM, New York (2012)

[5] Dice, D., Shalev, O., Shavit, N.N.: Transactional locking II. In: Dolev, S. (ed.) DISC 2006. LNCS, vol. 4167, pp. 194–208. Springer, Heidelberg (2006)

[6] Fatourou, P., Kanellou, E., Kosmas, E., Rabbi, M.F.: Wfr-tm: Knowledge of past, understanding of future, and perseverance in present. Tech. Rep. ICS-FORTH TR 449, Institute of Computer Science, Foundation of Research and Technology, Heraklion, Crete (November 2014)

[7] Felber, P., Fetzer, C., Marlier, P., Riegel, T.: Time-based software transactional memory. IEEE Transactions on Parallel and Distributed Systems 21, 1793–1807 (2010)

[8] Felber, P., Fetzer, C., Riegel, T.: Dynamic performance tuning of word-based software transactional memory. In: PPoPP '08: Proceedings of the 13th ACM SIGPLAN Symposium on Principles and Practice of Parallel Programming, PPoPP 008, pp. 237–246. ACM, New Yor (2008)

[9] Fernandes, S.M., Cachopo, J.A.: Lock-free and scalable multi-version software transactional memory. In: Proceedings of the 16th ACM Symposium on Principles and Practice of Parallel Programming, PPoPP 2011, pp. 179–188. ACM, New York (2011)

[10] Fraser, K., Harris, T.: Concurrent programming without locks. ACM Trans. Comput. Syst. 25(2) (May 2007)

[11] Gramoli, V., Harmanci, D., Felber, P.: Toward a theory of input acceptance for transactional memories. In: Baker, T.P., Bui, A., Tixeuil, S. (eds.) OPODIS 2008. LNCS, vol. 5401, pp. 527–533. Springer, Heidelberg (2008)

[12] Guerraoui, R., Henzinger, T.A., Singh, V.: Permissiveness in transactional memories. In: Taubenfeld, G. (ed.) DISC 2008. LNCS, vol. 5218, pp. 305–319. Springer, Heidelberg (2008)

[13] Guerraoui, R., Kapalka, M.: On the correctness of transactional memory. In: Proceedings of the 13th ACM Symposium on Principles and Practice of Parallel Programming, PPoPP 2008, pp. 175–184. ACM, New York (2008)

[14] Guerraoui, R., Kapalka, M.: The semantics of progress in lock-based transactional memory. SIGPLAN Not 44(1), 404–415 (2009)

[15] Herlihy, M., Luchangco, V., Moir, M., Scherer III, W.N.: Software transactional memory for dynamic-sized data structures. In: Proceedings of the 22nd ACM Symposium on Principles of Distributed Computing, PODC 2003, pp. 92–101. ACM, New York (2003)

[16] Herlihy, M., Moss, J.E.B.: Transactional memory: architectural support for lock-free data structures. SIGARCH Comput. Archit. News 21(2), 289–300 (1993)

[17] Keidar, I., Perelman, D.: On avoiding spare aborts in transactional memory. In: Proceedings of the 21st Symposium on Parallelism in Algorithms and Architectures, SPAA 2009, pp. 59–68. ACM, New York (2009)

[18] Matveev, A., Shavit, N.: Towards a fully pessimistic stm model. In: 7th ACM SIGPLAN Workshop on Transactional Computing (TRANSACT) (2012)

[19] Perelman, D., Byshevsky, A., Litmanovich, O., Keidar, I.: SMV: Selective multi-versioning STM. In: Peleg, D. (ed.) DISC 2011. LNCS, vol. 6950, pp. 125–140. Springer, Heidelberg (2011)

[20] Perelman, D., Fan, R., Keidar, I.: On maintaining multiple versions in stm. In: Proceedings of the 29th ACM Symposium on Principles of Distributed Computing, PODC 2010, pp. 16–25. ACM, New York (2010)

[21] Riegel, T., Felber, P., Fetzer, C.: A lazy snapshot algorithm with eager validation. In: Dolev, S. (ed.) DISC 2006. LNCS, vol. 4167, pp. 284–298. Springer, Heidelberg (2006)

[22] Shavit, N., Touitou, D.: Software transactional memory. In: Proceedings of the 14th ACM Symposium on Principles of Distributed Computing, PODC 1995, pp. 204–213. ACM, New York (1995)

[23] Tabba, F., Moir, M., Goodman, J.R., Hay, A.W., Wang, C.: Nztm: Nonblocking zero-indirection transactional memory. In: Proceedings of the 21st Symposium on Parallelism in Algorithms and Architectures, SPAA 2009, pp. 204–213. ACM, New York (2009)

On Developing Optimistic
Transactional Lazy Set

Ahmed Hassan, Roberto Palmieri, and Binoy Ravindran

Virginia Tech, ECE Department, Blacksburg VA 24061, USA
{hassan84,robertop,binoy}@vt.edu

Abstract. Transactional data structures with the same performance of highly concurrent data structures enable performance-competitive transactional applications. Although Software Transactional Memory (STM) is a promising technology for designing and implementing transactional applications, STM-based transactional data structures still perform inferior to their optimized, concurrent (i.e. non-transactional) counterparts. In this paper, we present OTB-Set, an efficient optimistic transactional lazy set based on both linked-list and skip-list implementations. We first provide general guidelines to show how to design a transactional (non-optimized) version of the highly concurrent lazy set with a minimal reengineering effort. Subsequently, we show how to make specific optimizations to the implementations of the OTB-Set for further enhancing its performance. We also prove that our OTB-Set provides linearizable individual operations and opaque transactions. Our experimental study on a 64-core machine reveals that OTB-Set outperforms competitors in most workloads.

Keywords: Software Transactional Memory, Semantic, Set Data Structure, Boosting.

1   Introduction

The increasing ubiquity of multi-core processors motivates the development of data structures that can exploit the hardware parallelism of those processors. The current widely used concurrent collections of elements (e.g., Linked-List, Skip-List, Tree) are well optimized for high performance and ensure isolation of atomic operations, but they do not *compose*. This is a significant limitation from a programmability standpoint, especially for legacy systems as they are increasingly migrated onto multicore hardware (for high performance) and must seamlessly integrate with third-party libraries.

Software transactional memory (STM) [20] can be used to implement transactional data structures (e.g., [7,12]), which makes them composable – a significant benefit. However, monitoring all of the memory locations accessed by a transaction while executing data structure operations is a significant (and often unnecessary) overhead. As a result, STM-based transactional collections perform inferior to their optimized, concurrent (i.e. non-transactional) counterparts.

M.K. Aguilera et al. (Eds.): OPODIS 2014, LNCS 8878, pp. 437–452, 2014.
© Springer International Publishing Switzerland 2014

As an alternative to STM, the transactional boosting methodology was introduced in [14] and further investigated in [11], to convert the highly concurrent data structures into transactional ones. Briefly, in [14], semantic locks are pessimistically acquired at early phases of the transaction to reduce false conflicts. For this reason, and following the trend in [11], we name this approach as *pessimistic transactional boosting* (or PTB). In contrast, the work in [11] lazily acquires the semantic locks, which motivates the name *optimistic transactional boosting* (or OTB). In both approaches, operations are saved in either semantic undo logs (in PTB) or semantic redo logs (in OTB) to correctly commit/abort transactions. As discussed in [11], OTB has benefits over PTB. First, OTB does not require defining inverse operations. Second, it uses the same phases of validation and commit, with the same semantics, as in STM systems, allowing an easy integration of OTB data structures with STM frameworks. Finally, it uses the underlying data structure as a white box, which allows further data structure-specific optimizations. Inspired by the general OTB's principles, in this paper we focus on set-based data structures providing an efficient transactional lazy set (called OTB-Set hereafter), which boosts the highly concurrent lazy set described in [13]. OTB-Set offers the implementation of linked-list and skip-list.

We split the design of OTB-Set into two phases. The first phase consists of: *i)* dividing each operation of the original lazy concurrent data structure into three steps (*traversal, validation*, and *commit*); *ii)* deferring the commit step to the end of the transaction; *iii)* modifying the validation step to guarantee opacity [9] rather than linearizability [17]. This phase is general and does not make any data structure-specific optimization as it provides guidelines independent from the actual implementation of the set. These optimizations are taken into account in the second phase, where we modify the previous OTB-Set design (i.e., the result of phase one) with the aim of further enhancing its performance. Here, we apply optimizations related to the implementation of the data structure rather than its semantic. Splitting the design in such a way allows the programmer to follow the same pattern for boosting more lazy data structures by first designing a general "non-optimized" transactional version using well-defined guidelines (phase one), and then adding optimizations to the specific data structure implementation, resulting in an "optimized", more performant, version (phase two).

We acknowledge that using concurrent data structures as "black boxes", as proposed by PTB, saves the effort for re-engineering them as transactional, however through OTB-Set we show that following our general guidelines it is not difficult to develop a transactional version of lazy set-based data structures, still retaining the advantage of enabling data structure specific optimizations.

We prove that OTB-Set (both the non-optimized and the optimized) provides individual linearizable operations and opaque transactions. To evaluate OTB-Set we compared its performance with both PTB [14] and lazy [13] sets. Our results show that OTB-Set's performance is closer to the highly concurrent lazy set than PTB set in most cases. Beyond the performance improvement, OTB-Set has an added benefit: it is easy to integrate with lazy STM frameworks without violating their correctness or progress guarantees. This way, as we showed in [10], programmer can execute transactions with mixed access types, namely classical memory accesses (managed by the STM framework) and data structure accesses (managed by OTB-Set), without suffering from the disadvantages (i.e., false-conflict) of using an STM on top of a data structure as explained above.

The PTB and OTB approaches take an orthogonal direction to other works in literature for allowing semantic conflict detection. Techniques like open nesting [18], elastic transactions [8], transactional collection classes [4], and transactional predication [3] are different alternatives to design transactional data structures by using STM frameworks more efficiently than the naive STM-based data structures' implementation. The distinguishing point in both PTB and OTB is that they are completely independent and decoupled from STM frameworks[1], and they focus more on digging into the design and the implementation of the highly concurrent data structures and optimize the specific implementation of each one of them. Along the same line of OTB, techniques like COP [1,2] and ParT [21] exploit the same idea of splitting data structures' operations into an unmonitored traversal phase and a speculated validation/update phase. While COP operations are only concurrent (non-transactional), which do not natively compose and cannot be integrated with traditional memory frameworks, ParT discusses how to compose operations by employing a set of validators. In this paper, OTB proposes more reliance on the semantics of the data structure (especially in the "optimized" versions), and provides more details on how to compose dependent operations. Despite their differences, the above trials, along with OTB, confirm the trend of moving towards more optimistic approaches for semantic validation.

Our lazy set is publicly available as JAVA library at http://www.hyflow.org/software.html.

2   Optimistic Transactional Boosting

Optimistic transactional boosting (OTB) [11] is a methodology to boost lazy data structures to be transactional. A common feature that can be identified in all lazy data structures is that they have an *unmonitored traversal* step, in which the object's nodes are not kept locked until the operation ends. To guarantee consistency, this unmonitored traversal is followed by a validation step before the last step that physically modifies the shared data structure. As described in [11], OTB modifies the design of these lazy data structures to support transactions. Basically, the OTB methodology can be summarized in three main guidelines.

(G1) *Each data structure operation is divided into three steps.* <u>*Traversal*</u>. This step scans the objects, and computes the operation's results (i.e., its postcondition) and what it depends on (i.e., its precondition). This requires us to define (in each transaction), what we call *semantic read-set* and *semantic write-set*, which store these information (*semantic write-sets* can also be called *semantic redo-logs*). <u>*Validation*</u>. This step checks the validity of the

---

[1] In fact, OTB does not use STMs, rather it has been designed to be easily integrated with existing STM frameworks.

preconditions. Specifically, the entities stored in the semantic read-set are validated to ensure that operations are consistent. *Commit*. This step performs the modifications to the shared data structure. Unlike concurrent data structures, this step is not done at the end of each operation. Instead, it is deferred to the transaction's commit time. All information needed for performing this step are maintained in the semantic write-sets during the first step (i.e., traversal). To publish the write-sets, a classical (semantic) two-phase locking is used. This semantic (or abstract) locking prevents semantic conflicts at commit.

(G2) *Data structure design is adapted to support opacity.* The correctness of transactional data structures does not only depend on the *linearization* of its operations (like concurrent data structures), but it also depends on the sequence of the operations executed in each transaction. Data structure design has to be adapted to guarantee this *serialization* part. OTB provides the following guidelines, which exploit the local read-sets and write-sets to guarantee opacity [9] [2] , the same consistency level of most STM algorithms [5,6,19,15]:

(G2.1) Each operation scans the local write-set first, before accessing the shared object. This is important to include the effect of the earlier (not yet published) operations in the same transaction.

(G2.2) The read-set is re-validated after each operation and during commit, to guarantee that each transaction always observes a consistent state of the system (even if it will eventually abort).

(G2.3) During commit, semantic locks of all operations are acquired before any physical modification on the shared data structure.

(G2.4) Operations are applied during the commit phase in the same order as they appeared in the transaction and, in case the outcome of an operation influences the subsequent operations recorded in the write-set, they are updated accordingly.

(G2.5) All operations have to be validated, even if the original (concurrent) operation does not make any validation (like `contains` operation in set). The goal of validation in these cases is to ensure that the same operation's result occurs at commit.

(G3) *Data structure design is adapted for more optimizations.* Each data structure can be further optimized according to its own semantic and implementation. For example, in set, if an item is added and then deleted in the same transaction, both operations eliminate each other and can be completed without physically modifying the shared data structure.

Unlike the first two guidelines, which are general for any lazy data structure, the third guideline varies from one data structure to another. It gives a hint to the developers that the data structures now are no longer used as black boxes, and further optimizations can be applied. It is important to note that the generality of the first two guidelines does not mean that they can be applied "blindly" without being aware of the data structure's semantics. OTB, like the former

---

[2] In section 4, we prove that those guidelines are sufficient to guarantee opacity.

techniques (including PTB) [1,2,21,14], performs better than the naive STM-based data structures only because it exploits semantics. However, we believe that OTB's guidelines make a clear separation between the general outline that can be applied on any lazy data structure (like validation, in *G2.2*, and commit, in *G2.4*, even if the validation/commit mechanisms themselves vary from one data structure to another) and the specific optimizations that are completely dependent on the data structures implementation.

In Section 3, we show in detail how those guidelines can be used to design OTB-Set, an efficient transactional set based on both linked-list and skip-list. In Section 3.2, we follow the first two guidelines to design a non-optimized transactional version of the lazy set. Then, in Section 3.3, we show how specific optimizations can be applied on our OTB-Set (according to the third guideline).

3   OTB-Set

3.1   Preliminaries

Set is a collection of ordered items, which has three basic operations: add, remove, and contains, with the familiar meanings [16]. No duplicate items are allowed (thus, add returns false if the item is already present in the structure). All operations on different items of the *set* are commutative – i.e., two operations add(x) and add(y) are commutative if $x \neq y$. Moreover, two contains operations on the same item are commutative as well. Such a high degree of commutativity between operations enables fine-grained semantic synchronization.

Lazy linked-list [13] is an efficient implementation of concurrent (non transactional) *set*. For write operations, the list is traversed without any locking until the involved nodes are locked. If those nodes are still valid after locking, the write takes place and then the nodes are unlocked. A marked flag is added to each node for splitting the deletion phase into two steps: the logical deletion phase, which simply sets the flag to indicate that the node has been deleted, and the physical deletion phase, which changes the references to skip the deleted node. This flag prevents traversing a chain of deleted nodes and returning an incorrect result. It is important to note that the contains operation in the lazy linked-list is wait-free and is not blocked by any other operation.

Lazy skip-list is, in general, more efficient than linked-list as it takes logarithmic time to traverse the *set*. In skip-list, each node is linked to multiple lists (i.e., levels), starting from the list at the bottom level (which contains all the items), up to a random level. Therefore, add and remove operations lock an array of *pred* and *curr* node pairs (in a unified ascending order of levels to avoid deadlock), instead of locking one pair of nodes as in linked-list. For add operation, each node is enriched with a *fullyLinked* flag to logically add it to the *set* after all levels have been successfully linked. Skip-list is also more suited than linked-list in scenarios where the overhead of rolling back (compared to execution) is dominating. In fact, for a linked-list (and especially a long linked-list), even if aborts are rare, their effect includes re-traversing the whole list again, in a linear time, to retry the operation. In a skip-list, the cost of re-traversal is lower (typically in a logarithmic time), which minimizes the overhead of the aborts.

The implementation of the PTB version of the *set* is straightforward and does not change if the *set* implementation itself changes. In fact, it uses the underlying concurrent lazy linked-list (or skip-list) to execute the *set* operations. If the transaction aborts, a successful **add** operation is rolled back by calling the **remove** operation on the same item, and vice versa (more details are in [14]).

Despite the significant improvement in the traversal cost and abort overhead, the implementation of OTB skip-list and OTB linked-list are very similar. Due to space constraints, and with the purpose of making the presentation clear, we focus on the linked-list implementation, and we highlight the main differences with respect to the skip-list implementation when necessary (the full implementation details of both linked-list-based and skip-list-based OTB-Set can be found in the source code).

3.2   Non-optimized OTB-Set

Following the first two guidelines (*G1* and *G2*) mentioned in Section 2, in this section we show how to boost the lazy *set* to design a transactional *set* without any specific optimization related to the details of its implementation. According to *G1*, we divide OTB-Set operations into three steps. The *Traversal* step is used to reach the involved nodes, without any addition to the semantic read-set. The *Validation* step is used to guarantee the consistency of the transaction and the linearization of the list. We define two different validation procedures: one is named *post-validation*, which is called after each operation, and the other is named *commit-time-validation*, which is called at commit time and after acquiring the semantic locks. The *Commit* step, which modifies the shared list, is deferred to transaction's commit. Following *G2*, we show how the usage of lazy updates, semantic locking, and post-validation guarantees opacity.

Similar to the lazy linked-list, each operation in OTB-Set involves two nodes at commit time: *pred*, which is the largest item less than the searched item, and *curr*, which is the searched item itself or the smallest item larger than the searched item[3]. To log the information about these nodes, with the purpose of using them at commit time, we adopt the same concept of read-set and write-set as used in lazy STM algorithm (e.g., [5,6]), but at the semantic level. In particular, each read-set or write-set entry contains the two involved nodes in the operation and the type of the operation. In addition, the write-set entry contains also the new value to be added in case of a successful **add** operation.

The only difference in skip-list is that the read-set and write-set entries contain an array of *pred* and *curr* pairs, instead of a single pair. This is because the searched object can be in more than one level of the skip-list.

Algorithm 1 shows the pseudo code of the linked-list operations. We can isolate the following four parts of each operation.

---

[3] Sentinel nodes are added as the head and tail of the list to handle special cases.

Algorithm 1. OTB Linked-list: add, remove, and contains operations

```
 1: procedure OPERATION(x)                         12:      read-set.add(rse)
                ▷ Step 1: search local write-sets  13:      if op is add or remove then
 2:     if x ∈ write-set then                      14:          wse = new WriteSetEntry(pred,curr,
 3:         ret = write-set.get-ret(op,x)                          op,x)
 4:         if op is add or remove then            15:          write-set.add(wse)
 5:             write-set.append(op,x)                                    ▷ Step 4: Post Validation
 6:         return ret                             16:      if ¬ post-validate(read-set) then
                          ▷ Step 2: Traversal      17:          ABORT
 7:     pred = head and curr = head.next           18:      else if Successful operation then
 8:     while curr.item < x do                     19:          return true
 9:         pred = curr                            20:      else
10:         curr = curr.next                       21:          return false
                ▷ Step 3: Save reads and writes    22: end procedure
11:     rse = new ReadSetEntry(pred,curr,op)
```

Local writes check (lines 2-6). Since writes are buffered and deferred to the commit phase, this step guarantees consistency of further reads and writes. Each operation on an item x checks the last operation in the write-set on the same item x and returns the corresponding result. For example, if a transaction previously executed a successful add operation of item x, then further additions of x performed by the same transaction must be unsuccessful and return false. In addition, if the new operation is a writing (i.e., add/remove) operation, it should be appended to the corresponding write-set entry (line 5). If there is no previous (local) operation on x in the write-set, then the operation starts traversing the shared linked-list as shown in the next step.

Traversal (lines 7-10). This step is the same as in the lazy linked-list. It saves the overhead of all unnecessary monitoring during traversal that, otherwise, would be incurred with a native STM algorithm for managing concurrency.

Logging the reads and writes (lines 12-15). At this point, the transaction records the accessed nodes, that are semantically relevant to the *set*, into its local read-set and write-set. All operations must add the appropriate read-set entry, while add/remove operations modify also the write-set (line 15). It is worth to note that having no entries in the write-set for contains operation means that it does not need to acquire locks during the commit phase. This way, although the contains operation is no longer wait-free, like its concurrent lazy version (because it may fail during the commit-time-validation), it still performs efficiently due to the absence of the semantic locks acquisition. We recall that, rather than OTB, PTB has to acquire semantic locks even for the contains operation to maintain consistency and opacity.

Post-Validation (lines 16-21). At the end of the traversal step, the involved nodes are stored in local variables (i.e., *pred* and *curr*). At this point, according to point *G2.2* and to preserve opacity [9], the read-set is post-validated to ensure that the transaction does not observe an inconsistent snapshot. The same post-validation mechanism is used at memory-level by STM algorithms such as NOrec [5]. More details about post-validation are discussed later in Algorithm 2.

As mentioned before, there is a difference between linked-list and skip-list regarding the add operation. In fact, in the skip-list the new node has to be linked to multiple levels, thus there could be a time window where the new node is only linked to some (and not all) levels. To handle this case in our OTB-Set, any concurrent operation waits until the *fullyLinked* flag becomes true, and then it proceeds.

Algorithm 2 shows the post-validation step. The validation of each read-set entry is similar to the one in lazy linked-list: both *pred* and *curr* should not be deleted, and *pred* should still link to *curr* (lines 6-8). According to *G2.5* of OTB guidelines, contains operation has to perform the same validation as add and remove, although it is not needed in the concurrent version. This is because any modification made by other transactions after invoking the contains operation and before committing the transaction may invalidate the returned value of the operation, making the transaction's execution semantically incorrect.

To enforce isolation, a transaction ensures that its accessed nodes are not locked by another writing transaction during validation. This is achieved by implementing locks as *sequence locks* (i.e., locks with version numbers). Before the validation, a transaction records the versions of the locks if they are not acquired. If some are already locked by another transaction, the validation fails. (lines 2-5). After the validation, the transaction ensures that the actual locks' versions match the previously recorded versions (lines 9-12).

Algorithm 2. OTB Linked-list: validation

```
1: procedure VALIDATE(read-set)          8:          return false
2:    for all entries in read-sets do    9:    for all entries in read-sets do
3:        get snapshot of involved locks 10:        check snapshot of involved locks
4:        if one involved lock is locked then  11:        if version of one involved lock is
5:            return false               changed then
6:    for all entries in read-sets do    12:            return false
7:        if pred.deleted or curr.deleted or  13:    return true
   pred.next ≠ curr then                 14: end procedure
```

Algorithm 3 shows the commit step of OTB-Set. Read-only transactions have nothing to do during commit (line 2), because of the incremental validation during the execution of the transaction. For write transactions, according to point *G2.3*, the appropriate locks are first acquired using CAS operations (lines 4-6). Like the original lazy linked-list, any add operation only needs to lock *pred*, while remove operations lock both *pred* and *curr*. As described in [13], this is enough for preserving the correctness of the write operations. To avoid deadlock, any failure during the lock acquisition implies aborting and retrying the transaction (releasing all previously acquired locks).

After the semantic lock acquisition, the validation is called, in the same way as in Algorithm 2, to ensure that the read-set is still consistent (line 7). If the commit-time-validation fails, then the transaction aborts.

The next step of the commit procedure is to publish writes on the shared linked-list, and then release the acquired locks. This step is not straightforward because each node may be involved in more than one operation of the same

Algorithm 3. OTB Linked-list: commit

```
 1: procedure COMMIT                              18:        n.next = curr
 2:    if write-set.isEmpty then                  19:        pred.next = n
 3:        return                                 20:        for all entries in write-sets do
 4:    for all entries in write-sets do           21:            if entry.pred = pred then
 5:        if CAS Locking pred (or curr if re-    22:                entry.pred = n
       move) failed then                          23:    else                            ▷ remove
 6:            ABORT                               24:        curr.deleted = true
 7:    if ¬ commit-validate(read-set) then        25:        pred.next = curr.next
 8:        ABORT                                   26:        for all entries in write-sets do
 9:    sort write-set descending on items         27:            if entry.pred = curr then
10:    for all entries in write-sets do           28:                entry.pred = pred
11:        curr = pred.next                        29:            else if entry.curr = curr then
12:        while curr.item < x do                 30:                entry.curr = curr.next
13:            pred = curr                         31:    for all entries in write-sets do
14:            curr = curr.next                    32:        unlock pred (and curr if remove)
15:        if operation = add then
16:            n = new Node(item)                  33: end procedure
17:            n.locked = true
```

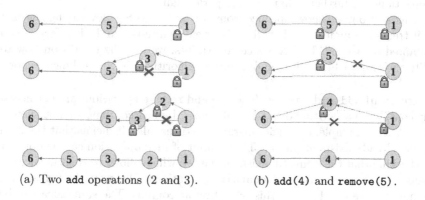

(a) Two **add** operations (2 and 3).   (b) **add**(4) and **remove**(5).

Fig. 1. Executing more operations that involve the same node in the same transaction

transaction. In this case, the saved *pred* and *curr* of these operations may change according to which operation commits first.

For example, in Figure 1(a), both 2 and 3 are inserted between the nodes 1 and 5 in the same transaction. During commit, if node 2 is inserted before node 3, it should be the new predecessor of node 3, but the write-set still records node 1 as the predecessor of node 3. In OTB guidelines, *G2.4* solves this issue. When node 2 is inserted, the operation scans the write-set again to find any other operation that has node 1 as its *pred* and replaces it with node 2. The same technique is used in the case of removal (Figure 1(b)). When node 5 is removed, any write-set entry that has node 5 as its *curr* replaces it with node 6, and any write-set entry that has node 5 as its *pred* replaces it with node 1. Lines 20-22 and 26-30 illustrate these cases.

It is clear that the inserted nodes have to be locked until the whole commit procedure is finished. Then they are unlocked along with the other *pred* and *curr* nodes (line 17). For example, in Figure 1(a), all nodes (1, 2, 3, 5) are locked and no transaction can access them until the commit terminates.

3.3   Optimized OTB-Set

One of the main advantages of OTB over the original PTB is that it uses the underlying (lazy) data structure as a white-box, which allows more data structures-specific optimizations.

In general, decoupling the boosting layer from the underlying concurrent data structure is a trade-off. Although, on the one side, considering the underlying data structure as a black-box means that there is no need to re-engineer its implementation, on the other side, it does not allow to customize its implementation and thus to exploit the new transactional specification, especially when the re-engineering effort can be easily achieved. For this reason, as showed in the previous section, we decided to split the re-engineering efforts (required by OTB) into two steps: one general (concluded in OTB guidelines *G1* and *G2*); and one more specific per data structure (concluded *G3*). We believe this division makes the re-engineering task easier and, at the same time, it allows specific optimizations for further enhancing the performance.

In this section, we show optimizations for our OTB-Set, leveraging the fact that it treats the underlying lazy linked-list as a white-box and, therefore, it can be adapted as needed. Due to space constraints, we put the details on how to modify the aforementioned "non-optimized" algorithms in the technical report.

Unsuccessful add and remove. The add and remove operations are not necessarily considered as writing operations, because duplicated items are not allowed in the *set*. For example, if an add operation returns false, it means that the item to insert already exists in the *set*. The commit of such operation can be done by only checking that the item still exists in the *set*, which allows to treat unsuccessful add operations as successful contains operations. This way, the transaction does not acquire any lock for this operation at commit. The same idea can be applied on the unsuccessful remove operation which can be treated as an unsuccessful contains operation during commit.

Accordingly, in our OTB-Set, both contains and unsuccessful add/remove operations are considered as read operations (which add entries only to the semantic read-set and do not acquire any semantic locks during commit). Only successful add and remove operations are considered read/write operations (which add entries to both the read-set and the write-set and thus acquire semantic locks during commit).

In the lazy linked-list, the add and remove operations acquire locks on the *pred* and *curr* nodes even if the operations are unsuccessful. PTB inherits this unnecessary lock acquisition because it uses the lazy linked-list as a black-box.

Eliminating Operations. As shown in Algorithm 1, each operation starts with checking the local writes before traversing the shared list. During this step, for improving OTB performance, if a transaction adds an item x and then removes the same item x, or vice versa, we allow those operations to locally eliminate each other. This elimination is done by removing both entries from the write-set, which means that the two operations will not make any physical modification on the shared list. No entry in the read-set is locally eliminated because, this way, the commit time-validation can still be performed on those operations in order to preserve transaction's correctness.

In PTB, due to the usage of the underlying lazy linked-list as a black-box, this scenario is handled by physically adding x to the shared *set*, and then physically removing it, introducing an unnecessary overhead.

Simpler Validation. In the case of successful `contains` and unsuccessful `add` operations, we use a simpler validation than the original validation of the lazy linked-list. In these particular cases, the transaction only needs to check that *curr* is still not deleted, since that is sufficient to guarantee that the returned value is still valid (recall that if the node is deleted, it must first be logically marked as deleted, which will be detected during validation). This optimization prevents false invalidations, where conflicts on *pred* are not real semantic conflicts.

The validation in the skip-list is similarly optimized because we leverage the rule that all items have to appear in the lowest level of the skip-list. For successful *contains* and unsuccessful *add* operations, it is sufficient to validate that *curr* is not deleted, which ensures that the item is still in the *set*. We can also optimize unsuccessful *remove* and *contains* by only validating the *pred* and *curr* in the lowest level to make sure that the item is still not in the *set*, because if the item is inserted by another transaction, it must affect this level. For successful *add* and *remove* operations, all levels need to be validated to prevent conflicts.

Optimized Commit. To ensure that the operations in Figure 1 are executed correctly, the write-set has to be re-scanned for each write operation (according to the OTB guideline *G2.4*), as we showed in Section 3.2. This overhead becomes significant if the write-set is relatively large. We optimize this routine and avoid the need of re-scanning the write-set by the following points. *(1)* The items are added/removed in descending order of their values, regardless of their order in the transaction execution. This guarantees that the *pred* of each write-set entry is always valid, non-deleted, and not touched by any previous operation in the transaction. *(2)* Operations resume traversal from the saved *pred* to the new *pred* and *curr* nodes. At this stage, the *pred* and *curr* nodes can only be changed because of some previous local operations. This is because the transaction already finished the lock acquisition and validation, which prevents any conflicting transaction from proceeding.

Using these two points, the issue in Figure 1(a) is solved without re-scanning the write-set. The first point enforces that node 3 is inserted first. Subsequently, according to the second point, when 2 is inserted, the transaction will resume its traversal from node 1 (which is guaranteed to be locked and non-deleted). Then, it will detect that node 3 is its new *succ*, and will correctly link node 2.

The removal case is shown in Figure 1(b), in which node 5 is removed and node 4 is inserted. Again, 5 must be removed as first (even if 4 is added earlier during the transaction execution), so that when 4 is added, it will correctly link to 6 and not to 5. Two subsequent `remove` operations follow the same procedure.

Skip-list uses the same procedure but at all levels. This is because each level is independent from the others, which means that the *preds* of the same node in two or more levels may be different. For this reason, the same procedure described above is repeated at each level, independently.

4   Correctness

In this section, we discuss the arguments that we use for assessing the correctness of OTB-Set, and, due to space constraints, we report the detailed correctness proof in the technical report.

The correctness of OTB-Set can be proved in two incremental steps. The first step is to show that, after the modifications needed for supporting the execution of transactions, each single operation on the *set* is still linearizable, like the lazy *set*. The second step consists of showing that the whole transaction is opaque [9].

A) Linearizability: Each operation traverses the *set* following the same rules as in the lazy *set*. After the traversal, we can distinguish between write and read operations' behavior. A write operation, instead of acquiring the locks on the involved nodes instantaneously after the traversal, it acquires the same locks, but at transaction commit time. Since the transaction is validated after the locks acquisition using the same validation done by the lazy *set*, the linearization points of each write operation is just shifted to the commit phase of the transaction (rather than after the operation as in the lazy *set*). We cannot use the same arguments for defining the linearization point of the read operations in our OTB-Set. In fact, in lazy *set*, a `contains` operation is wait-free, which implies that its linearization point is when the *curr* node is checked[4]. In OTB-Set, where `contains` operations are no longer wait-free, this point is replaced with the point when each operation is re-validated during the transaction commit.

B) Opacity: Considering the transaction as a whole, the combination of *lazy writes*, *post-validation*, and *commit-time-validation* is sufficient for guaranteeing opacity. In fact, this is the same approach used at memory level in many lazy STM algorithms such as NOrec [5] to enforce opacity. Specifically, all operations are linearized at the transaction's commit time and after acquiring all the semantic locks. This allows the committed transactions to appear as happened at a single indivisible point in time. Aborted transactions do not expose any write to other transactions, because, in general, transactions never write in the shared *set* unless they are sure that they will not eventually abort. Live transactions (whether they will eventually commit or abort) never observe an inconsistent state because they validate their entire read-set after each read (in the *post-validation* routine) and during the transaction commit (in the *commit-time validation* routine). Finally, the effect of interfering operations of the same transaction is preserved leveraging the points *G2.1* and *G2.4* of OTB guidelines.

---

[4] In some exceptional cases, discussed in [13], the linearization point of the unsuccessful *contains* operation becomes earlier. However, those special cases are not relevant when we discuss the correctness of our OTB-Set.

The optimizations described in Section 3.3 do not break opacity simply because they do not contradict with any of the previously mentioned evidences. It is also straightforward to prove that composing the operations on two different OTB-Set instances does not break the opacity of the transaction as a whole. This is because each read/write-set entry will be validated and/or published independently.

5 Experimental Evaluation

In this section we evaluate the performance of our OTB-Set's Java implementation equipped with the optimizations described in Section 3.3. We compared it with lazy set [13] and PTB set [14]. In order to conduct a fair comparison, the percentage of the writes in all of the experiments is the percentage of the successful ones, because an unsuccessful add/remove operation is considered as a read operation. Roughly speaking, in order to achieve that, the range of elements is made large enough to ensure that most add operations are successful. Also, each remove operation takes an item added by previous transactions as a parameter, such that it will probably succeed. In each experiment, the number of add and remove operations are kept equal to avoid significant fluctuations of the data structure size during the experiments.

The experiments were conducted on a 64-core machine, which has four AMD Opteron (TM) Processors, each with 16 cores running at 1400 MHz, 32 GB of memory, and 16KB L1 data cache. Threads start execution with a warm up phase of 2 seconds, followed by an execution of 5 seconds, during which the throughput is measured. Each plotted data-point is the average of five runs.

We use transactional throughput as our key performance indicator. Although abort rate is another important parameter to measure and analyze, it is meaningless in our case. Both lazy set and PTB set do not explicitly abort the transaction. However, there is an internal retry for each operation if validation fails. Additionally, PTB aborts only if it fails to acquire the semantic locks, which is less frequent than validation failures in the OTB-Set. We recall that the lazy set is not capable to run transactions at all (i.e., it is a concurrent data structure, not transactional). We only show it as a rough upper bound for the OTB-Set and PTB, but it actually does not support transactional operations.

We first show the results for a linked-list implementation of the set. In this experiments, we used a linked-list with 512 nodes. In order to conduct a comprehensive evaluation of OTB-Set's performance, in the first row of Figure 2 we show the results for four different linked-list workloads: read-only (0% writes and 1 operation per transaction), read-intensive (20% writes and 1 operation per transaction), write-intensive (80% writes and 1 operation per transaction), and high contention (80% writes and 5 operations per transaction). In both read-only and read-intensive workloads, OTB-Set performs closer to the (upper bound) performance of the lazy list than PTB-Set. This is expected, because PTB incurs locking overhead even for read operations. In contrast, OTB-Set, like lazy linked-list, does not acquire locks on read operations, although it still

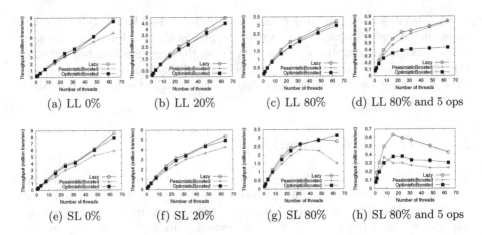

Fig. 2. Throughput of linked-list-based (LL) and skip-list-based (SL) set with 512 elements (labels indicate % write transactions). Four different workloads: read-only (0% writes), read-intensive (20% writes), write-intensive (80% writes), and high contention (80% writes and 5 operations per transaction).

has a small overhead for validating the read-set. For the write-intensive work-load, PTB starts to be slightly better than OTB-Set, and the gap increases in high contention workloads. This is also expected, because contention becomes very high, which increases abort rate (recall that aborts have high overhead due to re-traversing the list in linear time). In these high/very high contention scenarios, the "pessimism" of PTB pays off more than the "optimism" of OTB-Set. For example, in the high contention scenario, five operations are executed per transaction. In PTB, each operation (pessimistically) locks its semantic items before executing each operation and then it keeps trying to execute the operation on the underlying (black-box) concurrent data structure. On the other hand, OTB suffers from aborting the whole transaction even if the last operation of the transaction fails.

In the second row of Figure 2, the same results are shown for the skip-list-based *set* of the same size (512 nodes). The results show that OTB-Set performs better in all cases, including the high contention case. This confirms that OTB-Set gains because of the reduced overhead of aborts. Although the semantic contention is almost the same (for a *set* with 512 nodes, contention is relatively high), using a skip-list instead of a linked-list supports OTB-Set more than PTB. This is mainly because skip-list traverses less nodes of the set through the higher levels of the skip-list. Thus, even if the whole transaction aborts, re-executing skip-list operations is less costly than linked-list.

The last set of experiments (Figure 3), shows the performance when the contention is significantly lower. We used a skip-list of size 64K and measured throughput for the same four workloads. The results show that in such cases, which however are still practical, OTB-Set is up to 2× better, even in

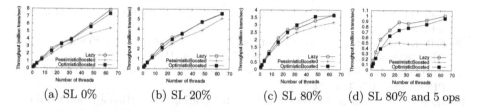

(a) SL 0%   (b) SL 20%   (c) SL 80%   (d) SL 80% and 5 ops

Fig. 3. Throughput of skip-list-based set with 64K elements (labels indicate % write transactions). Four different workloads: read-only (0% writes), read-intensive (20% writes), write-intensive (80% writes), and high-contention (80% writes and 5 operations per transaction).

write-intensive and high contention workloads. This is mainly because in the very low contention scenario, the PTB's eager locking mechanism becomes ineffective and a more optimistic algorithm, such as OTB-Set, is preferable.

6   Conclusions

In this paper we provided a detailed design and implementation of a transactional optimistic *set* data structure (OTB-Set). We presented two versions of OTB-Set: one "non-optimized", derived from the implementation of general guidelines; and one "optimized", which aims at further enhancing the performance. We also proved the correctness of the designed set and showed that OTB-Set operations guarantee opacity. Our evaluation revealed that the performance of OTB-Set is closer to highly concurrent (non-transactional) lazy *set* than the original transactional boosting version in most of the cases.

Acknowledgments. This work is supported in part by US National Science Foundation under grant CNS-1116190.

References

1. Afek, Y., Avni, H., Shavit, N.: Towards consistency oblivious programming. In: Fernàndez Anta, A., Lipari, G., Roy, M. (eds.) OPODIS 2011. LNCS, vol. 7109, pp. 65–79. Springer, Heidelberg (2011)
2. Avni, H., Kuszmaul, B.C.: Improving htm scaling with consistency-oblivious programming. In: TRANSACT (2014)
3. Bronson, N.G., Casper, J., Chafi, H., Olukotun, K.: Transactional predication: High-performance concurrent sets and maps for stm. In: PODC, pp. 6–15 (2010)
4. Carlstrom, B.D., McDonald, A., Carbin, M., Kozyrakis, C., Olukotun, K.: Transactional collection classes. In: PPOPP, pp. 56–67 (2007)
5. Dalessandro, L., Spear, M.F., Scott, M.L.: NOrec: streamlining STM by abolishing ownership records. In: PPOPP, pp. 67–78 (2010)
6. Dice, D., Shalev, O., Shavit, N.N.: Transactional locking II. In: Dolev, S. (ed.) DISC 2006. LNCS, vol. 4167, pp. 194–208. Springer, Heidelberg (2006)

7. Diegues, N.L., Romano, P.: Time-warp: Lightweight abort minimization in transactional memory. In: PPoPP, pp. 167–178 (2014)
8. Felber, P., Gramoli, V., Guerraoui, R.: Elastic transactions. In: Keidar, I. (ed.) DISC 2009. LNCS, vol. 5805, pp. 93–107. Springer, Heidelberg (2009)
9. Guerraoui, R., Kapalka, M.: On the correctness of transactional memory. In: PPOPP, pp. 175–184 (2008)
10. Hassan, A., Palmieri, R., Ravindran, B.: Integrating transactionally boosted data structures with stm frameworks: A case study on set. In: TRANSACT (2014)
11. Hassan, A., Palmieri, R., Ravindran, B.: Optimistic transactional boosting. In: PPOPP, pp. 387–388 (2014)
12. Hassan, A., Palmieri, R., Ravindran, B.: Remote invalidation: Optimizing the critical path of memory transactions. In: IPDPS, pp. 187–197 (2014)
13. Heller, S., Herlihy, M.P., Luchangco, V., Moir, M., Scherer III, W.N., Shavit, N.N.: A lazy concurrent list-based set algorithm. In: Anderson, J.H., Prencipe, G., Wattenhofer, R. (eds.) OPODIS 2005. LNCS, vol. 3974, pp. 3–16. Springer, Heidelberg (2006)
14. Herlihy, M., Koskinen, E.: Transactional boosting: A methodology for highly-concurrent transactional objects. In: PPOPP, pp. 207–216 (2008)
15. Herlihy, M., Luchangco, V., Moir, M., Scherer III., W.N.: Software transactional memory for dynamic-sized data structures. In: PODC, pp. 92–101 (2003)
16. Herlihy, M., Shavit, N.: The Art of Multiprocessor Programming, Revised Reprint. Elsevier (2012)
17. Herlihy, M., Wing, J.M.: Linearizability: A correctness condition for concurrent objects. ACM Transactions on Programming Languages and Systems 12(3), 463–492 (1990)
18. Ni, Y., Menon, V., Adl-Tabatabai, A.-R., Hosking, A.L., Hudson, R.L., Moss, J.E.B., Saha, B., Shpeisman, T.: Open nesting in software transactional memory. In: PPOPP, pp. 68–78 (2007)
19. Riegel, T., Felber, P., Fetzer, C.: A lazy snapshot algorithm with eager validation. In: Dolev, S. (ed.) DISC 2006. LNCS, vol. 4167, pp. 284–298. Springer, Heidelberg (2006)
20. Shavit, N., Touitou, D.: Software transactional memory. Distributed Computing 10(2), 99–116 (1997)
21. Xiang, L., Scott, M.L.: Composable partitioned transactions. In: WTTM (2013)

On the Mailbox Problem⋆

Uri Abraham and Gal Amram

Ben-Gurion University, Beer-Sheva, Israel
{abraham,galamra}@cs.bgu.ac.il

Abstract. The Mailbox Problem was described and solved by Aguilera, Gafni, and Lamport in [2] with a mailbox algorithm that uses two flag registers that carry 14 values each. An interesting question that they ask is whether leaner solutions exists. In addition to their algorithm, the authors of [2] proved that the mailbox problem cannot be solved with 1 bit flags. In this paper, we show that 2 bit flags suffice by presenting a mailbox algorithm with two bit flags. The space complexity of Aguilera et al. solution is $O(n \log n)$ and they conjectured that a solution with space complexity $O(\log n)$ exists. Our algorithm proves this conjecture. We also prove that there is no mailbox algorithm with a smaller space complexity.

Keywords: distributed algorithms, shared memory, synchronization, linearizability.

1 Introduction: The Mailbox Problem

The Mailbox Problem is a theoretical synchronization problem that arises from analyzing the situation in which a processor must cater to occasional requests from some device. The problem, as presented (and solved) in [2] requires the implementation of three operations: *deliver, check,* and *remove.* The device executes a *deliver* operation whenever it wants to get the processor's attention, and the processor executes from time to time *check* operations to find out if there are any unhandled device requests. After receiving a positive answer for its *check* operation the processor executes a *remove* operation to find-out the nature of the request and to clear the interrupt controller. In a serial execution, it is required that a *check* operation C returns a positive answer if and only if the number of *deliver* occurrences that precede C is strictly greater than the number of *remove* operations executed before C. The Mailbox Problem is to design a *deliver/check/remove* algorithm in which the *check* operation is as efficient as possible, namely that it employs bounded registers (called "flags") that are as small as possible.

In [2] the problem is presented first informally by means of a story about a *postman* (which is the device) and a *home owner* (the processor) in which the postman delivers its letters, and the owner removes them one by one every time she approaches the mailbox. The problem is to find an algorithm that ensures that the home owner approaches her mailbox if and only if it is nonempty. The *check* function returns a boolean value which tells the home-owner whether the mailbox is nonempty, and she approaches her

---

⋆ Research partially supported by the Frankel Center for Computer Science at Ben-Gurion University.

M.K. Aguilera et al. (Eds.): OPODIS 2014, LNCS 8878, pp. 453–468, 2014.

mailbox only after receiving a "nonempty" response. As noted in [2], depending on the assumptions made on the communication between the device and the processor, the mailbox problem can be extremely easy or surprisingly difficult. For example, a simple algorithm is suggested in [2] in which the postman and home-owner employ a flag at the mailbox. The postman can atomically (in a single step) deliver mail to the box and raise the flag, and the owner atomically removes mail from the box and lowers the flag. The mailbox problem becomes highly non-trivial when limitations are imposed on the communication devices. Specifically, Aguilera et al. require in [2], for efficiency reasons, that the mailbox solutions use only the simplest possible means, and the *check* operation (which is possibly invoked at higher frequency) should access only bounded registers. As formulated in [2], the mailbox problem asks for solutions that satisfy the following requirements.

1. Only registers with read/write actions can be employed.
2. Whereas the *deliver* and *remove* operations are allowed unbounded registers, the home-owner can only read bounded value registers in *check* operation executions. The point of this requirement, as explained in [2], is to make the *check* operation of the homeowner as light as possible.
3. Moreover, in her *check* operations the home-owner cannot use persistent local variables, that is variables that retain their values from one invocation of the operation to the following one (see the note at the end of this section).
4. The algorithms for the three operations (*deliver*, *check*, and *remove*) are bounded wait-free.

A solution is presented in [2] in which each of the two processes uses unbounded and bounded registers and the *check* operation decides on the value to return by reading only bounded registers (which are called 'flags'). The algorithm of [2] needs 14 values in each of the two flag registers, and a question is posed there if leaner solutions exist. Moreover, [2] contains a proof that there is no solution to the mailbox problem with single bit flags. In this article we describe a mailbox algorithm in which the flag registers of the postman and the home owner carry 4 values in each of the flag registers; that is, the mailbox problem can be solved with 2-bit registers. Furthermore, while the space complexity of the solution in [2] is $O(n \log n)$ (where n is the number of operation executions), the space complexity of our algorithm is $O(\log n)$. We also prove that there is no solution with space complexity $f(n) \in o(\log n)$ thus, under the problem constrains, our algorithm is optimal.

The safety property of the mailbox algorithm is expressed in [2] by first stating its sequential specification, and then requiring that a linearization exists which satisfies this sequential specification. This is the well-known approach to linearizability as defined by Herlihy and Wing in [5]. The following is the formulation in [2] for the sequential specification:

> If the homeowner and postman never execute concurrently, then the value returned by an execution of *check* is true if and only if there are more deliver than remove executions before this execution of *check*.

The specifications of a sequential (linear) mailbox are given by reference to a total ordering $\prec$ of the events. ($\prec$ is a *total ordering* of a set if it is a transitive and irreflexive relation so that for any two members a and b of the set $a \prec b$ or $b \prec a$.)

Definition 1. *Sequential mailbox specification:*

1. *The events are partitioned into deliver, check, and remove events, and are totally ordered by $\prec$. Every event is preceded by a finite number of events in this ordering. (The deliver events form the postman process, and the check and remove events form the home-owner process.) For every $n = 1, 2, \ldots$, if the n-th deliver event exists it denoted D_n.*
2. *For every check event C, the value that C returns, $Val(C)$, is in {true, false}. If $Val(C) = $ true then the first home-owner event after C is a remove event. For every remove event R there is a check event C such that $Val(C) = $ true, $C \prec R$ and R is the first home-owner event after C.*
3. *For every check event C let the removal number, removal_num(C), be the number of remove events R with $R \prec C$. And let deliver_num(C) be the number of deliver events D such that $D \prec C$. Then for every check event C,*

$$Val(C) = \text{``removal\_num}(C) < \text{deliver\_num}(C)\text{''}, \tag{1}$$

that is to say the boolean value of C is true iff the number of deliver events that precede C exceeds the number of letters that were removed by remove events that precede C.

As for the liveness requirements, [2] requires that the algorithm is bounded wait-free, which means (see [6] under the term loop-free, or [4]) that each operation completes before the process executing it has taken k steps, for some fixed constant k.

For communication, the Mailbox Problem as formulated in [2] requires atomic single-writer registers, but we prefer to present our algorithm with serial registers. Clearly, this simplifying assumption does not limit the applicability of our algorithm which works as well with atomic registers.

Definition 2. *For any serial register R we define a function μ over the read actions of register R, such that for any read r, $\mu(r)$ is the last write action on R that precedes r. That is, $\mu(r) < r$ and there is no write action w on R with $\mu(r) < w < r$.*

Then seriality of the register means that the read/write actions are linearly ordered and that r and $\mu(r)$ have the same value: $Val(r) = Val(\mu(r))$, where $Val(e)$ for a read/write action e, is the value that has been written/read at the action e. (To ensure that $\mu(r)$ is defined on all read actions, we have to assume an initial write event that precedes all read events.) Since the registers are serial it follows that if $r_1 < r_2$ are read actions of R then $\mu(r_1) \leq \mu(r_2)$.

An additional "access restriction" is made in [2] for efficiency's sake which requires that the *check* operation uses no persistent private variables. Namely, the owner's decision on whether to approach the mailbox or not should depend just on her readings of the flag values and not on any internal information sustained from some previous operation. One may argue that a small persistent variable would not harm the efficiency of the *check* operations. We do not argue otherwise, but we chose to accept all limitations as defined by the authors of [2], so our algorithm uses no persistent local variables in its *check* procedure. However, as we prove here, even if local variables are allowed during a *check* operation, the space complexity of our solution cannot be improved.

1.1 A Condition Equivalent to Linearizability

In simple cases, linearizability can be obtained by identifying in each operation execution a lower-level action (the linearization point of that operation execution) such that the operations "appear to take effect" instantaneously at these linearization points. In more complex cases no such fixed linearization actions exist, and in some cases the linearization point of an operation X can be found in some other operation execution Y that is concurrent with X (and belongs to a different process). The queue algorithm with which Herlihy and Wing present linearizability in [5] is an example of such a more complex algorithm, as is the mailbox algorithm that we present here. Our experience shows that in these complex cases a more natural proof of linearization can be obtained by finding an abstract intermediary property that implies linearizability, and then proving that any execution of the algorithm satisfies this intermediary property. In the case of the mailbox algorithm, we present a property that is in fact equivalent to linearizability of executions of mailbox algorithms.

As we have said, the two processes communicate by reading and writing serial registers, and we assume a linear ordering $<$ on all lower-level actions. Any operation execution consists of a set of lower-level actions, and so $<$ induces a partial ordering (still denoted $<$) on the operation executions: $A < B$ for operation executions A and B if, for every actions a in A and b in B, $a < b$ holds. *Linearization* as defined by [5] is the requirement that the partial ordering $<$ on the operation executions can be extended to a linear ordering $\prec$ that satisfies the linear mailbox specifications defined above in 1.

Let τ be an execution of a mailbox algorithm $\mathcal{A}$. Since the *postman* process is serial, we can enumerate the *deliver* operations of τ in increasing order $D_1 < D_2 \cdots$, and likewise the *remove* operations can be enumerated $R_1 < R_2 \cdots$. We do not claim that D_i or R_i exist for every i, but in case the i-th *deliver* (*remove*) operation exists then it is denoted D_i (R_i). If C is any *check* event, then *removal_num*(C) denotes the number of *remove* events that precede C in the ordering $<$ (which is a linear ordering on the *deliver* operation executions). Thus, for every $i > 0$, *removal_num*$(C) = i$ iff R_i is the last *remove* event that precedes C.

If C is a *check* event, we say that C is positive if C returns true and otherwise, C is said to be negative. Now we can present our intermediary properties.

Theorem 3. *Let $\mathcal{A}$ be a mailbox algorithm and let τ be an execution of $\mathcal{A}$. τ is linearizable iff for every check operation such that removal_num$(C) = k$ the following hold:*

P1. If C is positive then D_{k+1} exists and $\neg(C < D_{k+1})$.
P2. If C is negative, if D_{k+1} exists then $\neg(D_{k+1} < C)$.

We leave for the reader the straightforward proof for the "only if" direction of the theorem, and we focus on proving that the properties P1 and P2 imply linearizability. We fix an execution τ of a mailbox algorithm $\mathcal{A}$ that satisfies the two properties P1 and P2, and we shall prove that τ is linearizable. The proof is in three steps. First we define a relation $\lhd$ on some of the *check* and *deliver* events, and then we prove that the union $< \cup \lhd$ of the precedence relation $<$ with $\lhd$ has no cycles, and finally we note that any extension of this union into a linear ordering satisfies the linear mailbox specification given in definition 1.

$\lhd$ is defined as the union of the following two sets of ordered pairs:

1. $S_1 = \{\langle C, D_{k+1}\rangle : C$ is a negative *check* event, $rn(C) = k$ and D_{k+1} exists$\}$,
2. $S_2 = \{\langle D_{k+1}, C\rangle : C$ is a positive *check* event, and $rn(C) = k\}$.

Any relation is a set of ordered pairs, and hence the union $< \cup \lhd$ is meaningful. A cycle of length $n > 0$ in a binary relation R, is a sequence

$$(X_0, \ldots, X_n)$$

so that $X_i R X_{i+1}$ for each $i < n$ and $X_0 = X_n$. A binary relation R can be extended to linear order iff it admits no cycles.

We leave for the reader to prove that the following hold:

1. If $X \lhd Y$, then $\neg(Y < X)$.
2. If $X \lhd Y \lhd Z$, then $X < Z$.
3. If $X < Y \lhd Z < W$, then $X < W$.
4. If $X < Y \lhd Z$ or $X \lhd Y < Z$, then $X \neq Z$.
5. There are no cycles in $< \cup \lhd$ of length 1 or 2.

We conclude:

Lemma 4. *There are no cycles in* $< \cup \lhd$.

Proof. Assume for a contradiction that our lemma is false and consider a cycle of minimal length,

$$(X_0, X_1, \ldots, X_n).$$

By the minimality of n, since $<$ is transitive and by item 2, there are no two consecutive occurrences of $<$ and no two consecutive occurrences of $\lhd$. Therefore, we may assume w.l.o.g. that the sequence is of the form

$$X_0 < X_1 \lhd X_2 < \ldots X_{n-1} \lhd X_n = X_0.$$

Now, by the item 5, $n \geq 3$. Therefore, there is a prefix of the sequence of the form $X_0 < X_1 \lhd X_2 < X_3$. Hence, by item 3, $X_0 < X_3$ in contradiction to the minimality of n.   □

As $< \cup \lhd$ is an acyclic relation, it can be extended to a linear ordering. For completing the proof of theorem 3, we leave for the reader to verify that if $\prec$ is a linear ordering that extends $< \cup \lhd$, then $\prec$ satisfies the sequential specification of the mailbox object.

2 The 4/4 Mailbox Algorithm

In this section we define in Fig. 1 a mailbox algorithm with only two-bit flag registers: F_P and F_H. The *postman* process has three registers: D_num, T_P and F_P. D_num is of type $\mathbb{N}$, T_P is of type $\{0, 1, 2\}$ and F_P stores values from the set $\{0, 1, 2, \text{"stop"}\}$. The *home-owner*'s registers are R_num, T_H and F_H. R_num is of type $\mathbb{N}$, T_H is of type $\{0, 1, 2\}$ and F_H stores values from $\{0, 1, 2, \text{"go"}\}$. The initial value of F_P is "stop", and of all other registers is 0.

In addition to the registers, we have the FIFO queue Q which supports two operations: addition of a letter (executed by the *postman* process), and removal of a letter (executed by the *home-owner* when Q is nonempty)[1]. Q is initially empty.

deliver (*letter*):		*check*():	
1 add *letter* to Q;	// *enq*	1 $fh := F_H$;	// $r1$
2 $dn := dn + 1$;		if $fh =$ "go" return true;	
D_num $:= dn$;	// $w1$	2 $fp := F_P$;	// $r2$
3 $t := T_H$;	// $r1$	3 return "$fp = fh + 1 \mod 3$";	
4 $T_P := t + 1 \mod 3$;	// $w2$		
5 $rn := $ R_num;	// $r2$	*remove*():	
6 if $rn < dn$ then			
$F_P := t + 1 \mod 3$		1 remove one letter from Q;	// *deq*
else $F_P := $ "stop";	// $w3$	2 $rn := rn + 1$;	
		R_num $:= rn$;	// $w1$
		3 $t := T_P$;	// $r1$
		4 $T_H := t$;	// $w2$
		5 $dn := $ D_num ;	// $r2$
		6 if $rn < dn$ $F_H := $ "go"	
		else $F_H := t$;	// $w3$

Fig. 1. The 4/4 Mailbox Algorithm. Variables dn and rn are initially 0.

Each instruction line in the pseudocode of Fig. 1 is followed by the name of that instruction. So, for example, an execution of line 1 of the *deliver* operation is an *enq* action.

A *deliver* operation execution D is an execution of lines 1–6 of that code. It is a high-level event, namely a set of lower-level actions which are the executions of the code instructions. Variable dn (the *delivery number*) is initially 0, so that if D is the i'th *deliver* operation execution ($i = 1, 2, \ldots$) and $dn(D)$ denotes the value of dn after line 2 is executed in D, then $dn(D) = i$. Register D_num thus contains the current delivery number.

We shall use this sort of notation, $dn(D)$, for other variables as well. We note that in our algorithms any local variable is assigned a value in a unique instruction. So if v is a local variable and E some operation execution that assigns a value to v, then the notation $v(E)$ for that value that E assigns to v is meaningful and well defined. Likewise, if G is any register such that E contains a write into G then we denote with $G(E)$ the value of that write.

In line 3, register T_H is read into variable t and then $t + 1$ (mod 3) is written onto register T_P. We refer to the values that T_P writes as "colors". So, the *postman* is always changing the color obtained from the *home-owner* process, while the *home-owner* always copies the value obtained (see lines 3 and 4 in the *remove* code).

---

[1] As there is a unique process that enqueue elements into Q and a unique process that dequeue elements, it is not difficult to see that Q can be implemented with single-writer atomic registers.

There are two sorts of *check* operations. A "short" *check* C is one that returns true immediately after line 1 is executed. In this line, the *home-owner* process reads her own register F_H and returns true if that register's value is *"go"*. Note that line 1 is the only place in the algorithm where this flag is read, and hence F_H is dispensable and a local *home-owner* variable could replace it. The access restriction however prohibits persistent variables, and hence the need for this register which does nothing more than replacing a persistent local variable. In a "long" *check* operation C, the read of register F_H returns a value in $\{0, 1, 2\}$. Then C contains a read of F_P, and then C returns the truth value of $fp = fh + 1 \pmod{3}$. If fp is *"stop"*, then this equation cannot hold (as *"stop"* is not a number) and C returns false in this case.

A *remove* operation execution removes a letter from the queue, updates its removal number variable rn, and writes the new value on register R_num (line 2). Since the initial value of rn is 0, the value of R_num is the number of letters so far removed. In lines 3 and 4 the operation copies the value of T_P onto T_H, and in line 5 the value of D_num is read into variable dn. Then in line 6 the value of flag F_H is decided: if condition $rn < dn$ holds then the operation is positive and the value of the flag is "go"; otherwise the operation is negative and the value of the flag is t, which is also the value of register T_H.

There are two ways in which a *check* operation returns a positive answer, and their subtle combination is the essence of the algorithm. The first idea is that if during an execution of a *remove* event, the *home-owner* finds (in line 6) that condition $rn < dn$ holds (line 6), then surely there are some letters waiting to be removed from the queue and any additional actions by the *postman* can only increase their number. Thus, in this case *home-owner* writes "go" to F_P (line 6) and correspondingly, the following *check* event checks this value and returns true (line 1). A *check* event C that returns after executing line 1 solely, is named a "short" check event.

For the second idea, we want to consider the situation in which the *deliver/check/remove* operations are linearly ordered. That is, there is no interleaving and if A and B are any two operation executions then $A < B$ or $B < A$. Note that if C is any *check* operation that precedes all *deliver* operations then it must be negative, because the initial value of F_H is 0 (so it is not "go") and when C reads F_P it find the value "stop". Let D be the first *deliver* operation, then D writes 1 on F_P in executing line 6, and if C is any subsequent *check* operation that reads this value then (since the initial value of F_H is 0) condition $fp = fh + 1$ holds and so C is positive. The interplay with registers T_P and T_H, in which the *home-owner* tries to copy the value found in the *postman*'s register and the *postman* tries to have a newer value, is the second idea of the algorithm. If the *home-owner* sees a newer value at the F_P register then she is lead to believe that a new letter is waiting for her at the mailbox.

The reader may notice that if we assume seriality of the operation executions, then there is no need for the value "stop" at register F_P and the *postman* process may write to F_P, $t + 1 \pmod{3}$ while executing line 6 regardless of the value it read from R_num. We show that when operations may be executed concurrently, the value "stop" is required. For verifying this, we assume that *postman* writes to F_P, $t + 1 \pmod{3}$ while executing line 6 and we consider the following interleaved scenario of the linear operations algorithm. The first *deliver* operation $D1$ writes 1 on D_num and 1 on T_P and

F_P, and is followed by a positive *check* operation $C1$. Then a second *deliver* operation $D2$ begins, executes the write $w1(D2)$ (writes 2 on D_num) and stops for awhile. Now a *remove* operation $R1$ is prompted by the positive $C1$, it reads the value 2 in D_num, and writes "go" on F_H. The ensuing *check* operation $C2$ is short and positive. Then a *remove* operation $R2$ removes the message of D_2 and completes: it reads 1 in T_P and writes 1 on T_H and F_H. We let now $D2$ complete its operation: it reads 1 in T_H and writes 2 on T_P and F_P. The queue is now empty since the two messages were removed, and yet a *check* operation $C3$ is positive now because it reads 1 in F_H and 2 in F_P. Note that a corresponding scenario is not a problem for our mailbox algorithm of Fig. 1, because in the described scenario, D_2 writes "stop" to F_P and correspondingly $C3$ returns false after reading "stop" from F_P.

This explains the need for the "stop" value of the F_P register, but what about the three values of registers T_P and T_H, wouldn't two values suffice? Consider the variant of our algorithm in which mod 3 is replaced everywhere by mod 2. The following scenario shows that a *check* operation may return the value true even though the queue is empty. The first *deliver* operation $D1$ writes 1 on D_num and 1 on T_P and F_P and is followed by a positive *check* operation $C1$. Then a second *deliver* operation $D2$ begins, executes the write $w1(D2)$ (writes 2 on D_num) and stops for awhile. Now a *remove* operation $R1$ is prompted by the positive $C1$, it reads the value 2 in D_num and 1 in T_P, writes 1 to T_H and writes "go" on F_H. The ensuing *check* operation $C2$ is short and positive. So far we are as in the previous scenario. Now we let $D2$ continue with actions $r1$, $w2$, and $r2$. That is, $D2$ stops before the last write $w3$. Note that the value of $r1(D2)$ is 1, and the value of $w2(D2)$ is hence $0 = 1 + 1 \pmod 2$. Now comes $R2$ the second *remove* operation: it reads 0 in T_P, write 0 in T_H, and writes 0 in F_H. Finally, a positive *check* operation $C3$ reads 0 in F_H and 1 in F_P (as it obtains the value of $D1$). So $C3$ is positive despite the fact that the two letters were removed. Note that in the properly working algorithm of Fig. 1 this scenario is not problematic since $F_H(R2)$ would be 2. Lemma 8 proves that in a case as described by this scenario *check* operation $C3$ must be negative.

2.1   Correctness of the Algorithm

An *action* is an execution of an atomic instruction of the algorithm such as a read or a write of a register or a queue action. Since we assume that the registers are serial, and as the queue operations (to add or remove a letter) are also instantaneous, we have a total ordering $<$ on all actions. We write $a < b$ to say that a precedes b in this ordering.

An *operation execution* is an execution of the *deliver*, *check*, or *remove* algorithm. Every operation execution is a high-level event, namely a set of lower-level actions (also called lower-level events in [7]). The total ordering $<$ on the lower-level actions induces a partial ordering on the operation executions: for operation executions A and B we define that $A < B$ if $a < b$ for every $a \in A$ and $b \in B$. It is also very convenient to relate high-level events and lower-level actions: $A < x$ for a high-level event A and a lower-level event x means that $a < x$ for every a in A. And similarly $x < A$ is defined when $x < a$ for every a in A. The fact that we use the same symbol $<$ to denote both the total ordering relation on the actions and the resulting partial ordering relation on the high-level events should not be a source of confusion.

The aim of the correctness proof is to define a total ordering $\prec$ on the operation executions that extends the partial ordering $<$, and then to prove that properties $P1$ and $P2$ of Theorem 3 hold.

We assume two initial high-level events I_p and I_h by the *postman* and *home-owner* processes that determine the initial values of the registers and the initial values of the variables. These events precede all other high level events.

If a is any read/write action, then $[a]$ denotes that high level event to which a belongs. (Every low level action belongs to some operation execution, except for the assumed initial write actions which belong to the initial events I_h and I_p.)

The names of the read/write instructions appear in Fig. 1. If X is any operation execution and s a name of an instruction that X executes, then $s(X)$ denotes the corresponding action. That is, the execution of s in X. For example, if R is a *remove* operation, then $r2(R)$ is the read of register D_num in R.

If $D_1 < D_2$ are two postman operations (D_1 is possibly the initial I_p event) then D_2 is said to be the immediate successor of D_1 if there is no *postman* operation E such that $D_1 < E < D_2$. Similarly if $R_1 < R_2$ are two *remove* operations (or R_1 is the initial I_h event) then R_2 is an immediate successor of R_1 if there is no *remove* operation R such that $R_1 < R < R_2$.

It is convenient to define the "color" of operations. If E is a *deliver* or a *remove* operation then $color(E)$ is the value of the write action $w2(E)$. Thus, if D is a *deliver* operation, then $color(D) = T_P(D)$ (which is that value $c = 0, 1, 2$ that is written into register T_P when line 4 is executed in D) and if R is a *remove* operation then $color(R) = T_H(R)$ (which is also the value read from register T_P). The color of the initial events is 0.

The *deliver* and *remove* operations play some kind of ping-pong game with the values of registers T_H and T_P. The *home-owner* wants that these registers have the same value, but the *postman* wants that its register T_H is ahead of the *home-owner*'s register by 1 mod 3. As a result we have the following lemma.

Lemma 5. *If D_2 is an immediate deliver successor of D_1 then $color(D_2) = color(D_1)$ or else $color(D_2) = color(D_1) + 1$ (mod 3). Similarly, if R_2 is an immediate remove successor of R_1, then either $color(R_2) = color(R_1)$ or else $color(R_2) = color(R_1) + 1$ (mod 3).*

Proof. Suppose that the lemma does not hold and there are two operations $X_1 < X_2$ such that either X_2 is the immediate *deliver* successor of X_1 or else X_2 is the immediate *remove* successor of X_1, but $color(X_2) = color(X_1) + 2$ (mod 3). Since the write actions on the registers are assumed to be linearly ordered (and every action has only a finite number of actions that precede it) there is a minimal counterexample X_2 to the lemma, minimal in the sense that there is no counterexample $X_1' < X_2'$ such that $w2(X_2') < w2(X_2)$.

Case 1: the minimal counterexample $X_1 < X_2$ is a pair of *deliver* operations $D_1 < D_2$. The proof for this case is accompanied by Fig. 2.1 that illustrates our arguments. In this figure there is an arrow from a *deliver* event X into a *remove* event Y if *postman* reads from T_H in X the value that was written to T_H in Y (i.e. $Y = [\mu(r1(X))]$), and there is an arrow from a *remove* event X to a *deliver* event Y if *home-owner* reads from T_P in X the value that was written to T_P in Y (i.e. $Y = [\mu(r1(X))]$).

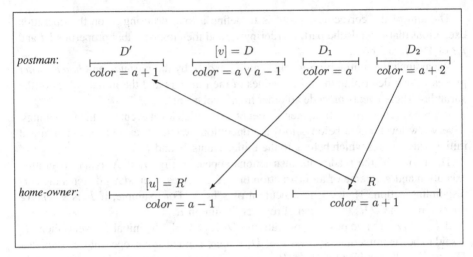

Fig. 2. A minimal counter example

Say a is the color of D_1 and $a + 2$ (mod 3) is the color of D_2. Since the *deliver* operation add 1 (mod 3) to the value read in T_H, $a = t(D_1) + 1$ (mod 3), and likewise $a + 2 = t(D_2) + 1$ (mod 3). So $t(D_1) = a - 1$ (mod 3), and $t(D_2) = a + 1$ (mod 3). Hence $t(D_1) \neq t(D_2)$, and this already implies that the read $r1(D_2)$ did not obtain the initial value of register T_H. Hence $w = \mu(r1(D_2))$, the corresponding write action on T_H is not the initial write. Thus there is a *remove* operation R such that $w = w2(R)$ is the write of value $a + 1$ (mod 3) that $t(D_2)$ obtains. The read $r1(R)$ of register T_P in R has the same value as that of w, namely $a + 1$ (mod 3). This implies that it is not the case that $w2(D_1) < r1(R)$, and thus $r1(R) < w2(D_1)$ (which implies that D_1 is not the initial event). Let v be the last write on T_P that precedes D_1. By minimality of the pair $D_1 < D_2$, the pair $[v] < D_1$ is not in contradiction to the lemma and hence the value of v is a or $a - 1$ (mod 3). This implies that $r1(R) < v$ (because the value of $r1(R)$ is $a + 1$ (mod 3)). Since $v < r1(D_1)$, $r1(R) < r1(D_1)$, and so $r1(D_1)$ obtains the value of u, the last write on T_H that is before R. So the value of u is $a - 1$ (mod 3), and the pair $[u] < R$ is also a contradiction to the lemma. But this is impossible by the minimality of D_2.

Case 2: the minimal counterexample is a pair of *remove* operations $R_1 < R_2$. This is argued in a similar vein.   □

We say that a *check* operation C is "positive" in case it returns the value true, and it is "negative" when it returns false. If E is a *deliver* or a *remove* operation execution then E is "positive" when condition $rn < dn$ holds when line 6 is executed (and E is "negative" otherwise). Equivalently, a *deliver* operation is negative when it writes "stop" on F_P, and a *remove* operation is positive when it writes "go" on F_H.

For any *remove* operation R, $rn(R)$ is the value of variable rn that is determined in executing line 2 and is written on register R_num. We also set $rn(I_h) = 0$ (and the initial value of variable rn is 0).

$rn(R)$ is called the "removal number"; it is the number of *remove* operations R' such that $R' \leq R$. Clearly, if $R_1 < R_2 < \cdots$ is the sequence of *remove* operations in increasing order, then $rn(R_i) = i$.

The *check* code does not contain a variable named rn, and so the number $rn(C) = $ *removal_num*(C) for a *check* operation execution C is defined directly as the number of *remove* operations R such that $R < C$.

Likewise, the "delivery number", $dn(D)$, of a *deliver* operation D is the value of variable dn as determined at line 2, and is equal to the number of *deliver* operations D' such that $D' \leq D$. Thus if $D_1 < D_2 < \cdots$ is the enumeration of the *deliver* operations in increasing order, then $dn(D_i) = i$. We also define $dn(I_p) = 0$.

Now we define two functions, pre_rem and ρ, on the *check* events.

Definition 6. *Let C be any check operation execution. Define pre_rem(C) as the last remove operation execution R such that $R < C$ if there is such a remove operation, and pre_rem$(C) = I_h$ is the assumed initial home-owner event otherwise.*

We note that a short *check* operation is positive, and that a *check* operation C is short if and only if pre_rem(C) is positive (it writes "go" on F_H). Since the assumed initial value of F_H is not "go", if C is short then pre_rem(C) is not the initial event– it is necessarily a *remove* operation execution.

The following is a key definition in our correctness proof. It relates every *check* operation C to $\rho(C)$ which is the *deliver* operation (or initial I_p event) that C considers in order to calculate which value (true or false) to return: we will prove that C is positive iff $rn(\rho(C)) < dn(\rho(C))$.

Definition 7. *For any check operation execution C we define $\rho(C)$ as follows. In case C is a short check operation let $R = $ pre_rem(C) (which is a remove operation execution as we noted) and then define $\rho(C) = [\mu(r2(R))]$. In case C is a long operation, define $\rho(C) = [\mu(r2(C))]$. (Recall that $r2(R)$ is the read of D_num in R, and $r2(C)$ is the read of F_P in C. For any read action r $\mu(r)$ is the corresponding write action, and $[\mu(r)]$ is the operation execution that contains this write action.)*

It is obvious that $\neg(C < \rho(C))$. We shall prove next some properties of the functions and predicates that we have defined. These properties will then be used to prove that the properties of theorem 3 hold.

Lemma 8. *Suppose that C is a long check operation and $D = \rho(C)$. If $S = $ pre_rem(C) is a remove operation such that $w2(D) < r1(S)$, then C is negative.*

Proof. Assumption $D = \rho(C)$ implies that $fp(C) = F_P(D)$. Assume for a contradiction that C is positive. So C finds that $fp = fh + 1 \pmod 3$ holds, and hence $fp(C)$ is not "stop", and this implies that D is positive (it did not write "stop" on F_P) and $T_P(D) = F_P(D)$ (consider line 6 in the *deliver* code). Since condition $fp = fh+1 \pmod 3$ holds in C, and as $fp(C) = F_P(D)$ and $fh(C) = F_H(S) = t(S)$, we get that

$$T_P(D) = F_P(D) = fp(C) = t(S) + 1 \pmod 3. \tag{2}$$

Since $w2(D) < r1(S)$ are a write and read actions on register T_P, $w2(D) \leq \mu(r1(S))$ and so we have two cases.

Case 1. $w2(D) = \mu(r1(S))$. Hence $Val(w2(D)) = Val(r1(S))$. That is, $T_P(D) = t(S)$, which is in contradiction to the above equation.

Case 2. $w2(D) < \mu(r1(S))$. Then $\mu(r1(S))$ is a write action on register T_P in some *deliver* operation D' such that $D < D'$ and

$$t(S) = T_P(D').$$

Hence (2) implies

$$T_P(D) = T_P(D') + 1 \quad (\text{mod } 3). \tag{3}$$

So $w2(D') < r1(S) < r2(C)$, and hence D' must be the immediate successor of D (or else we would have $D < \rho(C)$). But now, Lemma 5 says that $color(D') = color(D)$ or else $color(D') = color(D) + 1 \pmod 3$, and both cases negate (3) which says that $color(D) = color(D') + 1 \pmod 3$. So C is negative.

□

Proposition 9. *If C is a positive check operation and $\rho(C) = D$, then D is a deliver operation execution and*

$$rn(C) < dn(D). \tag{4}$$

Proof. Assume first that C is a short *check* operation and let $R = \text{pre\_rem}(C)$ be the previous *remove* operation, which necessarily has set its register F_H to "*go*" at line 6. So $rn(R) = rn(C)$, and inequality

$$rn(R) < dn(R) \tag{5}$$

holds (see line 6 in the *remove* code). Let $r = r2(R)$ be the read of register D_num which obtained the value $dn(R)$. By definition of $\rho(C)$ when C is short, $D = \rho(C) = [\mu(r2)]$, and $dn(D) = dn(R)$ follows. Since $dn(R) > 0$ follows from (5) and as $dn(I_p) = 0$, $D \neq I_p$ is concluded and necessarily D is a *deliver* operation execution and (4) follows.

Now suppose that C is a long *check* operation, and let $r2 = r2(C)$ be its read of register F_P. By definition of $D = \rho(C)$, $D = [\mu(r)]$, and $F_P(D) = fp(C)$. Since C is positive, it follows that "$fp = fh+1 \pmod 3$" holds in C, and hence $fp(D) = fp(C)$ is not "stop". Hence $rn < dn$ is evaluated to true when line 6 is executed in D. So

$$rn(D) < dn(D)$$

holds.

Define $R = [\mu(r2(D))]$. Then $rn(D) = rn(R)$. We shall prove that $R = \text{pre\_rem}(C)$. This will show that $rn(C) = rn(R)$, and hence that $rn(C) < dn(D)$ as required. It thus remains to prove that $R = \text{pre\_rem}(C)$. Note that $R < C$.

Suppose on the contrary that $R \neq \text{pre\_rem}(C)$, and then $R < \text{pre\_rem}(C)$ follows. Say $S = \text{pre\_rem}(C)$. Since $\mu(r2(D))$ is in R, $r2(D) < w1(S)$. But $w2(D) < r2(D)$. Hence $w2(D) < w1(S) < r1(S)$ and this implies by Lemma 8 that C is not positive, which yields a contradiction.

□

Proposition 10. *If C is a check operation and $D = \rho(C)$ is such that $rn(C) < dn(D)$, then C is positive.*

Proof. A short *check* operation is always positive, and hence we may assume that C is long. As $rn(C) < dn(D)$, $0 < dn(D)$ and D is thus a *deliver* operation and not the initial event. Say $R = \mathrm{pre\_rem}(C)$ and then $rn(R) = rn(C)$. Suppose first that $R = I_h$ is the initial event. So $rn(C) = 0$. In reading F_H, C obtains the initial value 0. So $fh(C) = 0$. We shall prove that $fp(C) = 1$, and hence that C returns true at line 3, as required.

By definition $D = \rho(C) = [\mu(r2(C))]$, and hence $D < r2(C)$. Now, the read of R_num in D returns the value 0 of I_h because $\mathrm{pre\_rem}(C) = I_h$. Hence $rn(D)$ is 0, but $dn(D) > 0$ implies that "$rn < dn$" holds in D when line 6 is executed, and hence the value of $w3(D)$ is $t(D) + 1 \pmod 3$. But $t(D) = 0$ because the initial value of T_H is 0, and hence the value of $w3(D)$ is 1. So $fp(C) = 1$.

So now we assume that R is a *remove* execution. In case $w1(D) < r2(R)$, $w1(D) \le \mu(r2(R))$ follows, and hence the read of D_num in R obtains the write in D or a later write. Hence $dn(D) \le dn(R)$. The fact that $rn(R) = rn(C)$ and our assumption that $rn(C) < dn(D)$ imply that $rn(R) < dn(R)$. So R is positive, it writes "go" on F_H and C is a short positive *check* operation.

So we may assume that $r2(R) < w1(D)$. It follows from this assumption that $w2(R) < r1(D) < w2(D) < r2(C)$, and hence that $w2(R) = \mu(r1(D))$. Say $c = color(R)$ (that is, by definition, the value of $w2(R)$, which is the write in T_H). Then c is also $F_H(R)$, and hence $c = fh(C)$. But as c is the value obtained in the read $r1(D)$, $F_P(D) = c + 1 \pmod 3$, and hence $fp = fh + 1 \pmod 3$ holds in C, and C is therefore positive. □

Now we can prove the correctness of our algorithm, relaying on theorem 3.

Proof. Let $D_1, \ldots,$ be the increasing order enumeration of the *deliver* operations. According to Theorem 3 we have to prove the following.

Let C be a *check* operation and suppose that $rn(C) = k$. Then:
1. If C is positive then D_{k+1} exists and $C \not< D_{k+1}$.
2. If C is negative then $D_{k+1} \not< C$.

For the first item suppose that C is positive and $rn(C) = k$. Define $D = \rho(C)$. By Proposition 9, D is a *deliver* operation such that $rn(C) < dn(D)$. That is, $k < dn(D)$, and hence $D = D_m$ for some $m \ge k + 1$. So $D_{k+1} \le \rho(C)$. Since it is not the case that $C < \rho(C)$, it is not the case that $C < D_{k+1}$.

Now suppose that C is negative and yet $D_{k+1} < C$. Then $D_{k+1} \le \rho(C)$. That is $\rho(C) = D_m$ for some $m \ge k + 1$. But $rn(C) = k < m$, and hence Proposition 10 implies that C is positive.

□

3   Space Lower Bound

Aguilera et al. [2] note that the space complexity of their mailbox solution is $O(n \log n)$ where n is the number of *deliver* and *remove* operations, and they conjecture that there is a solution with space complexity $O(\log n)$. Clearly, our algorithm proves this conjecture since the n-th *deliver* operation writes the number n and needs no more space than $4 + \log n$ (taking into account 12 values for the T_P and F_P registers), and the m-th *remove* operation writes the number m and needs no more space than $4 + \log m$. In this section we prove that there is no solution to the mailbox problem with space complexity $f(n) \in o(\log n)$.

Roughly speaking, this lower bound holds as the *postman* needs at least $\log(n)$ bits in order to inform the *home-owner* that n messages where delivered. For each $n > 0$ we consider the scenario in which the *postman* process delivers n letters in a solo run, and since the *home-owner* must distinguish between any two of theses scenarios, the *postman* must use at least $\log(n)$ bits in any of these executions for the *home-owner* to infer the number of messages that are to be removed. Note that this argument holds even if persistent local variables are allowed during a *check* operation, and even if the *check* operation may access unbounded registers. In fact, this result holds also if the shared registers support operations that are stronger than read/write actions. For the reader who wishes to see more details, we give below a detailed proof.

We consider a computation model which is much stronger than the model of read/write shared registers. In this model each of the two processes, *home-owner* and *postman*, has a single register, R_H and R_P respectively, that can hold words of any length. Any process can in a single atomic step read the two registers, perform an internal computation and write the computed value to its register. We use standard notations in our proof. A *process* is a state machine and a *configuration* is a tuple consisting of the local states of the processes and the values of the shared registers. An initial configuration is assumed. An *execution* is an interleaving sequence of configurations and actions that starts from some initial configuration. Two configuration C and C' are *indistinguishable* for a process P, if the local state of P is the same in both configurations and the two shared registers store the same values in C and in C'.

Aiming for a contradiction we assume that there is a wait free algorithm $\mathcal{A}$ that solves the mailbox problem, with space complexity $f(n) \in o(\log n)$. For each $k \geq 0$, we let τ_k be the execution in which the *postman* performs k deliver events in a solo run starting from the initial configuration. Let w_k denote the binary word that R_P stores at the end of τ_k. Since $f(n) \in o(\log n)$, it is not difficult to prove:

Lemma 11. *For some two natural numbers $m \neq k$, $w_m = w_k$.*

Our lower bound can now be proved.

Theorem 12. *There is no mailbox algorithm with space complexity $f(n) \in o(\log n)$.*

Proof. Assume for a contradiction that $\mathcal{A}$ is a wait free mailbox algorithm with space complexity $f(n) \in o(\log n)$. By lemma 11, there are some $k < m$ so that R_P stores the same value at the end of the executions τ_k and τ_m, executions that start from the initial configuration. τ_k leads to some configuration, C_1 and τ_m leads to some configuration ,C_2. Thus, C_1 and C_2 differ only by the state of the *postman* process and hence C_1 and C_2 are indistinguishable for the *home-owner*. As a result, if we let the *home-owner* process to take several steps starting from C_1, and starting from C_2, *home-owner* will reach the same local state and will write the same values to R_H.

We consider the execution $\tau_k \cdot \sigma$, namely the concatenation of τ_k and σ, where σ is a solo run by the *home-owner* in which, *home-owner* performs k-times *check* and *remove* events. Note that each *check* operation in σ returns true as there are k letters in the queue at the end of τ_k. The execution $\tau_k \cdot \sigma$ leads to a configuration C_1'. As C_1 and C_2 are indistinguishable for the *home-owner* process, the execution $\tau_m \cdot \sigma$ is valid and this execution leads to a configuration C_2'.

Now, since σ is an *home-owner* solo run, C_1' and C_2' are also indistinguishable for the *home-owner*. However, if we let the *home-owner* execute a *check* operation starting from C_1', this operation will return false as there are no messages in the queue. On the other hand, if we let the *home-owner* execute a *check* operation starting from C_2', this operation will return true since there are $m - k > 0$ messages in the queue at C_2'. This of course, contradicts our conclusion that C_1' and C_2' are indistinguishable for the *home-owner* process.   □

4   Conclusions

In [2], Aguilera, Gafni, and Lamport define the Mailbox problem, and present a solution in which the *check* operation reads two registers (the "flag" registers) that can carry 14 values each. Moreover, they prove that there is no solution to the problem with two single-bit flags. The authors of [2] asked if there is a solution with flags of smaller size and in this paper we provide a positive answer by presenting a much simpler solution to the Mailbox problem with two flags of two bits each.

Another problem from [2] is whether the space complexity of the mailbox algorithm presented in that paper can be improved. The algorithm of [2] uses $\Theta(n \log n)$ bits of shared memory, where n is the number of executions of deliver and remove, and the authors of [2] conjecture that there is a solution using logarithmic space. In addition to the 4-valued flags, our algorithm uses two registers D_num and R_num of width exactly $\log n$ bits for n procedure executions, and thus we answer this conjecture positively.

We prove in Theorem 12 that the space complexity of our algorithm is optimal. The question about the exact size of the flag registers however remains open. As we have mentioned, Aguilera et al. proved that there is no solution to the mailbox problem with two valued flags, and our solution needs two flags of size 4 each. We do not know if there is a solution with flags of size 3, or a solution in which one of the processes uses a flag of size 3 and the other process a flag of size 4.

The Signaling Problem is another interesting question with which [2] deals. The authors of [2] give a non-blocking solution and ask about the possibility of a bounded wait-free solution. The ideas developed in this paper have contributed to a solution of this wait-free Signaling Problem which was obtained by the second author [3]. These two algorithms, the one presented here and the algorithm of [3] have some ideas in common. However, the design of each algorithm among these two, also relay on unique ideas that fit the exact nature of each problem. Aguilera et al. also note in [2] that the mailbox problem can be solved by using a signaling algorithm. Accordingly, the signaling algorithm in [3] also provides a mailbox algorithm with 3 bit flags. Thus, the solution we present here is more efficient than the solution obtained by the reduction suggested in [2].

While introducing the mailbox problem, Aguilera et al. required that persistent local variables are not allowed during an execution of a *check* event. One may argue that using local variables of small size do not harm the efficiency of a mailbox algorithm. We chose to adapt this restriction and our algorithm does not use local variables during a *check* event. An interesting question asked by Aguilera et al. if there is a mailbox algorithm that use 1 bit flags in case that local variables are allowed during execution of *check* operations. We do not know the answer to this question. However, the $\Theta(\log n)$ space lower bound we present here, hold even when persistent local variables are allowed.

References

1. Aguilera, M.K., Gafni, E., Lamport, L.: The Mailbox Problem (Extended Abstract). In: Taubenfeld, G. (ed.) DISC 2008. LNCS, vol. 5218, pp. 1–15. Springer, Heidelberg (2008)
2. Aguilera, M.K., Gafni, E., Lamport, L.: The Mailbox Problem. Distributed Computing 23, 113–134 (2010)
3. Amram, G.: On the Signaling Problem. In: Chatterjee, M., Cao, J.-n., Kothapalli, K., Rajsbaum, S. (eds.) ICDCN 2014. LNCS, vol. 8314, pp. 44–65. Springer, Heidelberg (2014)
4. Herlihy, M.: Wait-free synchronization. ACM Trans. Program. Languag. Syst. 11(1), 124–149 (1991)
5. Herlihy, M., Wing, J.: Linearizability: A correctness condition for concurrent objects. ACM Transactions on Programming Languages and Systems 12, 463–492 (1990)
6. Lamport, L.: A new solution of dijkstraś concurrent programming problem. Communications of the ACM 17(8), 453–455 (1974)
7. Lamport, L.: On Interprocess Communication, Part I: Basic formalism, Part II: Algorithms. Distributed Computing 1, 77–101 (1986)

Distributed Universality

Michel Raynal[1,2], Julien Stainer[2], and Gadi Taubenfeld[3]

[1] Institut Universitaire de France
[2] IRISA, Université de Rennes 35042 Rennes Cedex, France
[3] The Interdisciplinary Center, PO Box 167, Herzliya 46150, Israel
{raynal,julien.stainer}@irisa.fr, tgadi@idc.ac.il

Abstract. A notion of a *universal construction* suited to distributed computing has been introduced by M. Herlihy in his celebrated paper *"Wait-free synchronization"* (ACM TOPLAS, 1991). A universal construction is an algorithm that can be used to wait-free implement any object defined by a sequential specification. Herlihy's paper shows that the basic system model, which supports only atomic read/write registers, has to be enriched with consensus objects to allow the design of universal constructions. The generalized notion of a k-*universal* construction has been recently introduced by Gafni and Guerraoui (CONCUR, 2011). A k-universal construction is an algorithm that can be used to simultaneously implement k objects (instead of just one object), with the guarantee that at least one of the k constructed objects progresses forever. While Herlihy's universal construction relies on atomic registers and consensus objects, a k-universal construction relies on atomic registers and k-simultaneous consensus objects (which are wait-free equivalent to k-set agreement objects in the read/write system model).

This paper significantly extends the universality results introduced by Herlihy and Gafni-Guerraoui. In particular, we present a k-universal construction which satisfies the following five desired properties, which are not satisfied by the previous k-universal construction: (1) among the k objects that are constructed, *at least* ℓ objects (and not just one) are guaranteed to progress forever; (2) the progress condition for processes is *wait-freedom*, which means that each correct process executes an infinite number of operations on each object that progresses forever; (3) if any of the k constructed objects stops progressing, all its copies (one at each process) stop in the same state; (4) the proposed construction is *contention-aware*, in the sense that it uses only read/write registers in the absence of contention; and (5) it is *generous* with respect to the *obstruction-freedom* progress condition, which means that each process is able to complete any one of its pending operations on the k objects if all the other processes hold still long enough. The proposed construction, which is based on new design principles, is called a (k, ℓ)-universal construction. It uses a natural extension of k-simultaneous consensus objects, called (k, ℓ)-simultaneous consensus objects ((k, ℓ)-SC). Together with atomic registers, (k, ℓ)-SC objects are shown to be necessary and sufficient for building a (k, ℓ)-universal construction, and, in that sense, (k, ℓ)-SC objects are (k, ℓ)-*universal*.

Keywords: Asynchronous read/write system, universal construction, consensus, distributed computability, k-simultaneous consensus, wait-freedom, non-blocking, obstruction-freedom, contention-awareness, crash failures, state machine replication.

M.K. Aguilera et al. (Eds.): OPODIS 2014, LNCS 8878, pp. 469–484, 2014.
© Springer International Publishing Switzerland 2014

1   Introduction

Asynchronous crash-prone read/write systems and the notion of a universal construction This paper considers systems made up of n sequential asynchronous processes that communicate by reading and writing atomic registers. Up to $n-1$ processes may crash unexpectedly. This is the basic $(n-1)$-resilient model, also called read/write *wait-free model*, and denoted here $\mathcal{ARW}_{n,n-1}[\emptyset]$. A fundamental problem encountered in this kind of systems consists in implementing any object, defined by a sequential specification, in such a way that the object behaves reliably despite process crashes.

Several progress conditions have been proposed for concurrent objects. The most extensively studied, and strongest condition, is wait-freedom. Wait-freedom guarantees that *every* process will always be able to complete its pending operations in a finite number of its own steps [14]. Thus, a *wait-free* implementation of an object guarantees that an invocation of an object operation may fail to terminate only when the invoking process crashes. The non-blocking progress condition (sometimes called lock-freedom) guarantees that *some* process will always be able to complete its pending operations in a finite number of its own steps [17]. Obstruction-freedom guarantees that a process will be able to complete its pending operations in a finite number of its own steps, if all the other processes "hold still" long enough [15]. Obstruction-freedom does not guarantee progress under contention.

It has been shown in [10,14,19] that the design of a general algorithm implementing *any* object defined by a sequential specification and satisfying the wait-freedom progress condition, is impossible in $\mathcal{ARW}_{n,n-1}[\emptyset]$. Thus, in order to be able to implement any such object, the model has to be enriched with basic objects whose computational power is stronger than atomic read/write registers [14].

Objects that can be used, together with registers, to implement any other object which satisfies a given progress condition *PC*, are called universal objects with respect to *PC*. Previous work provided algorithms, called *universal constructions*, based on universal objects, that transform sequential implementations of arbitrary objects into wait-free concurrent implementations of the same objects. It is shown in [14] that the *consensus* object is universal with respect to wait-freedom. A consensus object allows all the correct processes to reach a common decision based on their initial inputs. A consensus object is used in a universal construction to allow processes to agree –despite concurrency and failures– on a total order on the operations they invoke on the constructed object.

In addition to the universal construction of [14], several other wait-free universal constructions were proposed, which address additional properties. As an example, a universal construction is presented in [8], where "processes operating on different parts of an implemented object do not interfere with each other by accessing common base objects". Other additional properties have been addressed in [2,9].

From consensus to k-simultaneous consensus (or k-set agreement) in read/write systems. k-Simultaneous consensus has been introduced in [1]. Each process proposes a value to k independent consensus instances, and decides on a pair (x, v) such that x is a consensus instance ($1 \leq x \leq k$), and v is a value proposed to that consensus instance. Hence, if the pairs (x, v) and (x, v') are decided by two processes, then $v = v'$.

k-Set agreement [7] is a simple generalization of consensus, namely, at most k different values can be decided on when using a k-set agreement object ($k = 1$ corresponds to consensus). It is shown in [1] that k-set agreement and k-simultaneous consensus have the same computational power in $\mathcal{ARW}_{n,n-1}[\emptyset]$. That is, each one can be solved in $\mathcal{ARW}_{n,n-1}[\emptyset]$ enriched with the other[1]. Hence, 1-simultaneous consensus is the same as consensus, while, for $k > 1$, k-simultaneous consensus is weaker than $(k-1)$-simultaneous consensus.

While the impossibility proof (e.g., [14,19]) of building a wait-free consensus object in $\mathcal{ARW}_{n,n-1}[\emptyset]$ relies on the notion of valence introduced in [10], the impossibility to build a wait-free k-set agreement object (or equivalently a k-simultaneous consensus object) in $\mathcal{ARW}_{n,n-1}[\emptyset]$ relies on algebraic topology notions [5,16,26].

It is nevertheless possible to consider system models, stronger than the basic wait-free read/write model, enriched with consensus or k-simultaneous consensus objects. These enriched system models, denoted $\mathcal{ARW}_{n,n-1}[CONS]$ and $\mathcal{ARW}_{n,n-1}[k\text{-}SC]$ ($1 \leq k < n$), respectively, are consequently computationally strictly stronger than the basic model $\mathcal{ARW}_{n,n-1}[\emptyset]$.

Universal construction for k objects. An interesting question introduced in [12] by Gafni and Guerraoui is the following: what happens if, when considering the design of a universal construction, k-simultaneous consensus objects are considered instead of consensus objects? The authors claim that k-simultaneous consensus objects are k-*universal* in the sense that they allow to implement k deterministic concurrent objects, each defined by a sequential specification "with the guarantee that *at least one* machine remains highly available to all processes" [12]. In their paper, Gafni and Guerraoui focus on the replication of k state machines. They present a k-universal construction, based on the replication –at every process– of each of the k state machines.

Contributions. This paper is focused on *distributed universality*, namely it presents a very general universal construction for a set of n processes that access k concurrent objects, each defined by a sequential specification on total operations. An operation on an object is "total" if, when executed alone, it always returns [17]. This construction is based on a generalization of the k-simultaneous consensus object (see below). The noteworthy features of this construction are the following.

- On the object side, at least ℓ among the k objects progress forever, $1 \leq \ell \leq k$. This means that an infinite number of operations is applied to each of these ℓ objects. This set of ℓ objects is not predetermined, and depends on the execution.
- On the process side, the progress condition associated with the processes is wait-freedom. That is, a process that does not crash executes an infinite number of operations on each object that progresses forever.
- An object stops progressing when no more operations are applied to it. The construction guarantees that, when an object stops progressing, all its copies (one at each process) stop in the same state.

---

[1] This is no longer the case in asynchronous message-passing systems, namely k-simultaneous consensus is then strictly stronger than k-set agreement (as shown using different techniques in [6,24]).

- The construction is *contention-aware*. This means that the overhead introduced by using synchronization objects other than atomic read/write registers is eliminated when there is no contention during the execution of an operation (i.e., interval contention). In the absence of contention, a process completes its operations by accessing only read/write registers[2]. Algorithms which satisfy the contention-awareness property have been previously presented in [3,21,22,27].
- The construction is *generous*[3] with respect to *obstruction-freedom*. This means that each process is able to complete its pending operations on all the k objects each time all the other processes hold still long enough. That is, if once and again all the processes except one hold still long enough, then all the k objects, and not just ℓ objects, are guaranteed to always progress.

This new universal construction is consequently called a *wait-free contention-aware obstruction-free-generous (k, ℓ)-universal construction*. Differently, the universal construction presented in [12] is a $(k, 1)$-universal construction and is neither contention-aware, nor generous with respect to obstruction-freedom. Moreover, this construction suffers from the following limitations: (a) it does not satisfy wait-freedom progress, but only non-blocking progress (i.e., infinite progress is guaranteed for only one process); (b) in some scenarios, an operation that has been invoked by a process can (incorrectly) be applied twice, instead of just once; and (c) the last state of the copies (one per process) of an object on which no more operations are being executed can be different at distinct processes. While issue (b) can be fixed (see [25]), we do not see how to modify the construction from [12] to overcome drawback (c).

When considering the special case $k = \ell = 1$, Herlihy's construction is wait-free $(1, 1)$-universal [14], but differently from ours, it does not satisfy the contention-awareness property.

To ensure the progress of at least ℓ of the k implemented objects, the proposed construction uses a new synchronization object, that we call (k, ℓ)-simultaneous consensus object, which is a simple generalization of the k-simultaneous consensus object. This object type is such that its $(k, 1)$ instance is equivalent to k-simultaneous consensus, while its (k, k) instance is equivalent to consensus. Thus, when added to the basic $\mathcal{ARW}_{n,n-1}[\emptyset]$ system model, (k, ℓ)-simultaneous consensus objects add computational power. The paper shows that (k, ℓ)-simultaneous consensus objects are both *necessary and sufficient* to ensure that at least ℓ among the k objects progress forever.

From a software engineering point of view, the proposed (k, ℓ)-universal construction is built in a modular way. First a non-blocking $(k, 1)$-universal construction is designed, using k-simultaneous consensus objects and atomic registers. Interestingly, its design principles are different from the other universal constructions we are aware of. Then, this basic construction is extended to obtain a contention-aware $(k, 1)$-universal

---

[2] Let us recall that, in *worst case* scenarios, hardware operations such as compare&swap() can be 1000× more expensive that read or write.

[3] *Generosity* is a general notion. Intuitively, an algorithm is *generous* with respect to a given condition C, if, whenever C is satisfied, the algorithm does more than what it is required to do in normal circumstances. The condition C specifies the "exceptional" circumstances under which the algorithm does "more". These "exceptional" circumstances depend on the underlying system behavior.

construction, and then a wait-free contention-aware $(k, 1)$-universal construction. Finally, assuming that the system is enriched with (k, ℓ)-simultaneous consensus objects, $1 \leq \ell \leq k$, instead of k-simultaneous consensus objects, we obtain a contention-aware wait-free (k, ℓ)-universal construction. During the modular construction, we make sure that the universal construction implemented at each stage is also generous with respect to obstruction-freedom.

Roadmap The paper is made up of 5 sections. Section 2 presents the computation models and the specific objects used in the paper. Section 3 presents a non-blocking $(k, 1)$-universal construction. Then Section 4 extends it so that it satisfies contention-awareness, wait-freedom, and the progress of at least ℓ out of the k constructed objects. This section shows also that (k, ℓ)-simultaneous consensus objects are necessary and sufficient for the design of (k, ℓ)-universal constructions. Due to page limitation, (1) all proofs, (2) the presentation of an interesting simple variant of the general universal construction which is an obstruction-free $(1, 1)$-universal construction based on atomic registers only, and (3) definitions and notions which can be used to establish a (k, ℓ)-universality theory, are presented in [25].

2   Basic and Enriched Models, and Wait-Free Linearizable Implementation

2.1   Basic Read/Write Model and Enriched Model

The basic model presented in the introduction is the wait-free asynchronous read/write model denoted $\mathcal{ARW}_{n,n-1}[\emptyset]$ (see also [4,20,23]). The processes are denoted $p_1, ..., p_n$. Considering a run, a process is *faulty* if it crashes during the run, otherwise it is *correct*.

In addition to atomic read/write registers [18], two other types of objects are used. The first type does not add computational power, but provides processes with a higher abstraction level. The other type adds computational power to the basic system model $\mathcal{ARW}_{n,n-1}[\emptyset]$.

Adopt-commit object. The adopt-commit object has been introduced in [11]. An adopt-commit object is a one-shot object that provides the processes with a single operation denoted propose(). This operation takes a value as an input parameter, and returns a pair (tag, v). The behavior of an adopt-commit object is formally defined as follows:

- Validity.
  - Result domain. Any returned pair (tag, v) is such that (a) v has been proposed by a process and (b) $tag \in \{commit, adopt\}$.
  - No-conflicting values. If a process p_i invokes propose(v) and returns before any other process p_j has invoked propose(v') with $v' \neq v$, then only the pair $(commit, v)$ can be returned.
- Agreement. If a process returns $(commit, v)$, the only pairs that can be returned are $(commit, v)$ and $(adopt, v)$.
- Termination. An invocation of propose() by a correct process always terminates.

Let us notice that it follows from the "no-conflicting values" property that, if a single value v is proposed, then only the pair $(commit, v)$ can be returned. Adopt-commit objects can be wait-free implemented in $\mathcal{ARW}_{n,n-1}[\emptyset]$ (e.g., [11,23]). Hence, they provide processes only with a higher abstraction level than read/write registers.

k-Simultaneous consensus object. A k-simultaneous consensus (k-SC) object is a one-shot object that provides the processes with a single operation denoted propose(). This operation takes as input parameter a vector of size k, each entry containing a value, and returns a pair (x, v). The behavior of a k-simultaneous consensus object is formally defined as follows:

- Validity. Any pair (x, v) that is returned by a process p_i is such that (a) $1 \leq x \leq k$ and (b) v has been proposed by a process in the x-th entry of its input vector before p_i decides.
- Agreement. If a process returns (x, v) and another process returns (y, v'), and $x = y$, then $v = v'$.
- Termination. An invocation of propose() by a correct process always terminates.

Let $\mathcal{ARW}_{n,n-1}[k\text{-}SC]$ denote $\mathcal{ARW}_{n,n-1}[\emptyset]$ enriched with k-SC objects. It is shown in [1] that a k-SC object and a k-set agreement (k-SA) object are wait-free equivalent in $\mathcal{ARW}_{n,n-1}[\emptyset]$. This means that a k-SC object can be built in $\mathcal{ARW}_{n,n-1}[k\text{-}SA]$, and a k-SA object can be built in $\mathcal{ARW}_{n,n-1}[k\text{-}SC]$.

2.2   Correct Object Implementation

Let us consider n processes that access k concurrent objects, each defined by a deterministic sequential specification. The sequence of operations that p_i wants to apply to an object m, $1 \leq m \leq k$, is stored in the local infinite list $my\_list_i[m]$, which can be defined statically or dynamically (in that case, the next operation issued by a process p_i on an object m, can be determined from p_i's view of the global state). It is assumed that the processes are well-formed: no process invokes a new operation on an object m before its previous operation on m has terminated.

Wait-free linearizable implementation. An implementation of an object m by n processes is wait-free linearizable if it satisfies the following properties.

- Validity. If an operation op is executed on object m, then $op \in \cup_{1 \leq i \leq n} my\_list_i[m]$, and all the operations of $my\_list_i[m]$ which precede op have been applied to object m.
- No-duplication. Any operation op on object m invoked by a process is applied at most once to m. We assume that all the invoked operations are unique.
- Consistency. Any n-process execution produced by the implementation is linearizable [17].
- Termination (wait-freedom). If a process does not crash, it executes an infinite number of operations on at least one object.

Weaker progress conditions In some cases, the following two weaker progress conditions are considered.

- The *non-blocking* progress condition [17] guarantees that there is at least one process that executes an infinite number of operations on at least one object.
- The *obstruction-freedom* progress condition [15] guarantees that any correct process can complete its operations if it executes in isolation for a long enough period (long enough period during which the other processes stop progressing).

3   A New Non-blocking k-Universal Construction

As mentioned in the Introduction, the construction is done incrementally. In this section, we present and prove the correctness of a non-blocking k-universal construction, based on new design principles (as far as we know). This construction is built in the enriched model $\mathcal{ARW}_{n,n-1}[k\text{-}SC]$. In Section 4, we extend the construction, without requiring additional computational power, to obtain the contention-awareness property, and the wait-freedom progress condition (i.e., *each* correct process can always execute and completes its operations on any object that progresses forever). Then (k, ℓ)-SC objects are introduced (which are a natural generalization of k-SC objects), and are used to design a (k, ℓ)-universal construction which ensures that least ℓ objects progress forever. In Section 4, we also show that (k, ℓ)-SC objects are necessary and sufficient to obtain a (k, ℓ)-universal construction.

3.1   A new Non-blocking k-Universal Construction: Data Structures

The following objects are used by the construction. Identifiers with upper case letters are used for shared objects, while identifiers with lower case letters are used for local variables.

Shared objects

- $kSC[1..]$: infinite list of of k-simultaneous consensus objects; $kSC[r]$ is the object used at round r.
- $AC[1..][1..k]$: infinite list of vectors of k adopt-commit objects; $AC[r][m]$ is the adopt-commit object associated with the object m at round r.
- $GSTATE[1..n]$ is an array of SWMR (single-writer/multi-readers) atomic registers; $GSTATE[i]$ can be written only by p_i. Moreover, the register $GSTATE[i]$ is made up of an array with one entry per object, such that $GSTATE[i][m]$ is the sequence of operations that have been applied to the object m, as currently know by p_i; it is initialized to ϵ (the empty sequence).

Local variables at process p_i

- r_i: local round number (initialized to 0).
- $g\_state_i[1..n]$: array used to save the values read from $GSTATE[1..n]$.
- $oper_i[1..k]$: vector such that $oper_i[m]$ contains the operation that p_i is proposing to a k-SC object for the object m (as we will see in the algorithm, this operation was not necessarily issued by p_i).

- $my\_op_i[1..k]$: vector of operations such that $my\_op_i[m]$ is the last operation that p_i wants to apply to the object m (hence $my\_op_i[m] \in my\_list_i[m]$).
- $\ell\_hist_i[1..k]$: vector with one entry per object, such that $\ell\_hist_i[m]$ is the sequence of operations defining the history of object m, as known by p_i. Each $\ell\_hist_i[m]$ is initialized to ϵ. The function append() is used to add an element at the end of a sequence $\ell\_hist_i[m]$.
- $tag_i[1..k]$ and $ac\_op_i[1..k]$: arrays that, for each object m, are used to save the pairs $(tag, operation)$ returned by the invocation of $AC[r][m]$ of current round r.
- $output_i[1..k]$: vector such that $output_i[m]$ contains the result of the last operation invoked by p_i on the object m (this is the operation saved in $my\_op_i[m]$).

Without loss of generality, it is assumed that each object operation returns a result, which can be "ok" when there is no object-dependent result to be returned (as with the stack operation push() or the queue operation enqueue()).

3.2 Eliminating Full Object Histories

For each process p_i and object m, the universal construction uses a shared register $GSTATE[i][m]$ to remember the sequence of all the operations that have been successfully applied to object m, as currently known to p_i. We have chosen this implementation mainly due to its simplicity. While it is space inefficient, it can be improved as follows.

- Recall that we have assumed that all the operations are unique. This can be easily implemented locally, where each process attaches a unique (local) sequence number plus its id to each operation. The (local) sequence number attached can be the number of operations the process has invoked on the object so far. Now, instead of remembering (by each process) for each object m its full history, it is sufficient that each process p_i computes and remembers only the last state of m, denoted $\ell\_state_i[m]$, plus the sequence number of the last operation successfully applied to m by each process.
- As far as the function compute_output(op, h) used at line 9 and line 20 is concerned, we have the following, where $OUTPUT[1..n]$ is an array made up of one atomic register per process. Immediately after line 18, a process p_i executes the following statements, which replace lines 19-23.

> $output_i[m] \leftarrow$ compute_output($ac\_op_i[m], \ell\_state_i[m]$);
> **let** p_j **be** the process that invoked $ac\_op_i[m]$;
> **if** $(i = j)$ **then** lines 21-22
> > **else** $OUTPUT[j] \leftarrow output_i[m]$
>
> **end if**.

When executed by a process p_j, line 9 is replaced by $output_j[m] \leftarrow OUTPUT[j]$.

It is easy to see that these statements implement a simple helping mechanism that allow processes, which invoke append() at line 18, to pre-compute the operation results for the processes that should invoke compute_output(op, h) at line 9. Consequently, the distributed universal construction can be easily modified to use this more space efficient representation instead of the "full history" representation.

3.3 A New Non-blocking $(k, 1)$-Universal Construction: Algorithm

To simplify the presentation, it is assumed that each operation invocation is unique. This can be easily realized by associating an identity (process id, sequence number) with each operation invocation. In the following, the term "operation" is used as an abbreviation for "operation execution".

The function next() is used by a process p_i to access the sequence of operations $my\_list_i[m]$. The x-th invocation of $my\_list_i[m]$.next() returns the x-th element of this list.

Initialization The algorithm implementing the k-universal construction is presented in Figure 1. For each object $m \in \{1, ..., k\}$, a process p_i initializes both the variables $my\_op_i[m]$ and $oper_i[m]$ to the first operation that it wants to apply to m. Process p_i then enters an infinite loop.

Repeat loop: using the round r objects $kSC[r]$ and $AC[r]$ (lines 1-4) After it has increased its round number, a process p_i invokes the k-simultaneous consensus object $kSC[r]$ to which it proposes the operation vector $oper_i[1..n]$, and from which it obtains the pair denoted (ksc_obj, ksc_op); ksc_op is an operation proposed by some process for the object ksc_obj (line 2). Process p_i then invokes the adopt-commit object $AC[r][ksc\_obj]$ to which it proposes the operation output by $kSC[r]$ for the object ksc_op (line 3). Finally, for all the other objects $m \neq ksc\_obj$, p_i invokes the adopt-commit object $AC[r][m]$ to which it proposes $oper_i[m]$ (line 4). As already indicated, the tags and the commands defined by the vector of pairs output by the adopt-commit objects $AC[r]$ are saved in the vectors $tag_i[1..k]$ and $ac\_op_i[1..k]$, respectively. (While expressed differently, these four lines are the only part which is common to this construction and the one presented in [12].)

The aim of these lines is to implement a filtering mechanism such that (a) for each object, at most one operation can be committed at some processes, and (b) there is at least one object for which an operation is committed at some processes.

Repeat loop: returning local results (lines 5-13) Having used the additional power supplied by $kSC[r]$, a process p_i first obtains asynchronously the value of $GSTATE[1..n]$ (line 5) to learn an "as recent as possible" consistent global state, which is saved in $g\_state_i[1..n]$. Then, for each object m (lines 6-13), p_i computes the maximal local history of the object m which contains $\ell\_hist_i[m]$ (line 7). (Let us notice that $g\_state_i[i][m]$ is $\ell\_hist_i[m]$.) This corresponds to the longest history in the n histories $g\_state_i[1][m], ..., g\_state_i[n][m]$ which contains $\ell\_hist_i[m]$. If there are several longest histories, they all are equal as we will see in the proof. If the last operation it has issued on m, namely $my\_op_i[m]$, belongs to this history (line 8), some process has executed this operation on its local copy of m. Process p_i computes then the corresponding output (line 9), locally returns the triple $(m, my\_op_i[m], output_i[m])$ (line 10), and defines its next local operation to apply to the object m (line 11).

The function compute_output(op, h) (used at lines 9 and 20) computes the result returned by op applied to the state of the corresponding object m (this state is captured by the prefix of the history h of m ending just before the operation op).

Repeat loop: trying to progress on machines (lines 14-29) Then, for each object m, $1 \leq m \leq k$, p_i considers the operation $ac\_op_i[m]$. If this operation belongs to its

for each $m \in \{1, \ldots, k\}$ **do**
  $my\_op_i[m] \leftarrow my\_list_i[m].\text{next}(); oper_i[m] \leftarrow my\_op_i[m]$ **end for**.

repeat forever
(1) $r_i \leftarrow r_i + 1$;
(2) $(ksc\_obj, ksc\_op) \leftarrow kSC[r_i].\text{propose}(oper_i[1..k])$;
(3) $(tag_i[ksc\_obj], ac\_op_i[ksc\_obj]) \leftarrow AC[r_i][ksc\_obj].\text{propose}(ksc\_op)$;
(4) **for each** $m \in \{1, \ldots, k\} \setminus \{ksc\_obj\}$ **do**
       $(tag_i[m], ac\_op_i[m]) \leftarrow AC[r_i][m].\text{propose}(oper_i[m])$ **end for**;

(5) **for each** $j \in \{1, \ldots, n\}$ **do** $g\_state_i[j] \leftarrow GSTATE[j]$ **end for**;
     % the read of each $GSTATE[j]$ is atomic %
(6) **for each** $m \in \{1, \ldots, k\}$ **do**
(7) $\ell\_hist_i[m] \leftarrow$ longest history of $g\_state_i[1..n][m]$ containing $\ell\_hist_i[m]$;
(8)   **if** $(my\_op_i[m] \in \ell\_hist_i[m])$ % my operation was completed %
(9)   **then** $output_i[m] \leftarrow \text{compute\_output}(my\_op_i[m], \ell\_hist_i[m])$;
(10)   return $\{(m, my\_op_i[m], output_i[m])\}$ to the upper layer;
(11) $my\_op_i[m] \leftarrow my\_list[m].\text{next}()$
(12)   **end if**
(13) **end for**;

(14) $res \leftarrow \emptyset$;
(15) **for each** $m \in \{1, \ldots, k\}$ **do**
(16)   **if** $(ac\_op_i[m] \notin \ell\_hist_i[m])$ % operation was not completed %
(17)   **then if** $(tag_i[m] = commit)$ % complete the operation %
(18)   **then** $\ell\_hist_i[m] \leftarrow \ell\_hist_i[m].\text{append}(ac\_op_i[m])$;
(19)   **if** $(ac\_op_i[m] = my\_op_i[m])$ % my operation was completed %
(20)   **then** $output_i[m] \leftarrow \text{compute\_output}(ac\_op_i[m], \ell\_hist_i[m])$;
(21) $res \leftarrow res \cup \{(m, my\_op_i[m], output_i[m])\}$;
(22) $my\_op_i[m] \leftarrow my\_list[m].\text{next}()$
(23)   **end if**;
(24) $oper_i[m] \leftarrow my\_op_i[m]$
(25)   **else** $oper_i[m] \leftarrow ac\_op_i[m]$ % $tag_i[m] = adopt$ %
(26)   **end if**
(27)   **else** $oper_i[m] \leftarrow my\_op_i[m]$ % $ac\_op_i[m] \in \ell\_hist_i[m]$ %
(28)   **end if**
(29) **end for**;

(30) $GSTATE[i] \leftarrow \ell\_hist_i[1..k]$;   % globally update my current view %
(31) **if** $(res \neq \emptyset)$ **then** return res to the upper layer **end if**
end repeat.

Fig. 1. Basic Non-Blocking Generalized $(k, 1)$-Universal Construction (code for p_i)

local history $\ell\_hist_i[m]$ (the predicate of line 16 is then false), it has already been locally applied; p_i consequently assigns $my\_op_i[m]$ to $oper_i[m]$, where is saved its next operation on the object m (line 27).

If $ac\_op_i[m] \notin \ell\_hist_i[m]$ (line 16), the behavior of p_i depends on the fact that the tag of $ac\_op_i[m]$ is *commit* or *adopt*. If the tag is *adopt* (the predicate of line 17 is then false), p_i defines $ac\_op_i[m]$ as the next operation it will propose for the object m, which is saved in $oper_i[m]$ (line 25): it "adopts" $ac\_op_i[m]$. If the tag is *commit* (line 17), p_i adds (applies) the operation $ac\_op_i[m]$ to its local history (line 18). Moreover, if $ac\_op_i[m]$ has been issued by p_i itself (i.e., $ac\_op_i[m] = my\_op_i[m]$, line 19), p_i computes the result locally returned by $ac\_op_i[m]$ (line 20), adds this result to the set of results res (line 21), defines its next local operation to apply to the object m (line 22). Finally, p_i assigns $my\_op_i[m]$ to $oper_i[m]$ (line 24).

Repeat loop: making public its progress (lines 30-31) Finally, p_i makes public its current local histories (one per object) by writing them in $GSTATE[i]$ (line 30), and returns local results if any (line 31). It then progresses to the next round.

Theorem 1. *The algorithm of Figure 1 is a non-blocking linearizable $(k, 1)$-universal construction.*

Generosity wrt obstruction-freedom We observe that the construction of Figure 1 is also obstruction-free (k, k)-universal. That is, the construction guarantees that each process will be able to complete all its pending operations in a finite number of steps, if all the other processes "hold still" long enough. Thus, if once in a while all the processes except one "hold still" long enough, then all the k objects (and not "at least one") are guaranteed to always make progress.

4   A Contention-Aware Wait-Free (k, ℓ)-Universal Construction

4.1   A Contention-Aware Non-blocking k-Universal Construction

Contention-aware universal construction A *contention-aware* universal construction (or object) is a construction (object) in which the overhead introduced by synchronization primitives which are different from atomic read/write registers (like k-SC objects) is eliminated in executions when there is no contention. When a process invokes an operation on a contention-aware universal construction (object), it must be able to complete its operation by accessing only read/write registers in the absence of contention. Using other synchronization primitives is permitted only when there is contention. (This notion is close but different from the notion of *contention-sensitiveness* introduced in [27].)

A contention-aware non-blocking $(k, 1)$-universal construction A contention-aware $(k, 1)$-universal construction is presented in Figure 2. At each round r, it uses two adopt-commit objects per constructed object m, namely $AC[2r_i - 1][m]$ and $AC[2r_i][m]$, instead of a single one. When considering the basic construction of Figure 1, the new lines are prefixed by N, while modified lines are postfixed by M.

A process p_i first invokes, for each object m, the adopt-commit object $AC[2r_i - 1][m]$ to which it proposes $oper_i[m]$ (new line N1). Its behavior depends then on the

```
for each m ∈ {1, ..., k} do
      my_op_i[m] ← my_list_i[m].next(); oper_i[m] ← my_op_i[m] end for.

repeat forever
(1)    r_i ← r_i + 1;
(N1)   for each m ∈ {1, ..., k} do
            (tag_i[m], ac_op_i[m]) ← AC[2r_i − 1][m].propose(oper_i[m]) end for;
(N2)   if (∃ m ∈ {1, ..., k} : tag_i[m] = adopt) then
(2M)      (ksc_obj, ksc_op) ← kSC[r_i].propose(ac_op_i[1..k]);
(3M)      (tag_i[ksc_obj], ac_op_i[ksc_obj]) ← AC[2r_i][ksc_obj].propose(ksc_op);
(4M)      for each m ∈ {1, ..., k} \ {ksc_obj} do
               (tag_i[m], ac_op_i[m]) ← AC[2r_i][m].propose(ac_op_i[m]) end for
(N3)   end if;
lines 5- 31 of the construction of Figure 1
end repeat.
```

Fig. 2. Contention-aware Non-Blocking $(k, 1)$-Universal Construction (code for p_i)

number of objects for which it has received the tag *commit*. If it has obtained the tag *commit* for all the objects m (the test of the new line N2 is then false), p_i proceeds directly to the code defined by the lines 5- 31 of the basic construction described in Figure 1, thereby skipping the invocation of the synchronization object $kSC[r]$ associated with round r.

Otherwise, the test of the new line N2 is true and there is at least one object for which p_i has received the tag *adopt*. This means that there is contention. In this case, the behavior of p_i is similar to the lines 2-4 of the basic algorithm where, at lines 2 and 4, the input parameter $oper_i[m]$ is replaced by the value of $ac\_op_i[m]$ obtained at line N1 (the corresponding lines are denoted 2M and 4M). Moreover, at line 3, r_i is replaced by $2r_i$ (new line 3M).

Interestingly, for the case of $k = 1$, the above universal construction is the first known *contention-aware* $(1, 1)$-universal construction.

Theorem 2. *The algorithm of Figure 2 is a non-blocking contention-aware $(k, 1)$-universal construction.*

It is possible to still reduce the number of uses of underlying k-SC synchronization objects. by replacing the lines N1-N3 in Figure 2 as described in Figure 3. There is one modified line (N2M) and three new lines (NN1, NN2, and NN3). More precisely, if after it has used the adopt-commit objects $AC[2r_i − 1][m]$, for each constructed object m, p_i has received only tags *adopt* (modified line N2M), it executes the lines 2M, 3, and 4M, as in basic contention aware construction of Figure 2. Differently, if it has received the tag *commit* for at least one constructed object, it invokes $AC[2r][m]$ for all the objects m for which it has received the tag *adopt* (new lines NN1-NN3).

4.2 On the Process Side: From Non-blocking to Wait-Freedom

The aim here is to ensure that each correct process executes an infinite number of operations on each object that progresses forever. As far as the progress of objects is

(N1)	**for each** $m \in \{1, \ldots, k\}$ **do**
	$(tag_i[m], ac\_op_i[m]) \leftarrow AC[2r_i - 1][m].\text{propose}(oper_i[m])$ **end for**;
(N2M)	**if** $(\forall\, m \in \{1, \ldots, k\} : tag_i[m] = adopt)$ % $\forall\, m$ replaces $\exists\, m$%
(2M)	**then** $(ksc\_obj, ksc\_op) \leftarrow kSC[r_i].\text{propose}(ac\_op_i[1..k])$;
(3)	$(tag_i[ksc\_obj], ac\_op_i[ksc\_obj]) \leftarrow AC[2r_i][ksc\_obj].\text{propose}(ksc\_op)$;
(4M)	**for each** $m \in \{1, \ldots, k\} \setminus \{ksc\_obj\}$ **do**
	$(tag_i[m], ac\_op_i[m]) \leftarrow AC[2r_i][m].\text{propose}(ac\_op_i[m])$ **end for**
(NN1)	**else for each** $m \in \{1, \ldots, k\}$ **do**
(NN2)	**if** $(tag_i[m] = adopt)$ **then**
	$(tag_i[m], ac\_op_i[m]) \leftarrow AC[2r_i][m].\text{propose}(ac\_op_i[m])$ **end if**
(NN3)	**end for**
(N3)	**end if**.

Fig. 3. Efficient Contention-aware Non-Blocking $(k, 1)$-Universal Construction (code for p_i)

concerned, it is important to notice that it is possible that, in a given execution, several objects progress forever.

Going from non-blocking to wait-freedom requires to add a helping mechanism to the basic non-blocking construction. To that end, the following array of atomic registers is introduced.

– $LAST\_OP[1..n, 1..m]$: matrix of atomic SWMR (single-writer/multi-readers) registers such that $LAST\_OP[i, m]$ contains the last operation of my_list_i invoked by p_i. Initialized to $\bot$, such a register is updated each time p_i invokes $my\_list_i.\text{next}()$ (initialization, line 11n and line 22). So, we assume that $LAST\_OP[i, m]$ is implicitly updated by p_i when it invokes the function next().

Then, for each object m, the lines 24 and 27 where is defined $oper_i[m]$ (namely, the proposal for the constructed object m submitted by p_i to the next k-SC object) are replaced by the following lines ($|s|$ denotes the size of the sequence s).

(L1) $j \leftarrow |\ell\_hist_i[m]| \bmod n + 1$; $next\_prop\_m \leftarrow LAST\_OP[j, m]$;
(L2) **if** $next\_prop\_m \notin (\{\bot\} \cup \ell\_hist_i[m])$
(L3)   **then** $oper_i[m] \leftarrow next\_prop\_m$
(L4)   **else** $oper_i[m] \leftarrow my\_op_i[m]$
(L5) **end if**.

This helping mechanism is close to the one proposed in [14]. It uses, for each object m, a simple round-robin technique on the process identities, computed from the current state of m as known by p_i, i.e., from $\ell\_hist_i[m]$. More precisely, the helping mechanism uses the number of operations applied so far to m (to p_i's knowledge) in order to help the process p_j such that $j = |\ell\_hist_i[m]| \bmod n + 1$ (line L1). To that end, p_i proposes the last operation issued by p_j on m (line L3) if (a) there is such an operation, and (b) this operation has not yet been appended to its local history of m (predicate of line L2). This operation has been registered in $LAST\_OP[j, m]$ when p_j executed its last invocation of $my\_list_j[m].\text{next}()$. If the predicate of line L2 is not satisfied, p_i proceed as in the basic algorithm (line L4).

Theorem 3. *When replacing the lines* 24 *and* 27 *by lines* L1-L5, *the algorithms of Figure* 1 *and Figure* 2 *define a wait-free contention-aware linearizable* $(k, 1)$-*universal construction.*

Let us remark that requiring wait-freedom only for a subset of correct processes, or only for a subset of objects that progress forever is not interesting, as wait-freedom for both (a) all correct processes, and (b) all the objects that progress forever, does not require additional computing power.

4.3   On the Object Side: From One to ℓ Objects That Always Progress

Definition: (k, ℓ)-*Simultaneous consensus* Let (k, ℓ)-simultaneous consensus (in short (k, ℓ)-SC), $1 \leq \ell \leq k$, be a strengthened form of k-simultaneous consensus where (instead of a single pair) a process decides on ℓ pairs $(x_1, v_1), ..., (x_\ell, v_\ell)$ (all different in their first component). The agreement property is the same as for a k-SC object, namely, if (x, v) and (x, v') are pairs decided by two processes, then $v = v'$.

Notations Let (k, ℓ)-UC be any algorithm implementing the k-universal construction where at least ℓ objects always progress[4]. Let $\mathcal{ARW}_{n,n-1}[(k, \ell)\text{-}SC]$ be $\mathcal{ARW}_{n,n-1}[\emptyset]$ enriched with (k, ℓ)-SC objects, and $\mathcal{ARW}_{n,n-1}[(k, \ell)\text{-}UC]$ be $\mathcal{ARW}_{n,n-1}[\emptyset]$ enriched with a (k, ℓ)-UC algorithm.

A contention-aware wait-free (k, ℓ)*universal construction* A contention-aware wait-free (k, ℓ)-UC algorithm can be implemented on top of $\mathcal{ARW}_{n,n-1}[(k, \ell)\text{-}SC]$ as follows. This algorithm is the algorithm of Figure 2, where lines 24 and 27 are replaced by the lines L1-L5 introduced in Section 4.2, and where the lines 2M, 3M, and 4M, are modified as follows (no other line is added, suppressed, or modified).

- Line 2M: the k-simultaneous consensus objects are replaced by (k, ℓ)-simultaneous consensus objects, Hence, the result returned to a process is now a set of ℓ pairs whose first components are all distinct. It is denoted $\{(ksc\_obj_1, ksc\_op_1), ..., (ksc\_obj_\ell, ksc\_op_\ell)\}$. Let L be the corresponding set of ℓ different objects, i.e., $L = \{ksc\_obj_1, ..., ksc\_obj_\ell\}$. As already indicated, two different processes can be returned different sets of ℓ pairs.
- Line 3M: process p_i executes this line for each object $m \in L$. These ℓ invocations of the adopt-commit object (i.e., $AC[2r_i][ksc\_obj_x]$.propose(ksc_op_x), $1 \leq x \leq \ell$) can be executed in parallel, which means in any order. Let us notice that if several processes invokes $AC[2r_i][ksc\_obj_x]$.propose() on the same object ksc_obj_x, they invoke it with the same operation ksc_op_x.
- Line 4M: $AC[2r_i][m]$.propose($oper_i[m]$) is invoked only for the remaining objects, i.e., the objects m such that $m \in \{1, ..., k\} \setminus L$. As in the algorithm of Figure 2, the important point is that a process invokes $AC[2r_i][ksc\_obj_x]$.propose() first on the set L of the objects output by the (k, ℓ)-SC object associated with the current round, and only after invoke it on the other objects.

---

[4] It is possible to express (k, ℓ)-UC as an object accessed by appropriate operations. This is not done here because such an object formulation would be complicated without providing us with more insight on the question we are interested in.

Theorem 4. *With respect to the model $\mathcal{ARW}_{n,n-1}[\emptyset]$, (k,ℓ)-UC and (k,ℓ)-SC have the same computational power: (a) a (k,ℓ)-UC algorithm can be wait-free implemented in $\mathcal{ARW}_{n,n-1}[(k,\ell)$-SC$]$, and, reciprocally, (b) a (k,ℓ)-SC object can be wait-free built in $\mathcal{ARW}_{n,n-1}[(k,\ell)$-UC$]$.*

This theorem shows that (k,ℓ)-SC objects are both necessary and sufficient to ensure that at least ℓ objects always progress in a set of k objects. Let us remark that this is independent from the fact that the implementation of the k-universal construction is non-blocking or wait-free (going from non-blocking to wait-freedom requires the addition of a helping mechanism, but does not require additional computational power).

5 Conclusion

Our main objective was to build a universal construction for any set of k objects, each defined by a sequential specification, where at least ℓ of these k objects are guaranteed to progress forever. To that end, we have introduced a new object type, called (k,ℓ)-simultaneous consensus ((k,ℓ)-SC), and have shown that this object is both *necessary and sufficient* (hence optimal and universal) when one has to implement such a universal construction. We have related the notions of obstruction-freedom, non-blocking, and contention-awareness for the implementation of k-universal constructions. The paper has also introduced a general notion of *algorithm generosity*, which captures a property implicitly addressed in other contexts. The constructions presented in the paper can be seen as innovative generalizations of the universality notions introduced in [12,14]. More specifically, we have presented the following suite of constructions:

- A contention-aware construction, based on k-SC objects and atomic registers, which is both obstruction-free (k,k)-universal and wait-free k-universal (Section 3).
- A contention-aware (k,ℓ)-universal construction based on (k,ℓ)-SC objects which is both obstruction-free (k,k)-universal and wait-free (k,ℓ)-universal (Section 4).

Finally, a simple obstruction-free $(1,1)$-universal construction based on atomic registers only, and elements for a theory of (k,ℓ)-universality can be found in [25].

References

1. Afek, Y., Gafni, E., Rajsbaum, S., Raynal, M., Travers, C.: The k-simultaneous consensus problem. Distributed Computing 22(3), 185–195 (2010)
2. Anderson, J.H., Moir, M.: Universal constructions for large objects. IEEE Transactions of Parallel and Distributed Systems 10(12), 1317–1332 (1999)
3. Attiya, H., Guerraoui, R., Hendler, D., Kutnetsov, P.: The complexity of obstruction-free implementations. Journal of the ACM 56(4), Article 24, 33 (2009)
4. Attiya, H., Welch, J.L.: Distributed computing: Fundamentals, simulations and advanced topics, 2nd edn., p. 414. Wiley Interscience (2004) ISBN 0-471-45324-2
5. Borowsky, E., Gafni, E., Generalized, F.L.P.: impossibility results for t-resilient asynchronous computations. In: Proc. 25th ACM Symposium on Theory of Computing (STOC 1993), pp. 91–100. ACM Press (1993)

6. Bouzid, Z., Travers, C.: Simultaneous consensus is harder than set agreement in message-passing. In: Proc. 33rd Int'l IEEE Conference on Distributed Computing Systems (ICDCS 2013), pp. 611–620. IEEE Press (2013)
7. Chaudhuri, S.: More choices allow more faults: Set consensus problems in totally asynchronous systems. Information and Computation 105(1), 132–158 (1993)
8. Ellen, F., Fatourou, P., Kosmas, E., Milani, A., Travers, C.: Universal construction that ensure disjoint-access parallelism and wait-freedom. In: Proc. 31th ACM Symposium on Principles of Distributed Computing (PODC), pp. 115–124. ACM Press (2012)
9. Fatourou, P., Kallimanis, N.D.: A highly-efficient wait-free universal construction. In: Proc. 23th ACM Symposium on Parallel Algorithms and Architectures (SPAA), pp. 325–334. ACM Press (2012)
10. Fischer, M.J., Lynch, N.A., Paterson, M.S.: Impossibility of distributed consensus with one faulty process. Journal of the ACM 32(2), 374–382 (1985)
11. Gafni, E.: Round-by-round fault detectors: unifying synchrony and asynchrony. In: Proc. 17th ACM Symp. on Principles of Distr. Computing (PODC), pp. 143–152. ACM Press (1998)
12. Gafni, E., Guerraoui, R.: Generalized universality. In: Katoen, J.-P., König, B. (eds.) CON-CUR 2011. LNCS, vol. 6901, pp. 17–27. Springer, Heidelberg (2011)
13. Guerraoui, R., Kapalka, M., Kouznetsov, P.: The weakest failure detectors to boost obstruction-freedom. Distributed Computing 20(6), 415–433 (2008)
14. Herlihy, M.P.: Wait-free synchronization. ACM Transactions on Programming Languages and Systems 13(1), 124–149 (1991)
15. Herlihy, M.P., Luchangco, V., Moir, M.: Obstruction-free synchronization: double-ended queues as an example. In: Proc. 23th Int'l IEEE Conference on Distributed Computing Systems (ICDCS 2003), pp. 522–529. IEEE Press (2003)
16. Herlihy, M.P., Shavit, N.: The topological structure of asynchronous computability. Journal of the ACM 46(6), 858–923 (1999)
17. Herlihy, M.P., Wing, J.M.: Linearizability: A correctness condition for concurrent objects. ACM Transactions on Programming Languages and Systems 12(3), 463–492 (1990)
18. Lamport, L.: On inter-process communications, Part I: Basic formalism. Distributed Computing 1(2), 77–85 (1986)
19. Loui, M., Abu-Amara, H.: Memory requirements for agreement among unreliable asynchronous processes. Advances in Computing Research 4, 163–183 (1987)
20. Lynch, N.A.: Distributed algorithms, vol. 872. Morgan Kaufmann (1996)
21. Luchangco, V., Moir, M., Shavit, N.N.: On the Uncontended complexity of consensus. In: Fich, F.E. (ed.) DISC 2003. LNCS, vol. 2848, pp. 45–59. Springer, Heidelberg (2003)
22. Merritt, M., Taubenfeld, G.: Resilient consensus for infinitely many processes. In: Fich, F.E. (ed.) DISC 2003. LNCS, vol. 2848, pp. 1–15. Springer, Heidelberg (2003)
23. Raynal, M.: Concurrent programming: Algorithms, principles, and foundations, 515 p. Springer (2013) ISBN 978-3-642-32026-2
24. Raynal, M., Stainer, J.: Simultaneous consensus vs set agreement: A message-passing-sensitive hierarchy of agreement problems. In: Moscibroda, T., Rescigno, A.A. (eds.) SIROCCO 2013. LNCS, vol. 8179, pp. 298–309. Springer, Heidelberg (2013)
25. Raynal, M., Stainer, J., Taubenfeld, G.: Distributed universality. Tech Report, pages, IRISA, Université de Rennes, France (2014)
26. Saks, M., Zaharoglou, F.: Wait-free k-set agreement is impossible: the topology of public knowledge. SIAM Journal on Computing 29(5), 1449–1483 (2000)
27. Taubenfeld, G.: Contention-sensitive data structures and algorithms. In: Keidar, I. (ed.) DISC 2009. LNCS, vol. 5805, pp. 157–171. Springer, Heidelberg (2009)

A Practical Distributed Universal Construction with Unknown Participants*

Pierre Sutra, Étienne Rivière, and Pascal Felber

University of Neuchâtel, Switzerland

Abstract. Modern distributed systems employ atomic read-modify-write primitives to coordinate concurrent operations. Such primitives are typically built on top of a central server, or rely on an agreement protocol. Both approaches provide a *universal construction*, that is, a general mechanism to construct atomic and responsive objects. These two techniques are however known to be inherently costly. As a consequence, they may result in bottlenecks in applications using them for coordination. In this paper, we investigate another direction to implement a universal construction. Our idea is to delegate the implementation of the universal construction to the clients, and solely implement a distributed shared atomic memory on the servers side. The construction we propose is obstruction-free. It can be implemented in a purely asynchronous manner, and it does not assume the knowledge of the participants. It is built on top of *grafarius* and *racing* objects, two novel shared abstractions that we introduce in detail. To assess the benefits of our approach, we present a prototype implementation on top of the Cassandra data store, and compare it empirically to the Zookeeper coordination service.

1 Introduction

The management of conflicting accesses to shared data plays a key role in executing correctly and efficiently distributed applications. In general, strongly consistent operations on shared data are serialized either through a central server, or using the replicated state machine approach (e.g., with the Paxos consensus protocol [1]). These two techniques implement a wait-free universal construction, that is, a general mechanism to obtain responsive atomic objects [2]. It is however well-established that these two mechanisms are costly. This comes from the fact that in both cases a server serializes all updates emitted by the clients, creating a potential bottleneck in the system. Furthermore, central servers require human intervention to be constantly operational, and replicated state machines are known to be difficult to deploy and maintain.

In this paper, we propose to delegate the logic of strongly consistent operations to the client side, and to replace the central server/replicated state machine by a distributed shared memory. The resulting universal construction is dependable, while being conceptually simpler than state-machine-replication. Similar

---

* This work is sponsored in part by European Commission's Seventh Framework Program (FP7) under grant agreement No. 318809 (LEADS).

M.K. Aguilera et al. (Eds.): OPODIS 2014, LNCS 8878, pp. 485–500, 2014.

in spirit to [3,4], or more recently [5], we aim at bridging the gap that exists in practice between shared memory literature on universal constructions and their counterparts in distributed systems. Our approach is nonetheless different as we do not rely on a shared log to order all accesses, but instead make use of a distinct set of registers to implement each object used by the application. This leverages the intrinsic parallelism of the workload.

To achieve this, our first contribution is an obstruction-free universal construction on top of an asynchronous distributed shared memory that works even if the participants are unknown. We base our construction on two novel abstractions: a *grafarius* and a *racing*. A grafarius is close to the more common notion of ratifier, or adopt-commit object [6,7]. A racing object encapsulates the behavior of algorithms that repeatedly access new objects to progress. By combining these two abstractions, we devise an obstruction-free universal construction whose time complexity is optimal during contention-free executions.

Our previous solution makes use of an unbounded amount of memory to store the state of the object it implements. We solve this problem with a second contribution, in the form of a novel memory management mechanism. We formalize the notion of recycled objects then propose a mechanism to recycle all the base objects of our previous implementation. In a distributed system, the time complexity of the resulting universal construction during uncontended executions is constant, and it uses $O(k^2)$ shared registers, where k is the amount of processes that actually access the construction.

Our third contribution is a practical assessment of this approach. We present a prototype implementation on top of the Cassandra distributed data store [8] which we compare to Zookeeper, a state-of-the-art coordination service [9]. Several empirical results show that our system achieves results comparable to Zookeeper when clients rarely contend on shared objects, and that in addition, it exhibits a good scalability factor. For instance, with 12 servers and when the workload is completely parallel, our system is as dependable as a 3 servers deployment of Zookeeper, while being 3.2 times faster. This last property comes from the fact that our approach exhibits no bottleneck. Thus, the more it scales-out, the more likely operations that access distinct objects execute in parallel on the servers, improving performance.

Paper Outline. Section 2 surveys related work. In Section 3, we introduce the notions of grafarius and racing objects, and we present our first universal construction. We refine this construction to bound its memory footprint in Section 4. Section 5 describes a prototype implementation of our algorithm on top of Cassandra, and we evaluate it against Zookeeper. We close in Section 6. For the sake of conciseness, all the proofs are deferred to our companion technical report [10].

2   Related Work

Our work deals with the problem of transforming a sequential object into a concurrent strongly-consistent one. Such a mechanism is named in literature a *universal construction.* At core of this construction is consensus, an abstraction with which processes agree on the next state of the concurrent object. In a distributed system, the classical approach to implement consensus is the Paxos algorithm [1]. Due to the impossibility result of Fischer et al. [11], Paxos is indulgent [12]. This means that Paxos guarantees safety at all times but provides progress only under favorable circumstances. The alpha of consensus [13] captures the indulgent part of Paxos. This notion is close to the ranked-register object [14] which models the Disk Paxos algorithm of Gafni and Lamport [4].

Processes executing Paxos iteratively calls the alpha abstraction with a tuple (k, v), where k is a round number and v some (appropriately chosen) proposal value. Each such call translates the execution of a round in the original algorithm of Lamport [1]. A ratifier, or *adopt-commit*, object [6] is a one-shot object encapsulating the safety property of a round. Hence, from a high-level perspective, the alpha of consensus can be seen as successive (consistent) calls to adopt-commit objects (see [7] or [15, Fig.5]). In Section 3.3, we present the *racing* object that allows abstracting such iterative calls.

When there is no assumption on the proposed values, the result of Aspnes and Ellen [16] tells us that the solo time complexity of an adopt-commit belongs to $\Omega\left(\sqrt{\log n}/\log\log n\right)$. Surprisingly, some consensus algorithms exhibit constant solo decision time (e.g., [17]). This difference is explained by the convergence property of adopt-commit objects which requires processes to commit a value in case they all propose it. In Section 3.2, we introduce the notion of *grafarius* object. A grafarius can be seen as an adopt-commit object with a weak convergence property, namely a process has to commit its value only if it executes solo. As shown in Section 3.4, we can build an obstruction-free consensus with constant solo decision time on top of the grafarius and racing objects.

Some algorithms, e.g. [17,15] in shared-memory, or [4,14,3,5] in distributed systems, use strong synchronization primitives to implement consensus. On the contrary, our approach relies solely on a set of registers emulated by the underlying distributed system. As a consequence of this choice, our universal construction is obstruction-free. The work of Fich et al. [18] describes a practical transformation to convert an obstruction-free algorithm into a wait-free one. Jayanti et al. [19] proves an $\Omega(n)$ lower bound on the solo decision time and the space complexity of obstruction-free implementations.

During a step-contention free execution, processes do not contend on the base objects that implement the desired abstraction. The work of Attiya et al. [15] studies obstruction-free implementations that only make use of history-less primitives during step-contention free executions, but might rely on stronger ones under contention. The authors show that such implementation have $\Omega(n)$ space complexity, and that they exhibit $\Omega(\log n)$ time complexity in step-contention free executions.

The time complexity of the wait-free universal construction of Herlihy [2] is $O(n)$. Jayanti and Toueg [20] propose a variation of this construction which does not use unbounded integers. The space complexity of this last algorithm is $O(n^2)$. Attiya et al. [15] present an obstruction-free universal construction that employs an unbounded amount of memory. In Section 4.3, we describe a space-bounded universal construction that works in the case where processes participating to the construction are unknown. In a distributed system, it makes use of $O(k^2)$ registers, where k is the amount of processes that actually access the construction. To achieve this, we present a novel recycling mechanism in Section 4.2. At core of our mechanism is the observation that properly recycled grafarius objects can be concurrently accessed in different rounds.

3   The Construction

This section first introduces our system model. Then, it details the grafarius and the racing objects. Based on these two abstractions, we further depict a consensus algorithm that exhibits a constant time complexity in the contention-free case. This algorithm is our core building block to obtain an efficient universal construction. All the objects we present hereafter are depicted in the asynchronous shared-memory model, and they all support a bounded yet unknown amount of processes. These two assumptions reflect the message-passing system we target.

3.1   System Model and Notations

We consider an asynchronous message-passing system characterized by a complete communication graph where both communication and computation are asynchronous. Processes take their identities from some bounded set Π, with $n = |\Pi|$. The set Π is not accessible to processes for computation, but they may execute operations on the identities (e.g., equality tests).

During an execution, a process can fail-stop by crashing, but we assume that at most $\lceil \frac{n}{2} \rceil - 1$ such failures occur. There exists an implementation of an asynchronous shared-memory ($\mathcal{ASM}$) under such an assumption [21,22]. Consequently, we shall write all our algorithms in the $\mathcal{ASM}$ model where processes communicate by reading and writing to atomic multi-writer multi-reader (MWMR) registers.

In what follows, we detail how to implement higher level abstractions using the shared registers. Most of the objects we describe in this paper are linearizable [23]. An object is *one-shot* when a process may call one of its operations at most once. When there is no limit to the number of times a process may invoke the object's operations, the object is *long lived*. Besides, we shall be considering the following two progress conditions on the invocations and responses of operations [24]:(*Obstruction-freedom*) if at some point in time a process runs solo then eventually it returns from the invocation; and (*Wait-freedom*) a process returns from the invocation after a bounded number of steps.

In this paper, we are most interested in executions where processes rarely contend on shared objects. The canonical case of such an execution is the *solo* execution in which a single process executes computational steps. This class of execution is appropriate for one-shot objects but we need extending it for long-lived ones. To that end, we define the notion of *contention-free* execution that is an execution during which calls to the implemented shared object do not

---

Algorithm 1. Grafarius – code at process p

```
1:  Shared Variables:
2:      s                                      // A splitter object
3:      c                                      // Initially, false
4:      d                                      // Initially, ⊥
5:
6:  adoptCommit(u) :=
7:      if ¬ s.split() then
8:          c ← true
9:          if d ≠ ⊥ then
10:             return (adopt, d)
11:         d ← u
12:         return (adopt, u)
13:     d ← u
14:     if c then
15:         return (adopt, u)
16:     return (commit, u)
```

---

interleave. The *contention-free time complexity* of an algorithm is the worst case number of steps made by a process during such executions.

3.2   Grafarius

The first abstraction we employ in our construction is a shared object named *grafarius*. A grafarius is a one-shot object defined on a domain of values $\mathcal{V}$. It exports a single operation $adoptCommit(u \in \mathcal{V})$ that returns a pair $(flag, v)$, with $flag \in \{adopt, commit\}$ and $v \in \mathcal{V}$. During every history of a grafarius, and for every process p that invokes $adoptCommit(u)$, the following properties are satisfied:(*Validity*) If p adopts v, some process invoked the operation $adoptCommit(v)$ before. (*Coherence*) If p commits v, every process either adopts or commits v. (*Solo Convergence*) If p returns from its invocation before any other process invokes $adoptCommit$ then p commits u. (*Continuation*) If some process returns before p invokes $adoptCommit$, p adopts or commits a value proposed before it invokes $adoptCommit$.

The grafarius is closely related to the notion of adopt-commit object introduced by Gafni [6]. Nevertheless, the two abstractions are not comparable. On the one hand, the solo convergence property of a grafarius is weaker than the convergence property of an adopt-commit object. This avoids the lower bound $\Omega \left(\sqrt{\log n} / \log \log n \right)$ on the time complexity to execute $adoptCommit$ in $\mathcal{ASM}$ [16], while being sufficient to implement obstruction-free consensus. On the other hand, an adopt-commit object does not satisfy the continuation property, meaning that a process can return a value $(u, adopt)$ despite that such invocation follows a call which returned $(v, adopt)$. The continuation property improves convergence speed under contention. This also makes the grafarius a decidable object, which is needed by our memory management schema. We give further details regarding this last point in Section 4.

Algorithm 1 depicts a wait-free implementation of a grafarius. This algorithm makes use of a splitter object that detects a collision when two processes concurrently access the shared object. We first remind below how a splitter works, then we detail the internals of Algorithm 1.

Algorithm 2. Racing on $\mathcal{L}$ – code at process p

```
 1: Shared Variables:
 2:     L                                    // A map from Π to ℕ, initially ∅
 3:
 4: Local Variables:
 5:     F                                    // A function from ℕ to L
 6:     last                                 // Initially, 0
 7:
 8: enter() :=
 9:     L[p] ← last
10:     S ← codomain(L)
11:     let m = max(S)
12:     if last = m then
13:         last ← m + 1
14:     else
15:         last ← m
16:     return F(last)
```

The splitter object was first introduced by Lamport [25] then later formalized by Moir and Anderson [26]. A splitter is a one-shot shared object that exposes a single operation: $split()$. This operation takes no parameter and returns a value in $\{true, false\}$.[1] When a process returns $true$, we shall say that it *wins* the splitter; otherwise it *loses* the splitter. When multiple processes call $split()$, at most one receives the value $true$, and if a single process calls $split()$, this call returns $true$. Furthermore, when a process calls $split()$ after some other process returned, it necessarily loses the splitter. A splitter is implementable in a wait-free manner with atomic MWMR registers (see [26, Fig. 2] for further details).

Algorithm 1 works as follows. Upon calling $adoptCommit(u)$, a process p tries to win the splitter (line 7). If p fails, it raises the flag c to record that a collision occurred, i.e., the fact that two processes concurrently attempted to commit a value. Then, in case a decision was recorded, p adopts it; otherwise p adopts its own value (lines 9 to 12). On the other hand, if p wins the splitter, it writes its proposal u to the register d. Then, process p commits u if it detects no conflict, otherwise p adopts it (lines 14 to 16).

3.3 Notion of Racing

Many algorithms (e.g., [3,27]) repeatedly access new objects to progress. A *racing* is a long-lived object that captures such an iterative pattern. Its interface consists of a single operation $enter(p, l)$, defined on a countably infinite domain $\mathcal{L}$ of *laps*. During a history h, a process p *enters* lap l when $enter(p, l)$ occurs in h. Process p *leaves* lap l when l is the last lap entered by p and p enters a new one. The following invariant holds during every history of a racing: *(Ordering)* There exists a strict total order $\ll_h$ on the set of entered laps in h such that for every process p that enters some lap l, either(i) some process left l before p enters it, or (ii) the last lap left by p is the greatest lap smaller than l for the order $\ll_h$.

Let us consider an unbounded counter c at each process, and an indexing function F from ℕ to $\mathcal{L}$. Whenever a process p enters a new lap, suppose that

---

[1] A splitter is generally defined with the returned values $\{L, S, R\}$. Here, we make no distinction between L and R.

Algorithm 3. Consensus – code at process p

```
1: Shared Variables:
2:     R                                    // A racing on grafarius objects
3:     d                                    // Initially, ⊥
4:
5: propose(u) :=
6:     while true do
7:         if d ≠ ⊥ then
8:             return d
9:         o ← R.enter()
10:         (f, u) ← o.adoptCommit(u)
11:         if f = commit then
12:             d ← u
```

---

p increments c and then returns the object $F(c)$. This simple local algorithm implements a linearizable racing. However, because this construction does not bound the amount of laps a process has to retrieve before knowing the most recent one, it might be expensive when contention occurs.

Algorithm 2 presents a more efficient approach that allows a process to skip the laps it missed. This algorithm makes use of an initially empty shared map L from Π to $\mathbb{N}$. We map x to the value y via L when writing $L[x] \leftarrow y$; operation $codomain(L)$ returns the codomain of L. For some process p, the local variable $last$ stores the index of the last lap entered by p. When it calls $enter()$, process p stores the $last$ index in L (line 9). Then, p retrieves the content of L and computes the maximum element m in its codomain. Process p assigns $m + 1$ to $last$, if $last = m$ holds, and m otherwise (lines 12 to 15). The value of $F(last)$ is then returned as the result of the call.

Time Complexity. The adaptive collect object of Attiya et al. [28] can implement the shared map L used in Algorithm 2, without having the knowledge of Π. In such a case, the time complexity of Algorithm 2 is $O(k)$, where $k \leq n$ denotes the amount of processes that actually access the racing object.

3.4 Racing-Based Consensus

Using the racing abstraction introduced in the previous section, we now depict an obstruction-free implementation of consensus. Recall that consensus is a shared object whose interface consists of a single method *propose*. This method takes as input a value from some set $\mathcal{V}$ and returns a value in $\mathcal{V}$ ensuring both (*Validity*) if v is returned then some process invoked *propose*(v) previously, and (*Agreement*) two processes always return the same value.

Algorithm 3 describes our implementation of consensus. In this algorithm, processes compete on two shared abstractions: a racing R on grafarius objects, and a decision register d. When a process p suggests a value u for consensus, it attempts to commit u by entering the next grafarius object in R (line 9). Every time p executes *adoptCommit* on a grafarius object o, p updates its proposed value with the response returned by o (line 10). In case the grafarius returns a committed value, this value is stored in d as the result of the call to *propose* (lines 11 and 12).

Algorithm 4. Universal Construction – code at process p

```
 1: Shared Variables:
 2:     R                                    // A racing on consensus objects
 3:
 4: Local Variables:
 5:     C                                    // Initially, R.enter()
 6:     s                                    // Initially, s_0
 7:
 8: invoke(op) :=
 9:     while true do
10:         d ← C.d
11:         if d ≠ ⊥ then
12:             s ← d[1]
13:             C ← R.enter()
14:         else
15:             (s', v) ← τ(s, op)
16:             if s = s' then
17:                 return v
18:             d ← C.propose((p, s'))
19:             if d[0] = p then
20:                 return v
```

Time Complexity. The call to the splitter object in Algorithm 1 requires four computational steps [26]. Besides, the solo time complexity of the adaptive collect object of Attiya et al. [28] belongs to $O(1)$. It follows that Algorithm 3 solves consensus in $O(1)$ steps during solo executions.

This fast resolution of consensus allows us to implement a universal construction with a linear time complexity when no contention occurs. We detail our approach in the next section.

3.5 A Fast Obstruction-Free Universal Construction

A universal construction is a general mechanism to obtain linearizable shared objects from sequential ones. A sequential object is specified by some serial data type that defines its possible states as well as its access operations. Formally, a serial data type is an automaton defined by a set of states ($States$), an initial state ($s_0 \in States$), a set of operations (Op), a set of response values ($Values$), and a transition relation ($\tau : States \times Op \rightarrow States \times Values$). Hereafter, and without lack of generality, we shall assume that every operation op is $total$, i.e., $States \times \{op\}$ is in the domain of τ.

Algorithm 4 depicts our obstruction-free linearizable universal construction. The algorithm uses a single shared variable: a racing on obstruction-free consensus objects named R. When a process p invokes an operation via $invoke(op)$, p first checks the decision of the latest consensus object it entered (line 11). If a decision was taken, p updates its local variable s with the new state of the object. Then, p enters the next consensus (lines 12 to 13). Once p reaches the last consensus that was decided, variable s stores a state of the object that is older than the time at which p invoked op. At this point, process p executes tentatively the operation on s and stores the result in the pair (s', v). When s equals s', the invocation does not change the result of the object and p can immediately returns v. Otherwise, p proposes the pair (p, s') to change the state of the object to s'. If it successes, process p returns the response v (lines 19 and 20).

Time Complexity. As pointed out previously, the case we consider to be the most frequent one is the contention-free case, that is when multiple processes access the object but interleavings do not occur. In the worst case, a process freshly calling *invoke*() in a contention-free execution first retrieves the largest decided consensus, then it enters the next consensus and decides. From our previous time complexity analysis of Algorithms 2 and 3 and the lower bound result of Jayanti et al. [19], the contention-free time complexity of Algorithm 4 is optimal and belongs to $\Theta(k)$.

4   Managing Memory Usage

Every time the state of the object implemented by the universal construction changes, Algorithm 4 accesses a new consensus instance. This implies that the number of consensus instances is not bounded and may rapidly exhaust the available memory. In this section, we present a novel recycling technique that addresses this problem. To that end, we first introduce several definitions that capture the notion of recycled objects. Then, we depict a mechanism to recycle the objects used in Algorithm 4.

4.1   Preliminary Notions

Intuitively, every time an object is reused, it should behave according to its specification. We formalize this idea in the definitions that follow.

Definition 1 (*Round & Decomposition*). *Given some history h, a round r of h is a sub-history of h such that every invocation complete in h is complete in r. A decomposition of h is an ordered set of rounds $\{r_1.....r_{m \geq 1}\}$ satisfying $h = r_1.....r_m$.*

Definition 2 (*Recycled Object*). *Consider a history h of some object o. We say that o is a recycled object of type $\mathfrak{T}$ during h, when there exists a decomposition of h such that every round r in this decomposition is a correct history for an object of type $\mathfrak{T}$.*

In order to illustrate these definitions, let us consider two processes p and q, and a shared object o exporting an operation op. We can decompose history $h_1 = inv_{p,1}(op).inv_{q,1}(op).res_{q,1}(op)u.res_{p,1}(op)v.inv_{p,2}(op)$ in rounds $r_1 = inv_{p,1}(op).inv_{q,1}(op).res_{q,1}(op)u.res_{p,1}(op)v$ and $r_2 = inv_{p,2}(op)$. However, if we consider that $op = propose$ and $u \neq v$, there is no decomposition of h_1 for which o is a recycled consensus object.

The usual approach to recycle an object is to reset all its fields once the processes have stopped accessing it, that is once all the operations pending in a round have completed. The universal construction of Herlihy [2] implements this idea by provisioning for each process $O(n^2)$ cells, each cell storing the state of the implemented object. An array of $O(n)$ bits associated to each cell indicates when it can be reset by its owner.

Since the participants to the universal construction are unknown in our context, we cannot employ the previous approach. Instead, we propose to recycle the objects used in Algorithm 4 by signing each modification with the round at which it occurs. An operation that updates such an object will be oblivious to modifications made in prior rounds. If now the operation is in late, that is when a new round has started before it returns, the operation will observe the object in a state consistent with one of the rounds to which it is concurrent. We develop this idea in the next section, then apply it to Algorithm 4.

4.2   Recycling Objects

As a starter, let us remind the definition of a decidable object. This category of objects contains consensus, but also the splitter and the grafarius objects we described in Section 3.

Definition 3 (Decidable Object). *A decidable object o is a shared object whose state contains a* decision register d *taking its value in some set V, the domain of o, union $\bot \notin V$, and which initially equals $\bot$. The object is said* decided *when $d \in V$ holds. For every operation op of o, once o is decided, there exists some deterministic function f of d such that $f(d)$ is a valid response value for op.*

As an example of the previous definition, let us consider a grafarius object. We observe that when the decision register d does not equal $\bot$, the pair $(adopt, d)$ is a sound response for the call $adoptCommit$.

The first step of our recycling mechanism consists in recycling the MWMR registers that form the basic building blocks of our algorithms. We detail it below.

(Construction 1) Let $(\mathcal{T}, <)$ be a set of timestamps totally ordered by some relation $<$ and containing a smallest element $0 \in \mathcal{T}$. For every register x having some initial state s_0, we initialize x to $(0, s_0)$. Then, consider some timestamp t. When a value v is written to x, we write (t, v) to x. Now, upon reading from x, the value returned is the value u in the case where x contains (t', u) with $t \le t'$, and s_0 otherwise.

In a second step, we extend this technique to decidable objects as follows.

(Construction 2) For some decidable object o, a call to $recycle(o, t)$ returns a copy of o such that upon a call to an operation op of o,(i) if the object is decided then we return $f(d)$, and otherwise (ii) op is executed but read and write operations on the shared registers that implement o are replaced by the steps described in Construction 1 using timestamp t.

For some decidable object o, we shall write $recycle(o)$ the object obtained by proxying every call to the operations of o by corresponding calls to $recycle(o, t)$ for some timestamp t.

Proposition 1 establishes that, provided the timestamps are appropriate, $recycle(o)$ implements a recycled object of the same type as o.

Algorithm 5. Universal Construction – code at process p

```
 1: Shared Variables:
 2:     L                                       // A map from Π to ℕ, initially ∅
 3:
 4: Local Variables:
 5:     F                                       // A function from ℕ to consensus objects
 6:     s                                                            // Initially, s₀
 7:     last                                                         // Initially, 0
 8:     ts                                                           // Initially, 0
 9:     C                                                            // Initially, enter()
10:
11: invoke(op) :=
12:     while true do
13:         d ← C.d                          24: function free() :=
14:         if d ≠ ⊥ then                     25:     S ← codomain(L)
15:             (s, last, ts) ← (d[1], d[2], d[3])  26:     let (γ, Γ) = (min(S), max(S))
16:             C ← enter()                  27:     if γ > 0 then
17:         else                             28:         return γ − 1
18:             (s', v) ← τ(s, op)           29:     return Γ + 1
19:             if s = s' then
20:                 return v
21:             d ← C.propose((p, s', free(), ts + 1))  30: function enter() :=
22:             if d[0] = p then             31:     L[p] ← last
23:                 return v                 32:     return recycle(F(last), ts)
```

Proposition 1. *Consider a decidable object o of type $\mathfrak{T}$ and some history h of recycle(o) during which the following invariant holds:*

(P1) For any pair of operations op and op', executed respectively on recycle(o, t) and recycle(o, t') in h, if op' does not precede op in h and $t' < t$ holds, there exists an operation on recycle(o, t') that precedes op' in h and writes to the decision register d of o.

Then, recycle(o) implements a recycled object of type $\mathfrak{T}$ during history h.

4.3 Application

Algorithm 5 depicts our second obstruction-free universal construction. In comparison to Algorithm 4, we introduce two modifications: (i) processes now compete to decide which consensus will store the next state of the object, and (ii) consensus objects are recycled using the mechanism we presented in Construction 2.

With more details, Algorithm 5 works as follows. We implement a racing on consensus objects with variables L and F. When an operation changes the state of the implemented object, the calling process proposes to consensus the new state s' together with the index of the consensus object that will be used next and its associated timestamp (line 21). The index is determined by a call to the function *free*(). This function retrieves the codomain of L, and computes the smallest consensus index that is not currently accessed by a process (lines 25 to 29). In case all objects between 0 and Γ are busy, where Γ is the greatest index accessed so far, the index $\Gamma + 1$ is returned.

Algorithm 5 recycles the consensus objects in the codomain of F using the timestamping schema we introduced in Section 4.2. During an execution, for every object *recycle*(o) with $o \in codomain(F)$, the algorithm maintains the invariant P1 of Proposition 1. This ensures that accesses to variables L and F implement a racing on consensus objects, reducing Algorithm 5 to Algorithm 4.

Time and Space Complexity. The contention-free time complexity of Algorithm 5 is the same as for Algorithm 4, i.e., it belongs to $\Theta(k)$ in $\mathcal{ASM}$. From the code of function *free*(), Algorithm 5 employs at most $k+1$ consensus objects. In a distributed system, a quorum system can implement a collect object by emulating $O(k)$ shared registers. It results that in such a model the contention-free time complexity of Algorithm 5 measured in message delay is $O(1)$, and that its space complexity belongs to $O(k^2)$.

5   Empirical Assessment

To assess the practicability of our approach, we evaluate in this section a prototype implementation of Algorithm 5. This implementation is built on top of the Apache Cassandra distributed data store [8] which provides a distributed shared memory using consistent hashing and quorums of configurable sizes. In what follows, we describe the internals of our implementation then detail its performance in comparison to the Apache Zookeeper coordination service [9]. For the sake of reproducibility, the source code of our implementation, as well as the scripts we run during the experiments, are publicly available [29].

5.1   Implementation Details

Cassandra offers a data model close to the classical relational model at core of the database systems. The smallest data unit in Cassandra is a column, a tuple that contains a name, a value and a timestamp. Columns are grouped by rows, and a column family contains a set of rows. Each row is indexed by a key, and stored at a quorum of replicas (following a consistent hashing strategy). A client can read a whole row and write a column. The consistency of such operations is tunable in Cassandra. When the cluster running Cassandra is synchronized and both read and write operations operate on quorums, Cassandra provides an atomic snapshot model. This storage system also supports eventually consistent operations. When this consistency level is employed, a write operation accesses a quorum of replicas, while a read occurs at a single replica. Cassandra reconciles replicas via a timestamp-based mechanism in the background.

Prototype Implementation. Our implementation uses the Python programming language and it consists of the different shared objects we detailed in the previous sections (splitter, grafarius, consensus, and universal construction). The conciseness of Python allows the whole implementation to contain around 1,000 lines of code. Our implementation closely follows the pseudo-code of the algorithms. Each object corresponds to a row in a column family, and is named after the type of the object. When an object relies on lower-level abstractions, e.g., consensus employs multiple grafarius objects, the objects' keys at the low-level are named after the key at the higher one, e.g., *consensus:12:grafarius:3*. By changing the consistency of the decision register d in Algorithm 3, we can tune the consistency of our universal construction. When d is eventually consistent, the universal abstraction is sequentially consistent for read operations; otherwise it is linearizable. In Zookeeper, updates are linearizable while read operations are sequentially consistent. For that reason, when we compare the performance of our implementation to Zookeeper during the experiments, we use the sequentially consistent variation of our algorithm.

5.2   Evaluation

We conducted our experiments on a cluster of virtualized 8-core Xeon 2.5 Ghz machines running Gentoo Linux, and connected with a virtualized 1 Gbps switched network. Network characteristics, as measured by ping and netperf, are 0.3 ms for a round-trip and a bandwidth of 117MB/s. Each machine is equipped with a virtual hard-drive whose read/write (uncached) performance, as measured with `hdparm` and `dd`, is 246/200 MB/s. A *server machine* runs either Cassandra or Zookeeper. A *client machine* emulates multiple clients accessing concurrently the shared objects. During an experiment, a client executes 10^4 accesses to one or more objects. We used 1 to 20 clients machines, emulating 1 to 100 clients each, and 3 to 12 server machines. In all our experiments, the client machines were not a bottleneck.

Compare-and-swap. We first evaluate in Figure 1 the performance of our implementation when clients execute compare-and-swap operations, and the system is composed of 3 server machines. Recall that a compare-and-swap object exposes a single operation: $C\&S(u, v)$. This operation ensures that if the old value of the object equals u, it is replaced by v. In such a case, the operation returns *true*; otherwise it returns *false*. In Figure 1, we plot the la-

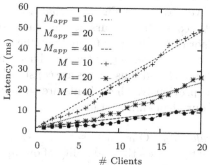

Fig. 1. Compare-And-Swap

tency to execute a compare-and-swap operation as a function of the number of clients and the arguments of the operations. The initial state of the compare-and-swap object is 0. Each client executes in closed-loop an operation $C\&S(k, l)$, where k and l are taken uniformly at random from the interval $[0, M[$ with M some maximum value.

When the size of the interval $[0, M[$ shrinks, each $C\&S()$ operation is more likely to success in transforming the state of the object; hence contention increases. Consequently, as observed in Figure 1, performance degrades. Contention between clients occur mainly on the splitter objects that form the building blocks of Algorithm 1. We briefly analyze how contention is related to performance next.

An operation $C\&S(u, v)$ is successful when the state is changed from u to v. Let us note ρ the ratio of successful operations, that is $1/M$, and λ_s (respectively λ_f) the latency to execute solo a successful (resp. failed) operation. In Figure 1, the light lines (M_{app}) plot for each value of M the curve $\lambda_f(1 - \rho) + \lambda_s \rho n$. This is a reasonable approximation where the term $\lambda_s \rho n$ follows Little's law [30] and translates the convoy effect [31] on successful operations.

Critical section In Figure 2, we compare the performance of our implementation and Zookeeper when clients access a critical section (CS). Such an object is not in line with the non-blocking approach, but it is commonly used in distributed applications. We implemented the CS on top of our universal construction using a back-off mechanism. For Zookeeper, we employed the recipe described in [9]. Figure 2 presents the average time a client takes to enter then leave the CS, and we vary the inter-arrival time of clients in the critical section according to a Poisson distribution.

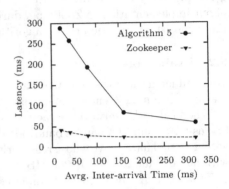

Fig. 2. Critical Section

We observe in Figure 2 that when the inter-arrival time is high, and thus little contention occurs, a client accesses the CS with Zookeeper in 20 ms. For Algorithm 5, it takes 60 ms, but the performance degrades quickly when clients access more frequently the CS. This comes from the fact that(i) we implemented a spinlock and thus clients are constantly accessing the system, and (ii) as pointed out previously, when clients are competing on splitter objects, the performance of our algorithm degrades.

Scalability Our last set of experiments assesses the scalability of our approach. To that end, we compute the maximal throughput of our prototype implementation when clients access different objects, precisely $C\&S()$ for $M = 10$. The amount of available server machines varies from 3 to 12 servers. In all cases, we implement a register with the help of a quorum of 3 servers. We compare our results to an instance of 3 Zookeeper machines. Zookeeper does

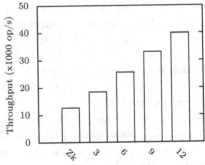

Fig. 3. Scalability

not implement natively a compare-and-swap operation. We devised the following implementation relying on the versioning mechanism exposed to the clients by Zookeeper. When a client executes $C\&S(u, v)$ it first retrieves the value w and the attached version k of the znode identifying the object. In case $w = u$, the client attempts writing v with version $k + 1$. If this write fails, the client re-executes $C\&S(u, v)$. In our experiments, a single client accesses each object. Thus it never retries and this implementation is optimal.

Figure 3 depicts the maximal throughput with 3 to 12 servers. With 3 servers, our system delivers 18.4K op/s and ZooKeeper 12.6K op/s. The bottleneck nature of the ZooKeeper leader which serializes all updates explains this gap.

Our prototype achieves 33K op/s when using 9 servers, and 40K op/s with 12. In this last case, our system is 3.2 times faster than Zookeeper on 3 machines.

6 Conclusion

This paper presents a novel algorithmic solution to implement a distributed universal construction when participants are unknown. Contrary to previous works, which mostly focus on state machine replication, our approach employs solely a distributed asynchronous shared memory, the logic of consistent operations being delegated to the client side. Hence, and as exemplified by our prototype, we can implement it in a client library that runs on top of an off-the-shelf distributed shared memory. To obtain this result, we introduce two novel shared abstractions: a grafarius and a racing, which we believe are of interest on their own. We also present a new mechanism to recycle the base objects at core of our construction.

References

1. Lamport, L.: The part-time parliament. ACM Trans. Comput. Syst. 16(2), 133–169 (1998)
2. Herlihy, M.: Wait-free synchronization. ACM Trans. Program. Lang. Syst. 13(1), 124–149 (1991)
3. Balakrishnan, M., Malkhi, D., Prabhakaran, V., Wobber, T., Wei, M., Davis, J.D.: Corfu: A shared log design for flash clusters. In: Proceedings of the 9th USENIX Conference on Networked Systems Design and Implementation, NSDI 2012, pp. 1–1. USENIX Association, Berkeley (2012)
4. Gafni, E., Lamport, L.: Disk paxos. In: Herlihy, M.P. (ed.) DISC 2000. LNCS, vol. 1914, pp. 330–344. Springer, Heidelberg (2000)
5. Balakrishnan, M., Malkhi, D., Wobber, T., Wu, M., Prabhakaran, V., Wei, M., Davis, J.D., Rao, S., Zou, T., Zuck, A.: Tango: Distributed data structures over a shared log. In: 24th ACM Symposium on Operating Systems Principles, SOSP (2013)
6. Gafni, E.: Round-by-round fault detectors (extended abstract): unifying synchrony and asynchrony. In: Proceedings of the Seventeenth Annual ACM Symposium on Principles of Distributed Computing, PODC 1998, pp. 143–152. ACM, New York (1998)
7. Aspnes, J.: A modular approach to shared-memory consensus, with applications to the probabilistic-write model. In: Proceedings of the 29th ACM SIGACT-SIGOPS Symposium on Principles of Distributed Computing, PODC 2010, pp. 460–467. ACM, New York (2010)
8. Lakshman, A., Malik, P.: Cassandra: A decentralized structured storage system. SIGOPS Oper. Syst. Rev. 44(2) (April 2010)
9. Junqueira, F.P., Reed, B.C.: The life and times of a ZooKeeper. In: PODC 2009: Proceedings of the 28th ACM Symposium on Principles of Distributed Computing, pp. 4–4. ACM, New York (2009)
10. Sutra, P., Rivière, E., Felber, P.: A practical distributed universal construction with unknown participants. CoRR abs/1309.2772 (2013)

11. Fischer, M.J., Lynch, N.A., Patterson, M.S.: Impossibility of distributed consensus with one faulty process. J. ACM 32(2), 374–382 (1985)
12. Guerraoui, R.: Indulgent algorithms (preliminary version). In: PODC 2000, pp. 289–297. ACM, New York (2000)
13. Guerraoui, R., Raynal, M.: The information structure of indulgent consensus. IEEE Trans. Comput. 53(4), 453–466 (2004)
14. Chockler, G., Malkhi, D.: Active disk paxos with infinitely many processes. In: Proceedings of the Twenty-first Annual Symposium on Principles of Distributed Computing, PODC 2002, pp. 78–87. ACM, New York (2002)
15. Attiya, H., Guerraoui, R., Hendler, D., Kuznetsov, P.: The complexity of obstruction-free implementations. J. ACM 56(4), 24:1–24:33 (2009)
16. Aspnes, J., Ellen, F.: Tight bounds for anonymous adopt-commit objects. In: 23rd Annual ACM Symposium on Parallelism in Algorithms and Architectures, pp. 317–324 (June 2011)
17. Luchangco, V., Moir, M., Shavit, N.N.: On the uncontended complexity of consensus. In: Fich, F.E. (ed.) DISC 2003. LNCS, vol. 2848, pp. 45–59. Springer, Heidelberg (2003)
18. Fich, F.E., Luchangco, V., Moir, M., Shavit, N.N.: Obstruction-free algorithms can be practically wait-free. In: Fraigniaud, P. (ed.) DISC 2005. LNCS, vol. 3724, pp. 78–92. Springer, Heidelberg (2005)
19. Jayanti, P., Tan, K., Toueg, S.: Time and space lower bounds for nonblocking implementations. SIAM J. Comput. 30(2), 438–456 (2000)
20. Jayanti, P., Toueg, S.: Some results on the impossibility, universality, and decidability of consensus. In: Segall, A., Zaks, S. (eds.) WDAG 1992. LNCS, vol. 647, pp. 69–84. Springer, Heidelberg (1992)
21. Attiya, H., Bar-Noy, A., Dolev, D.: Sharing memory robustly in message-passing systems. J. ACM 42(1), 124–142 (1995)
22. Lynch, N.A., Shvartsman, A.A.: Robust emulation of shared memory using dynamic quorum-acknowledged broadcasts. In: Proceedings of the 27th International Symposium on Fault-Tolerant Computing (FTCS 1997). IEEE Computer Society, Washington, DC (1997)
23. Herlihy, M., Wing, J.: Linearizability: A correctness condition for concurrent objects. ACM Trans. on Prog. Lang. 12(3), 463–492 (1990)
24. Herlihy, M., Shavit, N.: On the nature of progress. In: Fernàndez Anta, A., Lipari, G., Roy, M. (eds.) OPODIS 2011. LNCS, vol. 7109, pp. 313–328. Springer, Heidelberg (2011)
25. Lamport, L.: A fast mutual exclusion algorithm. ACM Trans. Comput. Syst. 5(1), 1–11 (1987)
26. Moir, M., Anderson, J.: Fast, long-lived renaming. In: Tel, G., Vitányi, P.M.B. (eds.) WDAG 1994. LNCS, vol. 857, pp. 141–155. Springer, Heidelberg (1994)
27. Guerraoui, R., Ruppert, E.: Anonymous and fault-tolerant shared-memory computing. Distributed Computing 20(3), 165–177 (2007)
28. Attiya, H., Fouren, A., Gafni, E.: An adaptive collect algorithm with applications. Distributed Computing 15(2), 87–96 (2002)
29. Sutra, P.: (2013), http://github.com/otrack/pssolib
30. Allen, A.O.: Probability, Statistics, and Queueing Theory with Computer Science Applications. Academic Press Professional, Inc., San Diego (1990)
31. Blasgen, M., Gray, J., Mitoma, M., Price, T.: The convoy phenomenon. SIGOPS Oper. Syst. Rev. 13(2), 20–25 (1979)

Author Index